PATTERN FORMATION IN COMPLEX DISSIPATIVE SYSTEMS

Advisory Board

H. R. Brand (Essen), P. E. Cladis (AT&T Bell), H. Furukawa (Yamaguchi), K. Kawasaki (Kyushu), L. Lam (San Jose), M. Matsushita (Chuoh), M. Sano (Tohoku), Y. Shiwa (KIT), R. Takagi (Tokyo), T. Yamada (KIT), Y. Yamazaki (KIT), W. Zimmermann (Juelich)

Supported by:

Kyushu Institute of Technology
Kitakyushu City
The Kajima Foundation
Shimadzu Science Foundation
The Society of Science of Form of Japan
The Physical Society of Japan
The Japan Fluid Mechanics Society

PATTERN FORMATION IN COMPLEX DISSIPATIVE SYSTEMS

FLUID PATTERNS, LIQUID CRYSTALS, CHEMICAL REACTIONS

KITAKYUSHU, JAPAN 18 – 20 SEPTEMBER 1991

EDITED BY

S. KAI

Department of Electrical Engineering
Kyushu Institute of Technology
Japan

World Scientific
Singapore • New Jersey • London • Hong Kong

Published by

World Scientific Publishing Co. Pte. Ltd.
5 Toh Tuck Link, Singapore 596224
USA office: 27 Warren Street, Suite 401-402, Hackensack, NJ 07601
UK office: 57 Shelton Street, Covent Garden, London WC2H 9HE

Library of Congress Cataloging-in -Publication Data
Pattern formation in complex dissipative systems : fluid patterns, liquid crystals, chemical reactions : Kitakyushu, Japan, 18–20 September, 1991 / edited by S. Kai.
p. cm.
ISBN-13 978-981-02-0879-0 -- ISBN-10 981-02-0879-0
1. Crystal growth -- Congresses. 2. Heat -- Convection -- Congresses. 3. Chemical reactions -- Congresses. 4. Turbulence -- Congresses.
I. Kai, S. (Shōichi), 1947–
QD921.P344 1992
530.4 -- dc20 92-14203
CIP

British Library Cataloguing-in-Publication Data
A catalogue record for this book is available from the British Library.

PREFACE

Pattern formation in nature in general has attracted the interest of scientists for a long time [see, for example, E. S. Hedges and J. E. Myers, *The Problem of Physico-chemical Periodicity*, Edward-Arnold Co., London, 1926]. Definition, quantization, characterization, formalization and classification of patterns are the areas to investigate in order to understand patterns as a science. The physical characterization of pattern formation has made considerable progress during the recent two decades and is now one of the major branches of nonlinear science. This is largely due to the development of nonlinear nonequilibrium physics. It is a branch of physics where in particular cases, for instance pattern formation in liquid crystals, a lively interaction between different disciplines, from applied mathematics to material science, exists. Fluid dynamics and turbulence, interface motion during solidification, chemical reaction and biological pattern formation have a similar impact on the conceptional development of the physics about systems far from thermal equilibrium.

For the closest way of one approach to understand a universality containing all these phenomena, better methodology and sophisticated mathematical tools should be found. One candidate is the so-called "reduction of information" which means that one only picks up spatially and temporally slow modes by elimination and renormalization of rapid modes. This is recently very successfully used to describe many macroscopic patterns and is frequently called amplitude and phase dynamics. Other currently developed conceptional tools are chaos and fractal which are in some sense also related to the above subject. All those are outstanding problems in the modern physics of pattern formation.

Originally, this workshop was planned as the second workshop on "Pattern Formation in Liquid Crystals", the first being organized by L. Kramer and W. Zimmermann at Bayreuth in Germany in 1989. The topics discussed at that time all concerned liquid crystal patterns. As described above, however, recent developments of research on patterns in liquid crystals show that they are closely related to other phenomena such as isotropic convective systems, crystal growth, interfacial motions between two phases, topological instabilities, phase transitions and so on. In order to really understand and solve these problems, we must look at their properties not only in liquid crystals but also in the related problems I have just mentioned. They have already been investigated for a rather long time and appear to be better understood than the field of liquid crystals. This was my motivation to organize a workshop with a more general scope. I hope there will be deep understanding and progress in the study of patterns by bringing together different points of view and new ideas across several fields.

In this workshop we had about 100 presentations, both in oral and in poster sessions on pattern formation phenomena and instabilities near convective onset, on turbulence and chaos, fractals, interface, liquid crystals, self-organized systems, pattern formation in chemical reaction systems and spatio-temporal phenomena. Several outstanding scientists from closely related fields also accepted our invitation to come to Japan. Although the fields covered were thus rather wide, there were many common grounds. The key words, therefore, characterizing the contents of this workshop on *Physics of Pattern Formation* were "Competition and Selection of Patterns, Instabilities, Bifurcations, Defects, Filaments, Interface, Turbulence, Chaos, Coherence, and Intermittence". The main goal of this workshop was to understand these concepts. These were discussed from several points of view and the important contributions are presented, especially those related to turbulence and spatio-temporal chaos.

It was the first international conference in the field of pattern formation in dissipative systems held in Japan and it would have been impossible to have this workshop without the support of many different organizations. Among them were, especially, The International Science Foundation of Kyushu Institute of Technology, The Kajima Foundation, The Simadzu Science Foundation and Kikakyushu City.

Shoichi KAI
Chairman of Organizing Committee
Kitakyushu, September 1991

CONTENTS

C: CONVECTIVE INSTABILITY

D: CHAOS AND TURBULENCE

E: OTHERS

F: ABSTRACTS

A: Interface and Crystal Growth

PATTERN FORMATION AT THE CHOLESTERIC-ISOTROPIC INTERFACE*

P. E. CLADIS

AT&T Bell Laboratories, Murray Hill, NJ 07974, USA

ABSTRACT

This is an account of our recent observations of the influence of a one-dimensional equilibrium length on the dynamics of a nonequilibrium process. The experimental tools are cholesteric liquid crystals prepared by adding a chiral impurity to a nematic liquid crystal, real-time image analysis and an automated directional solidification apparatus. The important new observation is: A system's response to perturbations of its equilibrium length when suitably oriented with respect to a moving phase transition boundary, is characterized by a frequency.

1. Introduction: Why is Macroscopic Pattern Formation Interesting?

Pattern formation, epitomized by beautiful snowflakes, has long been a mystery.[1] Indeed, in the words of Nicolis and Prigogine, *Virtually every structure in the natural world is the result of a long sequence of successive symmetry breaking instabilities due to nonlinear processes under nonequilibrium conditions.*[2] Despite their beauty, mystery and pervasiveness in nature, the appropriateness of a *scientific* investigation of pattern formation is questioned in some Physics Departments even today.[3]

Nevertheless, because of recent major advances in a macroscopic description of nature, a growing number of investigators believe that, indeed, pattern formation is an exciting new cross-disciplinary field for physical investigation. The link between disorder on molecular or atomic length scales and macroscopic order, the domain of statistical physics, is frequently difficult to forge. However, independent of the existence of this link, new ways of thinking about macroscopic patterns have extended the analytical framework for describing nonequilibrium structures sometimes revealing remarkable similarities among many different systems.[4] With each advance in the intellectual framework, new rules emerge to further guide the observer. This has been particularly striking in modern descriptions of temporal behavior under nonequilibrium conditions of simple systems.[5] Encouraged by these advances, interest has turned to 1-dimensional patterns exhibited by many nonequilibrium systems.[6] Patterns are spatial structures, usually characterized by a band of wavelengths, that replace a uniform state above a critical distance from equilibrium.

---

* In this work, partially supported by NATO CRG 890777, my collaborators are: H. R. Brand, P. L Finn and J. T. Gleeson. Support for this presentation from the KIT International Workshop on *The Physics of Pattern Formation* is gratefully acknowledged.

In parallel with theoretical advances in grappling with temporal phenomena, modern computing power now enables a precise quantification of a pattern's spatio-temporal evolution. Patterns are typically analyzed from pictures (images). While pictures are the broadest band of information humans can process, it is only recently that computer assisted image analysis has made it possible to compress such large amounts of information into physical knowledge. Indeed, with the new information age computing power, the old English adage, *A picture is worth a thousand words*, is turned upside-down. Picture files require large computer memory: the text file (TROFF) for our Phys. Rev. Letter[7] describing these experiments was only ~ 20kbytes, while its four picture files required ~ 2megabytes of memory!

Finally, the availability of well-characterized complex materials, thermotropic nematic liquid crystals,[8] has opened the door onto patterns potentially more amenable to a complete theoretical description near onset.[9] Because nematic liquid crystals (Fig. 1)

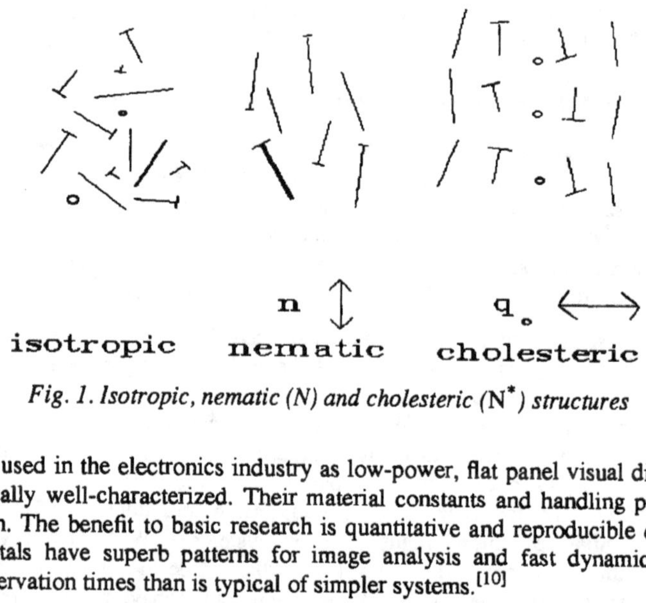

Fig. 1. Isotropic, nematic (N) and cholesteric (N) structures*

are widely used in the electronics industry as low-power, flat panel visual displays, they are chemically well-characterized. Their material constants and handling processes are well-known. The benefit to basic research is quantitative and reproducible data. Lastly, liquid crystals have superb patterns for image analysis and fast dynamics providing shorter observation times than is typical of simpler systems.[10]

2. Overview of the Experiment and the System Used

The experiment discussed here is directional growth at fixed speed, v, in a constant temperature gradient, **G** || **v**, of orientational order in a helix structure (the cholesteric liquid crystal state, N*), from the isotropic liquid that has continuous translational and rotational symmetry (Fig. 1). For growth rate, v, larger than a critical rate, v_c, the isotropic-cholesteric interface develops a cellular interfacial pattern of wavelength $2\pi/q$.

In the usual scenario,[1] the balance between impurity diffusion away from the interface into the isotropic liquid and impurity rejection by the growing cholesteric liquid crystal creates an impurity gradient in advance of a cholesteric-isotropic interface moving at speed v: the larger v is, the steeper the impurity gradient. When the nonequilibrium driving force is small enough, i.e. $v < v_c$, the impurity distribution across the interface is uniform and the interface is flat. The diffusion constant in the isotropic liquid is $D_I = 4\times10^{-7}\,cm^2/s$ for the system we use.[11] When v is larger than a critical speed, v_c, the impurity gradient in advance of the moving interface is sufficiently steep that the flat interface becomes unstable to a cellular pattern where impurities acquire a spatial distribution along the interface. Features of the cellular pattern are cell tips at a higher temperature and cell grooves extending to lower temperatures. Impurities are more concentrated in cell grooves and less concentrated at cell tips. However, conservation of impurities requires that impurity segregation in cell grooves be balanced by impurity depletion at cell tips.

The material we use is a cholesteric liquid crystal prepared by adding to a nematic liquid crystal (broken rotational symmetry along **n** but no helix structure, i.e. $q_o \equiv 0$ in Fig. 1) a fixed amount, c, of a chiral impurity to establish an equilibrium helix with pitch $p_o = 2\pi/q_o \sim 38\mu m$. The 1-d translational order ($\mathbf{q}_o$) associated with the helix is perpendicular to the orientational order, **n** (Fig. 1). The cholesteric state (broken rotational *and* translational symmetry) maintains a uniform q_o by orientational diffusion. The diffusion constant for this process is $D_o = 8\times10^{-7}\,cm^2/s$.[12]

The novel feature of the cholesteric system is that its helix pitch depends on the amount of impurity present. Thus, orientational diffusion competes against a pattern forming nonequilibrium driving force that creates spatial variations in impurity levels and, therefore, spatial variations around an equilibrium value in the helix pitch. The new insight we acquired is that a system with an equilibrium length has a frequency for response to perturbations of its equilibrium structure, even at the first bifurcation. As H. R. Brand noted during the course of these experiments: *The nonequilibrium force creates a spatial pattern of wavenumber q and the cholesteric fights back!*

When $q_o \neq 0$, then, both

an **intrinsic speed,**

$$v_{e\ell} = D_o/p_o \sim 2.1\mu m/s$$

and **frequency,**

$$\omega_{e\ell}/2\pi = D_o q_o^2 \sim 2s^{-1}$$

can be defined. We concentrate on a particular region of parameter space,[13] where the pattern wavelength is comparable to the equilibrium pitch: specifically, the region where $0.25q_o < q < q_o$. In this region, the **nonequilibrium observables**: q, pattern drift parallel to the interface (v_x), and pattern frequency, ω, when scaled by the **equilibrium** constants, q_o, $v_{e\ell}$ and $\omega_{e\ell}$, respectively, result in ratios that are **O**(1). We found that when the physics involved in the nonequilibrium and equilibrium processes are comparable and competitive, a "clock" (frequency) also characterizes the patterns.

3. Experimental Set-up

Sample Preparation: Samples are prepared by mixing c = 9% by weight of a chiral impurity (C15) with the well-known liquid crystal 8CB.[14] It was determined that the cholesteric-isotropic transition temperature, T_{Ch-I} and the equilibrium pitch, q_o depended linearly on the impurity content, c, according to

$$T_{Ch-I}(°C) = 40.36 - 0.22c(\%)$$

and

$$q_o(\mu m^{-1}) = 0.02c(\%).$$

For $c \approx 9\%$, $p_o = 2\pi/q_o \approx 38\mu m$. The higher the transition temperature, the smaller the impurity concentration and the larger the pitch.

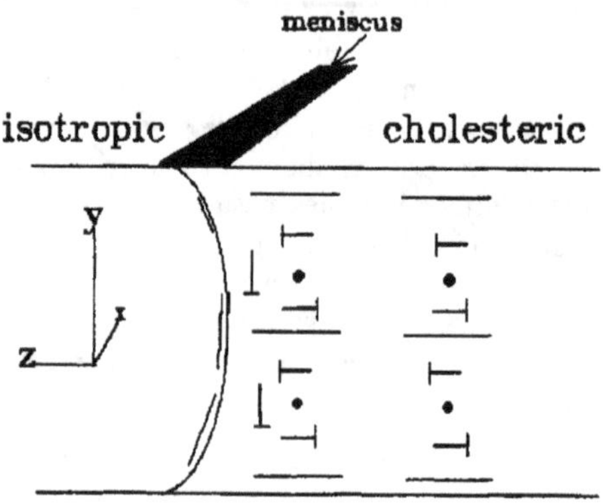

Fig. 2. Sample at rest.

The $\hat{x}-\hat{y}-\hat{z}$ sample dimensions were: 26mm×38μm×40mm. The glass slides containing the liquid crystal mixture were treated so that the direction of orientational order, **n**, was parallel to **G** (and v) at the top and bottom plates.[15] In addition, to minimize strains in the equilibrium structure, their spacing was chosen to be as close to p_o as possible. Fig. 2 shows the reference frame and a possible rest configuration of the sample with $\mathbf{q}_o \parallel \hat{y}$ except in the vicinity of the meniscus.

Directional Growth in a Temperature Gradient, G: In directional growth, temperature is mapped onto a spatial dimension, $\hat{z}$. Fig. 3 is a sketch of the experimental set-up.[16] In the experiment, a sample is placed in a temperature gradient, $\mathbf{G} \parallel \hat{z}$, and mechanically displaced ("pulled") from the hot contact towards the cold contact at speed v ($\mathbf{v} \parallel \mathbf{G}$).

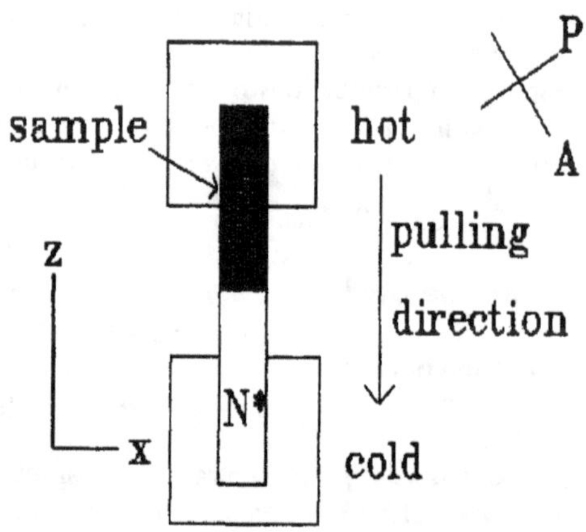

Fig. 3. Schematic of Experiment

The sample is viewed along $\hat{y}$ with a polarizing microscope interfaced to a computerized video-system with real-time image analysis. Polarizers are crossed at 45° to G so that, in an untreated image, the isotropic liquid is black (see Fig. 4a, an image of the system at rest). When $v < v_c \sim 19\mu m/s$, the interface is "flat", approximately an $\hat{x}-\hat{y}$

plane, i.e. an isotherm.

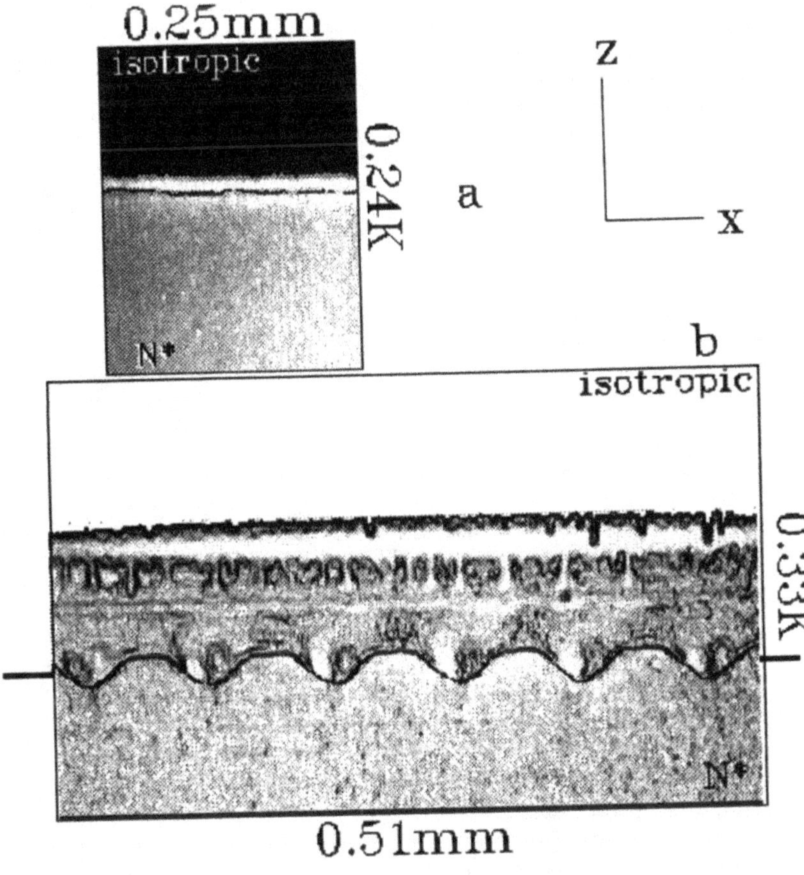

Fig. 4. In (a), the interface is at rest while in (b) $v > v_c$.

When $v > v_c$, the interface is unstable. A pattern forms (Fig. 4b) as impurity. diffusion overcomes the stabilizing influence of the temperature gradient, $\mathbf{G} \parallel_b \mathbf{v}$, at long length scales (~ 133μm), and capillary forces at short length scales (~ 2Å).[13] The contrast shown in Fig.4b is obtained by image enhancing an image with a contrast like Fig. 4a then reversing the contrast at each pixel making the isotropic liquid appear white between crossed polarizers. Dimensions on Figs. 4 refer to the size of the box enclosing the images.

Space-time diagrams are used to characterize the spatio-temporal behavior of the pattern. An x-t space-time plot is generated by capturing at fixed time intervals a single video line (such as a line between the two horizontal segments outside Fig. 4b's box) corresponding to a line of constant temperature that crosses the grooves of the pattern, then, replotting the lines in sequence (Fig. 5). Information on pattern wavelength and frequency is obtained from power spectra of x-t plots.

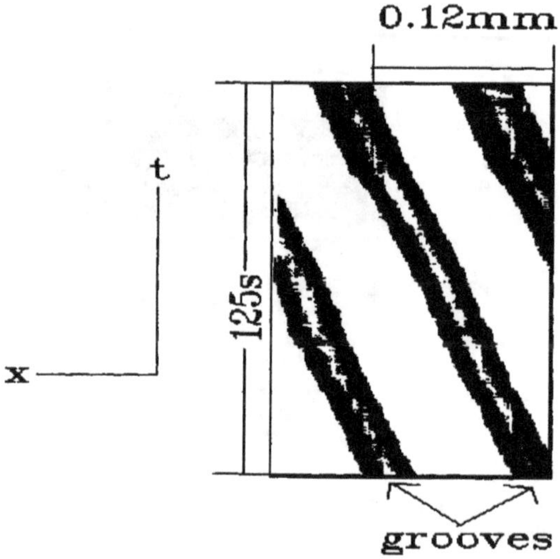

Fig. 5. Space-time Plot

4. Results

Pattern Wavelength: Our first observation is that the pattern's wavelength, defined as the distance between grooves, is always well-defined.[7] With increasing pulling speed, the pattern wavelength decreases from about 133μm at onset to close to the equilibrium pitch, 38μm, when a new instability takes place: the helix axis, $\mathbf{q}_o$, is parallel to the temperature gradient, **G**, and, there are no more cells. When the pattern wavelength is close to twice the equilibrium pitch, there is a second bifurcation to an instability that we call the "breathing mode". In the breathing mode, the grooves oscillate parallel to the interface: oscillations of neighboring grooves are in anti-phase while next-nearest neighbors are in phase.

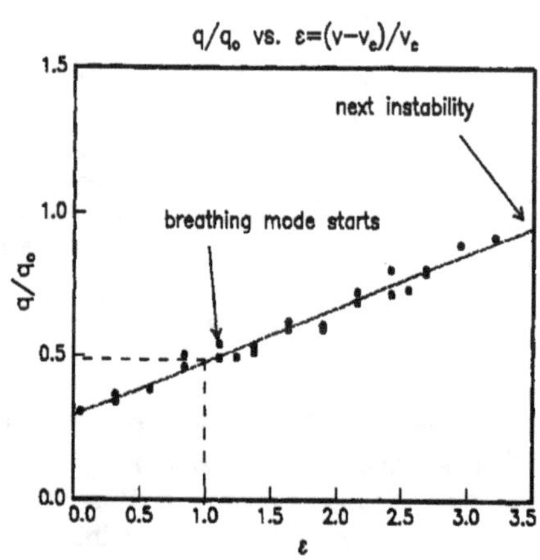

Fig. 6. Pattern wavenumber, q, vs. ε.

Defining $\varepsilon = (v - v_c)/v_c$, Fig. 6 shows that, to a good approximation, the pattern's wavenumber, q, is linear in ε for a large range of ε. The best fit line through the data points is: $q/q_o = 0.19\varepsilon + 0.29$.

Hopf Bifurcation: The pattern's drift parallel to the interface is determined from x-t plots such as Fig. 5. Both directions of propagation and standing wave patterns are observed. Even at the first instability, $v_x \neq 0$, characterizing this bifurcation as Hopf.[17]

In contrast to the smooth variation in q with ε, v_x shows two regimes. In the first regime, $\varepsilon < 2$, the helix axis is determined by boundary conditions: i.e. $\mathbf{q}_o \parallel \hat{\mathbf{y}}$ and

$$v_x/v_{el} = 0.32\varepsilon + 0.3.$$

The cholesteric texture that grows in this regime is unstructured and there are no relaxation processes observed in the material when pulling is stopped.

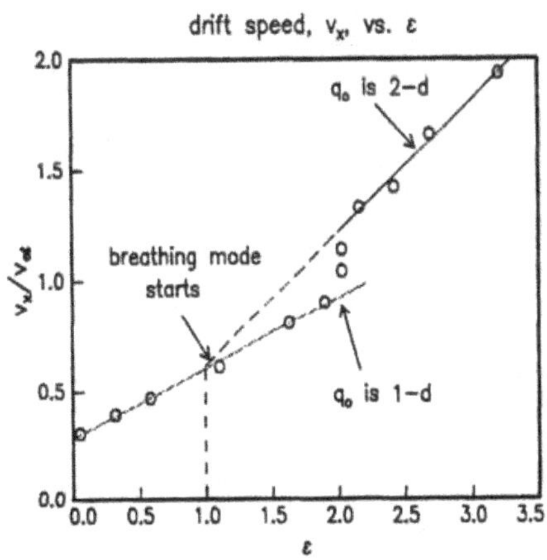

Fig. 7. Two regimes for v_x.

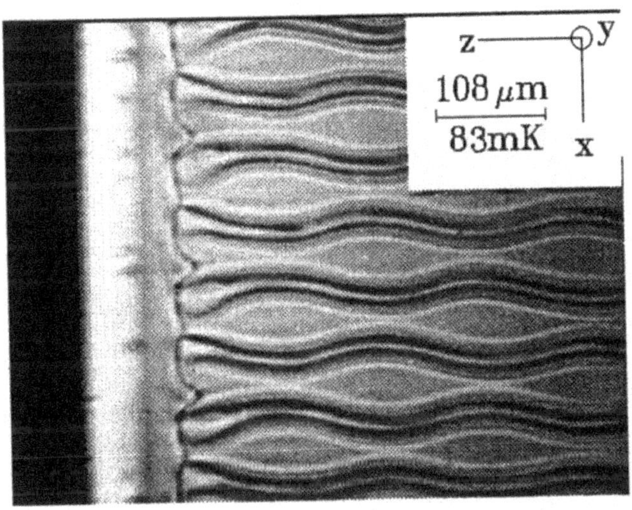

Fig. 8. Snapshot of Breathing Mode decorated by defect lines

In the second regime, $\varepsilon > 2$, $\mathbf{q}_o$ has two spatial degrees of freedom resulting in defect shedding at the interface (Fig. 8).[18] The texture frequently observed in this

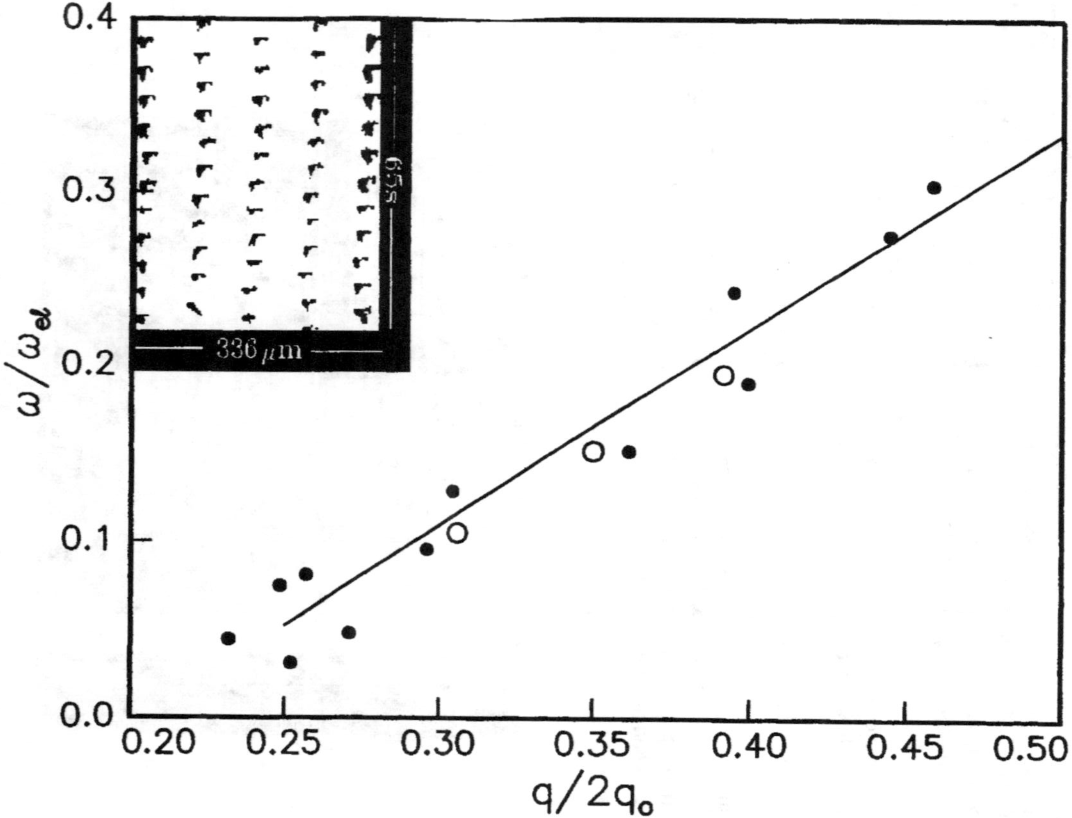

Fig. 9. Breathing mode dispersion curve and space-time crystal. (inset).

regime is a uniform array of line defects (one line per cell) that slowly relax when pulling is stopped. In this regime, v_x increases nearly twice as fast with increasing ε than in the first regime:

$$v_x/v_{e\ell} = 0.60\varepsilon + 0.03.$$

We note that, taking into account the scatter in the data, the intercept for this line could be zero. However, when this second line is extrapolated towards zero, we observe that it intersects the first line at $\varepsilon \approx 1$ where the breathing mode starts.[19]

Breathing Mode: A feature of the breathing mode is that alternate grooves grow while nearest neighbors shrink giving an alternating chain of larger and smaller grooves doubling the pattern wavelength (i.e. halving the number plotted in Fig. 6). A video-line crossing the the array of longer grooves but not the array of shorter ones, is captured every 0.3s then replotted in sequence. The binarized example in the inset of Fig. 9 shows the pattern in a standing wave. The dispersion relation is obtained by Fourier transforming x-t plots to obtain the structure factor, $S(q,\omega)$. In Fig. 9, q/2 is scaled by q_o and ω by $\omega_{e\ell}$. The best fit line through the data is

$$\omega/\omega_{e\ell} = -0.23 + 1.13q/2q_o.$$

The breathing mode is, therefore, nonlinear and nondispersive. Its group velocity, $v_G = 1.13\omega_{e\ell}/q_o = 1.13(2\pi)D_oq_o = (2\pi)^2 1.13v_{e\ell} \approx 95\mu m/s$.

5. Conclusions

The existence of an intrinsic length, $2\pi/q_o$, in cholesteric liquid crystals can give rise to a frequency for response of interfacial patterns of wavenumber q to perturbations from the equilibrium structure. After carefully choosing boundary conditions and sample dimensions, we explored the regime $q \sim q_o$. In this regime, the first instability to the cellular pattern is a Hopf bifurcation. A frequency is not associated with the cellular bifurcation when $q_o \equiv 0$. A second bifurcation to a nonlinear, nondispersive breathing mode is observed when $0.5q_o < q < q_o$.[7]

We pitted a nonequilibrium force against a comparable equilibrium one and found a "heartbeat".

REFERENCES

1. J. S. Langer, Rev. Mod. Phys., **52**, 1 (1980) and Science **243**, 1150 (1989). In his Science article, Langer notes that Kepler's monograph on snowflakes, *De Nive Sexangula* (1611), translated by C. Hardie (Clarendon, Oxford 1966), was the first time pattern formation was transferred from theology to science.

2. G. Nicolis and I. Prigogine, *Exploring Complexity*, (W. H. Freeman, New York 1989).

3. P. W. Anderson, Physics Today, **44**, 9, July 1991.

4. *See for example: Emerging Syntheses in Science* Vol. 1, David Pines, ed. (Addison-Wesley Publishing Company, Inc., New York 1987).

5. P. Bergé, Y. Pomeau and C. Vidal, *Order within Chaos* (John Wiley and Sons, New York 1984).

6. *See for example*: *Nonlinear Evolution of Spatio-temporal Structures in Dissipative Continuous Systems*, F. Busse and L. Kramer, eds., (Plenum Press, New York 1990).

7. P. E. Cladis, J. T. Gleeson, P. L. Finn and H. R. Brand, Phys. Rev. Lett. **67**, 3239 (1991).

8. P. G. de Gennes, *The Physics of Liquid Crystals*, (Clarendon, Oxford, 1982), 3rd edition.

9. J. M. Flesselles, A. J. Simon and A. Libchaber, Adv. Phys, **30**, 1 (1991).

10. *See for example*, W. Zimmermann, MRS Bulletin, **16**, 46 (1991).

11. M. Hara, H. Takezoe and A. Fukuda, Jpn. J. Appl. Phys. **25**, 1756 (1986). In the cholesteric liquid, these authors find an average diffusion constant in the cholesteric state, $D_{avg} \approx 2\times10^{-7} cm^2/s \sim D_I/2$.

12. $D_o = K/\gamma_1$ where K is an elastic constant and γ_1 is a rotational viscosity. For 8CB, $K = 2\times10^{-7}$dyn has been measured by M. J. Bradshaw, E. P. Raynes, J. D. Bunning, and T. E. Faber, J. Phys. (Paris), **46**, 1513 (1985) and $\gamma_1 = 0.25 dyn/cm^2/s$ by H. Kneppe, F. Schneider, and N. K. Sharma, J.Chem.Phys. **77**, 3203 (1982).

13. In these experiments, $G = 7.5 K/cm$. We measured the width of the two phase region for our mixture to be $\Delta T \approx 0.1 K$, thus the thermal length, $\ell_T = \Delta T/G \approx 133 \mu m$. It is interesting to note that the wavelength at onset is very close to ℓ_T. The dynamic diffusion length is, $\ell_D \equiv D_I/v < 2 \mu m$ as $v_c \approx 19 \mu m/s$.

14. 8CB = cyanooctylbiphenyl and C15 = cyano(methyl)butoxybiphenyl

15. J. S. Patel, T. M. Leslie and J. W. Goodby, Ferroelectrics **59**, 137 (1984).

16. K. A. Jackson and J. D. Hunt, Acta Metall. **13**, 1212 (1965) were the first to suggest this experiment to study solidification in transparent materials to develop understanding of the solidification process in metals.

17. This is in contrast to the forward bifurcation observed for the onset of the cellular pattern in nematic liquid crystals for which $q_o = v_{e\ell} = \omega_{e\ell} = 0$. See: P. Oswald, J. Bechhoefer, A. Libchaber and F. Lequeux, Phys. Rev. **A36**, 5832 (1987); J. Bechhoefer, A. J. Simon, A. Libchaber and P. Oswald, Phys. Rev. **A40**, 2042 (1989) and A. J. Simon and A. Libchaber, Phys. Rev. **A41**, 7090 (1990).

18. We identify the defects in Fig. 8 as λ defects, also called "double twist" by S. Meiboom, J. P. Sethna, P. W. Anderson and W. F. Brinkman, Phys. Rev. Lett. **46**, 1216 (1981), because the twist axis has two spatial degrees of freedom, i.e. along $\hat{x}$ and $\hat{y}$.

19. To underscore the generic aspect of the breathing mode, we note that a pattern similar to Fig. 8 has been observed by M. Zimmermann, A. Karma and M. Carrard, Phys. Rev. **42**, 833 (1990) in Al-Cu alloys close to the eutectic. A distinct advantage of our system over metallic systems is that, in our system, the pattern's time evolution can be followed.

EXPERIMENTAL STUDY OF THE PATTERN FORMATION OF ICE CRYSTAL GROWN IN SUPERCOOLED WATER

Y.FURUKAWA and W.SHIMADA
Institute of Low Temperature Science, Hokkaido University, Sapporo 060, Japan

ABSTRACT
In situ observations of ice crystals grown in supercooled water were conducted by using a new growth chamber combined with a Mach-Zehnder interferometer. As a result, the growth patterns of ice dendrites were analyzed in three-dimensional viewpoint. Several characteristics in the pattern formation of ice dendrite are summarized.

1.Introduction

It is well known that the morphology of ice crystal grown in the supercooled water sequentially changes with growth time from circular disks, to disks with perturbed periphery, and finally to developed dendrites with hexagonal symmetry[1,2,3]. A snow crystal which grows from the water vapor in the atmosphere also changes its morphology depending on the temperature and supersaturation. The growth process of both ice and snow crystals correspond to the physical problem of pattern formation under the non-equilibrium condition. Many researchers have studied both theoretically and experimentally the morphology of snow and ice crystal in such view point so far. In this paper, we concentrate to the problem of ice crystal growth from the supercooled water.

Komabayasi [4] conducted a simulation study such as the hexagonal pattern of snow dendrite is originated on the basis of Mullins-Sekerka morphological instability theory[5], which was applied to the diffusion field of water vapour around the snow crystal. It is to be emphasized that he introduced the anisotropy of surface tension of ice in the calculation of the melting temperature of ice surface. However, considering both the fact that the surface kinetic process is ignored in this calculation and the analogy between the diffusion fields of water vapor and heat, the patterns obtained by Komabayasi may be regarded as those of ice crystal grown from the supercooled water rather than the snow crystal[6]. Fujioka and Sekerka[7] also discussed the morphological stability of disk crystals growing at the water surface on the basis of M-S instability theory. Langer et al.[8] measured the growth velocities at the tip of ice dendrite growing in the supercooled water as a function of supercooling $\Delta T(=T_m-T)$ and indicated that the curve expressed in appropriate dimensionless variables is consistent with the prediction by the Langer-Müller-Krumbhaar theory[8]. Recently, Tirmizi and Gill[9,10] carried out the in-situ observation of ice crystal growing in supercooled water and discussed about the influence of free convection to ice crystal morphology and the scaling law in the development of dendritic patterns. It must be noted that all of these investigations have been done on the assumption that the ice dendrites are in two-dimensions(projected on a basal plane). Since the actual ice dendrites have three-dimensional morphologies, we can not conclude that the full understanding for the pattern formation of ice crystals have been already established.

In this paper, the results of in-situ observation for ice crystals growing in the supercooled pure water will be shown and the three-dimensional morphology of ice crystal will be preliminary discussed[12].

2. Experimental

An experimental apparatus was designed specifically for the purpose of observing the ice dendrites freely growing in the supercooled pure water. Figure 1 indicates a schematic illustration of both the growth cell and the optical system. A parallelepiped growth cell ($20x20x40mm^3$) made of optically flat glass-plates was immersed in a cubic water container ($70x70x70\ mm^3$) with glass windows. A glass capillary, whose diameter was smaller than 0.8mm, was inserted in the growth cell. Whole growth chamber was put in an insulated and temperature-controlled cold box.

After the water temperature has reached to a fixed supercooling, ice crystals are initially nucleated in the capillary by the method of inserting a chilled thin wire. Though they begin to grow inside the capillary, only one crystal whose a-axis is parallel to the capillary can be survived in the capillary. Because the the ice crystal can grow most rapidly along the direction of $<11\bar{2}0>$ axis[13]. It emerges at the tip of capillary and starts to grow freely in the supercooled water.

A growing ice crystal was observed through a Mach-Zehnder interferometer as shown in Fig.1. When an ice crystal developed in three-dimensions, interference fringes dependent on the ice crystal thickness were observed on the ice surface. During the experiment, the reference light beam of interferometer was periodically chopped using a chopper. When the chopper is closed, this optical system changes to the Schlieren system and then the outline of ice crystal can be observed more clearly. Results of in-situ observations were recorded using a TV camera-VTR system combined with an image analyzer.

Temperatures of supercooled water was measured by a thin thermocouple inserted in the growth cell. An accuracy of temperature measurements was absolutely within 0.01K in this experiment. The experiments were carried out in the temperature range between -0.05 and -1.00℃.

Figure 2 shows time-sequent pictures of an ice crystal grown at the supercooling $\Delta T=0.31K$, observed through the Schlieren system. It is to be noted that the perturbed branches extend radially with the hexagonal symmetry. The directions of the axes of dendrites are along the $<11\bar{2}0>$ axes[13].

Figure 3 shows the pictures observed through both the Schlieren system(left) and the Mach-Zehnder interferometer(right). The degrees of supercooling were given in the figure caption. Interference fringes are clearly observed on the ice crystal. In the next section, we will discuss about some interesting aspects obtained in this experiment.

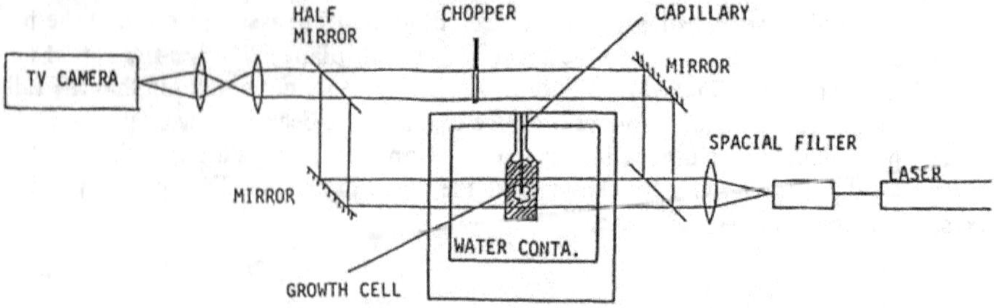

Fig.1 Schematic diagram of the growth cell and the optical system.

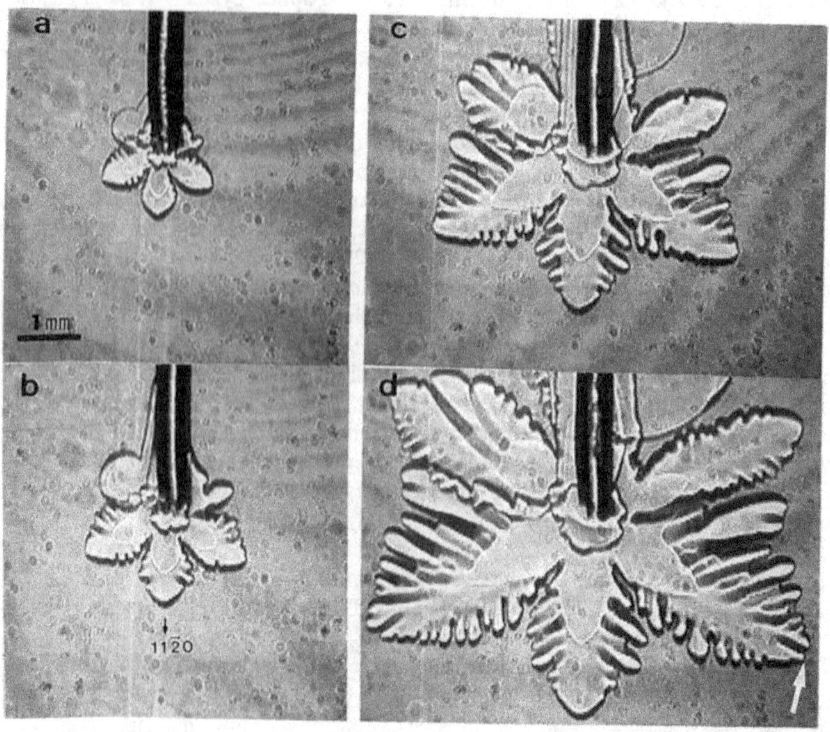

Fig.2 Time sequent pictures of an ice crystal observed through the Schlieren optical system. A tip splitting is observed at the tip of dendrite. Grown at the supercooling $\Delta T = 0.31K$. Growth time:(a)76sec, (b)143sec, (c)272sec and (d)560sec.

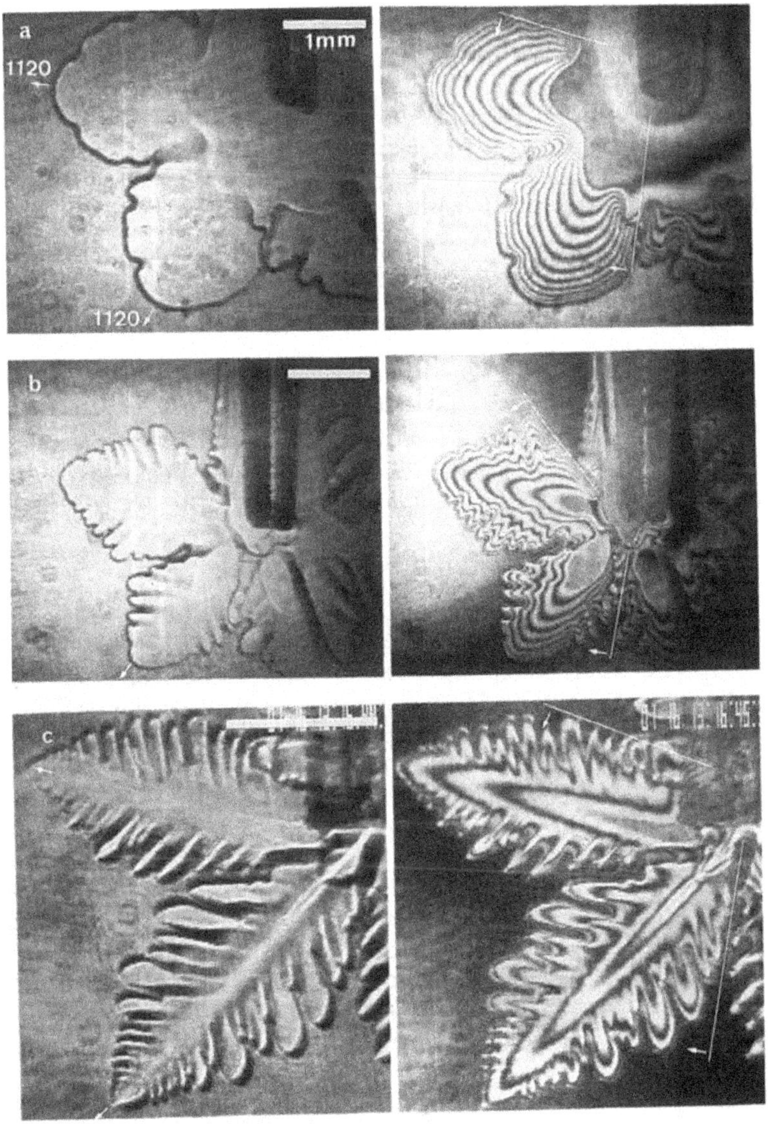

Fig.3 Ice crystals observed both the Schlieren system(left) and the Mach-Zehnder interferometer(right). Supercooling temperatures are (a) $\Delta T=0.12K$, (b) $\Delta T=0.37K$ and (c) $\Delta T=0.91K$, respectively.

3.Discussion

3.1 Anisotropy of ice-water interfacial tension

In order to understand the hexagonal symmetry of ice crystal dendrite, we first consider the anisotropy of ice-water interfacial free energy(interfacial tension) among the crystallographic faces parallel to the c-axis. It is well known that, in the equilibrium condition, the depression in melting temperature of ice-water interface due to Gibbs-Thomson effect is given by the following equation[14,15],

$$T_e(0) - T_e(K) = \frac{K}{\Delta S_f}\left(\gamma(\theta) + \frac{d^2\gamma(\theta)}{d\theta^2}\right), \qquad (1)$$

where $T_e(0)$ and $T_e(K)$ are the melting temperatures corresponding to the flat interface($K=0$) and the interface with curvature K, respectively, ΔS_f the fusion entropy and $\gamma(\theta)$ the interfacial tension at the interface with the polar angle θ of its normal. If $\gamma(\theta)$ is isotropic, the second derivative of (1) disappears. In order to the hexagonal symmetry of ice crystal, we assume that $\gamma(\theta)$ is given by the relation of

$$\gamma(\theta) = \gamma_0 (1 + \varepsilon_6 \cos 6\theta), \qquad (2)$$

where θ is the polar angle of interfacial normal, which is measured from $[1\bar{2}10]$ axis, and ε_6 the parameter indicating the degree of anisotropy. Six maxima of $\gamma(\theta)$ appear at the directions of crystallographic $<11\bar{2}0>$ axes, as shown in Fig.4(c). It is to be noted that this relation is reasonably presumed from the polar plot of surface energy $\nu(\theta)$ as shown in Fig.4(b). This polar plot was obtained from the ice crystal structure showed in Fig.4(a), assuming that the surface energy was proportional to the specific number density of hydrogen bonds intersecting to the ice surface, that is, the number density of uncompensated bonds .

If we can observe the equilibrium form of ice crystal, it is possible to determine the parameter of interfacial tension anisotropy ε_6 [16,17,18]. Though it was quite difficult to observe the ice crystal form at the exact equilibrium condition, we were able to observe the ice crystal growing at the condition close to the equilibrium. As the result, it was indicated[18] that its form is a nearly circular disk with weak hexagonal symmetry and the parameter of anisotropy ε_6 is given in the order of 10^{-2}.

3.2 Characteristic of tip-splitting growth of ice dendrite

It is usually known that the tip splitting at the dendritic growth of crystal does not occur at the tip of dendrite and occur at regular intervals from the dendrite tip. The pictures of ice crystal showed in Fig.2, however, indicate that the tip splitting can occur at the tip of dendrite in the case of ice crystal growth (at the portion pointed by an arrow). This phenomenon was usually observed when the ice crystals grew at the

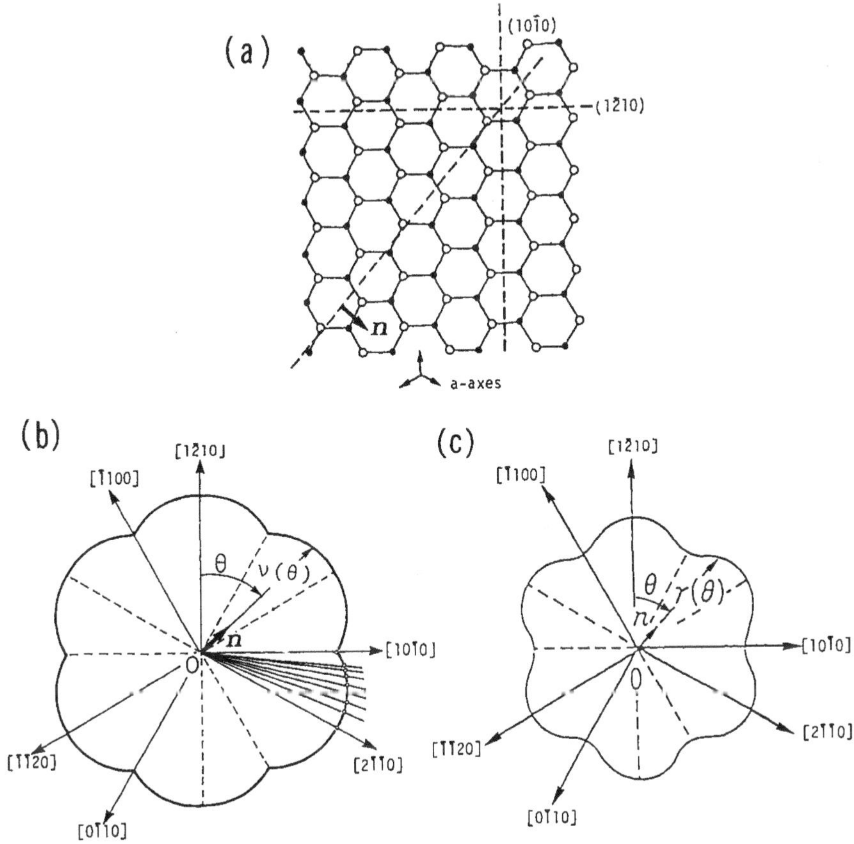

Fig.4 (a)Ice crystal structure: Open and solid circles indicate the oxygen atoms and solid lines connecting between circles the hydrogen bonds. Projected along [0001] axis; (b)Polar plot of the surface energy $\nu(\theta)$ of ice crystal; (c) Polar plot of surface tension of ice $\gamma(\theta)$, given by equation (2).

condition of supercooling lower than 0.6K. Saito et al.[19] has discussed the problem of tip splitting, taking the anisotropy of interfacial tension into account. They showed that the direction of the main axis of dendrite must coincide to the directions of the maxima of interfacial tension and the tip splitting never occur in this directions. Nevertheless, it is to be emphasized that the tip splitting is observed at the tip of ice dendrite, namely along the direction of maximum interfacial tension. A possible reason of this tip splitting may be originated from the higher order term in the anisotropy of interfacial tension (namely, $-\varepsilon_{12}\cos 12\theta$), which is ignored in the equation (2).

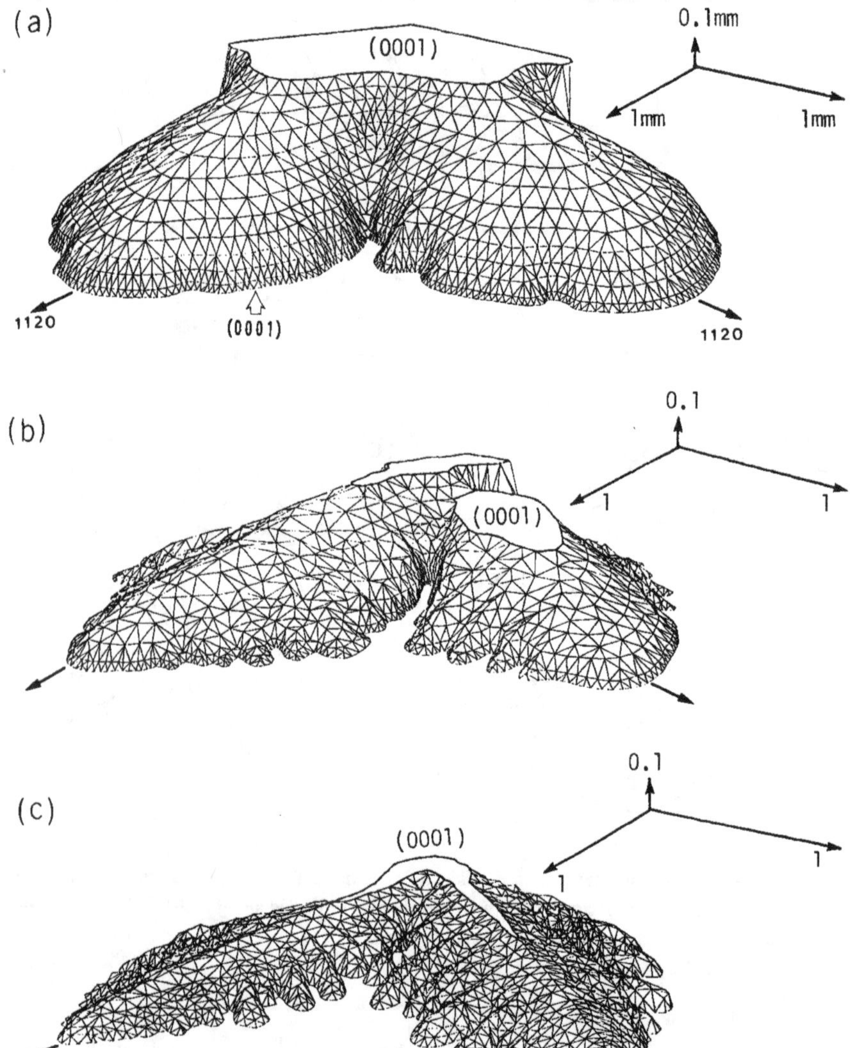

Fig.5 Three dimensional morphologies of the ice crystals. The ice crystal thickness was estimatedfrom the interference fringes. Figures (a), (b) and (c) correspond to the pictures of Fig.3.

3.3 Three-dimensional morphology of ice dendrite [12]

Using the interference fringes observed through the Mach-Zehnder interferometer, it is possible to estimate the thickness of ice crystal. The difference of ice crystal thickness between the neighboring interference fringes observed on the ice crystal is given as $d(m+1) - d(m) = \lambda / |n_i - n_w| = 25.4\ \mu m$, where λ (wave length) = 632.8nm, n_i (refractive index of ice)=1.3079 and n_w (refractive index of bulk water)=1.3327 at 0℃, respectively.

Figure 5 shows the three-dimensional illustrations of ice crystals showed in Fig.3. It is to be noted that the portion of dendritic form is constructed by the combination of a flat basal face and a rounded face. Though the results of detailed analysis of three-dimensional pattern formation will given in the other paper, we emphasize here that the previous theoretical considerations about the dendritic growth, in which the solid of revolution or two-dimensional pattern is usually assumed as the shape of dendrite tip, can not be applied to the problem of ice crystal pattern formation.

4.Conclusions

(1)In situ observations of ice dendrite were conducted by using a new optical technique and the analysis of the three-dimensional pattern formations of ice dendrite became possible.

(2)The hexagonal symmetry of ice dendrite was discussed on the basis of the anisotropy of ice-water interfacial tension.

(3)The tip splitting occurred at the tip of dendrite in case of ice crystal growth.

(4)Three-dimensional patterns of ice dendrites were shown illustratively, and it was to be emphasized that an ice dendrite is constructed by the combination of the flat basal face and the rounded interface.

5.Acknowledgments

This work was supported by Fund for Basic Experiments Oriented to Space Station Utilization (The Institute of Space and Astronautical Science, Japan), Grant No. U-4 and also partially supported by Grant-in Aid for Scientific Research on Priority Areas "Crystal Growth Mechanism in Atomic Scale" No.03243203.

6.References

1. K.Arakawa and K.Higuchi, *J.Fac.Sci.,Hokkaido Univ.*, Ser.II,4(1952)201.
2. K.Arakawa, *J.GLacio.*, *2*(1955)463.
3. P.V.Hobbs, in *Ice Physics*, (Clarendon Press, Oxford,1974)p572.
4. M.Komabayasi, *J.de Recherches Atmospheriques*, *6*(1972)307.

5. W.W.Mullins and R.F.Sekerka, *J.Appl.Phys., 34*(1963)323.
6. E.Yokoyama and T.Kuroda, *Phy.Rev., A41*(1990)2038.
7. T.Fujioka and R.F.Sekerka, *J.Cryst.Growth, 24/25*(1974)84.
8. S.Langer, R.F.Sekerka and T.Fujioka, *J.Cryst.Growth, 44*(1978)414.
9. J.S.Langer and H.Müller-Krumbhaar, *Acta Metall., 26*(1978)1681,1689,1696.
10. S.H.Tirmizi and W.N.Gill, *J.Cryst.Growth, 85*(1987)488.
11. S.H.Tirmizi and W.N.Gill, *J.Cryst.Growth, 96*(1989)277.
12. W.Shimada, Master's Thesis, Hokkaido Univ., Sapporo, (1992).
13. J.W.Glen and M.F.Perutz, *J.Glacio., 2*(1954)397.
14. T.Kuroda, in *Morphology and Growth Units of Crystal*, ed.I.Sunagawa, (Terra Sci.Pub.,Tokyo, 1989)577.
15. S.O'hara, L.A.Tarshis, W.A.Tiller and J.P.Hunt, *J.Cryst.Growth, 3/4*(1968)555.
16. A.Dougherty and J.P.Gollub, *Phy.Rev., A38*(1988)3943.
17. P.Oswald, *J.Phy.France, 49*(1988)1083.
18. Y.Furukawa and W.Shimada, in *Proceedings on Int.Sympo.Physics and Chemistry of Ice(PCI-91)*, (Hokkaido Univ. Press,Sapporo, 1992) in print.
19. Y.Saito,G.Goldbek-Wood and H.Müller-Krumbhaar, *Phy.Rev.Lett., 58*(1978)1541; *Phy.Rev., A38*(1988)2148.

COMPUTER SIMULATION OF MECHANICAL AND RHEOLOGICAL BEHAVIOR OF TWO-DIMENSIONAL RANDOM CELLULAR SYSTEMS

Kyozi KAWASAKI, Tohru OKUZONO, Toshihiro KAWAKATSU
Department of Physics, Kyushu University 33, Fukuoka 812, Japan

and

Tatsuzo NAGAI
Physics Department, Kyushu Kyoritsu University, Kitakyushu 807, Japan

ABSTRACT

Mechanical and rheological properties of foam systems which have random cellular structures are investigated by computer simulations. A simulation model describing the system is constructed taking effects of viscous flow inside the liquid films into account. The computer simulations using this model are performed on the case under simple shear flow. Results of the simulations suggest that the system shows non-Newtonian behavior with a yield stress, which is experimentally observed in oil-water emulsion systems.

1. Introduction

A foam under consideration consists of cells of gas or liquid separated by thin films of liquid of another kind. A large system composed of a great number of tiny cells exhibits quite interesting mechanical and rheological behavior. Under a small applied external stress the system behaves as an elastic body but exhibits plasticity under large stresses suggesting existence of a yield stress. A foam flow in a tube has a plug flow region in the middle and shows a finite slip at the tube wall and viscous shear flow in between with the shear-rate dependent viscosity. Thus macroscopically a foam system has the full complexity of a structured fluid. Experimental and theoretical investigations of foam systems have a long history dating back to the days of Plateau and Kelvin. The developments up to 1988 are reviewed by Kraynik.[1]

Random cellular systems exemplified are providing a class of most difficult many-body problems to treat theoretically. Here every cell of edge number n is in contact with its n neighbors and any change in the cell immediately gives rise to changes in its neighbors. Hence interactions among cells are always present and cannot be treated perturbatively in any sense. Furthermore, typical edge number n is not so large, about 6 in two dimensions. This precludes mean-field-like treatment which is sometimes valid when a body interacts with a large number of others as in polymer blend and Coulomb gas systems. In this sense our problem is similar to that of classical simple liquids where computer simulation has been an indispensable tool. This is also true here despite some attempts at analytical theory.[2] As far as mechanical and rheological properties are concerned, all of the existing theoretical studies are for regular cellular systems accompanied by periodic behavior[1] which is unrealistic for random cellular systems.

Turning now to computer simulations, previous efforts were mostly directed to the domain growth problem (coarsening).[3] The only exception is the works by Weaire and his coworkers[4,5] which dealt with elastic and plastic response under given deformation. So far, to our knowledge, no computer simulation exists to study nonequilibrium stationary state behavior of random cellular systems.

For the past several years we have been engaged in the task of developing a class of models for computer simulation of random cellular patterns, the so-called vertex models as described in Refs.6,7 and papers quoted therein.

2. Vertex Model

We consider an arbitrary cellular system specified by position $\mathbf{r}(a)$ of every element of cell boundaries parametrized by a. This automatically determines positions of intersections of cell boundaries denoted as $\mathbf{r}_i$, $i = 1, 2, \cdots$. A free energy functional $H(\{\mathbf{r}_i\}, \{\mathbf{r}(a)\})$ can be associated with each state where bulk free energies of cells can be also included if different cells are not energetically degenerate. If we assume purely dissipative dynamics we need the dissipation functional $R(\{\mathbf{v}_i\}, \{v(a)\}, \{\mathbf{r}_i\}, \{\mathbf{r}(a)\})$ which is one half of the energy dissipation rate. Here $\mathbf{v}_i$ is the velocity of the vertex i and $v(a)$ is the velocity of the cell boundary element at a along its normal direction $\mathbf{n}(a)$. Then the equations of motion take the following form,

$$\frac{\delta R}{\delta v(a)} = -\mathbf{n}(a) \cdot \frac{\delta H}{\delta \mathbf{r}(a)}, \tag{2.1}$$

$$\frac{\partial R}{\partial \mathbf{v}_i} = -\frac{\partial H}{\partial \mathbf{r}_i}, \tag{2.2}$$

if all the variables entering H and R were independent. Since this is not the case, Eqs.(2.1) and (2.2) must be suitably modified to incorporate constraints.

Numerical solution of this type of equation is computationally very expensive and cannot be readily extended to large systems. Thus we have chosen to develop computationally simpler models which hopefully capture essence of Eqs.(2.1) and (2.2). The idea is to focus only on the vertices $\{\mathbf{r}_i\}$ and eliminate other detailed information on cell boundary configurations contained in $\{\mathbf{r}(a)\}$. If we imagine fixing $\{\mathbf{r}_i\}$ artificially and let the system evolve with Eq.(2.1), eventually the system reaches a local equilibrium state with fixed positions of vertices. If different cells are energetically degenerate, this local equilibrium state is a state with minimal cell boundary surfaces (that is, surfaces with zero mean curvature). If cells are energetically non-degenerate, a boundary between two cells has a mean curvature proportional to the difference of the bulk free energy density of the two adjoining cells. In particular, for a two-dimensional system with degenerate bulk free energy densities, cell boundaries simply reduce to straight lines connecting two vertices. If we substitute this cell boundary configuration to the modified version of Eq.(2.2), we end up in a set of effective equations of motion involving vertices. In reality, however, this is justified only if motion of vertices is much slower than that of cell boundaries connecting pairs of vertices, which is generally not true. Hence the resulting vertex equations are better interpreted as new model equations for cellular systems.

The general vertex equation then has the form :

$$\sum_j \mathbf{D}_{ij} \cdot \mathbf{v}_j = \mathbf{f}_i \, , \tag{2.3}$$

where $\mathbf{f}_i = -\partial H_V / \partial \mathbf{r}_i$ is the force acting on the vertex i and H_V is the free energy of the vertex system in which $\mathbf{r}(a)$ is eliminated as above mentioned, $\mathbf{D}_{ij} = \mathbf{D}_{ji}^T$ is the friction tensor which in general still depends upon vertex configuration, where a superfix T stands for transpose. Eq.(2.3) implies that the vertex dissipation function is given by

$$R_V(\{\mathbf{v}_i\}, \{\mathbf{r}_i\}) = \frac{1}{2} \sum_{ij} \mathbf{v}_i \cdot \mathbf{D}_{ij} \cdot \mathbf{v}_j \, . \tag{2.4}$$

The concrete examples of vertex equations have been given elsewhere[6,7] and will not be repeated. Here we merely point out that anisotropy in the friction tensor $\mathbf{D}_{ij}$ which reflects local anisotropy in a cellular system is often quite important.

When cellular systems evolve the topology of cellular networks changes. It is described by the following two elementary processes when an edge becomes infinitesimal (In the computer simulation we have a cut-off length Δ.) : if the edge is that of a triangular cell the triangular cell changes to a vertex (T2-process), otherwise a neighbor switching occurs (T1-process). The details of the elementary processes are given in Refs.6 and 7.

3. Quasistatic Properties

Here we consider static response of the system under shape deformations, in particular, elongational deformations. Since a cellular system is not always in thermal equilibrium and degradation (or here coarsening) takes place over long time scales, our observation time must be limited to times much shorter than typical degradation times. Such degradation occurs by diffusion of fluid in a cell into another cell through cell boundaries, thereby changing areas of individual cells. We can thus stabilize the system by completely neglecting the diffusion process. Then our problem is to find a local free energy minimum under a given deformation starting from a suitable initial state.

Search for free energy minima can be most conveniently carried out by the relaxation method in which the simplest vertex equation like Eq.(2.3) is solved under the condition of fixed areas of all the cells until a stationary state is reached.

Here we have only the T1-process for topological change. The procedure and some preliminary results were reported in some detail in Refs.8,9 and so here we will be very brief.

The system shows elasticity for small deformations where no T1-process takes place, whereas it shows plasticity for larger deformations where T1-processes occur. The vertex models can be used to study mechanical properties of random cellular structures including dynamical problem like viscoelasticity as will be described in the next section.

4. Dynamic Properties

4.1 Vertex Model for Rheological Study

Here we consider a two-dimensional foam system with liquid films separating foams. It is now well understood[1] that the viscous dissipation occurring in liquid films near Plateau borders (vertices in our language) cannot be neglected. Thus, in addition to dissipation associated with diffusion of air across cell boundaries the viscous dissipation mentioned above must be considered.

The starting point of obtaining the latter dissipation is the following result[10] for energy dissipation rate occurring in the part of liquid film adjacent to a Plateau border :

$$\sigma^{1/3}(3\mu)^{2/3} I\,|U|^{5/3}\,, \tag{4.1}$$

where σ is the surface tension of the liquid film surface, μ the shear viscosity of the liquid in the film, U is the speed of the Plateau border relative to the liquid film ($U > 0$ and $U < 0$ for contracting and expanding liquid films, respectively). I is an integral of some function involving film thickness near the Plateau border having the following values,[11]

$$I = \begin{cases} I_e \equiv 1.2215 & U < 0 \\ I_c \equiv 1.1866 & U > 0 \end{cases} \tag{4.2}$$

The result Eq.(4.1) was derived using the so-called lubrication approximation where one assumes the Stokes flow and slow variation of film thickness with the distance from the Plateau border.

Our basic assumption is that the motion of a Plateau border is entirely due to ejection and withdrawal of liquid films. This is very natural in the framework of vertex model. Consider movement of a vertex in a fixed environment where all other vertices are motionless. The only way the vertex under consideration can move is by ejecting and/or absorbing liquid films connected to the vertex. Then, if the velocity of a Plateau border is $\mathbf{v}$, U in Eq.(4.1) is given by $-\mathbf{v}\cdot\hat{\mathbf{r}}$, where $\hat{\mathbf{r}}$ is the unit vector along the film directed toward the Plateau border. Thus, the dissipation function R_P associated with the viscous relaxation near Plateau borders is given using Eqs.(4.1) and (4.2) by

$$R_P = \frac{1}{2}\sigma^{1/3}(3\mu)^{2/3}\sum_i \sum_j^{(i)} |\mathbf{v}_i \cdot \hat{\mathbf{r}}_{ij}|^{5/3}[I_e\theta(\mathbf{v}_i \cdot \hat{\mathbf{r}}_{ij}) + I_c\theta(-\mathbf{v}_i \cdot \hat{\mathbf{r}}_{ij})] , \tag{4.3}$$

where $\hat{\mathbf{r}}_{ij}$ is the unit vector along $\mathbf{r}_i - \mathbf{r}_j$ and the summation with respect to j is over the vertices bonded to the vertex i. $\theta(x)$ is equal to 1 or 0 for positive and negative x, respectively.

We should also consider dissipation due to diffusion across cell boundaries. This was already given in Refs.6,7 but here a rough estimate of its size suffices. The corresponding dissipation function here denoted as R_D is the same as R_V, Eq.(2.4), where $\mathbf{D}_{ij}$ is of the order of $\sigma r_{av}/L$. Here r_{av} is the average linear size of cells and L is the kinetic coefficient for cell boundary motion due to diffusion. The same coefficient also appears in von Neumann's law : $dA_n/dt = L(n-6)$, where A_n is the area of an n sided cell. Thus we obtain the estimate

$$R_D \sim N\frac{\sigma}{L}r_{av}v^2 , \tag{4.4}$$

where N is the total number of cells and v the average speed of vertex motion. We now compare this with Eq.(4.3) which is roughly,

$$R_P \sim N\sigma^{1/3}\mu^{2/3}v^{5/3} . \tag{4.5}$$

We then find that R_D dominates whenever r_{av} exceeds r_c given by

$$r_c \equiv L\mu/\sigma . \tag{4.6}$$

Indeed in this case $v \sim L/r_{av}$ and hence we find

$$R_P/R_D \sim (\mu v/\sigma)^{2/3} \sim (r_c/r_{av})^{2/3} \ll 1 . \tag{4.7}$$

Since r_c is of microscopic size in general we are in the regime where R_D dominates. This implies that the cell boundary movement due to diffusion across cell boundary meets with strong resistance (friction) and will not likely to take place in short times, that is, areas of individual cells stay constant during the time interval in which rheological behavior is observed. Thus we adopt the vertex equation in which the dissipation function is R_P and the constraints of fixed areas of all cells are taken into account by introducing a Lagrangian multiplier λ_α for every cell α :

$$\frac{\partial R_P}{\partial \mathbf{v}_i} = -\frac{\partial}{\partial \mathbf{r}_i}[\sigma\sum_{\langle ij\rangle}|\mathbf{r}_{ij}| - \sum_\alpha \lambda_\alpha A_\alpha] , \tag{4.8}$$

where the two sums on the rhs. are over every pair of vertices connected by a bond and over every cell α with the area A_α, respectively. More explicitly,

$$\frac{5}{6}\sigma^{1/3}(3\mu)^{2/3}\sum_j^{(i)}|\mathbf{v}_i\cdot\hat{\mathbf{r}}_{ij}|^{2/3}\hat{\mathbf{r}}_{ij}[I_c\theta(\mathbf{v}_i\cdot\hat{\mathbf{r}}_{ij})-I_c\theta(-\mathbf{v}_i\cdot\hat{\mathbf{r}}_{ij})]$$

$$=-\sum_j^{(i)}\sigma\hat{\mathbf{r}}_{ij}-\frac{1}{2}\sum_\alpha^{(i)}\lambda_\alpha\hat{\mathbf{z}}\times\mathbf{r}_{jk}\,, \tag{4.9}$$

where the sum in the second term on the rhs. is over the cells which share the vertex i, $\hat{\mathbf{z}}$ is the unit vector perpendicular to the plane of cells, and j and k designate vertices of the cell α bonded to i such that (ijk) are arranged counterclockwise. The λ's are determined from the conditions

$$\frac{\mathrm{d}}{\mathrm{d}t}A_\alpha=-\frac{1}{2}\sum_i^{(\alpha)}\hat{\mathbf{z}}\times\mathbf{r}_{jk}\cdot\mathbf{v}_i=0\,, \tag{4.10}$$

where the sum is over the vertices belonging to the cell α. Incidentally we observe that the average vertex velocity in the Plateau border dominated dynamics $v\sim\sigma/\mu$ scales as the same way as the interface velocity in the spinodal decomposition of fluid mixtures.[13]

We now consider imposing shear strain with the constant rate of shear $\dot{\gamma}$ on the system. This introduces a macroscopic velocity field $\mathbf{u}(\mathbf{r})$ given by

$$\mathbf{u}(\mathbf{r})=\dot{\gamma}y\hat{\mathbf{x}}\,, \tag{4.11}$$

where $\hat{\mathbf{x}}$ is the unit vector along the x-axis. Then U now becomes $-[\mathbf{v}-\mathbf{u}(\mathbf{r})]\cdot\hat{\mathbf{r}}$, $\mathbf{r}$ in $\mathbf{u}$ being the position vector of a Plateau border. Thus $\mathbf{v}_i$ everywhere in Eq.(4.9) is to be replaced by $\mathbf{v}_i-\mathbf{u}(\mathbf{r}_i)$. One may also make the same replacement in Eq.(4.10) since the extra contribution involving $\mathbf{u}$ cancells because a shear deformation keeps an area fixed. Under shear strain the periodic boundary condition must be modified. For the x-direction the usual periodic boundary condition can be imposed. On the other hand, for the y-direction, the two boundaries move with different speeds in the x-direction due to the imposed constant strain rate. Thus the two points on the two boundaries connected by the periodic boundary condition also move with the same speeds as the two boundaries. Such a boundary condition is familiar in nonequilibrium molecular dynamics computer simulation[14], that is, a point (x,y) on the system is identified with a point $(x+m_xL_x+m_y\dot{\gamma}L_yt, y+m_yL_y)$ at time t, where m_x and m_y are integers and L_x and L_y are dimensions of the system along the x- and y-axes, respectively.

4.2 Simulation Results

We carry out computer simulations by the method given above. The units of length and time are chosen to be r_{av} and $2^{-3/2} \cdot r_{av}\mu/\sigma$, respectively. Here r_{av} is the average linear size of cells defined as $(A/N_c)^{1/2}$, where A is the total area of the system and N_c is the total number of the cells which is now a constant. The size of a time step δt at the time t is taken to be Δ/v_{max}, where v_{max} is the maximum value of $|\mathbf{v}_i|$ ($i = 1, 2, \cdots$) at that time and Δ is the cut-off length for the topological changes fixed at 0.01. (We permit small changes of cell areas of order $r_{av}\Delta$ when a topology change occurs.) The initial state is chosen to be a ripe state in the scaling regime obtained using the original vertex model (model O of Ref.6) and contains 459 cells. We perform 7 runs for this initial state with 7 different shear rates $\dot{\gamma}$ within the range $1.0 \times 10^{-3} \leq \dot{\gamma} \leq 1.6 \times 10^{-2}$.

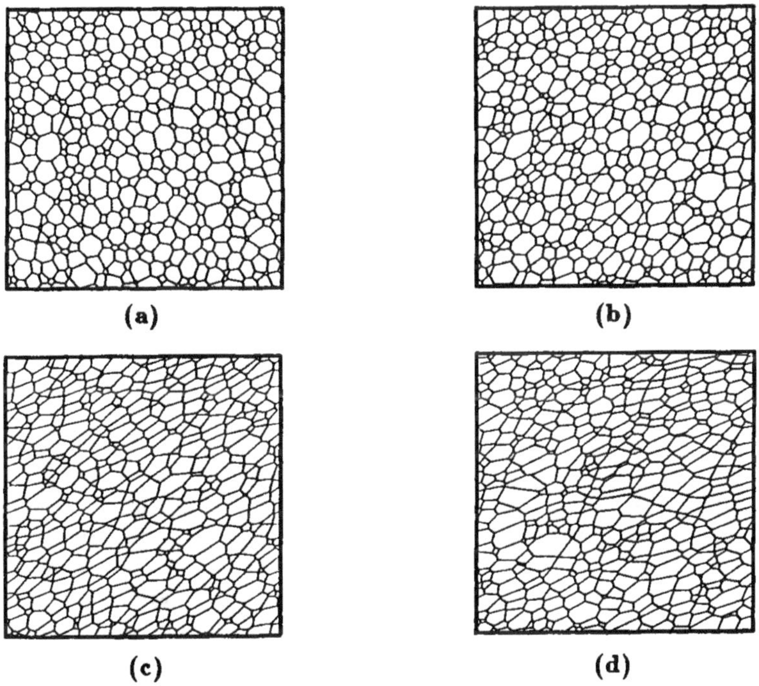

Fig.1 Sequential snapshots of cellular pattern during shear deformation : (a) $\dot{\gamma}t = 0.0$ (initial state), (b) $\dot{\gamma}t = 0.4$, (c) $\dot{\gamma}t = 1.2$ and (d) $\dot{\gamma}t = 4.0$; $\dot{\gamma} = 1.0 \times 10^{-3}$.

In Fig.1 we present some snap shots of the patterns during shear deformation for $\dot{\gamma} = 1.0 \times 10^{-3}$. In Fig.2 we show the shear stress $\tau_{xy}(t; \dot{\gamma})$ as a function of shear strain $\dot{\gamma}t$ for the three different values of $\dot{\gamma}$. The symbols □, △ and × in the figure correspond to $\dot{\gamma} = 1.0 \times 10^{-3}$, 4.0×10^{-3} and 1.6×10^{-2}, respectively. Here an $\alpha\beta$-component of stress tensor $\tau_{\alpha\beta}(t;\dot{\gamma})$ at time t is defined as

$$\tau_{\alpha\beta}(t;\dot{\gamma}) = \frac{2\sigma}{A} \sum_{\langle ij \rangle} |\mathbf{r}_{ij}| \hat{r}_{ij}^{\alpha} \hat{r}_{ij}^{\beta} , \tag{4.12}$$

where $\hat{r}_{ij}^{\alpha}$ is α-component of $\hat{\mathbf{r}}_{ij}$. The shear stress is an xy-component of this tensor and is measured in the unit of $2\sigma/r_{av}$ in the simulation. As one can see in Fig.2, the system shows elasticity in the early stage and appears to reach a steady state in the late stage in each run.

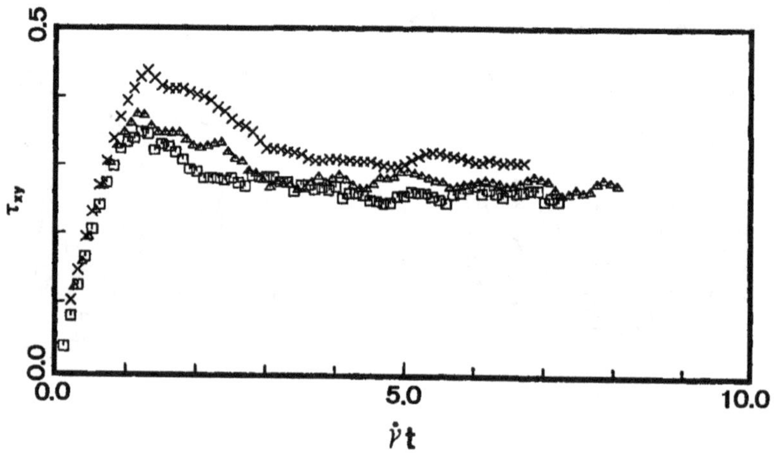

Fig.2 Shear stress $\tau_{xy}(t;\dot{\gamma})$ as a function of total shear strain $\dot{\gamma}t$ for the three different shear rates : 1.0×10^{-3} (□), 4.0×10^{-3} (△) and 1.6×10^{-2} (×).

In order to see what is actually happening in the system under shear flow we show in Fig.3 along side with the shear stress the frequencies of T1-processes per one cell divided by the shear rate $\dot{\gamma}$ plotted as functions of shear strain (time). We see that initially frequencies of T1-processes increase rapidly and seem to reach maxima at the peaks of shear stress, and then decrease towards constant values when steady states are reached. The time average of steady state frequencies of T1-processes per cell divided by the shear rate $\dot{\gamma}$ (a dimensionless quantity) is plotted in Fig.4. We see that this quantity is independent of $\dot{\gamma}$. This is the expected result since the system tries to maintain a steady state by balancing increasing strain at the constant rate $\dot{\gamma}$ by counteracting T1-processes with the rate proportional to $\dot{\gamma}$.

We regard the system in a regime $\dot{\gamma}t > 3.5$ to be in the steady state and obtain the steady state shear stress $\tau_{xy}(\dot{\gamma})$ by averaging $\tau_{xy}(t;\dot{\gamma})$ over that regime. We plot the steady state shear stress $\tau_{xy}(\dot{\gamma})$ versus shear rate in Fig.5.

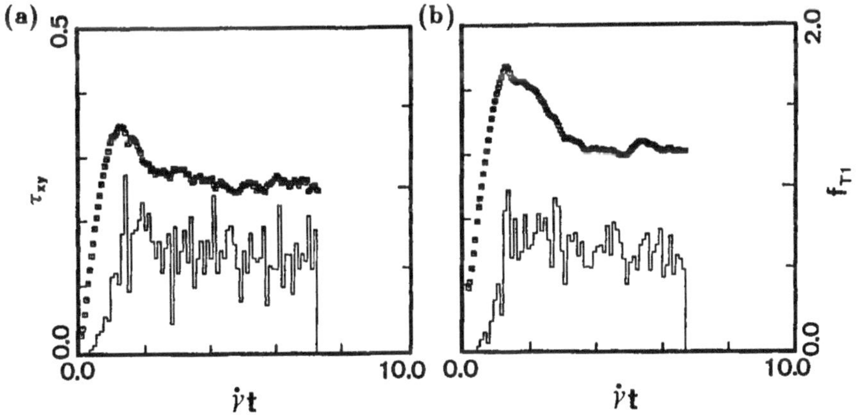

Fig.3 Frequencies of the T1-processes and the shear stress vs shear strain $\dot{\gamma}t$. Here f_{T1} is the frequency of the T1-processes per cell divided by $\dot{\gamma}$. The values of shear rates are (a) $\dot{\gamma} = 1.0 \times 10^{-3}$ and (b) $\dot{\gamma} = 1.6 \times 10^{-2}$.

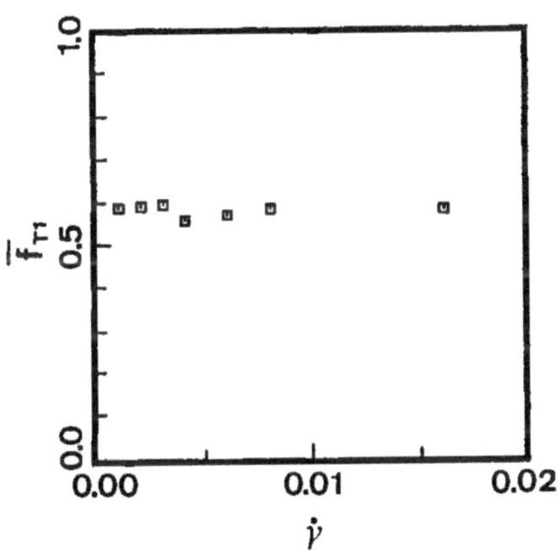

Fig.4 The time averages of f_{T1} in steady states $\bar{f}_{T1}$ where averages are taken for $\dot{\gamma}t \geq 3.5$ up to the maximum times of computation.

A recent experiment[12] on oil-water emulsion systems suggests the relation

$$\tau_{xy}(\dot{\gamma}) - \tau_0 = C\dot{\gamma}^{\alpha} \tag{4.13}$$

with $\alpha = 1/2$, where τ_0 is the yield stress (more precisely, the dynamic yield stress) and C is a positive constant. (Both τ_0 and C depend on a volume fraction of dispersed phase in general.) Based on the experimental fact, we assume the form Eq.(4.13) to fit our data. The solid line in Fig.5 is determined by the following procedure. Given τ_0, we plot $\log[\tau_{xy}(\dot{\gamma}) - \tau_0]$ versus $\log\dot{\gamma}$ and determine the best fitting line. We can obtain τ_0 to minimize the error sum of squares against the fitting line in the form of Eq.(4.13). Then we obtain the best fitting line drawn in Fig.5. In Fig.6 we show the doubly-logarithmic plot for the value 0.237 for τ_0 thus determined together with the fitting line. The value of α which is the slope of the line is 0.460. This value is close to the experimental value of 0.5.[12]

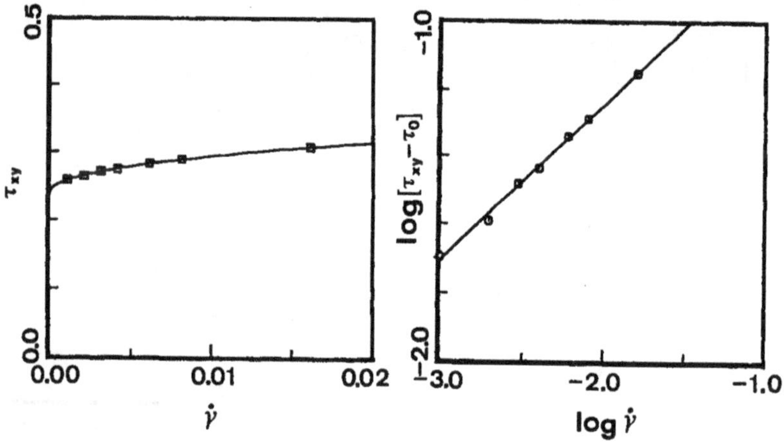

Fig.5(left) Shear rate dependence of shear stress $\tau_{xy}(\dot{\gamma})$ in steady states. Solid line is the best fitting one.

Fig.6(right) Plot of $\log[\tau_{xy}(\dot{\gamma}) - \tau_0]$ vs $\log\dot{\gamma}$ for $\tau_0 = 0.237$. Solid line is the best fitting one and the slope of the line is 0.460.

5. Concluding Remarks

In this paper we have proposed a new model to simulate rheological behavior of foam systems and performed computer simulations using the model. We have realized a steady state of the system under shear flow in the simulation and obtained the steady state shear stress as a function of shear rate. To our knowledge, this is the first computer simulation for the rheology of random cellular systems. Our results support the experimental data for oil-water emulsion systems[12] that gives the relation $\tau_{xy}(\dot{\gamma}) - \tau_0 \propto \dot{\gamma}^{1/2}$ together with a finite value of the yield stress τ_0. The exponent 1/2 in this relation, however, is different from the exponent 2/3 in an analytical model[10] where an ordered cell structure is assumed. This value also differs from the exponent 1 for a Bingham plastic. Such a behavior was observed for soap froth.[15] It is clear that the randomness of the structure is crucial even for the exponent α. On the other hand we have expected that our model is more appropriate for soap froth than for emulsions since viscous dissipation occurring within the dispersed phase, namely, in the interior of cells, may not be negligible.[16]

We should also be aware that our simulation was done for the values of shear rates considerably greater than the experimental values. Obviously further works are needed to clarify these points.

In the present work we have considered the limiting case where the volume fraction of dispersed phase (gas phase) ϕ is very close to unity although the viscous effects of the liquid films are taken into consideration. This quantity ϕ is an important parameter to understand foam rheology more completely. It is a future problem to study ϕ-dependence of rheological quantities, which is, however, beyond the present vertex model.

6. Acknowledgements

We would like to thank S.Ohta for valuable discussions. This work was supported by the Scientific Research Fund of the Ministry of Education, Science and Culture of Japan.

7. References

1. A.M.Kraynik, *Ann.Rev.Fluid Mech.* **20**, 325 (1988).
2. M.Hillert, *Acta Metall.* **13**, 227 (1965).
   V.E.Fradkov and D.G.Udler, in *Simulation and Theory of Evolving Microstructures*, eds. M.P.Anderson and A.D.Rollett, (TMS Publications, Warrendale, PA, 1990)
   J.Stavans, E.Domany and D.Mukamel, *Europhys.Lett.* **15**, 479 (1991).
3. See, e.g. articles by H.I.Frost *et al*, J.A.Glazier *et al* and E.A.Holm *et al* in *Simulation and Theory of Evolving Microstructures*, eds. M.P.Anderson and A.D.Rollett, (TMS Publications, Warrendale, PA, 1990)
4. D.Weaire and T-L.Fu, *J.Rheol.* **32**, 271 (1988).
5. D.Weaire and J.P.Kermode, *Phil.Mag.* **B50**, 379 (1984).
6. K.Kawasaki, T.Nagai and K.Nakashima, *Phil.Mag.* **B60**, 399 (1989).
7. K.Nakashima, T.Nagai and K.Kawasaki, *J.Stat.Phys.* **57**, 759 (1989).
8. K.Kawasaki, T.Nagai and T.Okuzono, in *Microscopic Aspect of Nonlinearity in Condensed Matter*, ed. V.Tognetti, (Plenum Press, 1991)
9. K.Kawasaki, T.Okuzono and T.Nagai, in *Research of Pattern Formation*, ed. R.Takaki, (KTK Scientific Publishers, Tokyo, 1991) in press
10. L.W.Schwartz and H.M.Princen, *J.Colloid.Interface Sci.* **118**, 201 (1987).
11. D.A.Reinelt and A.M.Kraynik, *J.Colloid.Interface Sci.* **132**, 491 (1989).
12. H.M.Princen and A.D.Kiss, *J.Colloid.Interface Sci.* **128**, 176 (1989).
13. E.D.Siggia, *Phys.Rev.* **A20**, 595 (1979).
14. M.P.Allen and D.J.Tildesley, *Computer Simulation of Liquids*, (Clarendon Press, Oxford, 1987)
15. S.A.Khan, C.A.Schnepper and R.C.Armstrong, *J.Rheol.* **32**, 69 (1988).
16. D.A.Reinelt and A.M.Kraynik, *J.Fluid Mech.* **215**, 431 (1990).

FILAMENTARY PATTERNS AND ROUGH SURFACES

L. LAM, R.D. FREIMUTH,* M.K. PON,
D.A. KAYSER, J.T. FREDRICK and R.D. POCHY

*Nonlinear Physics Group, Department of Physics,
San Jose State University, San Jose, CA 95192-0106, USA*

Experimental results from two types of filamentary growth patterns are presented, viz., dielectric breakdown in a thin layer of liquid under uniform electric field, and the dense linear morphology in copper electrodeposits. Active walker models (AWMs) consisting of interacting walker and landscape are used to simulate these and other filamentary patterns and rough surfaces. The AWMs are applicable to many other complex systems.

1. Introduction

In nonlinear science,[1,2] there are currently two directions of research being pursued intensely. One is the physics of pattern formation;[3,4] the other is the study of a landscape. The latter include the rough surfaces,[5] the sandpile model[6] and the surface growth model[7] of self-organized criticality,[8] and the rugged landscape model of evolutionary biology.[9] The former include both compact patterns (e.g., the interfacial studies in viscous fingering[10], directional solidification[11] and electrodeposits[12]) and filamentary patterns [e.g., those in diffusion limited aggregates (DLAs),[13] electrodeposits,[12] dielectric breakdowns,[14,15] and many other systems[16]].

Until recently, these two branches of research have been investigated separately. For example, in the study of filamentary pattern formation, random walkers (or their generalizations) are allowed to walk in a lattice -- a blank landscape; obviously one is not interested in the landscape. On the other hand, in the landscape study such as in the surface growth models,[17,7] even though random "walkers" are used to create and modify the landscape[18] -- the rough surface -- the attention is on the landscape and not on the tracks of the walkers.

A unified approach was proposed recently by Freimuth and Lam[18] (Ref. 18 will be referred as FL) in which the walker is capable of altering the landscape as it walks, and its steps are in turn influenced by the changing landscape. In other words, it is an *active* walker model with the walker and the landscape interacting with each other. The tracks

* Present address: Department of Physics, University of California at San Diego, La Jolla, CA 92093.

of the walker(s) form the filamentary pattern of interest if one is interested in pattern formation; the resulting landcape is the object of study if one is interested in the dynamics of landscape/rough surface. The AWM can be viewed as a new way of forming filamentary patterns which is quite different from the DLA model and, as shown in FL and below, presents more possibilities.

Essentially, in the AWM the landscape is represented by a scalar field, the "potential" ,$V(\mathbf{r}_i,n)$, defined at each site i (with position $\mathbf{r}_i$) on a lattice and at discrete time n (n =0,1,2,...). V is the sum of two parts: a possibly inhomogeneous, time-dependent background V_0 and an ever-updating part V_1 due to the influence of the active walker. Specifically, we assume that $V(\mathbf{r}_i,n) = V_0(\mathbf{r}_i,n) + V_1(\mathbf{r}_i,n)$, with $V_1(\mathbf{r}_i,0) = 0$ and

$$V_1(\mathbf{r}_i,n+1) = V_1(\mathbf{r}_i,n) + W(\mathbf{r}_i-\mathbf{R}(n-m)) \tag{1}$$

where W is an isotropic, time-independent potential (which may also be taken as anisotropic or time dependent if needed); for example, W could assume the form given in Fig. 1. In (1), m (= 0,1,2,...) is the time lag and W = 0 if n < m; $\mathbf{R}(n)$ is the position of the walker at time n. The walker will terminate if $V(\mathbf{R}(n),n+1)$ is lower than that at all its empty neighboring sites; otherwise, it will step onto one of the available sites -- those empty neighboring sites with potential lower than that at $\mathbf{R}(n)$. In the *deterministic* version (DAW), it will step onto the available site with the lowest potential; in the *probabilistic* version (PAW), onto one of the available sites with a probability proportional to the potential difference between that at $\mathbf{R}(n)$ and at the available site. [In the DAW model, if more than one neighboring sites have the same lowest potential, then one of these sites is chosen randomly to be the $\mathbf{R}(n+1)$.] Generalizations to other selection rules are straightforward. (See FL for more details.)

In this paper, experimental results on two examples of filamentary patterns are presented (Sec. 2 and 3). Computer results from the AWMs (in addition to those in FL and with these experimental patterns in mind) are given, which include patterns from single (Sec. 4) or multiple branching walkers (Sec. 5), and the scaling properties of rough surfaces produced as the landscapes (Sec. 6). A brief conclusion is included at the end (Sec. 7).

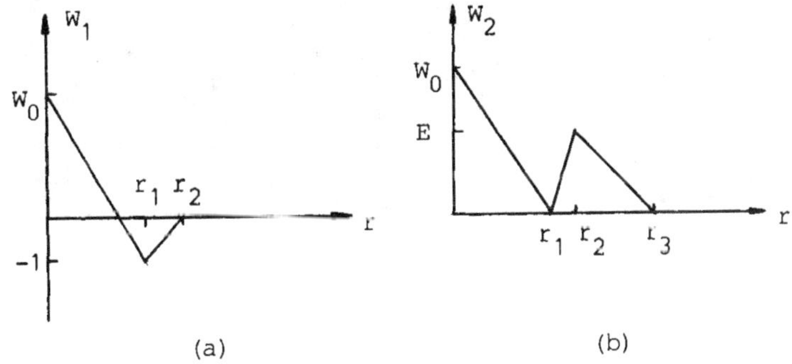

Fig. 1. Two possible isotropic W(r) functions. r is the magnitude of **r**.

2. Dielectric Breakdown in a Thin Layer of Liquid

Dielectric breakdown (DB) is an important topic in both basic and applied research. The usual investigation of DB in liquids is done by applying a high voltage between two electrodes immersed in the liquid.[19] In contrast, in Ref. 15 a new method of doing it is presented; the idea is to create and record a pattern of the DB tracks, and by placing a thin layer of the liquid between two conducting glass plates we are studying a quasi-two-dimensional system (under a uniform vertical electric field).

The same setup used in Ref. 15 is adopted here. Fig. 2(a) shows the tracks in a cell of mineral oil (NDC 0003-0559-33 from E.R. Squibb & Sons) when a d.c. voltage of 300 V is applied briefly. (The DB threshod voltage[15] is ~ 250 V.) For comparison, previous results[15] in a similar cell obtained by gradually increasing the voltage to slightly above the threshold is presented in Fig. 2(b). In both cases, a flash was observed during the BD which lasted a very short time (presumably in less than a msec which we did not measure). Apparently, the DB starts from a point in Fig. 2(a) and from multiple points in Fig. 2(b). What they share in common are: (i) the general features of individual tracks -- the *forward-bias* effect (the tendency to go forward), the *track-affinity* effect (winding when the track turns around), branching, nonoverlapping and termination. (ii) When enlarged, each track can be seen to consist of alternative thick, wrinkled strings and thin, smooth strings.[15,20] The physics of these DB patterns is complicated[18] and will not be addressed here. Intead, these patterns will be compared with those from the AWM below.

(a) (b)

Fig. 2. Dielectric breakdown pattern in a thin cell of mineral oil. Cell thickness is 25 μm. (a) Actual size is 1.1x0.6 cm^2. (b) Actual size is 2.5x2.1 cm^2. (See text.)

3. Dense Linear Morphology in Copper Electrodeposits

Electrodeposit experiments generate a large variety of patterns[12] which are still to be understood. In particular, the dense linear morphology (DLM) observed[21] in a linear cell is at once simple and intriguing because of the stability of a (almost) linear growth interface [Fig. 3(a)] and the existence of sharp morphological transition lines (MTLs) [Fig. 3(b)] as the pattern grows.[12,22] [The DLM is the counterpart of the dense radial morphology (DRM) observed in many different systems,[23] of which the stability is not yet understood.[24]] For both the DLM and the DRM, the current theoretical descriptions[25,26] assume a compact slab of the growing deposit by ignoring their filamentary and fractal (or self-affine) nature. Here, new experimental results on DLM in copper electrodeposit is presented and a phenomenological description is offered, which seems to agree surprisingly well with the experiments. The emphasis is on the time dependence of the current across the electrodes and the geometry of the deposit growth before the first sharp transition line.

The apparatus and experimental setup for a linear cell is described in Ref. 21. In the experiments described here, cell thickness d = 0.5 mm, cell width W = 4 cm, cell length L = 15 cm; the $CuSO_4$ solution is 0.05 M in concentration; the d.c. voltage between the two electrodes V is fixed at 30 V. During each run, the cell is photographed every 20 s; the current across the electrodes I(t) is sampled once every second. A typical I^{-1} is shown in Fig. 4. Note that I^{-1} increases first and then decreases monotonically; in each of the regions I to III, I^{-1} decreases linearly with time; the transition between regions II and III marks the MTL shown in Fig. 3(b). For region II, we propose three basic equations, viz., $\alpha I = dM/dt$, $M = \beta(x/W)^a$, and $R(x) = (W\rho_e/Ld)(1 - x/W)$ with $I \equiv V/R$. Here α, β and a are constants; M is mass of the aggregates; x(t) is (maximum) height of the aggregate filaments; R is the resistance of the cell, assuming to come completely from the electrolyte (of resistivity ρ_e) between the aggregates and the anode.

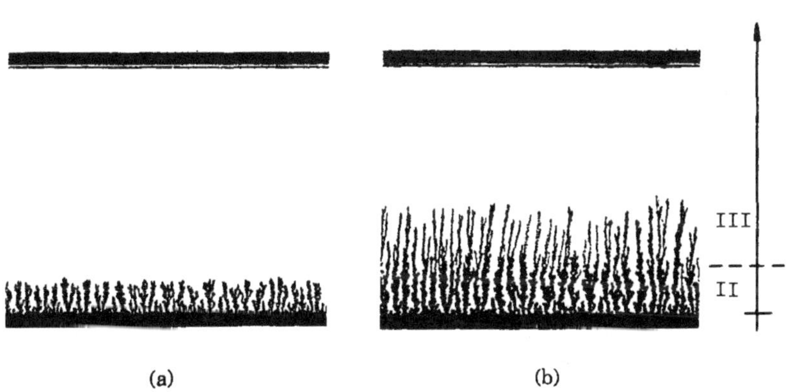

Fig. 3. Photographs of filamentary electrodeposit patterns of $CuSO_4$. Time t = 220 s (a); t = 480 s (b). Only the middle 1/3 of the cell is shown.

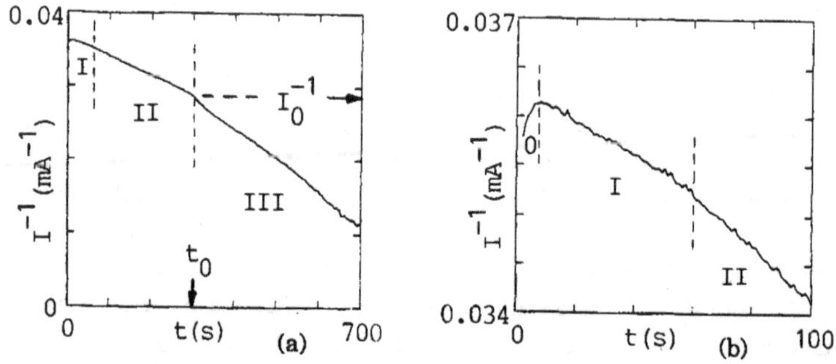

Fig. 4. Inverse current I^{-1} vs time t. (b) is an enlargement of (a) for small t.

Two immediate consequences are: (i) I^{-1} linear in t implies that the velocity of growth $v \equiv dx/dt = \text{const}$, which is verified from the series of photographs. (ii) $Q^{1/a} = c - eI^{-1}$, with $Q \equiv \int I dt$, $c \equiv (\alpha/\beta)^{-1/a}$ and $e \equiv cVLd/(W\rho_e)$. As shown in Fig. 5, $Q^{1/a}$ is indeed linear in I^{-1}, with $a = 1.21$ for the best fit.

From the scans of the photographs, we find $m \sim x^z$ with $z = 1.13$, where m is the number of pixels in each scan and is equal to the projection of the aggregate onto the two dimensional plane of the cell. The fact that $z < a$ confirms that the aggregate is not strictly two dimensional but quasi-two-dimensional, in consistent with the fact[12,21] that there is a sensitive dependence of the morphology on the cell thickness. Fig. 6 shows I^{-1} from repeated runs of the same experiment. There are slight quantitive differences between different runs due to the stochastic nature of the electrodeposit processes, but they are qualitatively the same. Further results are given in Ref. 27.

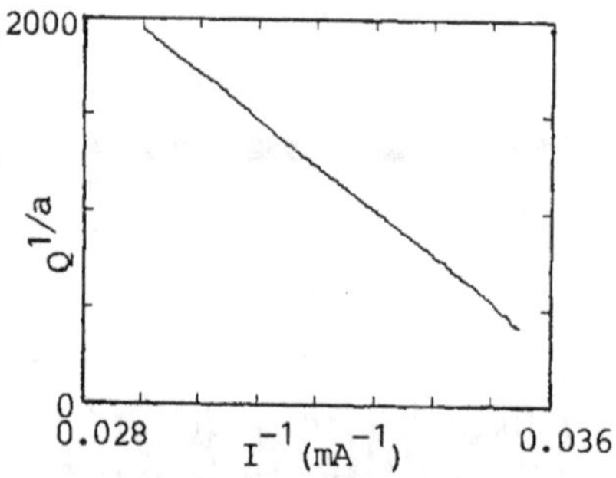

Fig. 5. Best fit of $Q^{1/a}$ vs I^{-1} with $a = 1.21$.

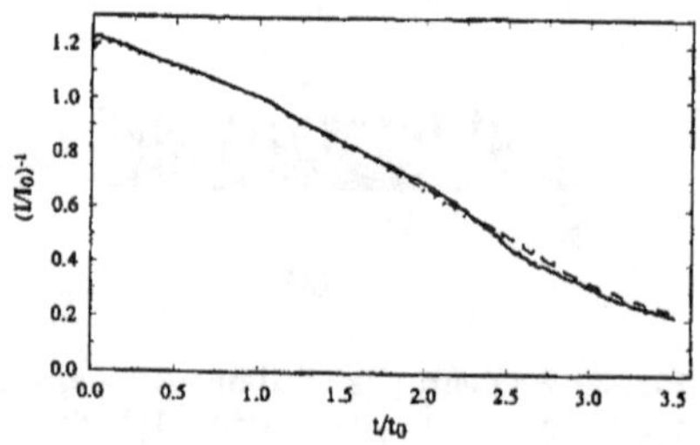

Fig. 6. Renormalized I^{-1} vs renormalized t for three different runs of the electrodeposit experiment (represented by solid, broken and dotted lines). I_0 and t_0 are defined in Fig. 4(a). The deviation from the mean of I_0 (t_0) is ~ 6% (~ 3%).

4. Filamentary Patterns I: Single Walkers

In Secs. 4-6, m = 0 and a square lattice is used. In Secs. 4 and 5, crossing of tracks is prohibited, and the walk is terminated if the walker hits the boundary of the finite lattice. In Figs. 7-9, single filaments from the PAW model with $V_0 = 0$ and $W = W_1$ ($W_0 = 15$, $r_2 = 20$, and $\rho = 0.1, 0.2, \ldots, 0.9$) are presented. Here, $\rho \equiv r_1/r_2$ and W is kept to one decimal point; grid size is 240x240. Compared to the case of m = 1 (see Fig. 4 of FL) there is more curling here. The ensemble averaged end-to-end length R_e and the radius of gyration R_g of the walk (before or until it naturally terminates) are found to vary with length (same as time) n as power laws [Fig. 8(a)]. The spatial distribution of the end points of repeated walks forms a fractal [Fig. 8(b)]. The overlapped picture of many walks varies interestingly as ρ is changed (Fig. 9).

(a) (b)

Fig. 7. Single filament from the PAW model. The dot represents the starting point at the center of the grid. (a) $\rho = 0.2$; (b) $\rho = 0.5$. Note that repeated runs could be different from each other in a PAW.

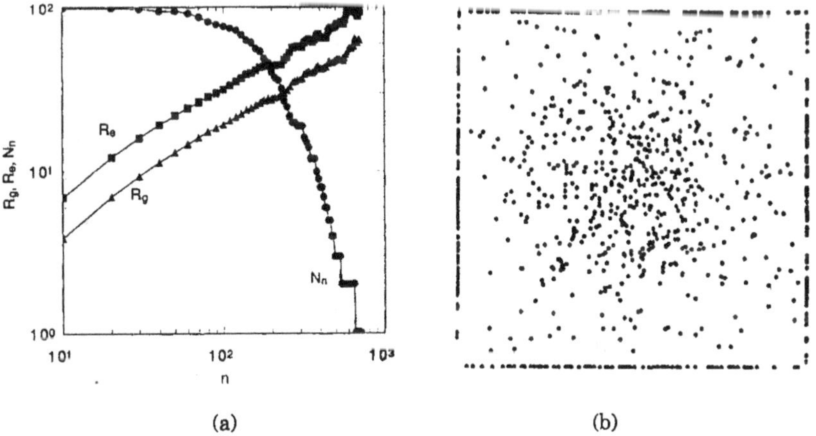

(a) (b)

Fig. 8. (a) Statistics of single filaments from 100 repeated runs. $\rho = 0.2$. $R_e \sim n^{0.63} \sim R_g$; N_n is the number of filaments with length larger than or equal to n. (b) Distribution of the end points of 1000 filaments, forming a fractal of dimension 1.82 (excluding the points on the edges of the grid). $\rho = 0.4$.

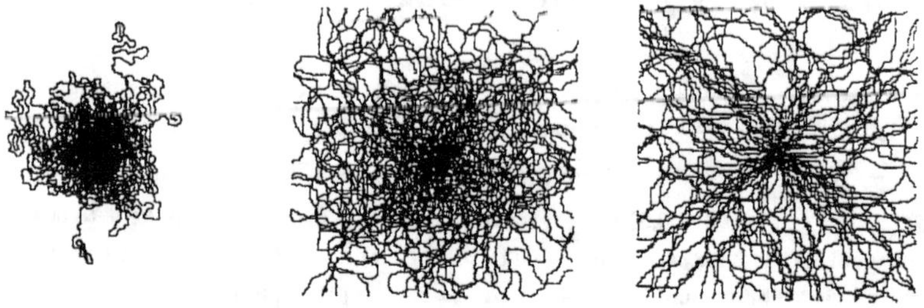

(a) (b) (c)

Fig. 9. Overlapped tracks of single walkers from 100 independent runs. (a) $\rho = 0.2$; (b) $\rho = 0.5$; (c) $\rho = 0.9$.

5. Filamentary Patterns II: Multiwalkers

The single walker is now allowed to branch. After the next site of the walker is chosen as described in Sec. 1, one looks at the next one among the set of available sites. If the potential difference of this next available site (with respect to the present location of the walker) is less than or equal to a factor of γ multiplying that of the chosen site, then this next available site is also occupied -- the walker branches into two walkers. Here γ is called the branching factor, a constant parameter.

For convenience, each walker is labeled with a seniority number, with the newly branched one assuming a new nember while the orginal walker continues to use the old number. The seniority number of a walker is assigned according to the order the walker is created, corresponding to its "age". At each step in time the walkers may be allowed to walk in sequence according to their seniority number, or be chosen randomly to walk irrespective of their seniority. The former (latter) version will be denoted by S (R). For example, SPAW (RPAW) stands for sequential (random) PAW, etc. Note that many other types of the seqential version may be easily created. As will be shown immediately, the morphology of some patterns may (see Fig. 10) or may not (see Fig. 11) depend on whether the S or R version is used.

When several walkers are used at n = 0, many different morphologies can be obtained. Some of them are shown in Figs. 10-12. In particular, the DRM is obtained (Fig. 11) where the filaments are more linear (i.e, less ramification) than those obtained from DLA-like models, in agreement with experiments (see, e.g., Ref. 12).

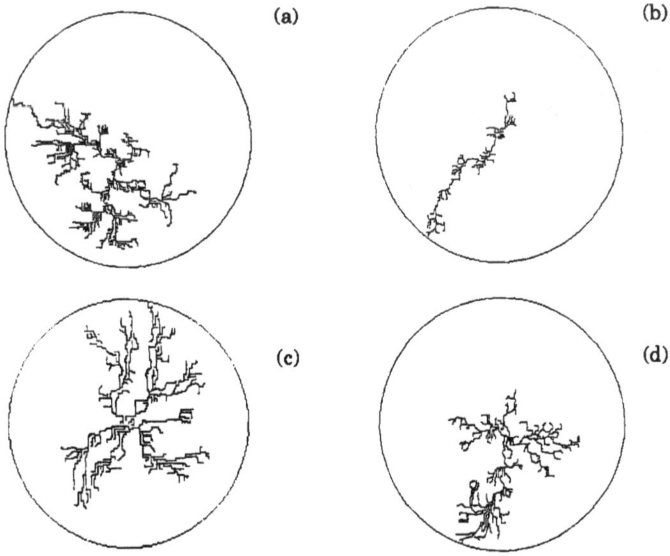

Fig. 10. Different fractal morphologies obtained from an initial set of four walkers placed at (0,±2) and (±2,0), where (0,0) is the center of the lattice. The seniority of the four is in the clockwise direction starting from the west; branching is allowed. V_0 is a cone with height V_c = 600 at (0,0) and reaching zero at a circle of radius L/2 (shown in each diagram), where LxL is the lattice size. $W = W_2$ with $W_0 = 5$, $r_2 = 10$ and $r_3 = 15$. Both V_0 and W are discretized, with integer values assumed. The box dimension is denoted by D; uncertainty is the standard deviation due to several runs. (a) RPAW; L = 252, $\gamma = 0.5$, $r_1 = 9$, E = 10; D = 1.666 ± 0.009. (b) RPAW; L = 252, $\gamma = 0.8$, $r_1 = 10$, E = 5; D = 1.29 ± 0.08. (c) RPAW; L = 175, $\gamma = 0.8$, $r_1 = 9$, E = 4; D = 1.42 ± 0.01. (d) SPAW; same parameters as in (b); D = 1.71. Note that the growth of the morphology can be either very much off center [as in (a), (b) and (d), similar to those found in electrodeposits in open cells[12]] or more symmetric with respect to the center [as in (c), similar to those from DLA].

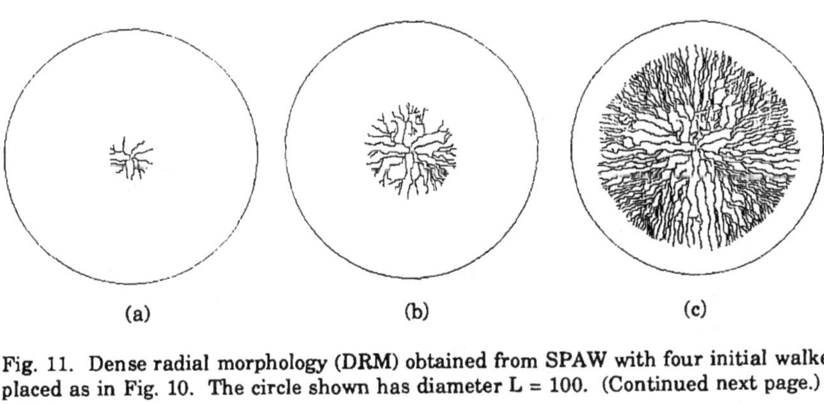

Fig. 11. Dense radial morphology (DRM) obtained from SPAW with four initial walkers placed as in Fig. 10. The circle shown has diameter L = 100. (Continued next page.)

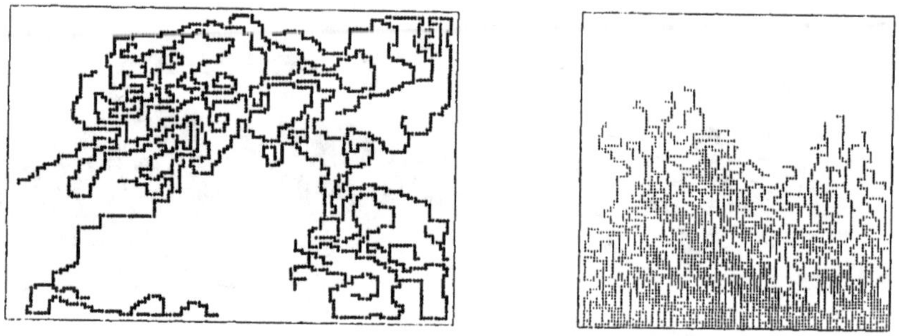

(a) (b)

Fig. 12. Filamentary patterns obtained from RPAW with $W = W_1$; $W_0 = 15$, $r_1 = 6$, $r_2 = 20$. The value of the potential is kept to one decimal point. (a) Seven initial walkers are placed randomly in a 100x125 lattice. $V_0 = 0$, $\gamma = 0.95$; number of pixels shown = 3363. Note the similarity with Fig. 2(b), including the existence of both large enclosed and open spaces. (b) 50 initial walkers placed evenly along the base line of a 100x100 lattice. V_0 is an upward inclined plane rising from zero at the base to 200 at the top line. $\gamma = 0.9$. This case fails to generate the filaments like that in Fig. 3.

6. Rough Surfaces

The landscape, the potential V(i,n) in our AWM, is a rough surface. (Here i stands for r_i, the lattice position.) In this section the simplest case of a one-dimensional (1D) lattice, a straight line of length L, is considered. Periodic boundary condition is used; $V_0 = 0$, i.e., the initial landscape is flat. A walker is dropped at random position along the line without changing the landscape at this first step. It then changes the landscape and walks as prescribed in Sec. 1, except that now self-crossing of the track *is* allowed, i.e., at every step it may move to left or right irrespective of whether that site has been visited before. At the end of this walk the walker is removed, and a new walker is introduced by dropping it randomly on the landscape created by the previous walker. The process is repeated.

In Figs. 13 and 14 we show two 1D rough surfaces created by our PAW model. The surface thickness $\sigma(L,h) \equiv \{\Sigma_i[V(i,N_s) - V_{av}(N_s)]^2/L\}^{1/2}$ where $V_{av}(N_s) \equiv \Sigma_i V(i,N_s)/L$, N_s is the total number of steps taken by all the walkers, and $h \equiv N_s/L$. The case in Fig. 13 does not conserve the total surface height, while that in Fig. 14 does. The log-log curves of $\sigma(L,h)$ vs h in both cases share the common property that there are four regions in

---

Fig. 11. (Continued.) Discretized $W = W_1$ with $W_0 = 5$, $r_1 = 12$ and $r_2 = 15$. Discretized V_0 is a cone with $V_c = -500$; $\gamma = 0.99999$. Number of pixels = 299 (a), 1475 (b), and 9452 (c). Note that the filaments are more sparse near the center; and the "defect," in the form of an opening, on the circular circumference appearing in the initial stages is always "healed" later -- the two phenomena often observed in electrodeposit DRM experiments [L. Lam (unpublished)]. The diamond shape at late stage is due to the square lattice used. Similar results are obtained with a RPAW.

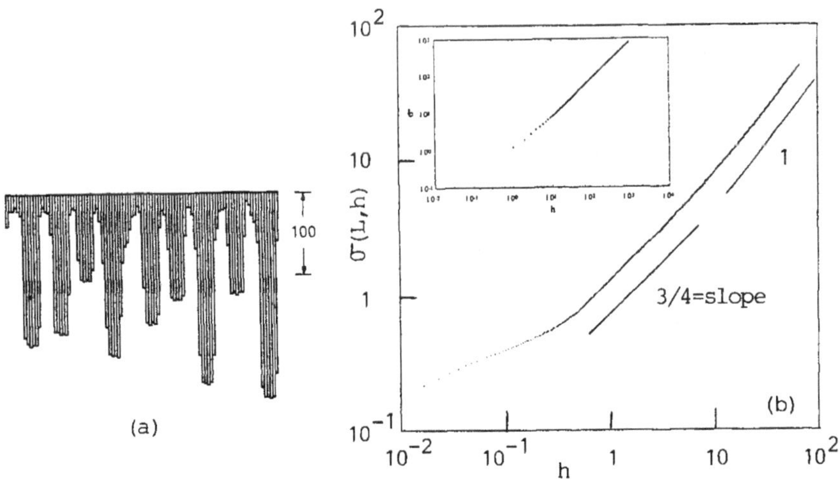

Fig. 13. (a) Rough surface generated by the PAW, with discretized $W = W_1$ ($W_0 = 1$, $r_1 = 1$ and $r_2 = 2$). $L = 100$; $N_s = 10^5$. All the heights are negative. (b) $\sigma(L,h)$ averaged over N_r runs vs h, with $N_r = 100$ and $N_s = 7000$. The insert is the same but with $N_r = 50$ and $N_s = 10^5$, where the h axis runs from 10^{-2} to 10^4.

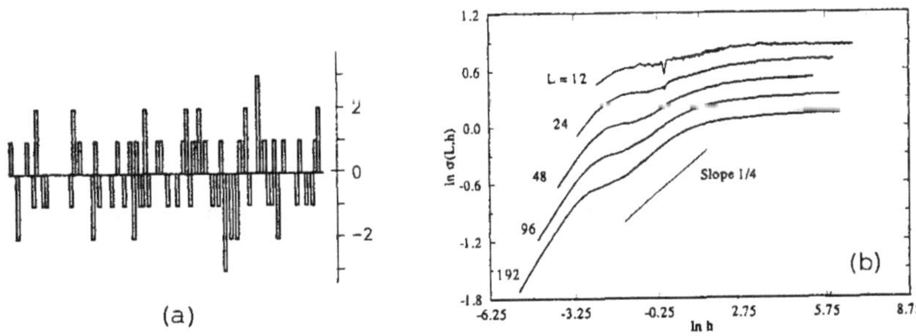

Fig. 14. (a) Rough surface generated by the PAW, with discretized $W = W_1$ ($W_0 = 2$, $r_1 = 1$ and $r_2 = 2$). $N_s = 10^5$. (b) Averaged Ln $\sigma(L,h)$ vs Ln h. $N_r = 10^3$. $N_s = 10^4$ for $L = 12$, 24 and 48. $N_s = 5\times10^4$ and 10^5 for $L = 96$ and 192, respectively. For clarity of presentation each curve, except the lowest one, is shifted 0.2 upward from the next lower one.

three of which a straight line (power law) exists. One is the usual initial region at small h, which is followed by a shoulder region, and then the two other straight line regions. The shoulder region is not found in the more conventional rough surface, e.g., the one created by surface diffusion[17]. We suspect that this is due to the existence of a new length scale in our case, viz., the end-to-end length of the active walker R_e.

Note that in Fig. 13, σ does not saturate, in contrast to the case in Ref. 17. Instead, the exponent α changes from 3/4 in the intermediate h region to 1 in the large h region, where α is defined by $\sigma(L,h) \sim h^{\alpha}$. The $\alpha = 1$ region corresponds to the steady state depicted in Fig. 13(a), corresponding to a almost periodic distribution of deep troughs in the landscape; this added "periodic" structure should be characterized by something apart from the scaling exponent α.

In Fig. 14, where V_{av} is conserved to be zero, $\alpha = 1/4$ which is the same as that in Ref. 17 but differs from the value (1/3) found in Ref. 28, another case with a conservation law. It looks like that $\sigma(L,h)$ will saturate at large h for fixed L, and $\sigma(L,\infty)$ seems to be independent of L for L large (i.e., $\beta = 0$), but more numerical data are needed to confirm these statements.

7. Conclusions

The study of formation of dielectric breakdown track patterns in a thin layer of liquid[15] is just one year old. More experiment and theory are needed. There is a longer history in the experimental study of electrodeposit patterns (at least since 1984), and the research is still going strong. But the physical understanding of these processes is just starting.

One of the great puzzles in Nature is the existence of many varieties of patterns which could be very different from (e.g., clouds and rivers) or very similar (e.g., rivers and tree branches) to each other. A mechanism which is simple but capable of producing very different patterns is called for. The AWMs proposed in FL and generalized here seem to have this important characteristic. The AWM is a local model close to the physical reality and offers much more possibilities than the DLA or the dielectric breakdown model. We have demonstrated that by varying the W function, very different patterns can be produced by the same model; we are able to provide here only a limited number of these demonstrations due to the lack of space and time. What remains to be seen is that when all the possible W's and walking rules are examined, if ever possible, will one be left with a small number of universal classes (such as the four cases found[29] in 1D cellular automata) of patterns and rough surfaces ? See FL for a discussion of the connection between our AWM and self-organized criticality.

On the biology front, recent experiments[30] demonstrate that the most common form of muscular dystrophy is caused by a gene that gradually grows bigger each time it is inherited. This seems to imply that the fitness landscape used in evolutionary biology[9] is not fixed but changes according to the location of the walker climbing the hills in the landscape. In other words, it is a situation of walker and landscape interacting with each other -- the case modeled by our active walker model.

This work is supported by the NSF-REU program and a Syntex Corporation grant from the Research Corporation.

8. References

1. *Nonlinear Physics for Beginners*, ed. L. Lam (World Scientific, River Edge, 1992).
2. *Introduction to Nonlinear Physics*, ed. L. Lam (Springer, New York, 1992).
3. *Nonlinear Structures in Physical Systems: Pattern Formation, Chaos and Waves*, eds. L. Lam and H.C. Morris (Springer, New York, 1990).
4. *Random Fluctuations and Pattern Growth: Experiments and Models*, eds. H.E. Stanley and N. Ostrowsky (Kluwer, Boston, 1988).
5. *Dynamics of Fractal Surfaces*, eds. F. Family and T. Vicsek (World Scientific, River Edge, 1991).
6. P. Bak, C. Tang and K. Wiesenfeld, Phys. Rev. Lett. **59**, 381 (1987); Phys. Rev. A **38**, 364 (1988).
7. L.M. Sander and H. Yan, Phys. Rev. A **44**, 4885 (1991).
8. P. Bak, Comput. Phys. **5**, 430 (1991).
9. *Molecular Evolution on Rugged Landscapes: Proteins, RNA and the Immune System*, eds. A.S. Perelson and S.A. Kauffman (Addison-Wesley, Redwood City, 1991).
10. P.G. Saffman, in Ref. 3.
11. E.A. Brener and V.I. Mel'nikov, Adv. Phys. **40**, 53 (1991); *Dynamics of Curved Fronts*, ed. P. Pelcé (Academic, San Diego, 1988); D. Kessler, J. Koplik and H. Levine, Adv. Phys. **37**, 255 (1988); J.S. Langer, in *Chance and Matter*, edited by J. Scouletie, J. Vannimenus and R. Stora (North-Holland, Amsterdam, 1987).
12. L. Lam, R.D. Pochy and V.M. Castillo, in Ref. 3.
13. T.A. Witten and L.M. Sander, Phys. Rev. Lett. **47**, 1400 (1981); L.M. Sander, in Ref. 2.
14. N. Niemeyer, L. Pietronero and H.J. Wiesmann, Phys. Rev. Lett. **52**, 1033 (1984).
15. L. Lam, R.D. Freimuth and H.S. Lakkaraju, Mol. Cryst. Liq. Cryst. **199**, 249 (1991).
16. *Kinetics of Nonhomogeneous Processes*, ed. G.R. Freeman (Wiley, New York, 1987).
17. F. Family, J. Phys. A **19**, L441 (1986).
18. R.D. Freimuth and L. Lam, in *Modeling Complex Phenomena*, eds. L. Lam and V. Naroditsky (Springer, New York, 1992).
19. I. Adamczewski, *Ionization, Conductivity and Breakdown in Dielectric Liquids* (Taylor and Francis, London, 1969).
20. L. Lam, in *Nonlinear and Chaotic Phenomena*, eds. W. Rozmus and J.A. Tuszynski (World Scientific, Teaneck, 1991).
21. M.A. Guzman, R.D. Freimuth, P.U. Pendse, M.C. Veinott and L. Lam, in Ref. 3.
22. V.M. Castillo, R.D. Pochy and L. Lam, in *Applications of Statistical and Field Theory Methods to Condensed Matter*, eds. D. Baeriswyl, A.R. Bishop and J. Carmelo (Plenum, New York, 1990); R.D. Pochy, A2. Garcia, R.D. Freimuth, V.M. Castillo and L. Lam, Physica **51**, 539 (1991).
23. E. Ben-Jacob and P. Garik, Nature **343**, 523 (1990).
24. The resistivity theory proposed by Grier et al (Ref. 25) to explain the stability of the DRM in electrodeposits seems to contain computational errors which invalidate their

conclusions. See M.K. Pon and L. Lam, in *Nonlinear and Chaotic Phenomena*, eds. W. Rozmus and J.A. Tuszynski (World Scientific, Teaneck, 1991); and unpublished. Moreover, the fact that the DRM is observed in systems other than electrodeposits points to the need of a much more general mechanism.

25. D.G. Grier, D.A. Kessler and L.M. Sander, Phys. Rev. Lett. **59**, 2315 (1987).
26. J.R. Melrose, D.B. Hibbert and R.C. Ball, Phys. Rev. Lett. **65**, 3009 (1990); D.P. Barkey and P.D. LaPorte, J. Electrochem. Soc. **137**, 1655 (1990); V. Fleury, M. Rosso, J.-N. Chazalviel and B. Sapoval, Phys. Rev. A **44**, 6693 (1991).
27. R.D. Freimuth, M.K. Pon and L. Lam, to be published.
28. T. Sun, H. Guo and M. Grant, Phys. Rev. Lett. **40**, 6763 (1989).
29. *Theory and Applications of Cellular Automata*, ed. S. Wolfram (World Scientific, Singapore, 1986).
30. *San Jose Mercury News*, Feb. 6, 1992, describing findings in three papers published in Nature by D. Housman and others.

DYNAMICS AND STRUCTURE OF AN AGGREGATION GROWN FROM A DIFFUSION FIELD

YUKIO SAITO
Department of Physics, Keio University,
3-14-1 Hiyoshi, Kohoku-ku, Yokohama 223, Japan

and

MAKIO UWAHA, SUSUMU SEKI
Institute for Materials Research, Tohoku University,
2-1-1 Katahira, Aoba-ku, Sendai 980, Japan

ABSTRACT

Structure and dynamics of an aggregation in a diffusion field is studied theoretically and the result is compared with Monte Carlo simulations of a lattice gas model. The model interpolates the diffusion-limited aggregation(DLA) and the Eden model. At low gas density and for short length scales, the structure of the aggregate is fractal and similar to the DLA, whereas for long length scales it is compact and uniform. The assumption that this crossover between the DLA and the compact structure happens at the scale of the diffusion length which is inversely proportional to the growth velocity can explain the density dependence of the growth rate in terms of the fractal dimension of DLA. By allowing solidification only on a large grid, the lattice anisotropy is enhanced. As the grid size increases, the growth pattern looks similar to dendrite, and the effective fractal dimension obtained by the $V - n_g$ relation decreases.

When the drift flow is introduced in the gas, a dynamical first order phase transition from the steadily growing state to the nongrowing one occurs on increasing the drift flow rate.

1. Introduction

A front moving in a diffusion field takes various patterns resulting from the competition among various destabilizing and stabilizing factors.[1] Typical example is a crystal interface growing from its melt or from a solution. When the crystal with a thermally rough interface grows in the diffusion field, a flat interface is unstable due to the Mullins-Sekerka instability. If the anisotropy in the interface tension or in the kinetic coefficient is sufficiently strong, the crystal has a dendritic form. Growth rate of the interface is shown to be determined by the microscopic solvability condition.[2] However, if the anisotropy is weak or the noise is large, we expect another mode of growth dynamics different from the regular dendrite, namely a growth of an irregularly formed crystal.[3] Our natural questions are then what determines the growth rate, and how the crystal structure looks like.

The simplest model to study the diffusion growth is a computer model called diffusion-limited aggregation(DLA)[4]. In the DLA model one places a solid seed at the origin of a square lattice. A gas atom is released far from the seed and migrates randomly on the lattice. When it happens to come next to the solid atom, it solidifies there. Then the second gas atom is released and migrates until it solidifies on some neighboring site of the solid. Repetition of this procedure produces a fractal aggregation. The shape of thus produced DLA is random fractal and almost isotropic. The number of solid atoms N changes in power of the linear size of the aggregate R:

$$N \sim R^D. \tag{1}$$

The exponent D here is called the fractal dimension of DLA, and smaller than the spatial dimension d. It also characterizes the power law decay of the correlation function of the solid density $n_s(\vec{r})$

$$< n_s(\vec{r}')n_s(\vec{r}' + \vec{r}) >\sim r^{D-d}. \tag{2}$$

The simulation of DLA shows that its fractal dimension is D =1.71 for $d = 2$.[5] The fractal pattern appears because the probability of finding a Brownian particle in the DLA model at a certain position follows the Laplace equation $\Delta n(\vec{r}) = 0$, which does not contain any characteristic length scale except the whole aggregate size.

In a general diffusional growth, the density of gas atoms obeys the time-dependent diffusion equation;

$$\frac{\partial n(\vec{r},t)}{\partial t} = \Delta n(\vec{r},t), \tag{3}$$

where the diffusion coefficient is set unity for convenience. The Laplace equation is merely an approximation for Eq.(3) when the growth is slow. A natural question then is the similarity and the difference of the DLA and the growth in a diffusion field.

2. Aggregation Growth From a Finite Density Lattice Gas

Our growth model in a diffusion field[6] is a generalization of the DLA model, such that an aggregation grows from a finite density lattice gas with the same solidification rule as that of the DLA model. The difference lies in the motion of a randomly chosen gas atom, such that it can move only to its *vacant* nearest neighbor site. We regard the lattice spacing as the length unit and assign, for each Monte Carlo trial, a time increment $\Delta t = (4N_g)^{-1}$, where N_g is the number of gas atoms in the system. With this choice of units, the gas density $n(\vec{r})$ satisfies the diffusion equation (3) with a unit diffusion coefficient.

For the study of the dynamics, it is important that our model has a physical time unit, which lacks in the DLA model. Also the DLA model contains only a single gas atom in the whole system at a time, and thus corresponds to the low density limit of our lattice gas model. In the other exteme of high density at $n_g = 1$, diffusion is no longer possible and the growth is limited by the process of picking up a "gas" atom at the solid-gas interface; the model now corresponds to the Eden model,[1] which is a well-studied local growth model. Our model thus naturally interpolates the Laplacian and the local growth models.

We study one-dimensional aggregation growth from a straight solid seed. In this geometry a steady state can be readily realized. Initially we provide a uniform gas with a density n_g in a large rectangular box and put a straight solid seed at the bottom of the box at $y = 0$. The periodic boundary condition is imposed in x direction such that $n(x + L) = n(x)$, with L being the width of the box.

Typical growth patterns with low gas densities are shown in Fig.1. There the structure of the "forest" may be characterized by a distance between main branches. The structure finer than this characteristic length has the similarity to the DLA. The characteristic length scale changes drastically with a small change of the gas density n_g as is apparent from Fig.1.

The dynamics is characterized by the growth rate V, which is defined as the velocity of an average interface position. On increasing the gas density, the growth rate increases. As is shown by the solid line in the logarithmic plot of the velocity V and the gas density n_g in Fig. 2, one clearly observes a power law relation between them as

$$V \approx 58 n_g^{3.50} \tag{4}$$

at low densities. We interpret the observation, in the following, by matching of the two characteristic length scales, which represent dynamics and the structure of the system.

On solidification gas atoms are absorbed by the solid and the gas density near the aggregate decreases, making the situation similar to the DLA formation. The thickness of the density depression layer is the diffusion length which is inversely proportional to the growth rate V as $l \sim 1/V$.

Even the aggregate looks like DLA for a short length scale, its average density must globally be the same as the average gas density n_g, because numbers of atoms should be conserved in the steady growth state. Thus at a crossover length ξ separating the DLA-like fractal regime from the uniform regime, the average density of the DLA should be n_g: $\xi^{D-d} \sim n_g$. We can rewrite the relation as

$$\xi \sim n_g^{-\nu}, \tag{5}$$

where the exponent ν is defined by

$$\nu = \frac{1}{d - D}. \tag{6}$$

We assume that these two length scales, the crossover length in the solid and the diffusion length in the gas, must be essentially the same. This assumption leads to

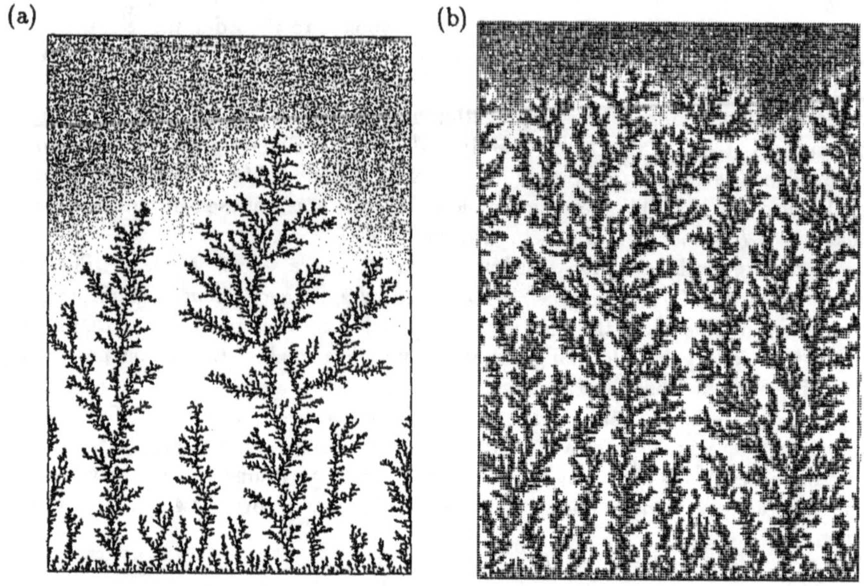

Fig.1. Growth patterns of aggregation without flow. Densities are (a) n_g=0.09 and (b) n_g=0.15.

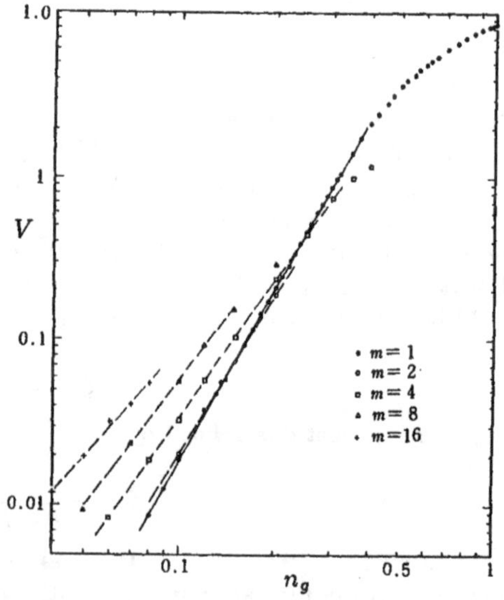

Fig.2. Growth velocity V versus gas density n_g for various grid sizes m.

an important prediction for dynamics of the growth, namely the growth velocity V is related to the gas density n_g as

$$V \sim \frac{1}{\xi} \sim n_g^{\nu}. \tag{7}$$

Relation (7) is valid at low densities where the diffusion controls the growth. Combining relation (7) with the simulation result (4), the fractal dimension D=1.71 is obtained, in very good agreement with the value obtained by the previous large-scale simulation.[5]

From an analysis of the distribution of gas atoms near growth points we found the exponential relaxation of gas density to the bulk value n_g.[6] The relaxation length defines a diffusion length, which was proven to change as gas density in accordance with Eq.(5). Also from a power law behavior of a solid density correlation function (2) at relatively short distances, an exponent $D = 1.68$ is obtained,[6] which is consistent to the value of the DLA.[5] At long distances the correlation function saturates to the bulk gas density n_g with the crossover length ξ.

We also performed a multifractal analysis[7] of the solid structure for scales smaller than ξ, and confirmed that the generalized dimension[8] D_q of the solid is constant, $D_q = 1.68 \pm 0.02$ for $q = 2 \sim 32$. This holds for solid samples with different gas densities, thus suggesting that the DLA is also a simple fractal.

(a) (b)

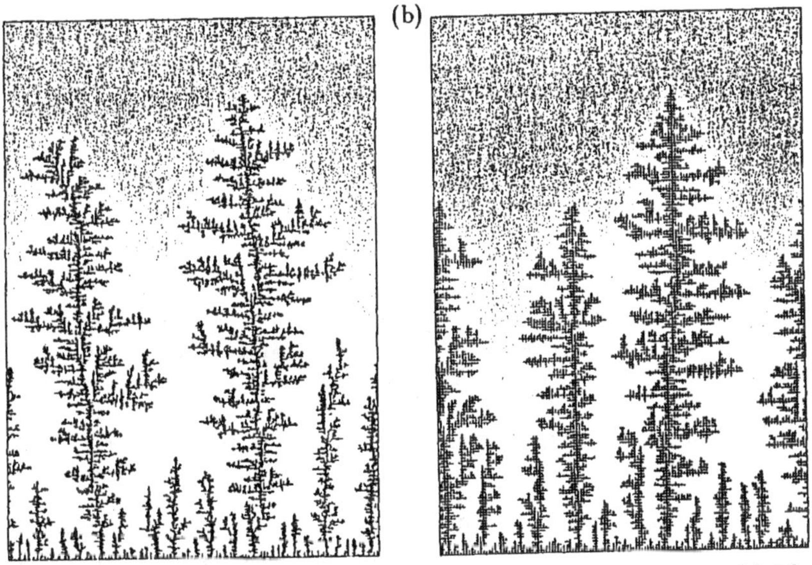

Fig.3. Anisotropy effect on the growth patterns with various grid sizes m. (a) The density is n_g=0.08 and m=4, and (b) n_g=0.07 and m=8.

3. Anisotropy Enhancement and Dendritic Growth

As is described in the introduction, anisotropy in interface tension or in kinetics is essential to the stable dendritic pattern.[2] Our system on a lattice naturally contains the anisotropy, but the strong fluctuation associated with the diffusion and the solidification dominates until the system size becomes enormously large. In order to observe the anisotropy effect in a relatively small DLA system, we apply here the grid method proposed by Matsushita and Kondo,[9] where the solidification is allowed on an $m \times m$ grid, namely only on every mth raw and column.[10] Once the crystal grows in one direction, it can grow in that direction for another $(m-1)$ steps, and thus the anisotropy is effectively enhanced.

The resulting pattern in Fig. 3 looks more regular and similar to the dendrite as the grid size m increases. The fractal dimension is expected to decrease to $D = 1.5$ asymptotically.[11] The logarithmic plot of the velocity V and the gas density n_g in Fig.2 clearly shows the power law relation similar to Eq.(7) at low densities. Apparent fractal dimension decreases when the grid size m increases as tabulated in Table.1.

There is another viewpoint that the m-dependence of the apparent fractal dimension is simply a crossover phenomena between the values of two fixed points, the DLA value $D = 1.71$ and the strong anisotropy limit $D = 1.5$.[12] To clarify the point, we have to explore less dense and larger systems, which is out of scope of the present capability.

Table 1: Apparent fractal dimension D versus grid size m.

m	1	2	4	8	16
D	1.71	1.69	1.64	1.61	1.56

4. Flow-Induced First Order Transition

The relation Eq.(7) of the characteristic length ξ and the gas density n_g reminds us that of the correlation length and temperature at a second order thermodynamic phase transition. Here the gas density plays the role of the temperature difference from the critical point, and the critical point of our model (or the fixed point in the renormalization group's sense) is the DLA model, that is $n_g \to 0$. In this sense the name[4] "kinetic critical phenomenon" may be appropriate. The exponent ν is determined by the geometrical fractal dimension D.

A finite density introduces a finite length scale ξ in the system. There is another way to introduce a length scale in the Laplacian field, namely via drift flow of the Brownian gas particles, which adds a term $\vec{U} \cdot \nabla n$ to the Laplace equation.[13,14] The question is then what would happen when the crystal grows from a finte density gas with a homogeneous drift flow.

We consider the same system as before and introduce a uniform gas flow in y direction.[15] Again, in this geometry a steadily growing state will be realized after some transient growth stage. In the steady state, the conservation of atoms implies

$$(n_s - n_g)V + n_g U = 0, \tag{8}$$

where n_s is the solid density deep in the solid, and U is the drift velocity of the gas in y-direction. The gas density n_g and the drift velocity U are the controlable quantities while the solid density n_s and the growth velocity V are unknown. Since the growth is still limited only by diffusion of the gas we can write an equation for the solidification flux:

$$n_s V = n_g \frac{a}{\xi}, \tag{9}$$

with a numerical constant a. The right hand side represents the diffusion current ∇n. Further we make the same assumption as before that the solid structure is DLA fractal up to a characteristic length ξ, which implies

$$\xi = b\, n_s^{-\nu}, \tag{10}$$

with b being another numerical constant, and $\nu = 1/(d - D)$ as before. With these three Eqs. (8), (9) and (10) we obtain an equation for ξ

$$1 - n_g(\frac{\xi}{b})^\nu = -U\frac{a}{\xi}. \tag{11}$$

In the DLA limit, $n_g = 0$, $\xi \approx 1/U$. For $U = 0$, Eq. (11) reduces to Eq. (7) and comparing the form with Eq. (4) we obtain $a/b \approx 58$. Once the characteristic length ξ is determined, the growth velocity V and the solid density n_s follow immediately.

The characteristic length ξ decreases when U is negative, that is for incoming flow, and increases when U is positive. As long as the flow velocity is small, the solid can adjust its density to the flowing gas by changing ξ. There is, however, a critical flow velocity U_c, above which there is no solution of Eq. (11). This implies that a steady growth ceases to exist and the solid stops growing. When the gas atoms flow away too fast from the interface, the growth is too slow to follow the gas flow. Therefore, the space of the two controlling parameters, n_g and U, is devided into two "phases", growing and nongrowing. It should be noted that ξ is finite at the critical point so that the transition resembles a first order phase transition.

Using Eq. (11) we can calculate various quantities at the critical point. These quantities satisfy simple relations to those without gas flow. For example, the critical flow velocity is related to the growth velocity V_0 without flow as

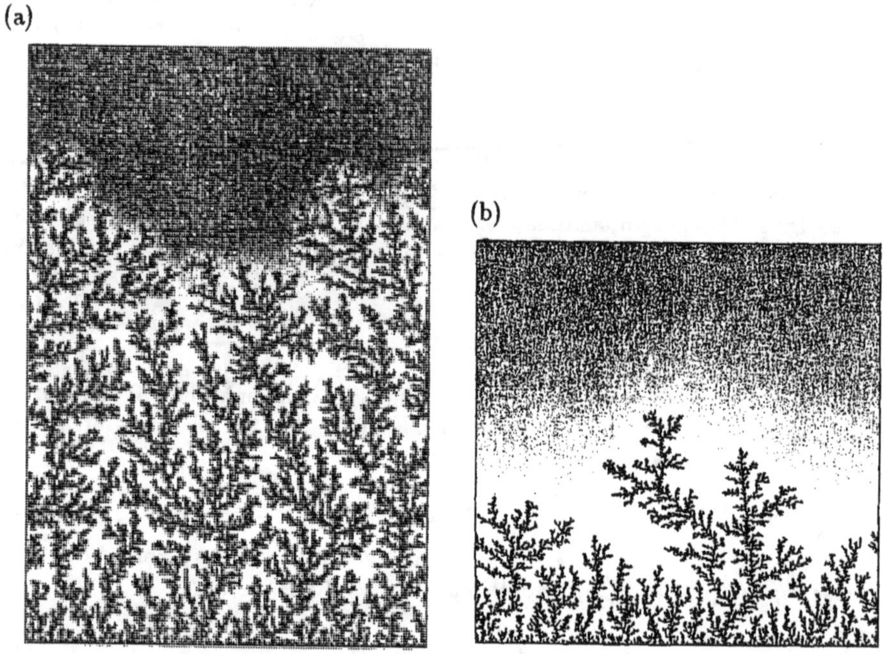

Fig.4. Growth patterns of aggregation with flow. Gas density is n_g=0.20 and the drift flow velocities are (a) U=0.025 and (b) U=0.032.

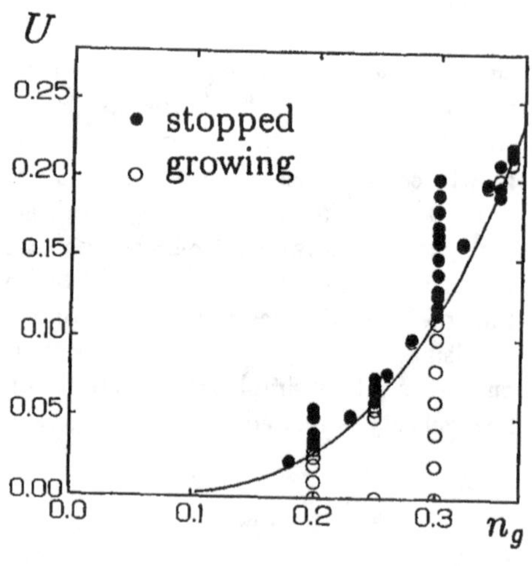

Fig.5. Phase diagram in the gas density n_g and the drift velocity U parameter space. The solid line represents a theoretical curve (12) with ν =3.30.

$$U_c = \frac{(\nu-1)^{\nu-1}}{\nu^\nu} V_0. \tag{12}$$

This relation does not include any unknown parameters and can be tested by a simulation.

In Fig.4 growth patterns with various drift velocities for the same gas density $n_g = 0.2$ are shown. The whole pattern of an aggregate looks the same with those under no drift flow shown in Fig.1. A structural analysis confirms this visual observation. Solid density and the growth velocity changes in a way predicted from Eq.(11).

The simulation with various combinations of a drift velocity U and a gas density n_g is summarized in a"phase diagram", Fig.5. The solid circles indicate that steadily growing solid and empty circles indicate growth has stopped. There is a well-defined phase boundary, which fits well to the curve given by Eq. (12) with $\nu = 3.3$. This value corresponds to the fractal dimension $D = 1.70$. Thus our simple theory works remarkably well.

5. Discussion

The lattice gas growth model provides a simple example of growth in a diffusion field. It reveals nature of the transition from a Laplacian growth model(DLA) to a local growth model(Eden model).

If a planer interface is growing in a diffusion field with a velocity V, the gas density relaxes exponentially to the bulk value within a diffusion length $l \sim 1/V$. In order to fix the diffusion length or to select the growth rate uniquely, however, one needs another intrinsic length scale. For the regular dendrite, this intrinsic length scale is given by the capillary length. For a system with strong noise or the one without surface tension, the local geometrical structure provides one characteristic length. Due to the absorption of gas atoms around the growing solid, the short-ranged structure of the solid is the same with its low density limit, the DLA, characterized by the fractal dimension D. On the other hand, the global mass conservation imposes the condition that in large scales the aggregate should be homogeneous with a finite density of the mother gas phase. The crossover of the fractal to the uniform structure takes place at a characteristic length ξ. Matching of the dynamical length l to the structural length ξ selects the growth velocity V. This is the velocity selection rule when diffusion is the only relevant physical process. The same rule also works when the gas has a drift velocity.

An aggregation pattern of two-dimensional electrochemical deposition[16] is very similar to that generated in the lattice gas models as well as in the DLA[17]. It is believed that certain electrochemical deposition is a realization of the DLA. Sawada and Hyosu[18] have made an attempt to test the growth law Eq.(7) in this system. At a constant applied voltage and with an increase of the solute concentration, the forest like aggregation pattern becomes denser and the distance of the main branches becomes smaller as expected. The growth velocity, however, turns out independent of the solute concentration, which is in complete disagreement with our simulation. The growth patterns of their experiment always have growth fronts with a very small height fluctuation, whereas the fluctuations of our pattern height is shown to increase indefinitely as the system width increases.[6] The disagreement may indicate that the experiment is not a diffusion-limited process, at least, in the investigated concentration and voltage range.

Inclusion of an evaporation process in the present model introduces the stabilizing effect of interface tension and kinetics, and brings about a series of growth stages:[19] a compact polyhedral shape, an open dendritic shape, and finally to the uniform structure, as observed in real crystals.

A part of the work is supported by the Grant-in-Aid for Scientific Research on Priority Areas by the Ministry of Education, Science and Culture with Grant No. 03243101. Also a partial support from IMR under the Visiting Researcher's Program is acknowledged.

References

1. T. Vicsek, in *Fractal Growth Phenomena,* (World Scientific, Singapore,1989)
2. J. S. Langer, in *Chance and Matter* ed. J. Souletie, J. Vannimenus and R. Stora (North Holland, Amsterdam, 1987).
3. H. Honjo, S. Ohta and M. Matsushita, J. Phys. Soc. Jpn. **55** (1986) 2487.
4. T. A. Witten and L. M. Sander, Phys. Rev. Lett. **47** (1981) 1400; Phys. Rev. B **27** (1983) 5686.
5. P. Meakin, in *Phase Transition and Critical Phenomena* Vol.12, eds. C. Domb and J. L. Lebowitz, (Academic, London, 1988).
6. M. Uwaha and Y. Saito, J. Phys. Soc. Jpn. **57** (1988) 3285; Phys. Rev. A **40** (1989) 4716.
7. M. Uwaha, S. Seki and Y. Saito, J. Phys. Soc. Jpn. **59** (1990) 1533.
8. H. G. E. Hentschel and I. Procaccia, Physica **8D** (1983) 435.
9. M. Matsushita and H. Kondo, J. Phys. Soc. Jpn. **55** (1986) 2483.
10. M. Uwaha and Y. Saito, J. Cryst. Growth **99** (1990) 175.
11. J. Nittmann and H. E. Stanley, Nature **321** (1986) 663.
12. E. Brener, H. Müller-Krumbhaar and D. Temkin, Europhys. Lett. (1991) to appear.
13. P. Meakin, Phys. Rev. B **28** (1983) 5221.
14. T. Nagatani, Phys. Rev. A **39** (1989) 438.
15. S. Seki, M. Uwaha and Y. Saito, Europhys. Lett. **14** (1991) 397.
16. M. Matsushita, Y. Hayakawa and Y. Sawada, Phys. Rev. A **32** (1985) 3814.
17. P. Meakin, Phys. Rev. A **27** (1983) 2616.
18. Y. Sawada and H. Hyosu, Physica D **38** (1989) 289.
19. Y. Saito and T. Ueta, Phys. Rev. A **40** (1989) 3408.

PHASE SEPARATION, CRYSTAL GROWTH AND MORPHOLOGICAL TRANSITIONS

Hiroshi Furukawa
Faculty of Education, Yamaguchi University, Yamaguchi 753, Japan

Abstract

We numerically performed inhomogeneous quenchings of two kinds of disordered systems, and examined subsequent ordering processes. The one system consists of two conserved components, and the other consists of infinite nonconserved components. The former is used for the phase separation of binary mixture, and the latter is used for the grain growth. The boundary between the stable phase and the unstable phase (referred to as quench boundary) is shifted with a constant velocity. According to the value of quench boundary velocity various types of domain morphology in ordering phase are observed. In the conserved binary system three typical domain morphologies are observed. For small, intermediate and large quench boundary velocities domain morphologies are columnar, lamellated, and random or turbulent, respectively. In the nonconserved system three typical grain morphologies are also observed, i.e., huge, elongated and compact ones, depending on the quench boundary velocity and system boundary conditions.

1. Introduction

When a binary mixture is suddenly quenched from a one-phase region to a state inside the miscibility gap a phase separation takes place. Almost all theoretical studies have been performed for the phase separation after a sudden and homogeneous quench. We here notice that the homogeneous quenching belongs to the category of the sudden quenching, no matter how the quench rate (cooling rate) is slow. This is because, once the system is brought into the unstable region, the process is completely irreversible. The counter process to the completely irreversible one is the quasi-static process. If the quenching process can be controlled by varying it from the quasi-static one to the completely irreversible one, then what happens during the phase separation process? This is a challenging problems in statistical physics of nonequilibrium state. There is a problem similar to this. That is the directional solidification from molten matter.[1] However, in the ordinary directional solidification the heat current flows in the direction of solidification, and the solidification is regarded as to be triggered by an interfacial instability.[2] We treat a system where the heat current flows in the reversal direction to the solidification or otherwise the material diffusion is much slower than the thermal diffusion: We mean that the disordered state is not a metastable stat. In such a case the instability is only the bulk one.[3]

We have performed a numerical simulation under the condition noted above using the Cahn-Hilliard equation. We set the following external condition, i.e., inside the quench boundary the phase separation proceeds and outside the quench boundary the system is kept to be disordered. The tránsient ordering patterns are then observed for various values of the shift velocity of the quench boundary. The same external condition is used also in the case of nonconserved order paramerter, which is the other subject here.

2. Conserved case[4]

The basic equation we use is the Cahn-Hilliard equation:[3]

$$\frac{d\psi(r,t)}{dt} = M\nabla^2\mu(r,t) \quad . \tag{1}$$

where μ is the first functional derivative of free energy with respect to the order parameter ψ, and we employed the following type:

$$\mu = -\frac{1}{2}\nabla^2\psi + \psi - \frac{\tanh\psi}{\tanh 1} \quad . \tag{2}$$

which was introduced by Oono and Puri,[5] who confirmed that any obtained results are equivalent to those by usual Gintzburg-Landau type free energy. As noted in §1 inside the quench boundary ψ

evolves following the equation (1), whereas outside the quench boundary ψ is kept disordered: $-\delta\psi_0 \leq 2\psi \leq \delta\psi_0$. Therefore, there are two control parameters, v (quench boundary velocity) and $\delta\psi_0$ (initial thermal fluctuation). Since the order parameter rapidly bilds up after quenching, we neglect the thermal fluctuation inside the quench boundary.

In Fig.1 we present three typical pattern for three values of quench boundary velocity v with the initial fluctuation $\delta\psi_0$=0.1. Here in each pattern the quench boundary starts from the upper system boundary and is now at the lower system boundary. The quench boundary is shifting. For larger values of v all patterns are random, whereas for smaller value of v all patterns are columnar. For intermediate value of v patterns are lamellated. But this lamellated pattern does not appear for values of $\delta\psi_0$ larger than about 1.0. In a real system therefore such a lamellated pattern appears possibly in polymer blend, where the effect of the thermal fluctuation is regarded as weak.

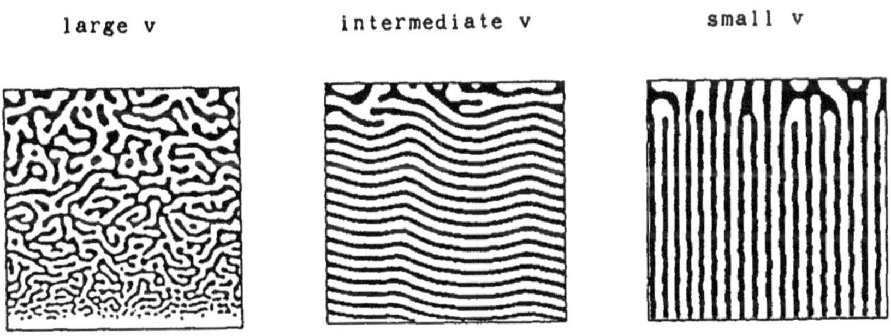

Fig.1 Case of conserved order parameter

3. Nonconserved case

Let us imagine the homogeneous material distinguished by interfaces such as the assembly of grains. The interfacial motion gives rise to the grain growth.[6] We introduce a lattice model which describes the interfacial motion in grain growth. Let σ be a spin variable with infinite number of spin states: $\sigma=1,2,..,\infty$. The value of σ represents the grain state such as the direction of crystal axis in a grain. The interaction between different spin states are equal to that of Potts model, i.e., the interacting energy between the same spin state is negative, whereas there is no interacting energy between different spin states. We use the square lattice and take the interaction up to the next nearest neighbor site. The spontaneous creation of new interface is inhibited. A suitable dynamics with the last restriction was proposed previously.[7] The Monte Carlo sampling is done only to shift the grain boundary. That is, a spin state changes into one of neighboring spin states, whenever the energy is reduced by this change. Notice that the dynamics is not of the Kawasaki type (spin exchange) but of the Glauber type (single spin change).

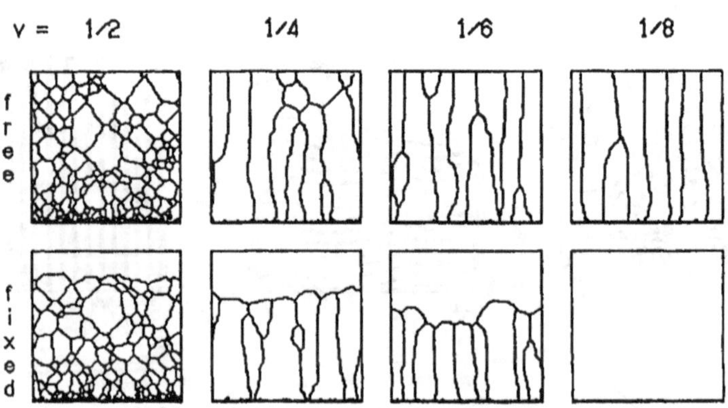

Fig.2 Case of nonconserved orderparameter

We performed the inhomogeneous quench of this system. Spins take random values before quenching. When spins comes inside the quench boundary, they obey the above dynamics yielding the grain growth. We observed three typical grain patterns according to the quench boundary velocity v and the system boundary condition. They are shown in Fig.2. Here in each pattern the quench boundary starts from the upper system boundary and is now at the lower system boundary. The quench boundary is shifting. The upper patters are those under the condition that the upper system boundary is free to evolve, whereas lower patterns are those under the condition that the upper boundary is fixed to be a single spin state. Under the free boundary condition grains become elongated for quench boundary velocities smaller than about 1/3 (Monte Carlo steps per spin). Under fixed boundary condition the system is occupied by a large single grain for quench boundary velocities smaller than about 1/8 (Monte Carlo steps per site).

5. References

1. D. Kessler, J. Koplik and H. Levine, Adv. Phys. 37 (1988) 255. J. S. Langer, Rev. Mod. Phys. 52 (1981) 1.
2. W.W Mullins and R.F. Sekerka, J. Appl. Phys. 35 (1964) 444.
3. J. W. Cahn and J.H. Hilliard, J. Chem. Phys. 28 (1958) 258; 31 (1959) 688.
4. H. Furukawa, Physica A 180 (1992) xxx.
5. Y. Oono and S. Purui, Phys. Rev. Lett. 58 (1987) 386; Phys. Rev. A 38 (1988) 434; A 38 (1988) 1542.
6. G. Grest, M.P. Anderson ang D. J. Srolovitz, in Proceedings of Nato Advanced Study Institute on *Time Dependent Effects in Disordered Materials*, edited by R. Pynn (Plenum, 1987). K. Kawasaki, T. Nagai and K. Nakamura, Phil. Mag. B60 (1989) 399.
7. H. Furukawa, in *Dynamics of Ordering Processes in Condensed Matter*, edited by S. Komura and H. Furukawa (Plenum, 1988).

DOMAIN GROWTH IN A SYSTEM WITH QUENCHED RANDOM IMPURITIES

HISAO HAYAKAWA and TOSHIYA IWAI
Department of Physics, Tohoku University, Sendai 980, Japan

ABSTRACT

Domain growth in a system with quenched random impurities is investigated using a cell-dynamics method. Our numerical simulation suggests (i) the growth rate of domains is reduced, and (ii) the dynamics is governed by the depinning of interfaces of domains in the late stage.

1. Introduction

The ordering dynamics after a quench from a disordered phase to an ordered phase below the critical temperature is a typical example of pattern evolution in nonequilibrium physics. Experiments and simulations show that the linear size of domains $l(t)$ obeys a power law as $l(t) \sim t^{1/2}$ for nonconserved systems of the scalar order parameter or $l(t) \sim t^{1/3}$ for conserved systems[1].

The effects of random impurities for ordering processes are drastic[2]. Interfaces between domains are pinned by impurities and the rate of domain growth is reduced. Huse and Henley[3] predicted the logarithmic growth to be $l(t) \sim (\log t)^4$ for two dimensional systems in the late stage of the ordering processes. Their prediction is independent of the existence of a conservation law of the order parameter, which is contrast to the usual case. Recently, numerical simulations of several groups[4–7] have obtained consistent results with the prediction by Huse and Henley[3] in the case of nonconserved systems. However, nobody has simulated the cases for conserved systems. In this report, we investigate the late stage of phase separations in systems with impurities both for conserved and for nonconserved cases.

2. The Method for Numerical Simulation

In this section we explain how to simulate phase separations in systems with random impurities. We use the cell-dynamics (CD) method proposed by Oono and Puri[8] for our simulations, because CD method is regarded as the most effective way for simulations of phase separations.

In our simulation, we assume that the system can be divided into two kinds of cells, the pure cells and impurities. The order parameter $S_{\vec{n}}(t)$ at the cell $\vec{n}$ is defined only in the pure cells. In each impurity which is immobile, $S_{\vec{n}}(t) = 0$ is always satisfied. We use a phenomenological effective Hamiltonian

$$H_{eff} = \frac{1}{2}\sum_{\vec{n}}[\sum_{\vec{n}'\in NN} J\theta(\vec{n},\vec{n}')(S_{\vec{n}} - S_{\vec{n}'})^2 + (b - \sum_{\vec{n}'\subset NN} J\theta(\vec{n},\vec{n}'))S_{\vec{n}}^2 + \frac{g}{2}S_{\vec{n}}^4], \quad (1)$$

where NN represents the nearest-neighbor cells of $\vec{n}$. $\theta(\vec{n},\vec{n}') = 1$ for the coupling between pure cells and $\theta(\vec{n},\vec{n}') = 0$ for other cases. The parameter b in (1) can be regarded as the temperature T, because the coefficient of $S_{\vec{n}}^2$ should be proportional to $T - T_0$ with the mean field critical temperature $T_0 = 2dJ$ (d: spatial dimension) without impurities. We choose the parameter $b = T$ and $\sqrt{(2dJ - T)/g} = 1$, because the equilibrium value of $S_{\vec{n}}$ without impurities is normalized to be 1.

The time evolution of the order parameter at a single cell dynamics is given by

$$S_{\vec{n}}(t+1) = f_{\vec{n}}[S(t)] \quad (2)$$

where $f_{\vec{n}}[S]$ can be regarded as a map function from time t to $t+1$. The expression of the map function is

$$f_{\vec{n}}[S(t)] = \frac{A^{\tau_{\vec{n}}/\tau_0}S(t)}{[1 + (\tau_0/\tau_{\vec{n}})S(t)^2(A^{2(\tau_{\vec{n}}/\tau_0)} - 1)]^{1/2}} \quad , \quad \frac{\tau_{\vec{n}}}{\tau_0} = \frac{(T/T_0) - \Theta(\vec{n})}{T/T_0 - 1}, \quad (3)$$

where $A = 1.3$ and $\Theta(\vec{n})$ is the existence probability of a pure cell in the nearest-neighbor cells of $\vec{n}$.

The time evolution of the system is the result of the coupling among cells. For a nonconserved system the equation is

$$S_{\vec{n}}(t+1) = D\Delta_{imp}S_{\vec{n}}(t) + f_{\vec{n}}[S(t)] + B\eta(\vec{n},t) \equiv \mathcal{F}[S(t)] + B\eta_{nc}(\vec{n},t) \quad (4)$$

with $D = 0.5$ and $B = 0.3\sqrt{2T/T_0}$. In our analysis we use a uniform random number (uniformly distributed between -1 and 1) for the thermal noise $\eta_{nc}(\vec{n},t)$. Δ_{imp} is the discrete Laplacian including the effects of impurities as

$$\begin{aligned}\Delta_{imp}S_{\vec{n}}(t) = &\frac{1}{6}\sum_{n'\in NN}\theta(\vec{n},\vec{n}')\{S_{\vec{n}'}(t) - S(\vec{n},t)\} \\ &+ \frac{1}{24}\sum_{n'\in NNN}\sum_{n''\in NN}\theta(\vec{n},\vec{n}'')\theta(\vec{n}'',\vec{n}')\{S_{\vec{n}'}(t) - S_{\vec{n}}(t)\},\end{aligned} \quad (5)$$

where NNN represents the second nearest-neighbor cells of $\vec{n}$.

For a conserved case, the time evolution is given by

$$S_{\vec{n}}(t+1) = S_{\vec{n}}(t) - \Delta_{imp}\mathcal{F}[S(t)] + \Delta_{imp}S_{\vec{n}}(t) + B\eta_c(\vec{n},t) \quad (6)$$

Since the thermal noise $\eta_c(\vec{n},t)$ is a conserved quantity, $\vec{\sigma}(\vec{n},t)$ defined by $\eta_c(\vec{n},t) = \nabla \cdot \vec{\sigma}(\vec{n},t)$ is a random variable. Note that $\vec{\sigma}(\vec{n},t) = 0$ should be satisfied on the surface of impurities to ensure the conservation law.

3. Results of Our Simulation

In this section we present the results of our simulation. Quenches are carried out at the center of the miscibility gap. The system size is 128×128 in the cell unit. The distribution of initial vales of the order parameter is randomly and uniformly distributed between ± 0.125. We use the periodic boundary condition. We perform simulations up to $t = 10000$ for nonconserved systems and $t = 160000$.

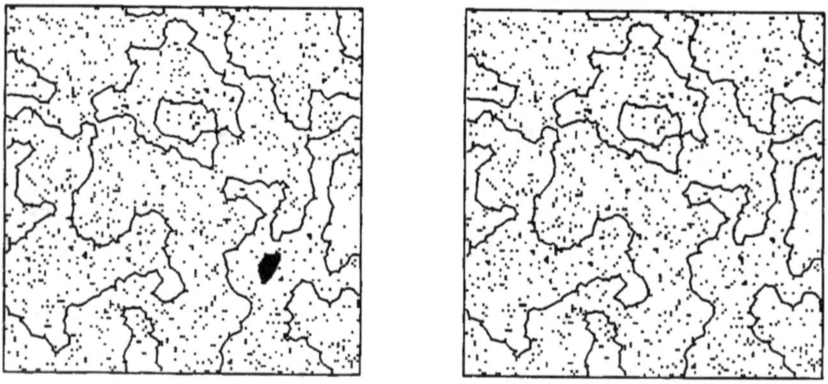

Fig.1: The configurations of the order parameter at $t = 5500$ (left) and $t = 6000$ (right) for $T/T_0 = 1/3$ and $c = 0.1$. Lines and spots represent interfaces of domains and impurities, respectively.

Let us show some interesting results of our simulation. In the middle stage of domain growths, interfaces between domains is driven by the curvature force. That is, the growth rate is essentially independent of the temperature. In the late stage, however, the motion of interfaces are dominated by the depinning process from a metastable state in which interfaces are pinned by impurities. Figures 1 and 2 are good examples of the domain growth by the depeinning process. Figure 1 shows the domain configurations at $t = 5500$ and $t = 6000$ for a nonconserved case with the concentration of impurities $c = 0.1$ and $T/T_0 = 1/3$. We find that the domain configuration is frozen except for disappearance of a shaded domain. Figure 2 shows a corresponding time evolution of the characteristic length, the domain size, defined by $l(t) \equiv \int_0^\infty I(q,t)dq / \int_0^\infty qI(q,t)dq$ where $I(q,t)$ and q are the scattering function and the wave number, respectively. We find that the growth of $l(t)$ is almost stopped after $t = 1000$ except for discontinuous growths of $l(t)$

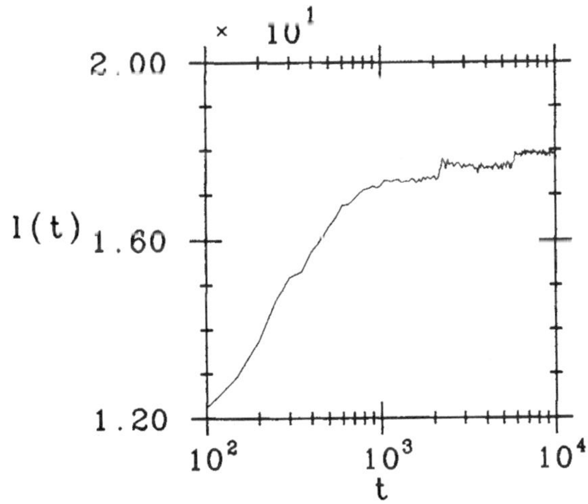

Fig.2: A plot of $l(t)$ versus t on double-logarithmic scales in the case corresponding to Fig.1.

about $t = 2000$ and $t = 5500$. These growths can be interpreted that the interface moves quickly by the curvature driven force after a depinning of the interface due to the thermal fluctuation. Note that such depinning processes could not be observed at $T = 0$.

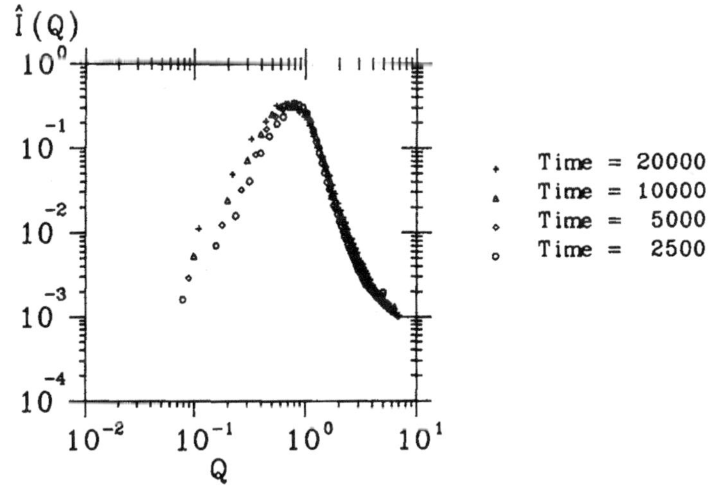

Fig.3: Plots of $\hat{I}(Q)$ versus Q on double-logarithmic scales.

Figure 3 shows a scaling plot of the scattering function $I(q,t) = l(t)^2\hat{I}(Q)$ versus the scaled wave number $Q = ql(t)$ for a conserved case with $c = 0.2$ and $T/T_0 = 1/3$. We performed 35 rus to eliminate the dependence of initial conditions, where we provide 5 configurations of impurities and perform 7 runs with the changes of both initial values of the order parameter and the thermal noise for each configuration of impurities. We can find that the system has a scaling property which is contrast to the case of the random field Ising model[9]. In addition, there is no shoulder which can be observed in simulation[10] and experiments[11] when systems do not contain impurities.

4. Concluding Remarks

We present a preliminary result of our simulation for phase separations in systems with quenched random impurities. We find that the depinning process of interfaces from impurities is most important in the late stage of phase separation. Our analysis suggests that even in conserved systems the scattering function satisfies a scaling law. We have not finished the quantitative analysis of our simulation. Details of our analysis will be published elsewhere.

References

1. J.D.Gunton and M.Droz, *Introduction to the Theory of Metastable and Unstable States*: Lecture Notes in Physics, *vol.***183** (Springer, Berlin, 1983).
2. H.Ikeda, Y.Endoh and S.Itoh, *Phys.Rev.Lett.* **64** (1990) 1266.
3. D.A.Huse and C.L.Henley, *Phys.Rev.Lett.* **54** (1985) 2708.
4. H.Hayakawa, *J.Phys.Soc.Jpn.* **60** (1991) 2492.
5. H.Hayakawa and T.Iwai, *to appear in* Proceedings of Slow Dynamics in Condensed Matter (AIP, New York, 1992).
6. S.Puri, D.Chowdhury and N.Parekh, *J.Phys.***A 24** (1991) L1087.
7. A.J.Bray and K.Humayun, *J.Phys.***A 24** (1991) L1185.
8. Y.Oono and S.Puri, *Phys.Rev.Lett.***58** (1987) 836, *Phys.Rev.***A38** (1988) 434. S.Puri and Y.Oono , *Phys.Rev. A* **38** (1988) 1542, *J.Phys.***A 21** (1988) L755.
9. S.R.Anderson, *Phys.Rev.* **B36** (1987) 8435.
   E.Oguz, A.Chakrabarti, R.Toral and J.D.Gunton, *Phys.Rev.***B42**(1990)704.
   E.T.Gawlinski, S.Kumar, M.Grant, J.D.Gunton and K.Kaski, *Phys.Rev.* **B32** (1985) 1575.
10. T.Koga and K.Kawasaki, *Phys.Rev.* **A 44** (1991) R817.
11. M.Takenaka, T.Izumitani and T.Hashimoto, *J.Chem.Phys.* **92** (1990) 4566.

Dense-branching morphology of an NH_4Cl dendrite

H.Honjo
College of General Education, Kyushu University,
4-2-1 Ropponmatsu Chuo-ku Fukuoka city, 810 Japan

and

S.Ohta
College of General Education, Kyushu University,
4-2-1 Ropponmatsu Chuo-ku Fukuoka city, 810 Japan

ABSTRACT

We have experimentally observed a dense-branching morphology(DBM) in an NH_4Cl dendritic crystal growth. The DBM forms a randomly and densely tip-splitted structure and the envelope of interface is stable. The measured interfacial velocity of our DBM decreases with a supersaturation. This behavior stabilizes the interface of DBM. The averaged length between tips corresponds to the diffusion length.

[Introduction]

A dendritic crystal growth represents a typical pattern formation in non-linear and non-equilibrium system[1]. The pattern in diffusion field changes from dendrite to diffusion-limited aggregation(DLA) or dense-branching morphology(DBM) by changing such a parameter as strength of anisotropy or diffusion field[2]. The DLA has self-similarity (fractal) with an opening and randomly branched structure. This pattern has been investigated actively by many authors[3]. The DBM forms a densely branched pattern and its envelope of many tips is stable. This pattern has been observed experimentally in electrochemical deposition[4], viscous fingering[5] and bacterial colonies[6]. However, the growing mechanism has not been understood yet.

In an NH_4Cl solution growth, a <100> dendrite changes its growth direction to <110> with an increasing supersaturation[7]. In the transition region, complicated growing behaviors are observed[8]. The growing crystal forms a randomly and densely tip-splitted structure, and the interfacial envelope is stable[9]. We suppose that this growing pattern corresponds to the DBM. We show an NH_4Cl dendritic DBM and discuss about the stability of the interface.

[Experiment]

We show the experimental setup in Fig.1. The temperature of system is controlled with both water-flow(±1mk) and heater using electric bridge-circuit independently. Analytical reagent grade NH_4Cl (99.0 %) is combined with distilled, deionized and filtered (a 0.2μm membrane filter) water. An NH_4Cl solution is sealed in the growth cell which has a 0.3 mm spacer(Fig.2). The

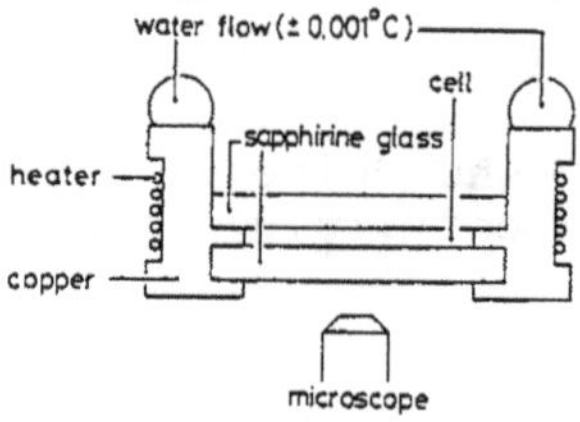

Fig.1

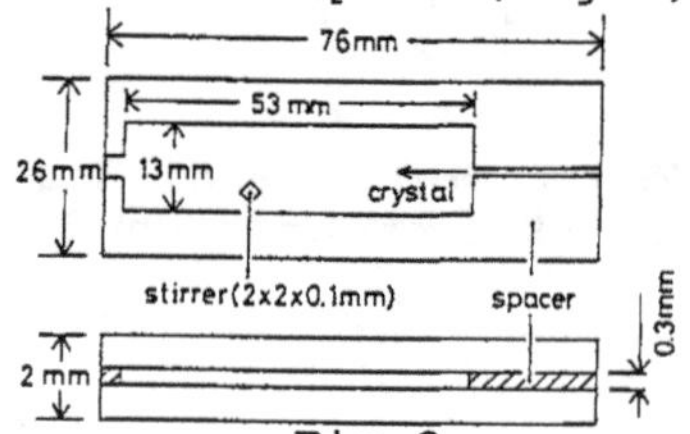

Fig.2

temperature corresponding to the NH_4Cl saturated concentration T_s is 313.15k(=40°C). We change the system temperature, that is, the effective supersaturation of the solution, which is defined as $\sigma=(T_s-T_\infty)/T_s$. Where, T_∞ is system temperature. A nucleation occurs at the narrow path in the cell and a dendrite grows along the path and into the wide region of the cell(Fig.2). We observe growing crystal with a microscope and choose the cases that dendrites grow pararell to the narrow path and the sidebranches grow pararell and perpendicular to the cell. We take the pictures of growing dendrites in the wide region into a video tape recorder and analyze them with image analysis technique; 512 x 480 pixels, 8-bit intensity resolution and 6.15μm pixel size. A stirrer(2x2x0.1mm) is included in the cell and the concentration of NH_4Cl solution is stirred to be uniform before a crystal grows. The stirrer is placed at the corner of the wide region when a crystal is growing.

[Results]

We show the relation between σ and crystal velocity v in Fig.3. The error bars of v are within the marks. In the case of $\sigma < \sigma_1(\simeq 0.04)$ in our experiment, a crystal grows as a <100> dendrite, which has four sidebranches because of four-fold symmetry of an NH_4Cl crystal. The tip velocity increases with σ. In the case of $\sigma_1 < \sigma < \sigma_2(\simeq 0.05)$, a regular tip splitting occurs. In this region, a <110> sidebranche, for example, grows and stops

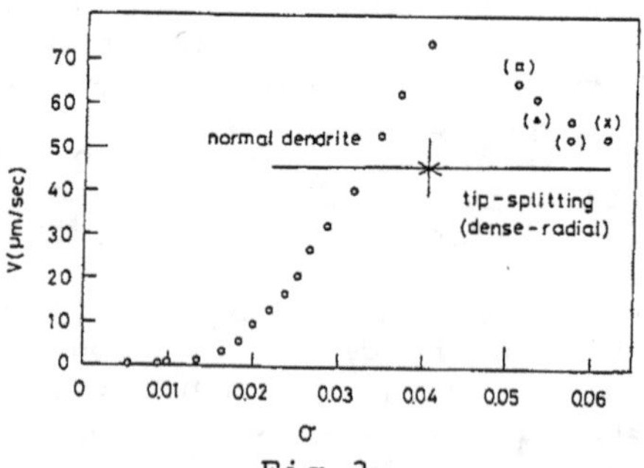

Fig.3

immediately because of the concentrational interaction with the tip. This phenomenon occurs in the vicinity of the tip, then it is observed as if the tip were splitted. This behavior repeats spatially in regular[8]. In the case of $\sigma > \sigma_2$, the tip-splitting occurs irregularly and the <110>sidebranch does not stop but immediately changes its direction to <100> unlike the former case. This phenomenon occurs many times irregularly and many new tips are produced consequently. We show the typical pattern (σ=0.0574) in Photo 1. This pattern grows repeating tip-

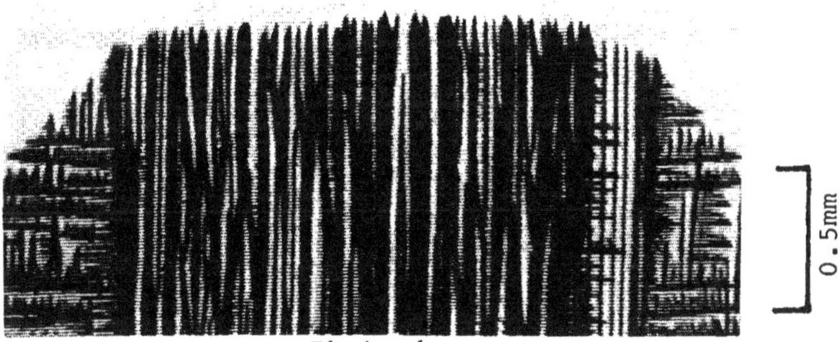

Photo 1

splittings from the dendrite that has one tip. We call this type as a dense-branching morphology(DBM) corresponding to the patterns observed in electrochemical deposit system. The velocity of this type is measured as the one of the averaged interfacial line of the tips and decreases with σ. We measure about four cases of DBM, σ=0.0511, 0.0532, 0.0574 and 0.0614. The accuracy of σ is $\pm$ 0.0001 resulting from the temperature deviation of the system. As shown in Photo 1, the DBM is divided into two regions, one is a dense-branching region(DBR) and the other is a side-branching region(SBR) where normal dendrites grow.

We show the interfacial length L and the number of tips N of the DBM against time t in Fig.4 and Fig.5 (log-log

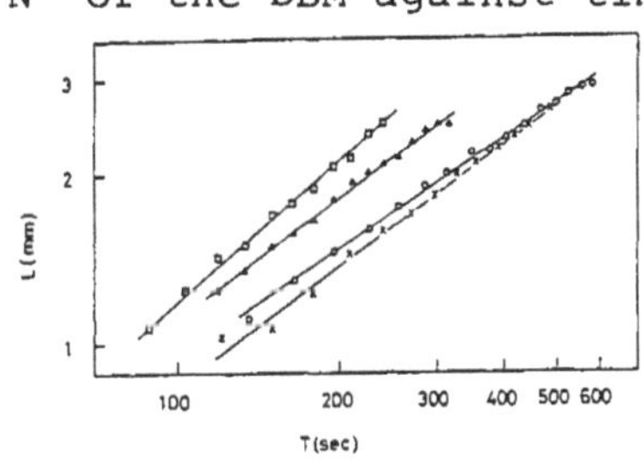

Fig.4

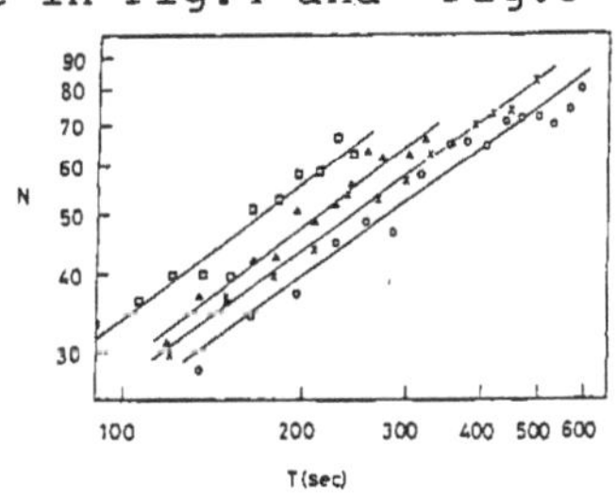

Fig.5

plots) respectively. The marks □, △, ○ and × corresponds to the cases of σ=0.051, 0.0532, 0.0574 and 0.0614 respectively. The interfacial position of the crystal slightly depends on the level of brightness detecting crystal in an image and is averaged one by changing the level. The positions of the tips are defined as the places where the digitized interfacial curve has minimun values, where the growing direction of the interface is the negative axis. The dependences of L and N against t are $L \sim t^{\alpha}$, and $N \sim t^{\beta}$, and these exponents are given in Table 1. The difference of time extences in the data results from the one of the scales of patterns which appear at the entrance of the wide region of the cell.

σ	0.0511	0.0532	0.0574	0.0614
v(μm/sec)	65.3	61.5	56.6	52.4
α	0.848±0.021	0.733±0.014	0.653±0.010	0.718±0.014
β	0.735±0.066	0.758±0.055	0.696±0.037	0.705±0.023
ℓ(μm)	41.53	38.40	36.78	31.87
ℓ_d(μm)	79.67	84.54	91.86	99.24

Table 1

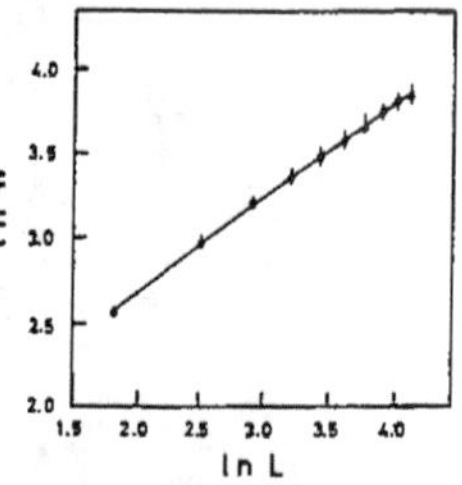

Fig.6

In Fig.6, we show the typical relation between the surface thickness W and the length scale L in the case of σ=0.0511. Where, $W \equiv \langle [h(x)-h_0]^2 \rangle^{1/2}$ and h_0 is the averaged interfacial position of the DBM. The relation is a power law and the value of the exponent γ is 0.555±0.004.

[Discussion]

The spatial dimension of our system depends on the relation between the diffusion length ℓ_d=2D/v and the thickness of spacer d. Where, D is the diffusion constant of an NH_4Cl crystal in water and 2.6 x 10^{-5} $(cm^2/sec)^{10}$. Our DBM system is three dimensional because $\ell_d < \ell$; ℓ_d < 100 μm(see Table 1) and ℓ = 300 μm.

The growth of many dendrites are influenced with the upper and bottom walls of the cell. By observation of growing behavior, about two or three overlapped dendrites can grow together. We analyze the two dimensional patterns projected from such three dimensional patterns. There happen the following two cases; one is that a bottom tip is completely blind by a upper tip, and the other is that some dendrites are overlapped and observed as one larger dendrite. These cases influence to the estimation of N. However, we suppose it is a few cases.

As given in Table 1, the dependences of L and N on t are power-laws. The values of α and β are roughly equal for each σ, and it means that the length between tips is constant independently on time t. We also give the values of the length between tips ℓ=L/N and the diffusion length in Table 1. The value of ℓ at a higher σ is less than the case of a lower σ. Considering the case of σ=0.0511, which is a relatively less supersaturation, ℓ is nearly eaqual to ℓ_d(=79.67 μm) neglecting 2 factor of ℓ_d. With an increasing σ, ℓ_d increases because of the decreasing interfacial velocity. If the velocity of the only one <100> dendrite without tip-splitting could be measured, the tip velocity might increase and ℓ_d decrease with σ. Then ℓ might correspond to ℓ_d in a higher σ. Brener has shown that with an increasing σ the grwoth velocity of a dendrite without tip-splitting increases and that there exists a transition of its velocity between two growing mechanisms[11]. We conclude that the length between tips nearly corresponds to the diffusion length. S.-K. Chan et al. have measured the relation between v and σ in the tip-splitting region, and have shown that v is constant within the error bars for only a dendrite[7]. Our experiment is about collective growing behavior of dendrites and there exists a concentrational interaction among them.

As shown in Fig.3, the interfacial velocity decreases with σ. This behavior suggests the reason why the interface of our DBM is stable. If one tip sticks out from the interface, the consentration converges into the tip and the local σ near the tip increases. Then many tip-splittings occur and their tip-velocities decrease. The other tips meanwhile catch up with them and the whole interface becomes stable. On the other hand, if the interfacial velocity of our DBM is constant independently on σ, such a sticking tip grows with the same velocity with the others. Then its sidebranches screen the other tips and the whole interface becomes unstable.

We suppose the above decreasing velocity mechanism as follows. The distance between tips which form the interface of DBM is determined with the diffusion length as the above discussion. Many sub-tips and sidebranches grow or stop in the space between the tips. There exists the concentrational interaction among them, which decreases the tip velocities. In the case of a higher σ, the number of sub-tips and sidebranches is larger, then the interfacial velocity decreases with σ. This discussion implies that it is necessary that another information should be measured about the density of sub-tips and sidebranches within the tips. We suppose that such an information may be related with the surface thickness[12]. As shown in Fig.6, our DBM is a self-affine fractal. In this case, the diffusion constant ℓ_d is 50.13 μm and ln(ℓ_d)=3.91, then the surface of the DBM within the scale length ℓ_d is a self-affine fractal structure. However, the value of the exponent(γ=0.555) is less than the experimental values of Rubio et. al. and Vicsek et. al.(0.7∿0.8)[12]. Our exponent γ measured for various patterns in the case of σ=0.0511 is roughly 0.5∿0.6. In the case of a larger linear scale length ($L > \ell_d$), we suppose $\gamma \simeq 1$ because the interface of the DBM is stable.

In conclusion, we have shown the dense-branching morphology in an NH_4Cl dendritic crystal growth. The averaged spacing between the tips corresponds to the diffusion length. The stability of the interface results from the decreasing interfacial growth velocity with an increasing supersaturation.

[Acknowledgements]

We thank M.Matsushita and M.Sano for fruitful discussions and suggestions.

[References]

1.J.S.Langer, *Rev.Mod.Phys.* **52** (1980) 1.

2.E.Ben-Jacob, R.Godbey, N.D.Goldenfeld, J.Koplik, H.Levine,T.Mueller and M.Sander, *Phys.Rev.Lett.* **55** (1985) 1315; H.Honjo, S.Ohta and M.Matsushita,*Phys.Rev.A* **36** (1987) 4555.

3.T.A.Witten and L.M.Sander, *Phys.Rev.Lett.***47** (1981) 1400. See also, e.g., T.Vicsek, *Fractal Growth Phenomena*(World Scientific, Singapore, 1989), and references therein.

4.Y.Sawada, A.Dougherty and J.P.Gollub, *Phys.Rev.lett.* **56** (1986)1260;D.Grier,E.Ben-Jacob,R.Clarke and L.M.Sander, *Phys.Rev.Lett.* **56** (1986) 1264; D.G.Grier, D.A.Kessler and L.M.Sander,*Phys.Rev.Lett.* **59** (1987) 2315; J.R.Melrose, D.B.Hibbert and R.C.Ball,*Phys.Rev.Lett.* **65** (1990) 3009.

5.E.Ben- Jacob, G.Deutscher, P.Garik, N.D.Goldenfeld and Y.Lareah, *Phys.Rev.Lett.* **57** (1986) 1903.

6.H.Fujikawa and M.Matsushita, *J.Phys.Soc.Jpn.* **58** (1989) 3875; H.Fujikawa and M.Matsushita, *J.Phys.Soc.Jpn.***60** (1991) 88.

7.S.-K.Chan, H.-H.Reimer and M.Kahlweit, *J.Cryst.Growth* **32** (1976) 303.

8.H.Honjo,S.Ohta and Y.Sawada,*Phys.Rev.Lett.***55** (1985) 841.

9.A Similar pattern is reported in reference E.Raz, S.G.Lipson and E.Ben-Jacob,*J.Cryst. Growth***108**(1991) 637.

10.A.Tanaka and M.Sano, preprint.

11.E.A.Brener,*J.Cryst.Growth***99**(1990) 165, and references therein. See also F.Liu and N.Goldenfeld, *Phys.Rev.A* **42** (1990) 895.

12.M.A.Rubio, C.A.Edwards, A.Dougherty and J.P.Gollub, *Phys.Rev.Lett.* **63**, 1685 (1989); T.Vicsek, M.Cserzo and V.K.Horvath,*Physica A***167** (1990) 315. See also M.Kardar, G.Parisi and Y.-C.Zhang, *Phys.Rev.Lett.* **56** (1986) 889.

COMPUTER SIMULATIONS OF SPINODAL DECOMPOSITION IN BINARY FLUIDS

Tsuyoshi KOGA and Kyozi KAWASAKI
Department of Physics, Kyushu University 33, Fukuoka 812, Japan

ABSTRACT

Results of a computer simulation of spinodal decomposition in binary fluids are presented. The growth law of domains which is peculiar to viscous fluids is obtained in the late stage of spinodal decomposition. The dynamical scaling is satisfied in this time region and the functional form of the scaling function is in good agreement with the experimental results of polymer blends.

1. Introduction

The dynamics of ordering processes in quenched systems has been studied in various systems.[1] Recently, spinodal decomposition in polymer blends has been extensively studied and these experiments provide us precise information on the late stage dynamics of the spinodal decomposition.[2,3] Since the hydrodynamic effects play an important roll in the late stage of spinodal decomposition in such fluid systems,[4] it is necessary to take the hydrodynamic effects into account in order to study the late stage dynamics in these systems.

We have proposed a computationally efficient method to simulate ordering processes in fluid systems with the long-range hydrodynamic interaction and reported some results of the computer simulation of spinodal decomposition in binary fluids.[5,6] In this paper, we present results of the computer simulation of spinodal decomposition in binary fluids using a large three-dimensional system and compare our results with the experimental data of polymer blends.

2. Model and Simulation Methods

Here we consider binary fluids, where we assume that the fluid is incompressible. In order to describe the spinodal decomposition of binary fluids, we introduce a local order parameter $S(\boldsymbol{r})$,e.g. the local concentration, and the transverse component of the local velocity $\boldsymbol{v}(\boldsymbol{r})$. When the relaxation of the local velocity is more rapid compared to that of the local order parameter, we can eliminate the local velocity by assuming that the local velocity instantaneously follows the changes in the local order parameter.[6–8] This approximation may be valid in many experiments of fluid systems such as critical fluids and polymer blends, because, in critical fluids, the relaxation of concentration fluctuations is very slow due to critical slowing-down and, in polymer blends, the viscosity is very large due to the entanglement effects of polymer chains. As a result, we obtain the following time-dependent Ginzburg-Landau (TDGL) equation for binary fluids[6-8] :

$$\frac{\partial}{\partial t}S(\boldsymbol{r},t) = L\nabla^2\mu(\boldsymbol{r}) - \nabla S(\boldsymbol{r})\cdot\int \boldsymbol{T}(\boldsymbol{r}-\boldsymbol{r}')\cdot\nabla' S(\boldsymbol{r}')\mu(\boldsymbol{r}')d\boldsymbol{r}', \tag{2.1}$$

where $\mu(\boldsymbol{r}) \equiv \delta H\{S\}/\delta S(\boldsymbol{r})$ is the chemical potential, $H\{S\}$ is the Ginzburg-Landau type free-energy functional, L is the Onsager kinetic coefficient and $\boldsymbol{T}(\boldsymbol{r})$ is the Oseen tensor given by

$$\boldsymbol{T}(\boldsymbol{r}) = \frac{1}{8\pi\eta}\left(\frac{\boldsymbol{1}}{|\boldsymbol{r}|} + \frac{\boldsymbol{r}\boldsymbol{r}}{|\boldsymbol{r}|^3}\right), \tag{2.2}$$

where η is the shear viscosity and $\boldsymbol{1}$ is the unit tensor. The second term on the right-hand side of Eq.(2.1) describes the long-range hydrodynamic interaction between the order parameter fluctuations. We have ignored the thermal noise in Eq.(2.1) because we are interested in the dynamics in the very late stage of spinodal decomposition.

To integrate Eq.(2.1) numerically, we employ the cell-dynamical method.[9] Since a great deal of computational resources is required to compute the integral in the second term in real space, we calculate this integral in Fourier space using Fast Fourier Transform. Details of this method have been reported elsewhere.[5] The system used is a three-dimensional cubic lattice of size 128^3 with periodic boundary conditions. The results are averaged over two independent runs and the other parameters are the same as those in Ref.5.

In order to study the growth law of domains and the dynamical scaling in this system, we calculate the spherically averaged structure factor $I_k(t)$, which is obtained from the normalized structure factor $I_{\mathbf{k}}(t)$ defined by

$$I_{\mathbf{k}}(t) \equiv < S_{\mathbf{k}}(t)S_{-\mathbf{k}}(t) > / \sum_{\mathbf{k}} < S_{\mathbf{k}}(t)S_{-\mathbf{k}}(t) >, \tag{2.3}$$

where $S_{\mathbf{k}}$ is the Fourier component of $S(\boldsymbol{n})$ and $\boldsymbol{n}$ is the position vector of a lattice point.

3. Results

3.1 *Time Dependence of Characteristic Length*

In this study, the characteristic wave number, which is proportional to the inverse of a characteristic length of random domains, is defined as the first moment of $I_k(t)$ denoted as $k_1(t)$.

Figure 1 shows the time dependence of $k_1(t)$ and the negative slope in this figure gives the exponent z in the power law fit i.e. $k_1(t) \sim t^{-z}$. According to earlier theoretical studies,[4,7] the domain growth in the late stage of spinodal decomposition in fluid systems is driven by the flow induced by the surface tension and, in this case, the normal component of the velocity of the interface v_n is given by

$$v_n \sim \frac{\sigma}{\eta}, \tag{3.1}$$

where σ is the surface tension. By assuming that at the late stage of spinodal decomposition there is one characteristic length scale $\ell(t)$ associated with domains, we find $\ell(t) \sim \sigma/\eta\ t$ from (3.1), which gives the exponent $z = 1$. Our results in Fig.1 shows that the value of z increases rapidly after about 100 time steps and approaches to its expected value 1. The value of z, however, is somewhat smaller than 1 even at the late stage, which is possibly due to the fact that we have not yet reached the regime where $z = 1$.

On the other hand, if the domain growth proceeds by the evaporation-condensation mechanism which is another coarsening mechanism of domains in this system,[7] the value of z should be 1/3. The value of z in Fig.1, however, is apparently bigger than 1/3. Therefore, it is evident that the hydrodynamic interaction almost dominate the domain growth after about 200 time steps, although z has not yet reached its asymptotic value 1.

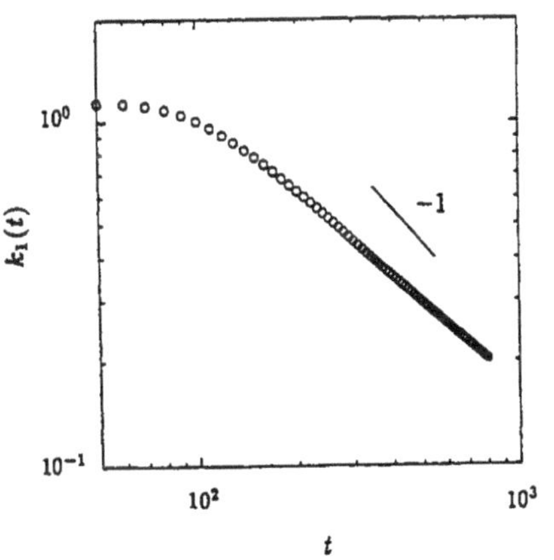

Fig.1 Plot of $k_1(t)$ vs t on a double-logarithmic scale. (From Ref.6)

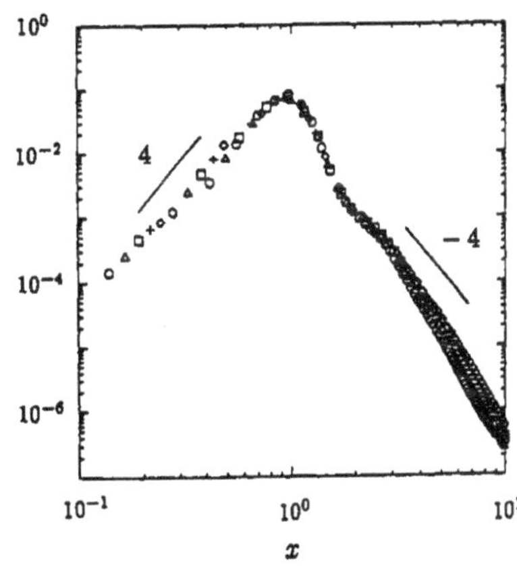

Fig.2 Plot of $F(x)$ vs x on a double-logarithmic scale. The symbols o, Δ, $\Box$, + and $\Diamond$ correspond to t = 400, 500, 600, 700 and 800 time steps, respectively. (From Ref.6)

3.2 *Scaling function*

We study the scaling law for $I_k(t)$: $I_k(t) = k_1(t)^{-3}F(x)$, in the time region in which the exponent z is nearly 1, where $F(x)$ is the scaling function and $x \equiv k/k_1(t)$ is the scaled wave number. The scaling plot for I_k at $t = 400 - 800$ time steps is presented in Fig.2 on double-logarithmic scales. This figure shows that the dynamical scaling law is satisfied in this time region. In this figure, we also find the following three features :

(i) For large x, $F(x)$ approaches the form $F(x) \sim x^{-4}$, which is indicated by the solid lines in this figure. This result corresponds to Porod's law.[10]

(ii) The asymptotic form of $F(x)$ in the limit $x \to 0$ is $F(x) \sim x^4$ as is indicated by the solid lines.[11–13]

(iii) In recent experiments on polymer blends, the "shoulder" is observed at $k \sim 3k_1$,[3] which can also be observed in Fig.2.

These features are consistent with the recent experimental data of polymer blends,[3] in which the exponent $z \simeq 1$ has been obtained and the hydrodynamic effects seem to be the dominant mechanism of the domain growth.

In Fig.3(a), a comparison between the simulation data and the experimental data is presented. We find that our results are in good agreement with the experimental data, except at the large $x \geq 3$. We can remove this difference at the large x by replacing the order parameter profile by that for infinitely sharp interfaces as shown in Fig.3(b).[6,14]

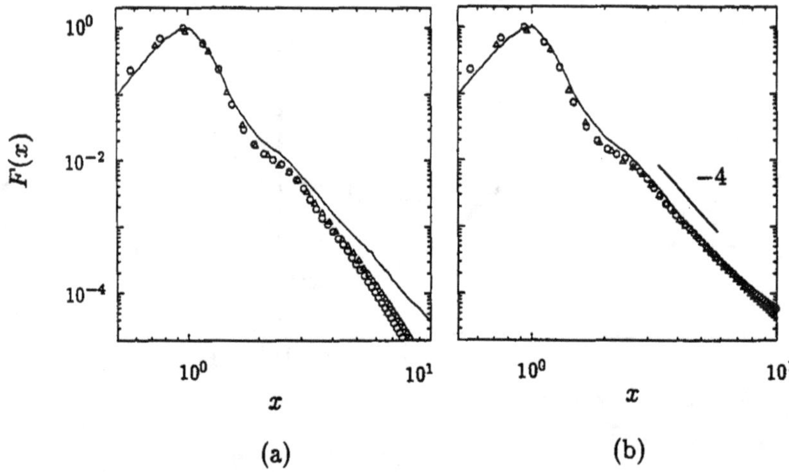

Fig.3 Comparison of our data with experimental data of polymer blends. The symbols o and Δ correspond to $t = 600$ and 800 time steps, respectively. Data in Fig.3(b) is obtained from the transformation: $S \to S_e \mathrm{sgn}(S)$, where S_e is the equilibrium value of S. The solid line is the experimental data. (From Ref.6)

4. Conclusion

The results of the computer simulation of spinodal decomposition in binary fluids were presented. The growth law of domains which is peculiar to viscous fluids was obtained. The scaling function obtained in the late stage, where the hydrodynamic interaction dominates the domain growth, agrees well with the experimental results of polymer blends.

5. Acknowledgments

We would like to thank Dr. T. Kawakatsu, Prof. T. Hashimoto and Mr. M. Takenaka for valuable discussions. This work was mainly supported by the Scientific Research Fund of the Ministry of Education, Science and Culture of Japan and also partially supported by the Joint Program of the National Institute for Fusion Science.

6. References

1. *Dynamics of Ordering Processes in Condensed Matter*, edited by S. Komura and H. Furukawa (Plenum, New York, 1988).
2. F. S. Bates and P. Wiltzius, J. Chem. Phys. **91**, 3258 (1989).
3. T. Hashimoto, M. Takenaka and H. Jinnai, J. Appl. Cryst. **24**, 457 (1991); M. Takenaka and T. Hashimoto, (submitted to J. Chem. Phys.).
4. E. D. Siggia, Phys. Rev. A **20**, 595 (1979).
5. T. Koga and K. Kawasaki, Phys. Rev. A **44**, R817 (1991).
6. K. Kawasaki, T. Koga and T. Kawakatsu, (to be published in MRS Fall meeting Proceedings).
7. K. Kawasaki and T. Ohta, Physica **118A**, 175 (1983).
8. K. Kawasaki, Prog. Theor. Phys. **57**, 826 (1977); K. Kawasaki and T. Ohta, *ibid.* **59**, 362 (1978).
9. Y. Oono and S. Puri, Phys. Rev. A **38**, 434 (1988).
10. G. Porod, in *Small Angle X-ray Scattering*, edited by O. Glatter and O. Kratky, (Academic, New York, 1982).
11. C. Yeung, Phys. Rev. Lett. **61**, 1135 (1988).
12. H. Furukawa, J. Phys. Soc. Jpn **58**, 216 (1989).
13. P. Fratzl and J. L. Lebowitz, Acta Metall. **37**, 3245 (1989); P. Fratzl, J. L. Lebowitz, O. Penrose and J. Amar, Phys. Rev. **A44**, 4794 (1991).
14. A. Shinozaki and Y. Oono, Phys. Rev. Lett. **66**, 173 (1991).

Flow Effects on DLA I

Sasuke MIYAZIMA, Kaneyasu MARUYAMA, Kouichi OKUMURA and Yasuhito OOTA

Department of Physisc, Chubu University,
Kasugai, Aichi 487, JAPAN

Abstract

Diffusion limited aggregation (DLA) is investigated in a Helle-Shaw cell with flow field. The morphology of the aggregation strongly depends on the flow. In general, the patterns with low fractal dimension are obtained in the highest ratio of flow, because sticking to the aggregation occurs more often at the tip.

1. Introduction.

Recently fractal growth phenomena have become one of central fields in the stastistical physics. Basic concept of fractality and multi-fractality was introduced by Mandelbrot[1] into physics. A typical example is diffusion limited aggregation (DLA) which was first discussed by Witten and Sander.[2] Meakin[3] progressed more details of fractal behaviors by very large scale simulations. Stimulated by these pioneering works, many models with various types of modifications have been proposed corresponding to various experiments.

The examples are cluster-cluster aggregation,[4] ballistic model,[5] Eden model,[6] epidemics[7] and so on.[8] Bunde and one of the present authors also proposed further extensions by introducing a lifetime of growing elements such as a lifetime of cell in Eden model. They found the intersting phenomena i.e. cross-over behaviors in the morphology and dynamical phase transition[9)–11].

There are several interesting experimental investigations such as electrodeposition,[12] viscous fingering,[13] aggregation in colloids, [14] fracture[15] and so on. Electrodeposition is seemed to be the best experimental example of DLA simulation with the fractal dimension 1.7 in the two dimension, even though it contains much more physics, e.g. multifractal behavior, than many physicists considered at initial. Simulational investigation plays an important role initially, now we need more details of knowledges of experimental works.

It may be one way in order to inspect an essential mechanism of growth that one tries to invent a simpler experimental situation. It may be another way in order to obtain more informations that one add another external force to the system. By the additional force, some portion of interactions become clear and might be stressed, and other portions are depressed. Therefore we consider here that a flow is a candidate for it.

We have found that the morphology depends strongly on the flow rate. In the Helle-Shaw cell, we have already obtained a usual DLA growth for the case of no fluid flow. We can conclude that the fractal dimension decreases with the flow rate. In the highest ratio of flow, we obtain a needle type growth, and the number of needles from the center decreases with the flow rate.

2. Experimental apparatus.

In our setup shown in Fig.1, the flow directs outward from the center. In order to avoid the eddy flow-in to the narrow gap between two acrylic plates, we make a bump cavity before the entrance to the gap from the bottom of acrylic plate. The hole size is 2 mm in diameter, and the cathode which is made of carbon rod and used for the pencil is at the center of the top of acrylic plate. The anode is just the Zn-plate which surrounds the center with the radius of 5 cm. The electric voltage between two electrodes is 6 volts.

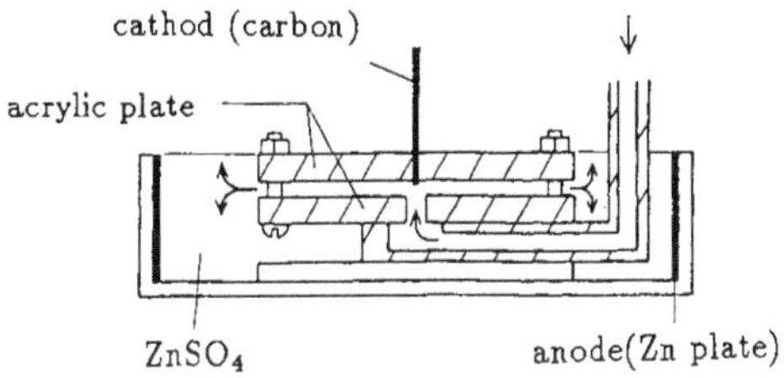

Fig.1. Radially symmetric flow of 2 normal ZnSO4 fluid is produced between two acrylic plates.

The flow rate is caused by the potential difference, where the bottle of fluid source is set at 1 m of height and controled by the gap width of the tube. The flow rate is varied from 0 to 200 milli-litre per minute. Although one keeps the fluid quantity at the tube constant, the flow rate along the radial direction changes with proportion to $1/r^2$, therefore the flow rate decreases rather rapidly towards the outer portion.

3. Experimental results and discussions

We have found the usual pattern of DLA with the fractal dimension of about 1.7 at the zero flow rate at first. The morphology changes and the growth patterns become to the one dimension from 1.7 as the flow rate. Global fractal dimension depends on the rate of flow as shown in Fig.2.

From Fig.2 we can notice several interesting points as follows.

1. The number of branches decreases untill 2, i.e. one-dimensional growth (Fig.3).
2. The length of needle increses (Fig.4).
3. After the needle growth, we have usual DLA growth with the fractal dimension 1.7.

4. The global fractal dimension changes from 1.7 to 1.1 (Fig.5).
The data are averaged over about 30 samples for Fig.3 to Fig.5.

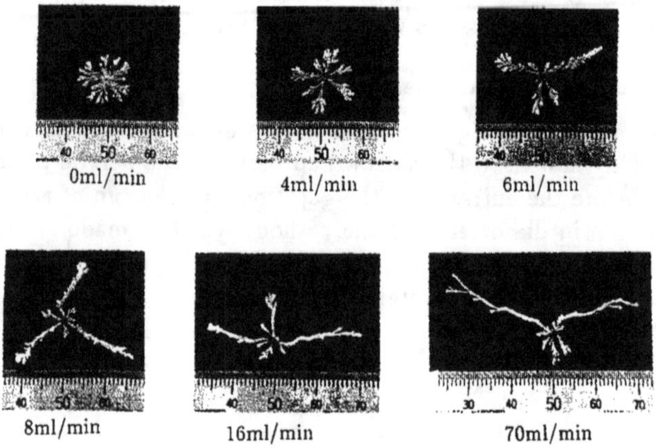

Fig.2. Typical growth pattern for various fluid flow.

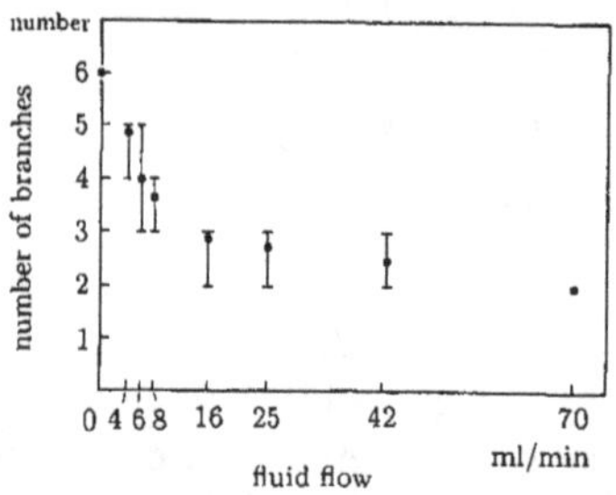

Fig.3. The number of branches agaist fluid flow.

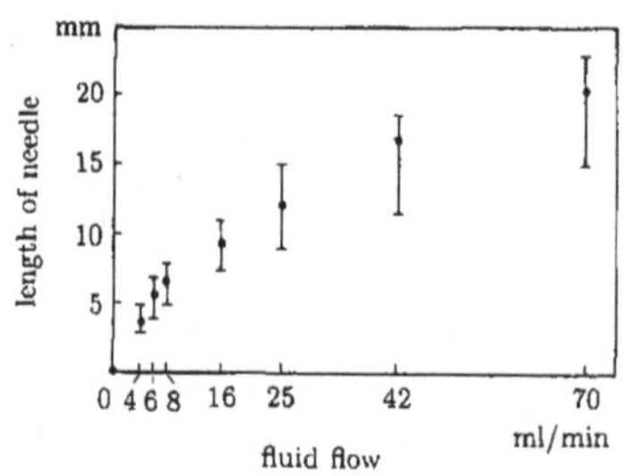

Fig.4. The length of needle gainst fluid flow.

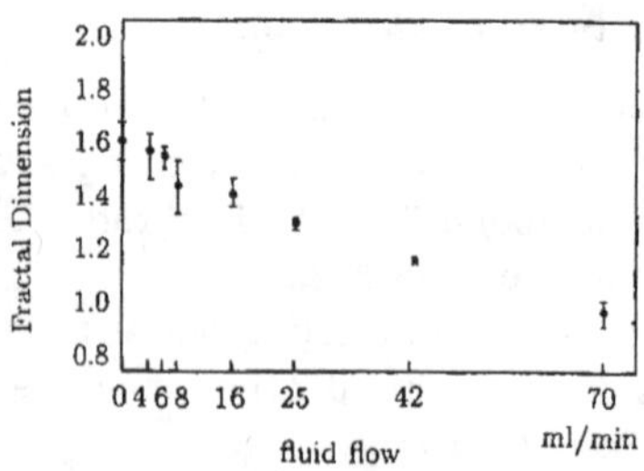

Fig.5. The global fractal dimension decreases depending on the fluid flow.

In this radial symmetric aparatus, however, the flow ratio decreases outwards as $1/r^2$, where r is the distance from the center. The fluid flow near the center is rather quick, and stronger than the electric field toward the center. On the other hand, the flow velocity decreases rather rapidly at the outer region, then the electric field becomes stronger and grows as the usual DLA. It is an interesting fact that the distance for the growth manner change also increases, depending on the ratio flow.

As for morphology of DLA, several parameters have been considered, such as concentration of fluid, strenght of electric field. Mainly two groups investigated the effect of the two factors onto the morphology of DLA, however, they obtained quite different phase diagrams, respectively. We need more detail investigations on the same investigation.

Here we have proposed an experiment with another parameter to DLA. The new parameter which is relevant for DLA is the flow of fluid in the Helle-Shaw cell. In the highest flow field we have obtained rather simple pattern of growth. The growth mechanism is still not simple. We tried several different simulations, but we make no success to form one-dimensional growth. We can expect this simulation might give more details of growth manner of DLA.

Now we consider the case of opposite direction of flow, i.e. the fluid gathers to the center from the outside. Therefore the acceleration of Zn ion to the aggregation at the center increases by fluid flow adding to the accleration by electric field between anode and cathode. In this case the morphology of DLA does not depend on the fluid flow as seen in Fig 6.

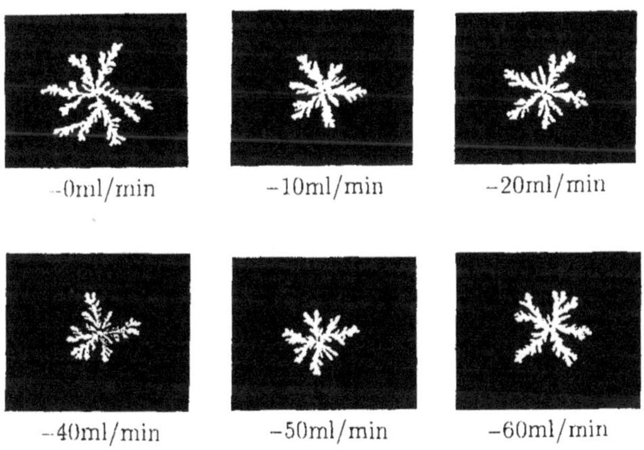

Fig.6. Typical patterns of DLA with flow to the center.

References

1. B. Mandelbrot, *Fractal Geometry in Nature*,(Freeman, New York, 1982).
2. T. Witten and Sander, Phys. Rev. Lett. **47** (1981), 1400.

3. P. Meakin, in *Phase Transition and Critical Phenomona* Vol.12 edited by C.Domb and J.L.Lebowitz (academic Press, New York, 1987).

4. P.Meakin, Phys. Rev. Lett. **51**. (1983) 1119., M. Kolb, R. Botet and R. Jullien, Phys. Rev. Lett. **51** (1983) 1123.

5. P. Ramanlal and L. M. Sander, Phys. Rev. Lett. **54** (1985) 1828.

6. M. Eden, *Proc. Fowth Berkeley Symp. on Mathematical Statistics and Probability*, Vol.IV edited by F. Negman (Univ. of California Press, Berkeley, 1961), P223.

7. G. Mackay and N. Jan, J. Phys. **A17** (1984) L759.

8. P. L. Leath Phys. Rev. B14 (1976) 5046. S.Miyazima, P. Meakin and F. Family, Phys. Rev. **A36** (1987) 1421.

9. S. Miyazima, A. Bunde, S. Havlin and H. E. Stanley, J. Phys. **A19** (1986) 1159.

10. S. Miyazima, A. Bunde, and H. E. Stanley, J. Stat. Phys. **47** (1987) 1.

11. S. Miyazima, Y. Hsaegawa, A. Bunde and H. E. Stanley, J. Phys. Soc, Japan **57** (1988) 3376.

12. M. Matsusita, M. Sato, Y. Hayakawa, H. Honjo and Y. Sawada, Phys. Rev. Lett. **53** (1984) 286.

13. P. G. Saffman and G. I. taylor, Proc. Roy. Soc. London Ser. **A245** (1988) 312.

14. D. A. Weity and M. Oliveria, Phys. Rev. **B31** (1985) 2993.

15. E. Charlaix, E. Guyon and N. River, Geol. Mag. **122** (1985) 157.

MORPHOLOGICAL CHANGES IN CONVECTION-DIFFUSION LIMITED AGGREGATIONS

TAKASHI NAGATANI
College of Engineering, Shizuoka University, Hamamatsu 432, Japan

ABSTRACT

The effect of flow on the diffusion-limited aggregation(DLA) is investigated by using a Monte Carlo model. The growth process of aggregates is governed by the convection and the diffusion. The convective diffusion field is simulated by the biased-random walker. With varying types of flows, the diffusion-limited aggregation changes into distinct morphologies. The morphological changes are shown by Monte Carlo simulation.

1. Introduction

There has recently been increasing interest in a variety of nonequilibrium aggregations such as the ballistic model and the diffusion-limited aggregation(DLA) model[1-3]. A variety of computer simulations and experiments have been carried out to investigate the relationships between the cluster geometry and the growth mechanism.The structure of the aggregates strongly depends on the dynamics of the growth process. The ballistic deposition model provides a basis for understanding the vapor-deposition process. The DLA model presents a prototype of the pattern formation of diffusive systems including electrochemical deposition, crystal growth, viscous fingering, dielectric breakdown, and bacterial colony.

In this paper, we study the morphological changes of the DLA by introducing convective flows into diffusion field. We simulate the morphological changes under various flows by a Monte Carlo model. The simulation patterns are shown under the following types of flows: (A) the radial flow[4], (B) the rotating flow[5], (C) the horizontal flow[6], (D) the shear flow[7], (E) thermal convection[8], and (F) the flow on a wedge.

2. Basic equation and simulation

We consider the aggregation process growing at a point seed and the deposition process growing from the plate in two dimensions. The pattern formation is controlled by the convective diffusion field of a component moving outward from the growing aggregate without surface kinetics. The concentration C of the diffusing component under hydrodynamic flow satisfies the convective diffusion equation under the quasi-stationary approximation

$$u_x\frac{\partial C}{\partial x} + u_y\frac{\partial C}{\partial y} = D\frac{\partial^2 C}{\partial x^2} + D\frac{\partial^2 C}{\partial y^2} , \qquad (1)$$

where u_x and u_y are the x and y components of the velocity and D is the diffusion constant. The boundary condition on the surface of the aggregate is given by

$$(1-p_s)\partial C/\partial n - p_s C = 0. \qquad (2)$$

Now we consider the simulation of the aggregation process following Eq.(1). The convective diffusing particles undergo random walks except for a bias resulting from the convection. The probability p(i,j) that the biased random walker will visit the site(i,j) satisfies the equation

$$\begin{aligned} p(i,j) = &\left[|P_x|\{1-\theta(P_x)\} + \frac{1-|P_x|-|P_y|}{4}\right]p(i+1,j) + \left[|P_x|\theta(P_x) + \frac{1-|P_x|-|P_y|}{4}\right]p(i-1,j) \\ &\left[|P_y|\{1-\theta(P_y)\} + \frac{1-|P_x|-|P_y|}{4}\right]p(i,j+1) + \left[|P_y|\theta(P_y) + \frac{1-|P_x|-|P_y|}{4}\right]p(i,j-1) \end{aligned} \quad , (3)$$

where $P_x=(u_x a/D)/\{1+(u_x a/D)+(u_y a/D)\}$,

$P_y=(u_y a/D)/\{1+(u_x a/D)+(u_y a/D)\}$,

and a is the lattice spacing. The boundary condition(2) can be simulated by introducing the sticking probability p_s.

3. Simulation result

We show the simulation results for various flow conditions. The simulation of the aggregation in the presence of flows is carried out with the use of a simple square lattice.

(A) We consider the convection-diffusion limited aggregation(CDLA) under a radial flow. Particles perform a radially biased random walk with a probability P_r toward the radial direction. Figures 1(a)-(d) show the typical patterns for attractive drift $|P_r|=0.1$, no drift $P_r=0$, and repulsive drifts $|P_r|=0.009$, 0.018. When the direction of the drift changes from an attractive drift to a repulsive drift, a structure transition occurs from a dense structure, through the DLA fractal, to a single needle structure.

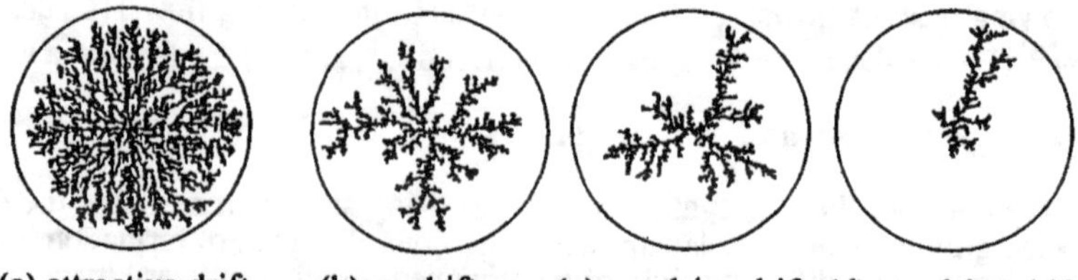

(a) attractive drift (b) no drift (c) repulsive drift (d) repulsive drift

FIG.1. Typical clusters under radial drifts.

(B) We consider CDLA under a rotating flow. Particles perform a tangentially biased random walk with P_t toward the tangential direction. Figures 2(a)-(d) show the typical patterns obtained with increasing P_t. As the tangential drift P_t becomes higher, the overall shape of the cluster becomes a less ramified structure and changes finally into a single-arm structure which is oppositely vortical to the tangential flow. The rotating flow induces a strong screening. The screening effect prevents particles from going into the inner part of the cluster.

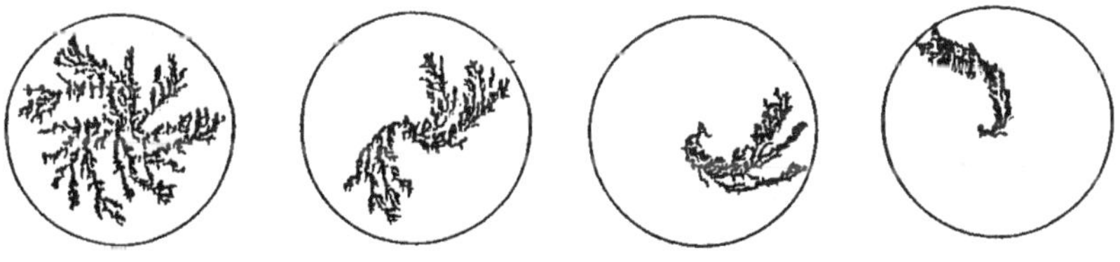

(a) P_t=0.1 (b) P_t=0.2 (c) P_t=0.5 (d) P_t=0.8

FIG.2. Typical clusters grown under tangential drifts.

We also study the effect of the tangential drift upon the DLA in the presence of the constant radial drift. Figures 3(a)-(d) show, respectively, the clusters for P_t=0.1, 0.2, 0.5, and 0.8 under the constant radial drift P_r=0.2. In the presence of pure radial drift(see FIG.1(a)), the pattern is the dense-radial pattern. As the tangential drift P_t becomes higher, the overall shape of the cluster changes into the multiarm structure and the width of the arm decreases.

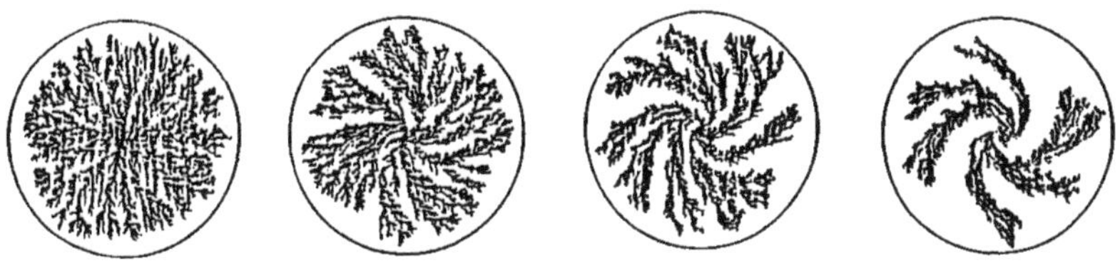

(a) P_t=0.1,P_r=0.2 (b) P_t=0.2,P_r=0.2 (c) P_t=0.5,P_r=0.2 (d) P_t=0.8,P_r=0.2

FIG.3. Typical clusters grown under both tangential and radial drifts.

(C) We consider the convection-diffusion limited deposition(CDLD) on the plate in the presence of the horizontal flow. Particles perform a horizontally biased random walk with P_x to the deposition plate. Figures 4(a)-(d) show the typical patterns obtained with increasing P_x. By introducing the horizontal flow, the overall shape of the deposit changes into a single tree. As the horizontal drift increases, the shape of the deposit becomes a less ramified structure and changes finally into a single needle inclined against the horizontal flow. Since the horizontal flow induces a strong screening effect, the tip of the deposit grows mainly.

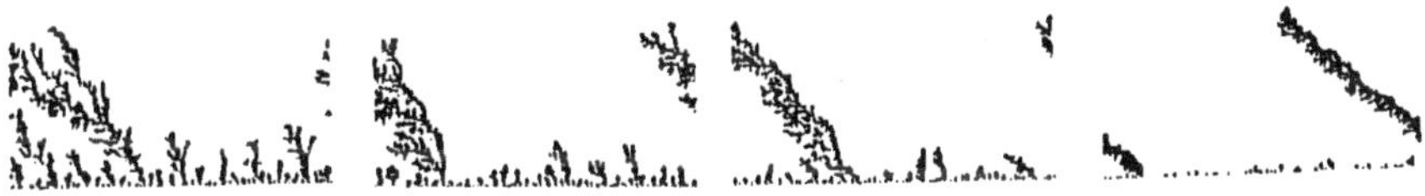

(a) P_x=0.05 (b) P_x=0.1 (c) P_x=0.2 (d) P_x=0.5

FIG.4. Typical deposits grown under horizontal flows.

We also study the effect of both horizontal and vertical drifts upon the DLA. Figures 5(a)-(d) show, respectively, the deposits for P_x=0.1, 0.3, 0.5, and 0.7 under the constant vertical drift P_y=0.1. With increasing horizontal drift under the constant vertical drift, the pattern of the deposits changes into the columnar morphology. The width and the number of the columns decrease with increasing horizontal drift.

(a) P_x=0.1,P_y=0.1 (b) P_x=0.3,P_y=0.1 (c) P_x=0.5,P_y=0.1 (d) P_x=0.7,P_y=0.1

FIG.5. Typical deposits grown by varying the horizontal drift under the constant vertical drift.

(D) We consider the CDLD on the plate under a shear flow. Figure 6(a) shows a schematic representation of the model under the Couette flow. In the shear flow, u_y=0 and $u_x=u_{max}y/H$ where u_{max} is the maximum velocity on the upper plate and H is the distance between the lower plate and the upper plate. The drift P_x varies vertically from position to position. Figures 6(b) and 6(c) show, respectively, the typical patterns of deposits obtained by varying the sticking probability p_s under the constant $P_{x,max}$=0.5. The deposits lead to the formation of a characteristic needle morphology, the needle inclines to the shear flow, and the width of the needle increases with decreasing sticking probability.

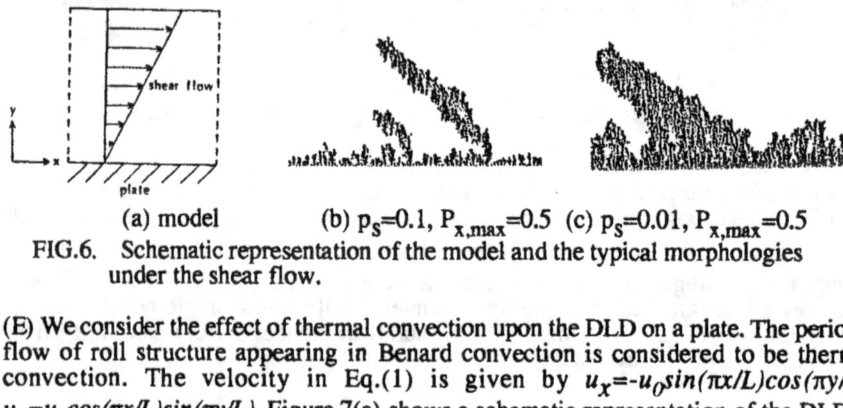

(a) model (b) p_s=0.1, $P_{x,max}$=0.5 (c) p_s=0.01, $P_{x,max}$=0.5

FIG.6. Schematic representation of the model and the typical morphologies under the shear flow.

(E) We consider the effect of thermal convection upon the DLD on a plate. The periodic flow of roll structure appearing in Benard convection is considered to be thermal convection. The velocity in Eq.(1) is given by $u_x=-u_0 sin(\pi x/L)cos(\pi y/L)$, $u_y=u_0 cos(\pi x/L)sin(\pi y/L)$. Figure 7(a) shows a schematic representation of the DLD in the presence of Benard convection. Particles perform a biased random walk following the thermal convection.

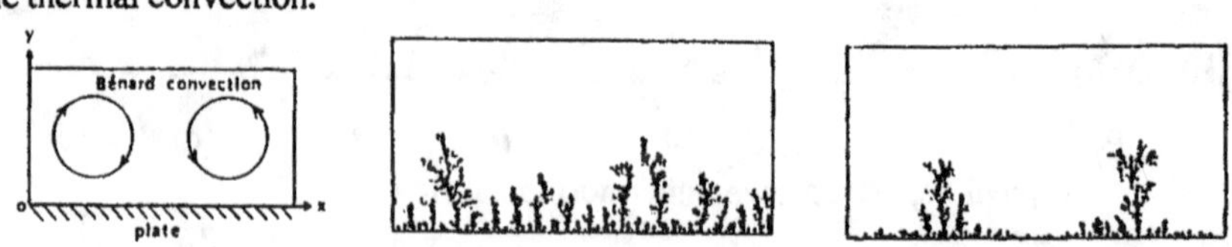

(a) model (b) $P_{x(or\ y),max}$=0.001 (c) $P_{x(or\ y),max}$=0.01

FIG.7 Schematic representation of the model and the typical morphologies under the thermal convection.

Figures 7(b) and 7(c) show the typical deposits obtained by varying the strength of thermal convection. In figures 7(b) and 7(c), the maximum values $P_{x(or\ y),max}$ of the drift resulting from thermal convection are given, respectively, by 0.001 and 0.01. With increasing thermal convection, the pattern of the deposit changes into a periodic columnar morphology. Each column grows toward the center of the roll. The morphological change appears even with very weak thermal convection.

(F) We consider the deposition process on a wedge. Figure 8(a) shows the geometry of the model. Flow is vertical to the wedge where the angle of the apex is 30°. Figures 8(b), (c), and (d) show, respectively, the typical patterns for the sticking probabilities p_S=1, 0.1, and 0.01 under constant drift P_y=0.2. The dendrite structure appears with increasing vertical flow. The dendrite structure will be constructed by the screening effect being similar to the case(C)(see FIG.5). With decreasing sticking probability, the width of branches increases. The effect of the sticking probability is similar to the case(D)(see FIG.6). The branches on the wedge are not symmetric.

(a) model (b) p_S=1 (c) p_S=0.1 (d) p_S=0.01

FIG.8. Schematic representation of the model and the typical morphologies grown on the wedge for the vertical flow P_y=0.2.

4. References

1. H.E.Stanley and N.Ostrowsky, *On Growth and Form* (Nijhoff, The Hauge, 1985).
2. P.Meakin, in *Phase Transitions and Critical Phenomena*, eds. C. Domb and J.L.Lebowitz (Academic, New York, 1988), Vol.12, p.336.
3. T.Vicsek, *Fractal Growth Phenomena* (World-Scientific, Singapore, 1989).
4. T.Nagatani, *Phys.Rev.*A **39** (1989) 438.
5. T.Nagatani and F.Sagues, *J.Phys.Soc.Jpn.* **59** (1990) 3447.
6. T.Nagatani and F.Sagues, *Phys.Rev.*A **43** (1991) 2970.
7. T.Nagatani, *J.Phys.Soc.Jpn.* 60 (1991) 2700.
8. T.Nagatani, *J.Phys.Soc.Jpn.* **60** (1991) 1181.

OSTWALD RIPENING OF PRECIPITATES IN OPEN SUBSYSTEMS

Akio Nakahara, Toshihiro Kawakatsu and Kyozi Kawasaki

Department of Physics, Kyushu University 33, Fukuoka 812, Japan

ABSTRACT

Ostwald ripening in open cells shows new scaling laws when the total volume of precipitates in the cell decreases through the diffusion of the solute.

1. Introduction

As compared to coarsening processes in homogeneous systems which are characterized by the universal scaling laws, coarsening processes of precipitates in non-uniform systems show various different features that depend on the inhomogeneity of the system, such as the spatial distribution of precipitates[1,2]. The Ostwald ripening in spatially uniform systems leads to the results that physical quantities in the system obey the universal scaling laws, i.e., the supersaturation of the solution, $\sigma(t)$, the mean radius of precipitates, $\bar{r}(t)$, and the total number of precipitates per unit volume, $n(t)$, at time t obey the universal power laws in time as $\sigma(t) \sim t^{-1/3}$, $\bar{r}(t) \sim t^{1/3}$, $n(t) \sim t^{-1}$, and, the size distribution function of precipitates with radius r, $f(r,t)$, can be described as $f(r,t) \sim t^{-4/3}F^{\rm LSW}(z)$ by using the Lifshitz-Slyozov-Wagner (LSW) universal scaling function[3,4]

$$F^{\rm LSW}(z) \sim \begin{cases} \left[z^2/(2z_0+z)^{7/3}(z_0-z)^{11/3}\right] \times \exp\left[-z_0/(z_0-z)\right] & (0 \le z < z_0) \\ 0 & (z \ge z_0) \end{cases} \tag{1.1}$$

Here, $z_0 = (3/2)^{1/3}$ and the scaled radius z is given by $z = r/t^{1/3}$. On the other hand, Miyazaki[1] investigated the Ostwald ripening of garnets in metamorphic rocks, and found that, when garnets are distributed uniformly in space, the size distribution of garnets is consistent with the LSW scaling function, while the size distribution becomes wider than the LSW scaling function if the garnets show heterogeneous spatial distribution, such as clusters and layers. Kai and Müller[2] performed experiments on the spatial patterns of precipitates and reported that the size distribution of precipitates is much more symmetric and much flatter than the LSW scaling function. Therefore, the Ostwald ripening of precipitates in non-uniform systems is characterized by new features, which are strongly related to the inhomogeneity of the system.

2. Basic Formulae of Ostwald Ripening in Open Subsystems

As a first step to understand the Ostwald ripening in the presence of spatial inhomogeneity, we consider the situation that the space is separated into two regions, i.e., the cell a where many larger precipitates exist and the cell b where only smaller precipitates are found. These cells (subsystems), each of which is labeled by the suffix i ($i = a, b$), are coupled with each other by the diffusion of the solute. The size distribution function in cell i, $f_i(r,t)$, obeys the continuity equation[3–7]

$$\frac{\partial}{\partial t} f_i(r,t) + \frac{\partial}{\partial r}[v_i(r,t) f_i(r,t)] = 0, \tag{2.1}$$

where $v_i(r,t)$ is the growth rate of precipitates in cell i. $v_i(t)$ is determined from the Gibbs-Thomson condition which is, in some units, expressed as

$$v_i(r,t) = \frac{1}{r}\left[\sigma_i(t) - \frac{1}{r}\right]. \tag{2.2}$$

Here, $\sigma_i(t)$ is the supersaturation of the solution in cell i. The total number and the total volume of precipitates per unit volume in cell i, denoted as $n_i(t)$ and $u_i(t)$, are written as

$$n_i(t) = \int_0^\infty dr f_i(r,t), \tag{2.3}$$

$$u_i(t) = \frac{4\pi}{3}\int_0^\infty dr r^3 f_i(r,t). \tag{2.4}$$

Since the mass transport takes place due to the difference in the supersaturation between the two cells, the time evolution of the total amount of both the solute and the precipitates, $q_i(t) = \sigma_i(t) + u_i(t)$, is given by[6,7]

$$\frac{d}{dt} q_i(t) = D[\sigma_j(t) - \sigma_i(t)], \tag{2.5}$$

where $j \neq i$ and D is the coupling constant between the two cells. Equation(2.5) satisfies the mass conservation in the whole system

$$q_a(t) + q_b(t) = \text{constant}. \tag{2.6}$$

In the following sections, we investigate the Ostwald ripening of precipitates in open subsystems by using dynamical scaling assumptions and by performing numerical simulations.

3. Dynamical Scaling in Open Subsystems

Since the difference in the supersaturation between the two cells becomes asymptotically negligible, the whole system behaves asymptotically as a closed system, and the supersaturation in each cell, $\sigma_i(t)$, obeys the LSW scaling law, i.e., $\sigma_i(t) \sim t^{-1/3}$ for large t. However, the evolution of the total volume of precipitates, $u_i(t)$, depends on the initial conditions. When there are many large precipitates in the cell as compared to the other cell at the initial time t_0, the total volume of precipitates $u_i(t)$ approaches a finite value in the limit $t \to \infty$. On the other hand, when there are only few large precipitates in the cell as compared to the other cell at the initial time t_0, the total amount of both the solute and the precipitates, $q_i(t)$, flows out into the other cell, and the total volume of precipitates $u_i(t)$ vanishes in the limit $t \to \infty$. Assuming that this quantity $u_i(t)$ behaves as $u_i(t) \sim t^{-x_i}$ with a non-negative parameter x_i for cell i, we found that the size distribution function of precipitates obeys a new scaling law[7] $f_i(r,t) \sim t^{-(4/3+x_i)} F^{x_i}(rt^{-1/3})$. Here, $F^x(z)$ is a scaling function for open subsystems which depends on the parameter x as,

$$F^x(z) \sim \begin{cases} \left[z^2 / (2z_0 + z)^{1+(4/3)(1+x)} \, (z_0 - z)^{2+(5/3)(1+x)} \right] \times \exp\left[-\, (1+x)\, (z_0/(z_0 - z)) \right] & (0 \le z < z_0) \\ 0 & (z \ge z_0) \end{cases} , \quad (3.1)$$

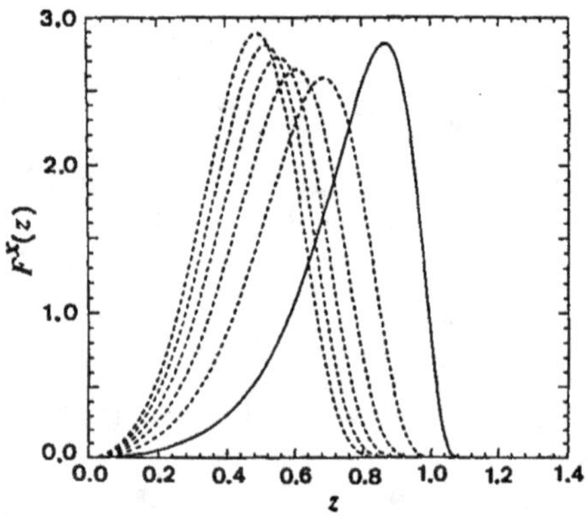

Fig.1 The LSW scaling function $F^{LSW}(z)$, denoted by the solid line, and the scaling functions $F^x(z)$ in open subsystems, denoted by dotted lines, are shown as functions of the scaled size z, where the values of the parameter x in $F^x(z)$ are $x = 2, 4, 6, 8$ and 10, from the larger z side to the smaller z side. As x increases, the peak position of $F^x(z)$ shifts to the smaller z side and the functional form of $F^x(z)$ becomes more symmetric (From Ref.7).

where $z_0 = (3/2)^{1/3}$. Figure 1 shows $F^x(z)$ as a function of the scaled size z. We note that, for vanishig x, i.e., when the total volume of precipitates $u_i(t)$ approaches a finite value like the one in the closed system, $F^x(z)$ coincides with the LSW scaling function $F^{LSW}(z)$. On the other hand, as x increases, i.e., as the total volume of precipitates $u_i(t)$ decreases more rapidly through the diffusion of the solute, the peak position of the scaling function $F^x(z)$ shifts to the smaller z side and the functional form of $F^x(z)$ becomes more symmetric. These results on the functional form of the size distribution agree with Miyazaki's investigation showing that the size distribution of garnets becomes wider than the LSW scaling function when the garnets are not distributed uniformly in space. The total number of precipitates $n_i(t)$ is also found to obey a scaling law as $n_i(t) \sim t^{-(1+x_i)}$. Thus, as the total volume of precipitates $u_i(t)$ decrease more rapidly, the number of precipitates which still survive decreases faster than the LSW prediction. As for a mean radius $\bar{r}_i(t)$, it obeys the same power law as the LSW scaling law, i.e., $\bar{r}_i(t) \sim t^{1/3}$.

4. Application to an Example

We consider a situation where the initial size distribution function of large precipitates takes the form $f_i(r, t_0) \sim \exp[-(r/s_i)^{m_i}]$ for large r, where m_i and s_i are positive constants characterizing the value of the initial size distribution function of large precipitates of cell i.[7] For the case $m_a = m_b = m$ and $s_a > s_b$, we obtain $x_a = 0$ and $x_b = (s_b/s_a)^{-m} - 1 > 0$ by using the arguments in the preceding section. Therefore, we find that, when there are more larger precipitates in cell a as compared to cell b at the initial time t_0, more precipitates survive at the later time t in cell a than that in cell b, and the total amount in cell b, $q_b(t)$, flows out into cell a. For the case $m_a = m_b = m$ and $s_a = s_b$, we obtain $x_a = x_b = 0$. Thus, when there are as many large precipitates in cell b as in cell a at the initial time t_0, physical quantities in each cell obey the LSW scaling laws.

Next, we present the results of our numerical simulations[7]. We take $m_a = m_b = 6$, $s_a = 10$ and $s_b = 10 \times (3/4)^{1/6}$. In this case, the values of x_i are expected to be $x_a = 0$, $x_b = 1/3$. Figures 2(a) and 2(b) show the time evolution of the size distribution of precipitates in cell a and cell b, respectively. Here, we use the normalized size distribution function $G_i(z,t)$ in the scaled size space z defined by $dzG_i(z,t) \equiv drf_i(r,t)/\int_0^\infty drf_i(r,t)$. We see that $G_a(z,t)$ approaches the LSW scaling function $F^{\mathrm{LSW}}(z)$, while $G_b(z,t)$ approaches the scaling function $F^x(z)$ with $x = 1/3$.

5. Conclusion

We studied the Ostwald ripening in open systems where two homogeneous cells are coupled with each other by the diffusion of the solute. We found that the supersaturation in each cell asymptotically obeys the LSW scaling law. On the other hand, as the total volume of precipitates in one cell decreases through the diffusion of the solute into another cell, the size distribution function of precipitates becomes more symmetric and wider than the LSW scaling function. These results coincide with the investigation showing that the size distribution of garnets in non-uniform systems is wider than the LSW scaling function[1].

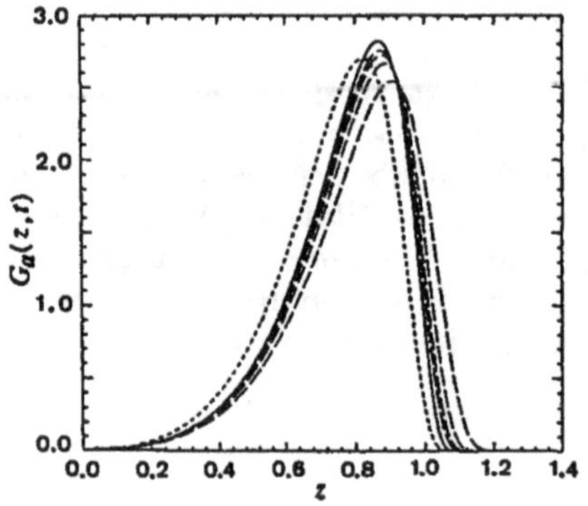

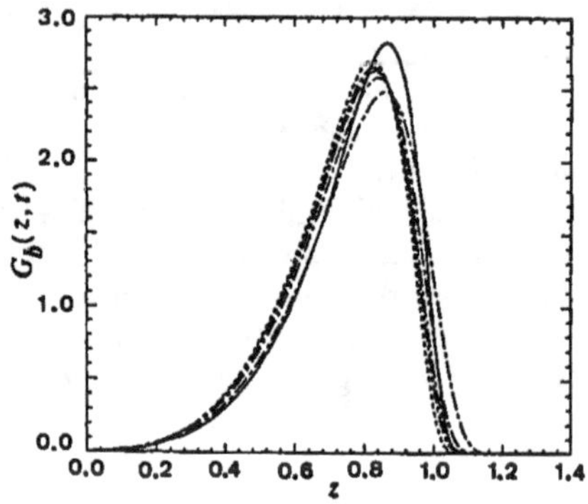

Fig.2(a) The evolution of the normalized size distribution function of precipitates in cell a, $G_a(z,t)$, is shown as a function of the scaled size z. Broken lines denote $G_a(z,t)$, where the time is, from right to left, $\log_{10}t = 5.0, 6.0, 7.0$ and 8.0, respectively. The solid line denotes the LSW scaling function $F^{LSW}(z)$. The dotted line denotes the scaling function $F^x(z)$ with $x = 1/3$. Note that $G_a(z,t)$ approaches $F^{LSW}(z)$ for large t. (b) Same as Fig.2(a) but for cell b, with dash-dotted lines denoting $G_b(z,t)$. Note that $G_b(z,t)$ approaches $F^x(z)$ with $x = 1/3$ for large t. (From Ref.7).

Acknowledgements

One of the authers (A.N.) thanks Dr.K.Sekimoto, Dr.H.Hayakawa and Mr. N.Suematsu for valuable discussions. He also thanks Dr.T.Nishiyama and Dr.K.Miyazaki for introducing him to their works on the pattern formation in metamorphic rocks. This work was supported by the Scientific Research Fund of the Ministry of Education, Science and Culture of Japan.

References

1. K.Miyazaki, *Contrb.Mineral.Petrol.* **108**, (1991) 118.
2. S.Kai and S.C.Müller, *Sci.Form* **1**, (1985) 9.
3. I.M.Lifshitz and V.V.Slyozov, *J.Phys.Chem.Solids* **19**, (1961) 35.
4. C.Wagner, *Z.Elektrochem.* **65**, (1961) 581.
5. J.A.Marqusee and J. Ross, *J.Chem.Phys.* **79**, (1983) 373.
6. G.Venzl, *Phys.Rev.* **A31**, (1985) 3431.
7. A.Nakahara, T.Kawakatsu and K.Kawasaki, *J.Chem.Phys.* **95**, (1991) 4407.

HOMOGENEOUS DIFFUSION-LIMITED AGGREGATION INCLUDING SURFACE-DIFFUSION PROCESS

Shonosuke Ohta
Physics Department, College of General Education,
Kyushu University, Ropponmatsu, Fukuoka 810, Japan

ABSTRACT

Morphology of aggregates including surface-diffusion process is studied by Monte Carlo simulations on a two-dimensional square lattice. The anisotropy of pattern changes between <10> and <11> directions depending on the surface-diffusion mechanism. Various types of crystal growth such as hopper shape, and side branching and tip-splitting dendrites are simulated. A homogeneous diffusion-limited aggregation grows in the crossover domain of the two anisotropies under the adequate noise.

Much attention has been paid to the morphology of the pattern form growing in the diffusion field[1,2]. Many attractive forms such as dendrite, diffusion-limited aggregation (DLA), and dense-branching morphology (DBM) has been experimentally observed in various systems of physical and chemical fields. These results show the universality for the growth mechanisum of the pattern formation in diffusion systems. While in a unitary system, we can observe the variously different patterns. A typical example is the denditic crystal growth of NH_4Cl in the supersaturated solution. The preferd ditection of NH_4Cl dendrite changes from <100> to <110> and then changes into <111> with increase of supersaturation. The growth morphologies of the stable tip type with side branching, regular tip-splitting type, and complicated tip-splitting type appear[3]. Moreover, the DLA pattern of NH_4Cl crystal has been observed in the system with the additional roughness[4]. It is clear that such morphological changes depend on the boundary conditions on the crystal surface. In fact, the effects of surface tension which depends on the crystalline anisotropy has been studied for the boundary condition of dendritic growth, and also the importance of kinetic term has been pointed out[5]. On the other hand, it is well known that the surface kinetic process plays an important part in the growth of snow crystal with faceted surfaces. Under this process the regular, hopper, and dendritic crystals grow with six-fold symmetry. We supporse the study of the pattern formation based on the surface kinetic process is an interesting problem in order to understand the

Fig. 1. Hopper-like pattern simulated by $\underline{m}$=100, τ=10000, γ=1, and N=520000.

growth mechanisms of the dendrite, DLA, and DBM.

The Laplace field is assumed for the surroundings of the pattern in the same manner as the DLA model[6] on a two-dimensional square lattice, which stands for the growth velocity of the pattern is negligible. The surface kinetics of the particles take account of the surface diffusion and of the solidification by means of the step growth and the surface nucleation. The evaporation process of the particles located on the pattern surface is ignored. Simplified simulation procedures are as follows: A random-walker released from a distant boundary is moving on a square lattice. After it reaches an unoccupied nearest-neighbor (NN) site of the pattern particles, called the surface site, further random walking along the surface site is performed. The hopping site of the walker is chosen from among the surface sites of n_1 on the NN and of n_2 on the next-nearest-neighbor (NNN). Assuming the ratio γ of the hopping probability for a NNN surface site (P_2) to that for a NN one (P_1), hopping site is decided from the relations $P_2=\gamma P_1$ and $n_1P_1+n_2P_2=1$ by using a random number less than unity. Such a

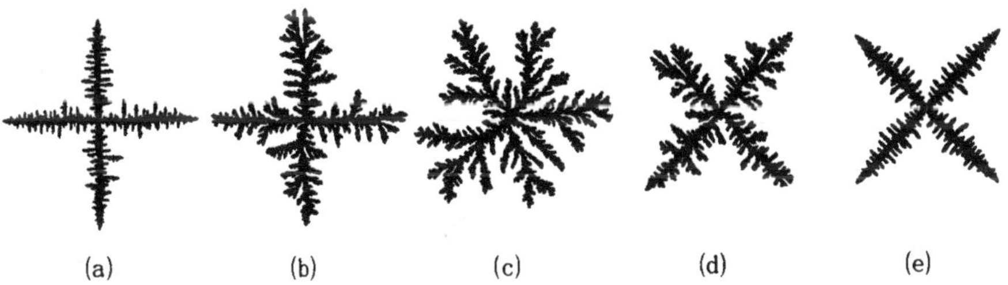

Fig.2. Simulated patterns for $\underline{m}$=5 and τ=10 with radius of 600 lattice units. The hopping parameters are γ=(a)0.05, (b)0.175, (c)0.23, (d)0.3 and (e)0.6.

random walk is repeated until the walker visits a kink site or until the trial of the surface walk becomes τ times. Here, the kink site is defined as where the two NN sites neighboring each other and a NNN site between them are occupied by the pattern. The number of walkers stuck to the surface site is counted by its own counter. If its count reaches a given number $\underline{m}$ then the surface site becomes part of the pattern and the new surface sites are born. Another random walker is released from a distant boundary and the procedure iterated. If we imagine the three-dimensional system constituted from the planes of $\underline{m}$ layers, the pattern growth at the kink site corresponds to the step growth on the step whose length is $\underline{m}$ layers. While on the flat surface without step, a surface nucleus composed of $\underline{m}$ particles is formed by increase of the concentration of surface-diffusion particles.

In case of τ>>1 and γ=1, the effect of surface diffusion is emphasized, and thus the lateral growth caused by the step and nucleation growths constructs the regular pattern as shown in Fig.1. When the pattern size is larger than the surface diffusion length $\sqrt{\tau}$, the pattern changes into the hopper-like pattern with giant steps. Such a pattern grows toward the <11> direction. Dendritic patterns in small τ are shown in Fig.2, whose anisotropy changes from the <10> direction to the <11> one with increase of γ. Since the surface diffusion is limited within a plane surface in γ<<1, the <10> dendrite is formed in Fig.2(a). At the crossover domain of two anisotropies a randomly ramified structure appears in Fig.2(c), which is a homogeneous DLA of the fractal dimension d_f=1.715 as indicated from the measurements of the pattern anisotropy and from the analysis of the scaling properties[7].

A morphological diagram for the simulated patterns is shown in the γ-τ space of Fig.3. The <10> dendrites (a1-a3) and the <11> dendrites (b1-b3) simulated by $\underline{m}$=100 have strong anisotropy and form the side branches. The crossover points of two anisotropies are dotted for $\underline{m}$=1,5,100. The diagram clarifies that the anisotropy of pattern tends to the <11> direction with increase of γ, τ, and $\underline{m}$. It is reasonable because the pattern growth at kink sites is enhanced. The increasing $\underline{m}$ makes the averaging effects of rondomness grow intense, and therefore the growth probability of kink site enlarges. In the case of $\underline{m}$=100, the crossover pattern formes a coexistent type (c1) and an intermediate type (c2,c3) on the <10> and <11> anisotropies. Thus, the homogeneous DLA which appers in the crossover domain of small $\underline{m}$ (d1) needs noise. The growth mechanism of such

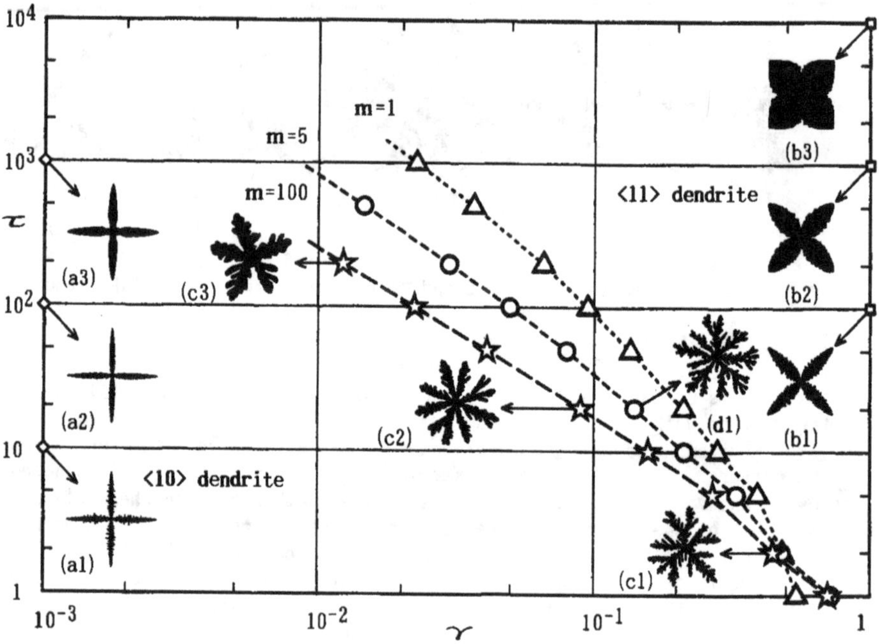

Fig.3. Morphology on the γ-τ plane. Pattern radius is 400 lattice units.

a DLA is the random tip-splitting type, whose hottest tips expand and then randomly bifurcate caused by the interfacial instability and by the noise as shown in Fig.4.

The above simulations reveal various and realistic patterns growing in the diffusion field. However, the growth velocity of pattern is neglected. In the system far from the equilibrium this effect should be considered. The tip front of growing pattern experiences the flow of the field particles. According to this consideration we imposed the drift motion of field particles toward the center. The results simulated by the same conditions of γ, τ, and $\underline{m}$ as Fig.2 are shown in Fig.5. Here, the ratio of the flight

Fig.4. Random tip-splitting process in the hottest tips of DLA.

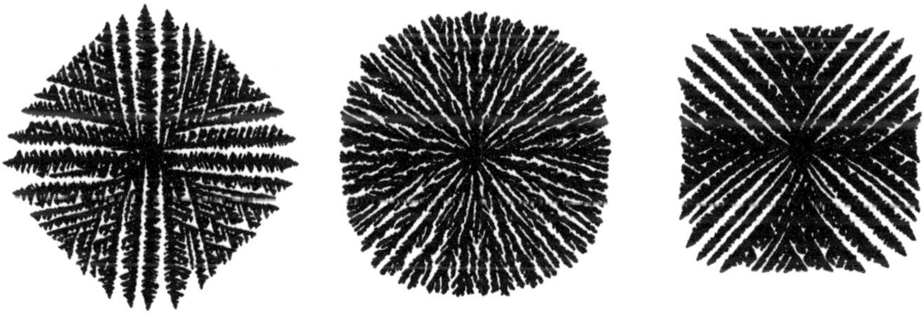

(a) (b) (c)

Fig.5. DBM-like patterns simulated in the drift of field particles at $\underline{m}$=5, τ=10, γ= (a) 0.05, (b) 0.23, (c) 0.6, and the radius of 600 lattice units.

length for the drift motion to that for the Brownian motion is a quarter. A dense-radial-like pattern (b) is obtained from the same conditions as the DLA shown in Fig.2(c). From these figures it is obvious that the pattern homogeneity is due to the random tip-splitting process of the hottest tips. Thus, we can understand the pattern formation of DBM from a standpoint of the growth velocity.

In summary our simulations reveal the realistic patterns of the dendrite, DLA, and DBM. Results show that the side-branching dendrite, DLA, and DBM are caused by the anisotropy, by the isotropy and noise, and by the growth velocity respectively. In the case of low noise the interesting growth forms appear around the crossover domain of the anisotropies. The present results remark that the surface kinetics plays an important role in the pattern formation especially in the dendritic crystal growth.

References

1. T.Vicsek, Fractal growth phenomena, (World Scientific,Singapore,1989).
2. J.S.Langer, Rev. Mod. Phys. 52 (1980) 1; Science **243** (1989) 1150.
3. S.K.Chan, H.H.Reimer, and M.Kahlweit, J. Cryst. Growth **32** (1976) 303; H.Honjo, S.Ohta, and Y.Sawada, Phys. Rev. Lett. **55** (1985) 841.
4. H.Honjo, S.Ohta, and M.Matsushita, J. Phys. Soc. Jpn. **55** (1986) 2487; S.Ohta and H.Honjo, Phys. Rev. Lett. **60** (1988) 611.
5 E.A.Brener, J. Cryst. Growth, **99** (1990) 165.
6. A.Witten and L.M.Sander, Phys. Rev. Lett. **47** (1981) 1400; P.Meakin, in On Growth and form, ed. H.E.Stanly and N.Ostrowsky, (Martinus Nijhoff,Hague,1986).
7. S.Ohta and H.Honjo, Phys. Rev. **A44** (1991) 8425.

EFFECT OF HIGH MAGNETIC FIELDS ON FRACTAL GROWTH OF METAL LEAVES

S. Okubo, I. Mogi, and Y. Nakagawa

Institute for Materials Research, Tohoku University, Sendai 980, Japan

ABSTRACT

Growth patterns of silver metal-leaves were investigated in high magnetic fields up to 8 T as an experimental approach to the diffusion-limited aggregation (DLA) with the lateral drift of particles. Silver metal-leaves were grown from a $AgNO_3$ aqueous solution by means of the electrodeless reduction of silver ions. A growth pattern was a typical DLA structure in the absence of a magnetic field, while it approached a spiral and dense branching structure with increasing magnetic fields. The spiral direction was reversed by the reversal of the applied field. The fractal dimension D_f, which was estimated by a box-counting method, was 1.62 at zero field. At 8 T the dimension crossover from 1.69 to 2 was observed with increasing length scale.

1. Introduction

In the last decade, much attention has been paid to a random pattern formation in the nature. In particular, a pattern formation of irreversible aggregation in nonequilibrium systems, such as crystal growth in a diffusion field, has been investigated theoretically and experimentally. The diffusion-limited aggregation (DLA) model was introduced by Witten and Sander[1,2] to reproduce the random fractal pattern in irreversible aggregation. Computer simulations showed that the DLA model produces irregular and self-similar patterns. A typical DLA pattern in a real physical system was observed by Matsushita *et al.*[3] in two-dimensional zinc metal-leaves grown by electrodeposition. They estimated the fractal dimension D_f of the metal leaves from the density-density correlation function; D_f=1.66, which is in good agreement with the theoretical result[4] of D_f=5/3. On the other hand, Uwaha and Saito[5] demonstrated that a DLA-like fractal pattern is reproduced from a lattice gas of finite density.

Recent computer simulation studies about DLA have dealt with the aggregation pattern in the presence of a particle drift. Meakin[6] studied the influence of unidirectional particle drifts on the DLA pattern with a Monte Carlo method and showed that the aggregation pattern with the particle drift had the crossover from a fractal structure on short length scales to a uniform structure on long length scales. Nagatani and Sagúes[7] simulated DLA patterns with rotating and radial flows and found that the DLA pattern changes into a single-arm structure in a rotating flow and a multiarm structure in the presense of the both flows.

The purpose of this paper is to investigate the effect of a lateral particle drift on DLA in a real physical system. We have observed the growth patterns of two-dimensional silver metal leaves in high magnetic fields up to 8 T, where the Lorentz force affects the diffusive motion of ions. In our previous paper[8], we reported dense radial growth of a silver metal-leaf at 8 T. In this paper, we will show the dependence of the growth pattern on both Ag^+ concentration and magnetic field.

2. Experimental

Two-dimensional silver metal-leaves were grown by sandwiching a piece of copper metal ($3 \times 3 \times 0.02$ mm^3) in a silver nitrate aqueous solution between two glass plates. The distance between the two plates was approximately 0.06 mm.

The experiments with high magnetic fields were carried out using a superconducting magnet SM-3 in the High Field Laboratory of Tohoku University. This magnet produced fields of up to 8 T in a large bore of 220 mm at room temperature. Magnetic fields were applied perpendicularly to the glass plates which were placed on a bakelite holder within the magnet. Temperature within the magnet was controlled at 25±0.2 °C by a water-circulating system. Silver metal-leaves were photographed and analyzed digitally with an image-scanner and a personal computer. A fractal dimension of the metal-leaves were estimated by means of the box-counting method.

3. Results and Discussion

Silver metal-leaves were grown from a $AgNO_3$ aqueous solution by the electrodeless reduction of Ag^+ ions. The growth pattern depends on the Ag^+ concentration, as shown in Fig. 1, in which all the metal-leaves were grown for 150 minutes in the absence of magnetic fields; a square at the center of the leaf is a piece of a copper metal 'seed'. The metal leaf (a) grown from a 0.20 M (M=mol·dm^{-3}) $AgNO_3$ solution has a dense and mossy structure. The metal-leaf (b) from a 0.050 M solution shows a typical DLA pattern, in which some short branches cease to grow owing to the growth of longer branches. Such a screening effect is a prominent feature in DLA[1,2]. In the growth pattern (c)of a 0.025 M solution, which also shows a DLA feature, there appears the dependence of pattern on the shape of the copper metal seed; the main four branches develop at the four corners of the square seed.

The chemical reaction to form a silver metal-leaf is a simple redox (reduction-oxidation) reaction of

$$Cu + 2Ag^+ \rightarrow Cu^{2+} + 2Ag \downarrow .$$

The difference of the redox potentials between Cu/Cu^{2+} and Ag/Ag^+ causes this reaction to proceed. Silver ions are brought to a copper piece by a diffusive motion. They are reduced by the copper metal and deposited around it. The electron transfer

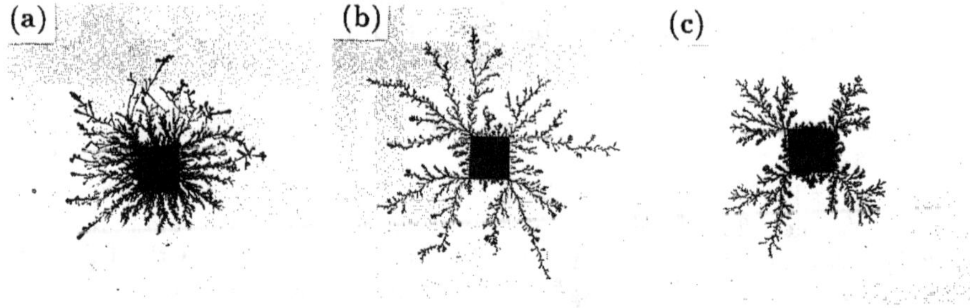

Fig. 1. Growth patterns of silver metal leaves from $AgNO_3$ solutions of (a)0.2 M, (b)0.05 M and (c)0.025 M.

process is, generally, faster than the diffusive mass-transfer process, so that DLA is realized.

Uwaha and Saito simulated aggregation patterns from lattice gases with various densities and reproduced the pattern change from a DLA fractal to a compact structure with increasing the gas density. A similar pattern change is observed in Fig. 1. The aggregation pattern depends on a diffusion length which corresponds to the distance between ajacent main branches. The diffusion length decreases with increasing Ag^+ concentration, hence, the growth pattern becomes compact.

Figure 2 shows the magnetic field dependence of the growth pattern of silver metal-leaves from a 0.075 M $AgNO_3$ solution. A typical DLA pattern is observed in the absence of a magnetic field (a). In magnetic fields the growth pattern becomes dense and spiral (b,c). When the applied field is reversed, the spiral growth is in the opposite direction (d). The metal-leaf at 8 T has a dense-branching morphology with a circular periphery (e).

The fact that the spiral direction depends on the field direction indicates that the pattern change is caused by the Lorentz force acting on the motion of silver ions. When a magnetic field is applied to an electrolytic solution, the Lorentz force induces lateral drifts on the diffusive motion of ions. This is well known as the magnetohydrodynamic (MHD) effect.[9] According to the DLA simulation with a rotating flow by Nagatani and Sagúes,[7] a tangential drift causes a spiral DLA pattern, which changes into a single-arm structure with increasing tangential-drift probability. However, a dense branching morphology has not emerged in their simulations. This is probably because the DLA simulation does not include the term of the particle concentration, so that the dependence of the diffusion length on the tangential drift has not been considered.

In electrochemistry the MHD effect has been intensively studied[9]; it has been found that the MHD convection makes a diffusion layer thinner in electrode reactions and increases a current. The decrease in the diffusion-layer thickness or the diffusion length is also observed in the metal-leaf growth in high magnetic fields, even though the reaction is electrodeless. The distance between ajacent main branches decreases with increasing fields, as can be seen in Fig. 2.

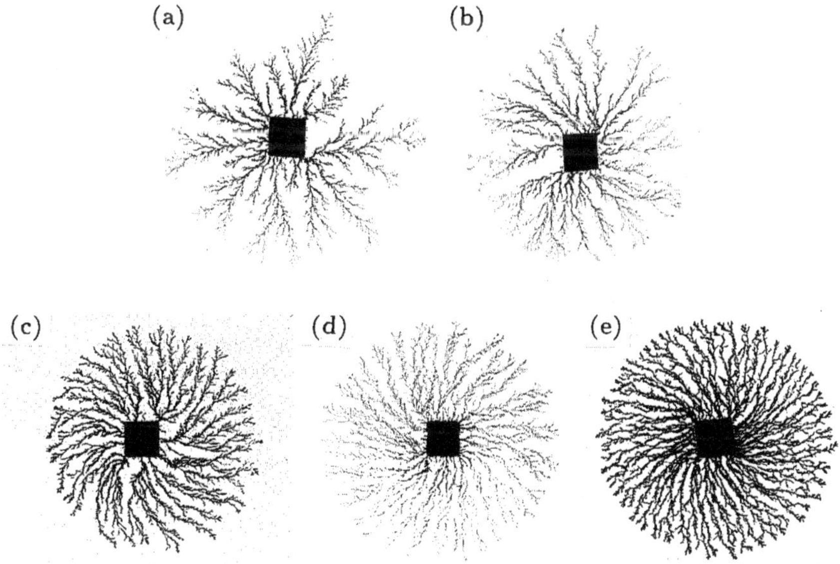

Fig. 2. Growth pattern of silver metal leaves from 0.075 M $AgNO_3$ aqueous solution in magnetic fields (a)0 T, (b)2 T $\odot$, (c)4 T $\odot$, (d)4 T $\otimes$, (e)8 T $\odot$.

In order to estimate the fractal dimension D_f by the box-counting method, the numbers of boxes including parts of the metal leaf $N(r)$ are plotted as a function of the box size r in Fig.3, where the copper seed was excluded from counting. The plots at zero field show a self-similar pattern with D_f=1.62, which is in good agreement with that of the two-dimensional DLA model[3] D_f=5/3. In the plots at 8 T a dimension crossover is seen with increasing the box size; D_f=1.69 at short scales and $D_f \sim 2$ at long scales.

The crossover in the fractal dimension from D_f=5/3 on short scales to D_f=2 on longer scales was reported by Meakin[3] in Monte Carlo simulations of DLA with unidirectional particle drifts. He showed that the fractal structure with D_f=5/3 remains on short scales in a dense pattern grown with particle drifts. A similar result was observed in the metal-leaf at 8 T. The log-log plots of $N(r)$ vs r on short scales are on a straight line with the descending slope of 1.69, as can be seen in Fig.3. This indicates that the DLA like fractal structure is present on short scales even in the dense branching pattern.

In conclusion, the drastic change in the growth pattern of silver metal-leaves in magnetic fields is ascribed to the MHD effects: The spiral structure is caused by the lateral drifts of the Lorentz force acting on the diffusive motion of silver ions. On the other hand, the MHD drifts make the diffusion length shorter and reduces the screening effect. As a result, the metal-leaf grows into a dense branching morphology. The analysis of the fractal dimension indicates that the DLA-like fractal structure remains on short scales even in the dense branching structure.

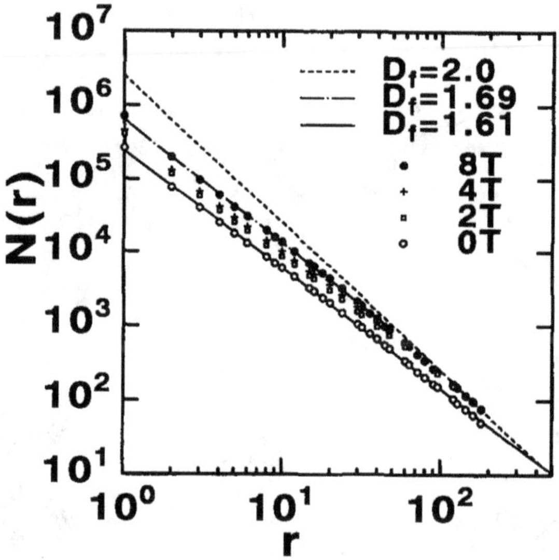

Fig.3. The log-log plots of the number of boxes supporting the metal leaf $N(r)$ vs the box size r. The unit length of the box size (r=1) is approximately 0.01 mm. The fractal dimension D_f was estimated from the slope of the linear regression line (solid lines). The dashed line represents D_f=2.

Acknowledgments

The authors wish to thank Dr. Uwaha of IMR Tohoku University for valuable discussion. They are also indebted to the members of the High Field Laboratory and the Cryogenic Center of Tohoku University for operating the superconducting magnet.

References

1. T.A.Witten and L.M.Sander, Phys. Rev. Lett. **47** (1981) 1400.
2. T.A.Witten and L.M.Sander, Phys. Rev. B **27** (1983) 5686.
3. M.Matsushita, M.Saono, Y.Hayakawa, H.Honjo and Y.Sawada, Phys. Rev. Lett. **53** (1984) 286.
4. M.Tokuyama and K.Kawasaki, Phys. Lett. **100A** (1984) 337.
5. M.Uwaha and Y.Saito, Phys. Rev. A **40** (1989) 4716.
6. P.Meakin, Phys. Rev. A **27** (1983) 1495.
7. T.Nagatani and Sagúes, J. Phys. Soc. Jpn. **59** (1990) 3447.
8. I.Mogi, S.Okubo and Y.Nakagawa, J. Phys. Soc. Jpn. **60** (1991) 3200.
9. T.Z.Fahidy, J. Appl. Electrochem. **13** (1983) 553.

A simulation of pattern formation by erosion

Hideki Takayasu and Hajime Inaoka
Department of Earth Sciences , Kobe University
Kobe 657 , Japan

ABSTRACT

The time evolution of river pattern and earth's relief is simulated on lattice by modeling the process of water erosion. Starting from a randomly given surface the river pattern and earth's relief develops simultaneously. The river pattern becomes stationary after all lakes are vanished. In the stationary state the river pattern shows some fractal properties such as power law distribution of drainage basin area and Horton's laws.

Landscapes and river patterns are familiar to everybody but they do not have attracted much physicists' interest so far. Mandelbrot firstly showed that the coastlines can be characterized by fractal dimensions[1]. He also graphically demonstrated that the earth's reliefs are fractals by proposing fractional Brownian surfaces as landscapes. Although the resulting surfaces look natural for the first sight, there is an obvious flaw, that is, no river exists on the surface, and the evolution of landscapes by erosion cannot be introduced in the model.

River patterns are also known to be typical fractals[1,2]. Scheidegger proposed a lattice model of rivers which is defined on a slope where water on a site flows randomly either to the left down site or right down site[3]. The model is known to show critical behaviors automatically like self-organized criticality model[4], for example, the system is proved to converge to a steady-state where the distribution of drainage basin area rigorously follows a power law[5]. It is anticipated that a large portion of fractals in nature may be created by such kind of self-organization.

For the creation of real river patterns and landscapes we believe that the effect of water erosion should play the central role. By this motivation we model the erosion process on lattice and simulate the formation of river patterns.

We consider the situation that rain is falling uniformly and constantly on a land. We ignore the effects of underground water and evaporation, so the water fallen on the surface can either flow out from the boundary or accumulates making lakes. The elevation of the surface never increase because we also neglect sedimentation of sand.

Our model is defined on a two dimensional triangular lattice. Every site has three variables, $h(x,y)$, the height of the land, $s(x,y)$, the water flow intensity, and $w(x,y)$, the thickness of water accumulation. The height of water surface, $h_w(x,y)$, is defined as $h_w(x,y)\equiv h(x,y)+w(x,y)$. Note that $s(x,y)$ does not contribute to the thickness, so $h_w(x,y)$ is equal to $h(x,y)$ when $w(x,y)$ is zero. The site where water accumulation is not zero is called "lake". For a given initial landform, $h(x,y)$, $w(x,y)=0$, and $s(x,y)=0$, the time evolution is performed by repeating the following 3 procedures.

1: Determination of the water flow direction

For every site (x,y) we find the site (x',y') which has the lowest height of

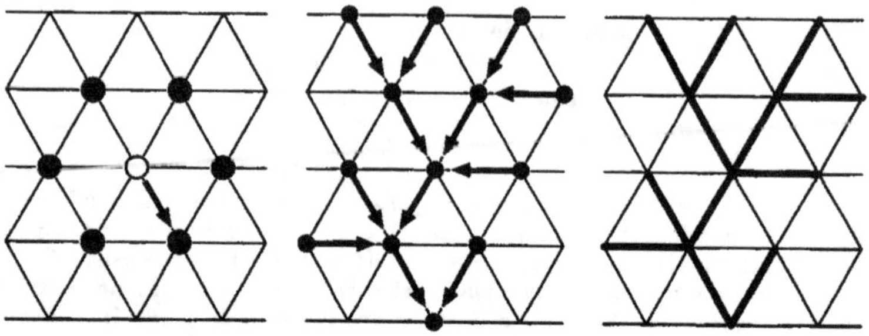

Fig.1 (a) (b) (c)

water surface in the six neighbors. When $h_w(x,y)>h_w(x',y')$ the water at (x,y) flows out to the site (x',y') (Fig.1(a)). However, if $h_w(x,y)\leq h_w(x',y')$, it is unphysical for the water at (x,y) to flow out. In that case the water accumulates and we define the water thickness by $w(x,y)\equiv h_w(x',y')-h(x,y)+\varepsilon$ to raise the water surface, where ε is a very small positive number. On the other hand, it is natural to assume that the water accumulation is zero if $h(x,y)$ is higher than $h_w(x',y')$. So we make $w(x,y)$ zero at such a site. We can draw global water flow pattern which we call a river pattern when the water flow directions for all sites are determined (Fig.1(b)(c)).

2: Water flow and rain fall

The value of $s(x,y)$ is renewed by the sum of a constant $s_0(=1)$ and the water flow intensities which flow into (x,y) in the previous time step. This s_0 corresponds to the uniform rain fall. Water is supplied only by rain and is conserved in the flow process.

3: Water erosion

By the effect of water erosion $h(x,y)$ is decreased by $\delta h(x,y)=F(J(x,y))\cdot\{h_w(x,y)-h_w(x',y')\}$, where $J(x,y)$ denotes the water power which is defined as $J(x,y)\equiv s(x,y)\cdot\{h_w(x,y)-h_w(x',y')\}$. The function F is a positive monotonically increasing function characterizing the strength of erosion $(0\leq F<1)$. In the following simulation $F(J)=C_1\cdot J/(C_2+J)$, where C_1 and C_2 are positive constants $(C_1<1)$.

An example of time evolution starting from a nearly flat surface with small random fluctuation is shown in Fig.2. Here, we are observing the system of size 100×100 with periodic boundary for the left and right edges and with fixed boundaries for the top and bottom edges. The bottom edge is much lower than the top edge, so water flows out only from the bottom edge. Accordingly river patterns and contours grow from the bottom to the top. We find the growth of river patterns from the early disconnected stage to the fully developed space-filling stage (Fig.3).

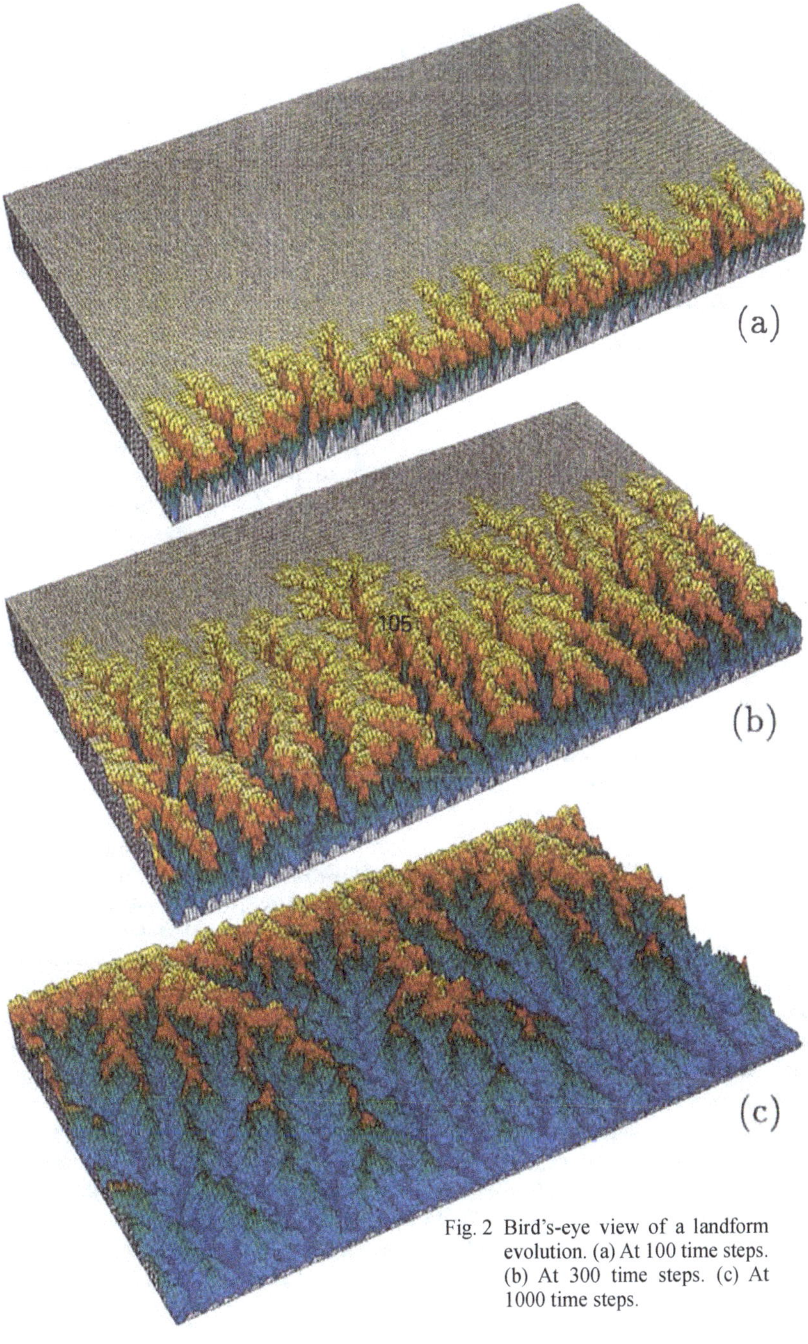

Fig. 2 Bird's-eye view of a landform evolution. (a) At 100 time steps. (b) At 300 time steps. (c) At 1000 time steps.

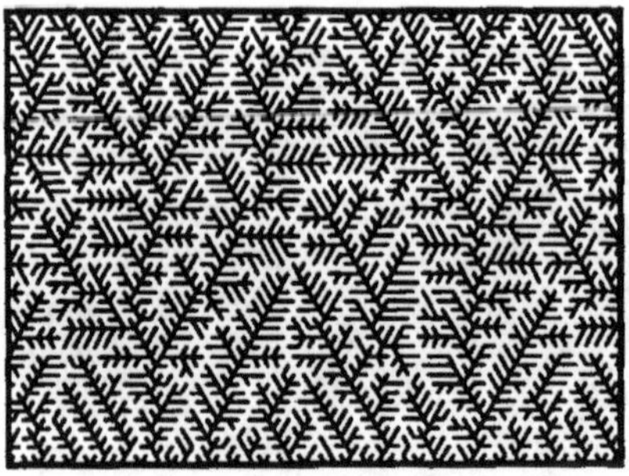

Fig.3 An example of steady-state river pattern.

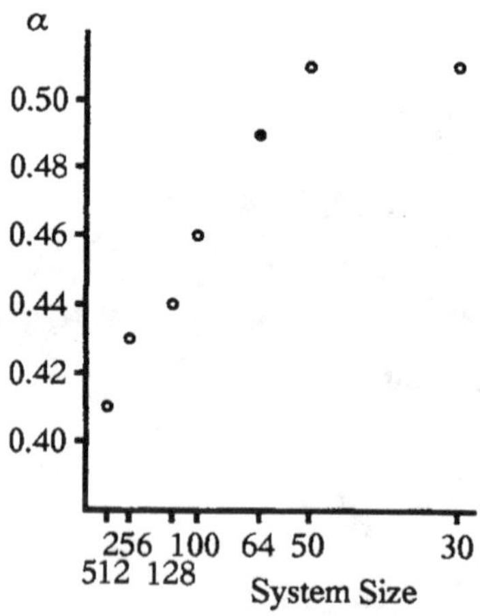

Fig.4 The inverse of the system size vs. α. The dot (•) shows the real data.

In the beginning almost all sites are lakes due to the initial condition. The number of lake-sites decreases nearly constantly and no lake remains when the whole surface is covered by valleys. At this time all rivers are percolated to the boundary and the whole rain water is drained out without any accumulation. Water flow directions are unstable in and around a lake especially when water begins to flow out from the lake's edge. The number of unstable flows which change flow directions in one time step decreases also monotonically and it becomes negligibly small about the time that all lakes are vanished, namely, the river patterns on the surface become frozen[6]. We focus our attention on this stationary state and analyze its statistical properties.

In Scheidegger's river model, it is theorically shown that the drainage basin area converges to a power law distribution which is characteristic of fractal geometry. In the present model the distribution of drainage basin area follows

$$P(\geq n) \propto n^{-\alpha} \tag{1}$$

where $P(\geq n)$ denotes the probability that the drainage basin area of a randomly chosen site is larger than n. The exponent α seems to depend on the system size of the model. Fig.4 shows the finite size effect of the exponent. One of the points shows the exponent from real data degitized from the map of the mountainous region of Ryohaku in Japan. All the points are roughly on a smooth line, which indicates that our model does not contradict to the real system. We estimate the exponent in an infinite system as α=0.40. This exponent is significantly larger than the value for Scheidegger's model[5],1/3.

Real river patterns are known to satisfy Horton's laws[7], so we check the validity of the laws for the steady state river patterns in our model[8]. In Fig.5 (a) and (b) we plot $N(\omega)$, the number of branches of order ω and $L(\omega)$, the averaged length of branches of order ω, in semi-log scale. Here the system size is 512×512 with time steps 5000. The points are clearly on straight lines, respectively. The estimated values are R_B=5.3 for the bifurcation ratio, and R_L=2.7 for the stream length ratio, which are in the range of values for natural drainage basins ($2<R_B<6$ and $1.5<R_L<3.5$)[9].

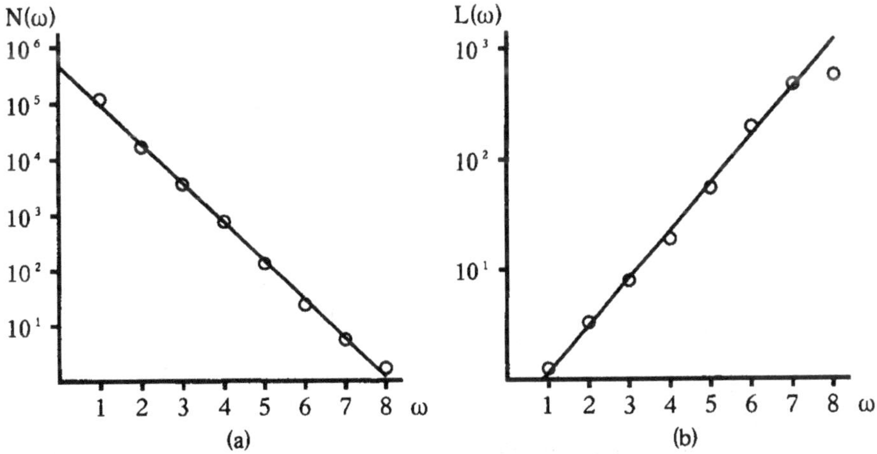

Fig.5 (a) Log N(ω) vs. ω. (b) Log L(ω) vs. ω.

Although the river patterns are stationary in the steady-state, the contours keep changing because the surface is always eroded by water flows. In the steady-state the snap-shot shape of contours can be characterized by the fractal dimension. By using the box counting method we confirm power law scalings over nearly two decays of scale length. The values of contours' fractal dimension are estimated as $D=1.60\pm0.15$[6], which seems to depend slightly on both height and time steps.

We would like to acknowledge Rikei Corporation for making an opportunity to use a Massive Parallel SIMD computer, MasPar MP-1204.

References

1. B.B.Mandelbrot, *The Fractal Geometry of Nature* (W.H.Freeman and Company, San Francisco, 1982).
2. H.Takayasu, *Fractals in the physical sciences* (Manchester Univ. Press, 1990).
3. A.E.Scheidegger, *Bull.I.A.S.H.* **12**,(1), 15 (1967).
4. P.Bak, C.Tang, and K.Wiesenfeld, *Phys.Rev.Lett.* **59**, 381 (1987); *Phys.Rev.A* **38**, 364 (1988).
5. G.Huber, *Physica* **A170**, 463 (1991); H.Takayasu, M.Takayasu, A.Provata, and G.Huber, *J.Stat.Phys.* **65**, 725 (1991).
6. H.Takayasu, and H.Inaoka, Preparing for publication.
7. R.E.Horton, *Bull.Geol.Soc.Am.* **56**, 275 (1945).
8. H.Inaoka, and H.Takayasu, Preparing for publication.
9. P.La Barbera, and R.Rosso, *Water Resour.Res.* **25**, 735 (1989).

PATTERN FORMATION IN VISCO-ELASTIC MATERIALS BY FLUID INJECTION

TAKASHI YOSHINO and TAKAYUKI HIRATA
Institute of Applied Physics, University of Tsukuba,
Tsukuba, Ibaraki 305, Japan

ABSTRACT

Systematic experiments of fluid (air) injection into visco-elastic materials were carried out. A rectangular Hele-Shaw cell (250 mm x 250 mm) spaced by 2mm was used. Agar gels of 0.1, 0.2, and 0.3 weight percent solutions were used as visco-elastic model materials. Shapes of intruding fluid show different patterns that depend on both the agar concentrations and the injection pressure of fluid. These growing patterns can be classified into three types: type I) a single plane cracking, type II) a visco-elastic fingering, and type III) a viscous fingering. These patterns will be changed from type III) to type I) via type II) with increasing agar concentration. Type II) pattern was a new pattern that is distinguished from viscous fingering by tip growth. A visco-elastic fingering is a new class pattern formation that includes both characters of viscous flow and crack propagation.

1. Introduction

A fluid intrusion into materials shows various patterns. Hydrofracturing is an example of a fluid intrusion into brittle materials (see, e.g., Spence and Turcotte; 1985), and a viscous fingering is an example of a fluid intrusion into viscous fluid (see, e.g., Bensimon *et al.*; 1986). Although these phenomena are different in governing mechanism (a crack propagation in hydrofracturing and a flow obeying the Laplace equation in viscous fingering), they can be treated as a similar problem from the viewpoint of interfacial growth. In these phenomena, an instability growth at the interface is a dominant mechanism to generate complex patterns.

The interfacial growth by fluid injection into visco-elastic materials shows complex patterns because the material has both properties of viscous fluids and elastic materials. Pattern formation in visco-elastic materials is an important phenomenon not only in applied physics (e.g., melting down of a nuclear reactor and dynamics of magma transportation) but also in pattern formation (it will give us new class patterns). In spite of its importance, there are few studies relating to these phenomena. In this paper, we carried out experimental studies of a fluid injection into visco-elastic materials by using Hele-shaw cell.

2. What will be happened in visco-elastic materials?

Let us consider the stress transportation in elastic solids, viscous fluids, and visco-elastic materials. Conservation of momentum leads to the equations: the Navier's equation and the Navier-Stokes equation, for an isotropic elastic material and for a Newtonian fluid, respectively. The former is,

$$\frac{D^2 S}{Dt^2} = \frac{\lambda + \mu}{\rho} \, grad\, div\, S + \frac{\mu}{\rho} \Delta S + K \,, \tag{1}$$

and the latter is,

$$\frac{Du}{Dt} = -\frac{1}{\rho} grad\, P + \frac{\lambda' + \mu'}{\rho} grad\, div\, u + \frac{\mu'}{\rho} \Delta u + K \,. \tag{2}$$

In viscous fingering, the Navier-Stokes equation is approximated to the Laplace equation for pressure (Saffman and Taylor; 1958),

$$\Delta P = 0 \,. \tag{3}$$

Roughly speaking, the pattern growth obeying the Eqs. 1 and 3 have different growth rates by the order. A crack in elastic materials grows at the order of t and a viscous finger grows at the order of $t^{1/2}$. Here, t denotes time.

The governing equations for visco-elastic materials can be derived in the same way. In visco-elastic materials, however, they will be more difficult due to including combination of strain, strain rate and stress history. We will show some simple examples of visco-elastic materials: e.g., a Maxwell fluid and a Voigt solid. The governing equations for a Maxwell fluid and a Voigt body are written as follows:

$$\begin{aligned} \frac{D^2 S}{Dt^2} = \frac{\mu'}{\rho} & \left[\int_0^t exp\left(-\frac{\mu'}{\mu} (\tau - t) \right) \frac{\partial}{\partial \tau} \left(\Delta S + grad\, div\, S \right) d\tau \right] \\ & + exp\left(-\frac{\mu'}{2\mu} t \right) \cdot sinh\left(\frac{\mu'}{2\mu} t \right) \cdot \frac{\mu'}{\rho} \left[\frac{\lambda}{\mu} grad\, div\, S + \frac{\lambda'}{\mu'} grad\, div\, u \right] + K \,, \end{aligned} \tag{4}$$

for a Maxwell fluid and,

$$\frac{D^2 S}{Dt^2} = \frac{\lambda + \mu}{\rho} grad\, div\, S + \frac{\lambda' + \mu'}{\rho} grad\, div\, u + \frac{\mu}{\rho} \Delta S + \frac{\mu'}{\rho} \Delta u + K \,, \tag{5}$$

for a Voigt body. As already seen in the above equations, it is difficult to construct an interfacial growing master equation even for a Maxwell fluid or a Voigt solid. Furthermore, in our study, the pattern formation relates to "fracturing". Fracturing is a highly nonlinear phenomenon and has discontinuous growth. The interfacial growth master equation will lose its meaning in such a system. Therefore, the experimental studies are needed.

3. Experiments

Figure 1(a) shows a schematic diagram of the experimental system. A horizontal rectangular Hele-Shaw cell that consisted of two 250 mm x 250 mm acrylic boards spaced by 2 mm was used. The Hele-Shaw cell was filled with a visco-elastic material. Air was injected into the visco-elastic materials from the hole that located at the center of the upper acrylic board. The hole diameter was 1.8 mm. Growing patterns were recorded using a CCD video camera/recorder.

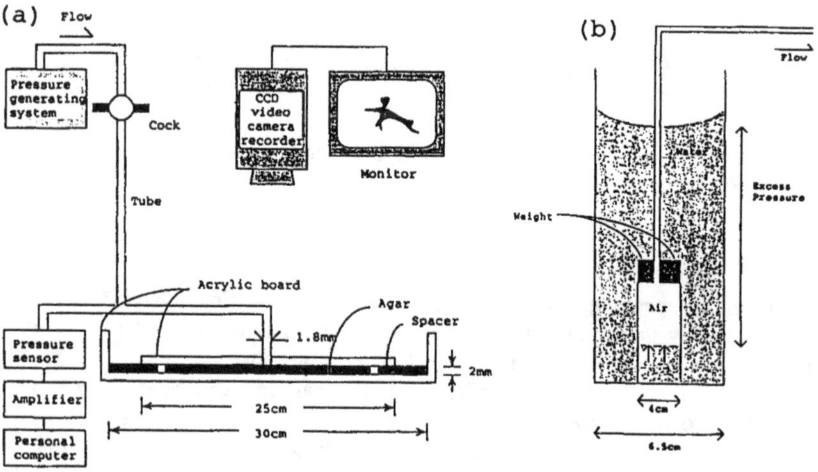

Figure 1. (a) Schematic diagram of experimental system. An rectangular Hele-Shaw cell (250 mm x 250 mm x 2 mm) is placed under the CCD video camera/recorder that records a growing pattern. The change of pressure is measured by a diffusion type semiconductor pressure sensor. The data are stored in a computer via 12 bits AD converter. (b) Schematic sketch of the pressure generating system. The hydrostatic pressure applies the air in the cup. The pressure is controlled by changing the depth of the cup.

Agar gels of 0.1, 0.2 and 0.3 weight percent solutions were used as model materials that have a visco-elastic property at a room temperature. The constant pressure generated by the water height, hydrostatic pressure, as shown in Fig. 1(b). It covers the pressure range from 0 to 4.5 kPa. Strictly speaking, although the pressure is continuously decreasing with the fluid injection, the decrease of the pressure can be negligible as shown in Fig. 2. Figure 2 is a time vs. pressure plot. The pressure change during the experiments is smaller than 1%. Therefore, we can say that the patterns grow under a constant injection pressure.

4. Results

Shapes of the intruding fluid show various patterns. We classified the observed patterns into three typical types according to their shapes: type I) a single plane cracking, type II) a visco-elastic fingering that we newly termed, and type III) a viscous fingering. Figure 3 is a schematic illustration of these typical patterns. These patterns

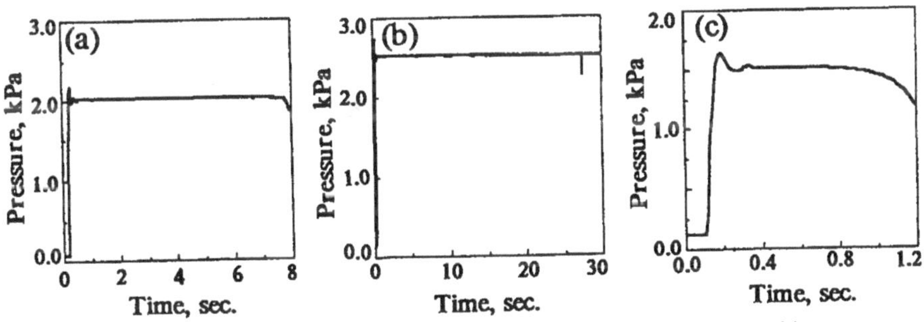

Figure2. The pressure of injecting fluid is plotted against time for typical three patterns: (a) single plane crack, (b) visco-elastic fingering, (c) viscous fingering. When a tip of finger reaches to the edge of the region covered with the upper acrylic board, the abrupt pressure decrease occurred.

will change from type I) to type III) via type II) with increasing agar concentration. The visco-elastic fingering is a new pattern that is characterized by the shape and growth rate of tips. We can distinguish the visco-elastic fingering from the viscous fingering by the shape of growing finger tips. A tip of visco-elastic fingering is acute, in other words, very large curvature, in compare with one of viscous fingering. Abrupt growth occurs at the tips. The growth of visco-elastic fingering has several stages. Tensile crack opening occurs at the first stage. Swelling of interface occurs at the second stage in order to prepare for next cracking. Tip splitting occurs at the last stage. The pattern of visco-elastic fingering is generated by the iteration of this sequence.

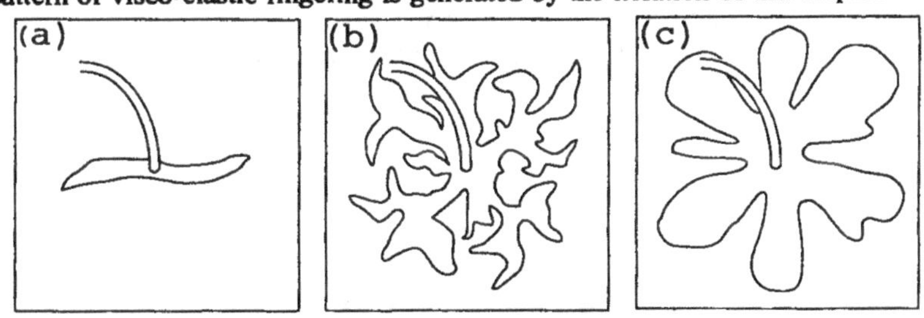

Figure 3. Schematic illustration of three typical patterns: (a) a single plane cracking, (b) a visco-elastic fingering, and (c) a viscous fingering.

Figure 4 is a morphological phase diagram. If the injection pressure is lower than the material strength (K_{IC}), there is no pattern growth. An increasing injection pressure transform the patterns from the single plane cracking to the visco-elastic fingering except the case of 0.1 wt.% agar concentration. Only viscous fingering is observed when 0.1 wt.% agar solution is used. Visco-elastic fingering is not observed under a low pressure.

5. Discussion

Three typical patterns, i.e., single plane cracking, visco-elastic fingering, and

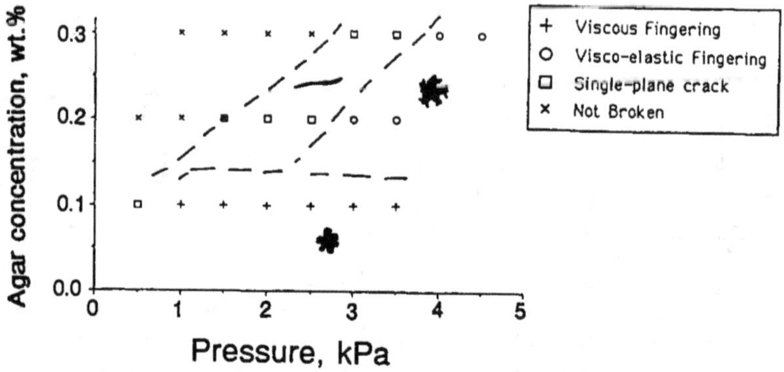

Figure 4. Morphological phase diagram of the growth patterns. The parameters are the pressure and the agar concentration.

viscous fingering, were observed by changing degree of the agar concentration. As the extreme cases of visco-elastic materials, they are approximated to elastic solids and viscous fluids. Both the single plane cracking and the viscous fingering were observed in such cases. Since an agar gel with high concentration behaves as a solid, fracturing is a main mechanism to control the pattern growth. On the other hand, as a low concentration agar gel is a fluid, viscous flow determines the pattern. At the intermediate concentration, these two mechanisms contribute to the pattern formation. A visco-elastic fingering was observed as a result of two competitive mechanisms.

A pattern formation of visco-elastic fingering has two different stages in growth process. The first step is tensile crack opening and the second one is swelling. Tensile crack opening has the growth rate of the order of t and viscous fingering has the order of $t^{1/2}$. These stages will appear as a result of competition of two mechanisms with different growth rates. The visco-elastic fingering may be a new universal class in pattern formation that is concerned with pattern growths of the order of t and $t^{1/2}$.

Visco-elastic fingering has not only interests in pure physics but also applications to applied physics. A magma migration in the Earth's mantle and a magma eruption are examples. To study the visco-elastic fingering will bring a new field of pattern formation to us.

6. Acknowledgements

We would like to thank Dr. T. Ogawa and Mr. Y. Nakashima for helpful discussions. This study was partially supported by a grant-in-aid from the Ministry of Education, Science and Culture of Japan (No. 03232202).

7. References

1. D. A. Spence and D. L. Turcotte, *Jour. Geophys. Res.* **90** (1985) 575
2. D. Bensimon, L. P. Kadanoff, S. Liang, B. I. Shraiman and C. Tang, *Rev. Mod. Phys.* **58** (1986) 977
3. P. G. Saffman and G. I. Taylor, *Proc. Roy. Soc.* **A245** (1958) 312

A MODEL OF PATTERN FORMATION IN AGGREGATION PROCESSES

Satoshi Tsurumi
Gunma College of Technology, Maebashi, 371

Katsuya Honda
Department of Applied Physics, Faculty of Engineering, Nagoya University, Nagoya, 464
JAPAN

Abstract

To clarify the pattern formation mechanism in precipitation, we investigate a computer-simulated lattice model, based on the diffusion-limited aggregation (DLA) model. Taking account of evaporation processes, it is shown that the present model forms periodic band patterns("Liesegang band" like) and irregular patterns. Our results suggest that the movement of reaction front plays an important role in the band formation and the decision of the band location.

1 Introduction

Precipitation pattern formation is an important problem of reaction-diffusion systems. When a concentrated salt solution such as lead nitrate diffuses into a gel containing, for example, potassium iodide, it is known that precipitants produce circular rings in a petri dish, or periodic bands parallel to the surface of the diffusion front in a test tube (called "Liesegang phenomenon"). Many experimental studies have been made on this phenomenon and different varieties of structure have been reported, for instance, a wave-like pattern, a network pattern, and a helicoidal pattern[1]. The recent studies lead to the conclusion that precipitation patterns are a postnucleation phenomenon[2]. The experimental results are summarized as follows: (1)In the presence of concentration gradients the periodic patterns are appeared. (2)In the absence of gradients the irregular patterns are appeared. (3)Some empirical laws are known, for instance, the spacing law, the temporal law, and the band width law [2][3][4].

On the other hand, these have been investigated by a large number of theoretical approaches. The recent theoretical analyses with a set of diffusion-reaction equations suggest that the Ostwald ripening, which often occurs by a condensation-evaporation mechanism, is important for macroscopic pattern formation[5]. In spite of previous extensive study of the Liesegang phenomenon, the formation mechanism of the precipitation patterns is not understood completely yet.

In this study, we present a computer-simulated lattice model that can form various types of aggregation pattern including the Liesegang band-like one. The present model is based on the diffusion-limited aggregation (DLA) model taking account of evaporation processes. As the presence of concentration gradients is dominant in the band formation, we may regard both band and ring formations as a one-dimensional reaction-diffusion phenomenon. Therefore we assume that the diffusion of reactants is one dimensional. But, the aggregation and evaporation are assumed to be occurred in two-dimensional space.

2 Computer model and results

We start from the reactions

$$\begin{cases} A+B \longrightarrow C \\ C \longleftrightarrow D. \end{cases}$$

A and B are reacting electrolytes and C is a colloid. D is a precipitate that is formed by a colloidal growth, and weakly soluble. C and D are chemically identical, but D cannot diffuse in the gel.

For the sake of simple simulation, we deal with these reactions as two essential processes. That are an activation process of seed particles (corresponding to the production of C) and an aggregation-evaporation process of the activated particles (corresponding to the growth and dissolution of D). Simulations are carried out on a $L \times L$ two dimensional lattice space.

First of all let N seed particles be located randomly on the lattice. When a reaction front moves along the axis of the tube (denotes the coordinate x), the seed particles included in this area are activated completely. Then the activated particles start one-dimensional random walks along the x axis. If any of these particles touch each other, they stick together and form aggregates. Note that the free (activated) particles may aggregate two-dimensionally each other. The aggregate that is consisted of more than two particles cannot move. The aggregated particles can evaporate in all directions with a given rate (probability P). If the evaporated particle is free, i.e. have no nearest-neighbor particles, it starts a one-dimensional random walk. Then the above procedures are repeated.

We have some assumptions on this simulation. First, let us assume that boundaries are periodic but both ends of the x axis are a wall. Secondly, in the evaporation processes, if there are more than three nearest-neighbors around a particle, it cannot evaporate. Although two nearest-neighbors, a particle whose two sides are occupied cannot evaporate. Finally, a "one step" in the simulation is when the reaction front moves by a lattice unit. After the front arriving at the end of axis, the step is continued counting.

Figure.1 shows the result of a simulation under the conditions of $L = 100$, $N = 2000$ and $P = 0.1$. Top figure indicates the evolution of band pattern for four stages. Bottom indicates the density distribution ($D(x)$) for aggregated particles.

As the result show, it is found that four clear bands are formed gradually, but they grow nonequivalently. Intervals between bands are also nonequivalent. We should notice that the bands are not always formed at the same location on the other trials.

Figure.2 shows the result of a simulation on the condition of no reaction front. Parameters and procedures are the same as Figure.1, except that we assume all N particles are activated initially. As shown in the figure, the early bands are formed homogeneously in the space. Then the final three bands grow equivalently.

Figure.3 shows the result of a simulation corresponding to the case of gradient free. The simulation is started from the same initial condition with Figure.2. However, the initial activated particles can move any direction randomly (two-dimensionally). This is a particular case of the cluster-cluster aggregation (CCA) model. Parameters used are also the same as in Figure.2. As the result shows, the random aggregation pattern is formed. It is considerable that changing the evaporation mechanism, the present model can form the network-pattern which is observed in a supercooled solution.

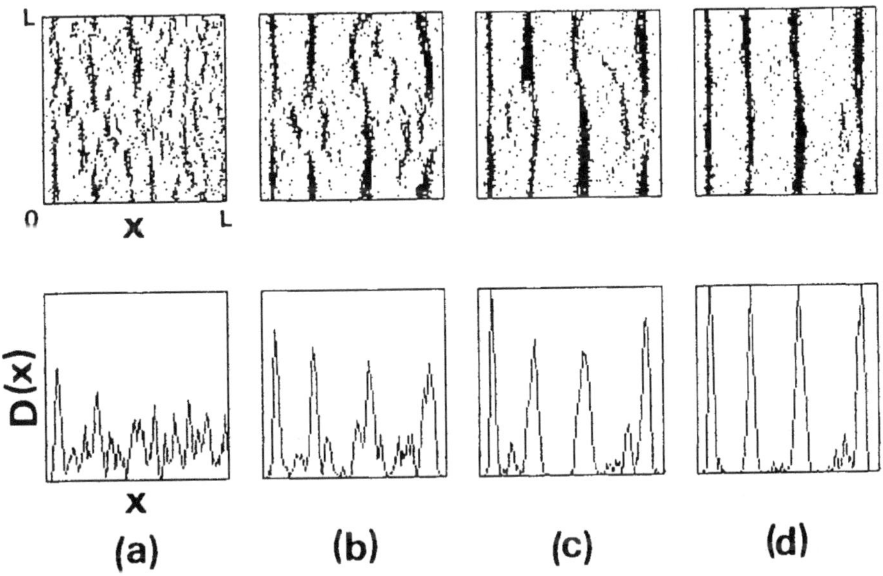

Figure 1: Four stages of a band formation process. Top figure shows the evolution of band pattern. Bottom shows the density distribution of aggregated particles. Parameters are $L = 100$, $N = 2000$, and $P = 0.1$. The aggregation-evaporation processes are repeated 30 times every step. (a), (b), (c), and (d) are corresponding to 100, 300, 500, 700 steps, respectively.

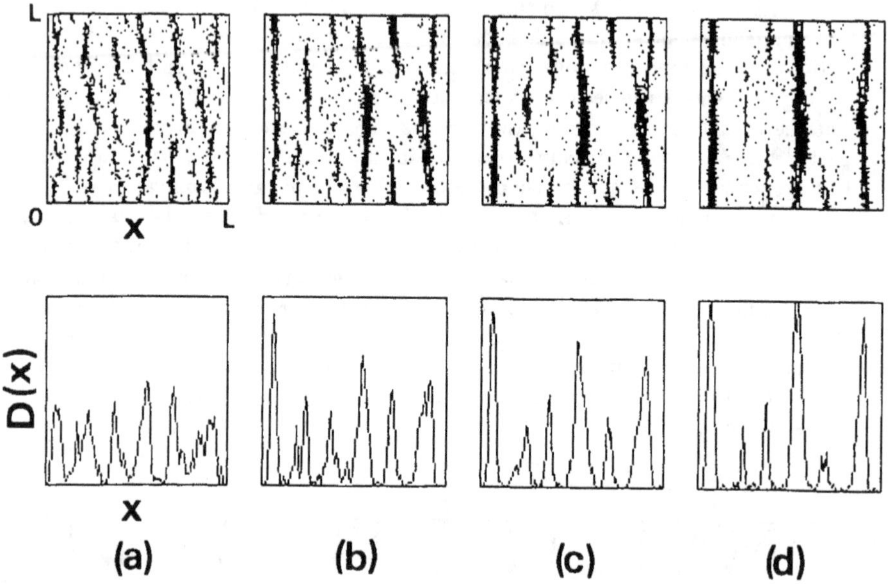

Figure 2: Four stages of a simulation on the condition of no reaction front. Parameters and conditions are the same as Figure.1, except that the all initial particles are active.

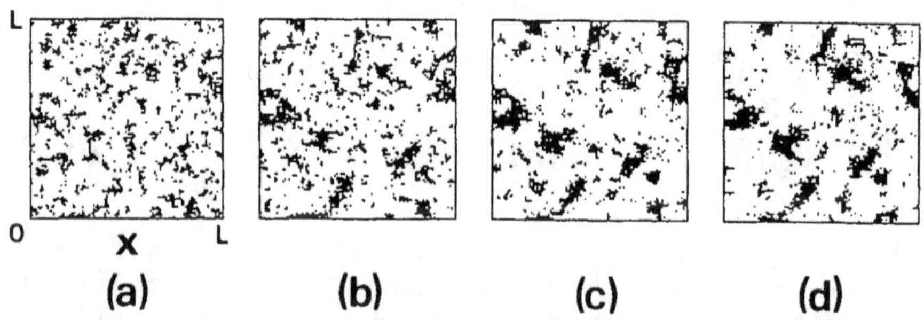

Figure 3: Four stages of a simulation on the case of gradient free. Parameters used are the same as Figure.2. This simulation is started from the same initial state as Figure.2, however, the particles can move any direction randomly.

3 Discussion

We have examined a simple pattern formation model by use of a computer simulation technique. In the presence of reaction front, the band patterns are formed with nonequivalent intervals. In the absence of reaction front, the intervals between bands are equivalent. These results are understood as follows. The reaction front forms early bands (small clusters) inhomogeneously in space. The earlier clusters are formed, the faster they grow. As a competition between early bands decides the location of final bands, the intervals between them become nonequivalent. Since the movement of reaction front is controlled by the number of aggregation-evaporation loops, it is obvious that the parameter influences the location of bands. Thus we found that the reaction front plays an important role for the band formation. Furthermore, it is noted that the evaporation rate P is an important factor for the band formation though it is small(for example, 0.1).

On the other hand, as shown in Figure.3, in the case of gradient free, we have observed the random pattern that is similar with the network-pattern in the supercooled solution. It means that the cluster-cluster aggregation (CCA) mechanism is essential for this kind of patterns. Taking account of various types of evaporation mechanism, it is considerable that the present model can describe the Liesegang phenomenon synthetically.

Finally, we comment that our model assumed a one-dimensional reaction-diffusion process, nevertheless, it is sufficient to describe the two-dimensional band formations. As far as the concentration gradients are dominant, we expect our model can be applied to other reaction-diffusion problems in the system.

References

[1] D.Feinn,P.Ortoleva,W.Scalf,S.Schmidt,and M.Wolff, *J.Chem.Phys.*, **69** (1978)27

[2] S.Kai,S.C.Müller,and J.Ross, *J.Chem.Phys.*, **76**(1982)1392

[3] S.C.Müller,S.Kai,and J.Ross, *J.Phys.Chem.*, **86**(1982)4078

[4] S.C.Müller,S.Kai,and J.Ross, *J.Phys.Chem.*, **87**(1983)806

[5] S.Kai,S.Higaki,H,Yamazaki,and T.Yamada, *T.IEE.Japan*, **107**C(1987)1011

B: Chemical Reaction

SIMULATIONS OF THREE DIMENSIONAL DENDRITES

Ryo Kobayashi

Department of Applied Mathematics and Informatics,
Ryukoku University, Seta, Ohtsu 520-21, Japan

ABSTRACT

Three dimensional dendritic growth was simulated by a phase field model for melt growth. The results represent many features which are found in real physical systems. In particular, the axisymmetry is first broken near the tip of growing parabolloidal revolution into a four fold fin-like structure, and then sidebranch instability takes place on the fins as was observed in real systems. The results would hopefully encourage theoretical works to understand three dimensional dendritic growth.

1. Introduction

Dendritic crystal growth has been studied in supercooled melt of succinonitrile[1], Helium4[2], Krypton[3] and in supersaturated solution of Anmonium bromide[4-6]. Experimentally the shape of the growth tip was found to be of parabolloidal revolution and the radius of curvature was nearly propotional to square of the supercooling or supersaturation.

Theoretically, geometric model[7], boundary layer model[8] and a fully nonlinear model[9-15] were investigated extensively. It has been found that the dendrite growth is stabilized by a crystal anisotropy and the curvature of the growth tip were selected by the presence of surface tension. Numerical simulation was carried out for the fully nonlinear model, and the anisotropy dependence of growth velocity was investigated[16]. The simulation, however, based on the fully nonlinear model is limited at present computational power to the two dimensional case.

It is, therefore, important to provide some means to simulate three dimensional dendritic growth, as there are some interesting features specific in real three dimensional case which cannot be represented by any simplified two dimensional models. In this paper I report three dimensional simulation based on a phase field model[17-22].

2. Mathematical Modeling

In this model, the evolution of the crystal shape is represented by the function $p(\boldsymbol{r},t)$. The $p(\boldsymbol{r},t)$ is a ordering parameter at the position $\boldsymbol{r}$ and the time t, where $p=0$ means liquid phase and $p=1$ means solid phase. The solid/lquid interface is represented by the steep internal layer between the regions $p=0$ and $p=1$. Temperature field is represented by $T(\boldsymbol{r},t)$ which is non-dimensionalized such that $T=0$ corresponds to the characteristic cooling temperature and $T=1$ to the equilibrium temperature. Using the pair $(p(\boldsymbol{r},t),\ T(\boldsymbol{r},t))$, we can describe the crystal growth from the melt. If there is no anisotropy, the dimensionless model equations are the following. The equation for p is

$$\tau\frac{\partial p}{\partial t}=\varepsilon^2\nabla^2 p+p(1-p)(p-1/2+m(T)) \tag{1}$$

where τ and ε are small positive parameters and $m(T)$ is some decreasing continuous function satisfying $m(1)=0$ and $|m(T)|<1/2$. The thickness of the internal layer of p is determined by ε. The equation for T is

$$\frac{\partial T}{\partial t}=\nabla^2 T+K\frac{\partial p}{\partial t} \tag{2}$$

where K is a dimensionless latent heat or equally the inverse number of the dimensionless supercooling $\Delta=c\Delta T/L$ (c: heat capacity per unit mass, ΔT: initial supercooling and L: latent heat per unit mass) in the case of supercooling solidification. The space variables are non-dimensionalized by the characteristic size of the dendrites, and the time variable is non-dimensionalized such that the diffusion constant in Eq.2 equals to 1. Eq.1 is an isotropic interface equation and Eq.2 is a heat conduction equation including the term which describes the release of the latent heat along the advancing interface. Anisotropy is introduced into Eq.1 by assuming that τ and $m(T)$ depend on the direction. So we adopt the following equation as an interface equation.

$$\tau(\nu)\frac{\partial p}{\partial t}=\varepsilon^2\nabla^2 p+p(1-p)(p-1/2+m(T,\nu)) \tag{3}$$

where $\nu=-\nabla p$ which means an outer normal vector at the interface. In the following simulations, I took $\tau(\nu)$ as a constant $\bar{\tau}$ and

$$m(T,\nu)=\frac{\alpha}{\pi}Tan^{-1}(\gamma(\nu)(1-T)) \tag{4}$$

$$\gamma(\nu)=\bar{\gamma}\sigma(\nu) \text{ and } \sigma(\nu)=1-\delta(1-|\nu|_4^4/|\nu|_2^4) \tag{5}$$

where α and γ are positive constants satisfying $\alpha<1$, $|\cdot|_h$ is h-norm of the vector, and δ is a constant. Arctangent form in Eq.4 guarantees $|m(T,\nu)|<1/2$. And $\gamma(\nu)$ represents the sensitivity of the interfacial velocity to the direction. The value of $\gamma(\nu)$ equals to $\bar{\gamma}$ for the direction (1,0,0), which is the maximum, $\bar{\gamma}(1-\delta/2)$ for (1,1,0), and $\bar{\gamma}(1-2\delta/3)$ for (1,1,1) which is the minimum for positive δ. Therefore we can regard the parameter δ as a strength of anisotropy. Note that the form of Eq.5 is not intended to reflect exactly the structure of molecule arrays in crystals or kinetic process. It only describes the rough tendency that the crystal grows more easily to the (1,0,0) (also (0,1,0) and (0,0,1)) direction than the other directions. In two dimensional case, it gives four-fold anisotropy such as $\sigma(\theta)=1-\delta(1-\cos 4\theta)/4$. You may feel unnatural that anisotropy is fixed to the space coordinate system in Eq.3. But we can interpret it as follows. The model describes the growth of monocrystal and therefore the directions of anisotropy are determined at the nucleation stage and unchange through the growth process. Therefore we can determine the space coordinates from this directions as in Eq.3.

The parameter ε can be regarded as a parameter which makes a sequence of the interface equations. By putting $\tau(\nu) = a(\nu)\varepsilon^2$ and $m(T,\nu) = \varepsilon f(T,\nu)/\sqrt{2}$ in Eq.3, we obtain the ε-sequence of the equations

$$a(\nu)\varepsilon^2 \frac{\partial p}{\partial t} = \varepsilon^2 \nabla^2 p + p(1-p)(p - 1/2 + \varepsilon f(T,\nu)/\sqrt{2}) \tag{6}$$

This ε-sequence has a limiting interface equation with $\varepsilon \downarrow 0$,

$$a(\nu)V = f(T_i,\nu) - (n-1)\kappa \tag{7}$$

where V is a velocity of the interface to the outer normal direction ν, T_i denotes a temperature at the interface, κ is a mean curvature of the interface and n is a spatial dimension. Therefore the interfacial motion coming from Eq.6 must approximates the one of Eq.7 if ε is sufficiently small, and ε is interpreted as a parameter that indicates how well Eq.6 approximates Eq.7. So ε is not a physical parameter but a computational parameter. As stated before, it determines the thickness of the internal layer of p, and it's impossible for this model to describe a structure with the curvature radius less than the thickness of the layer. Therefore we must take ε rather smaller than the characteristic length of the finest structure to be simulated. But to take ε small demands the small space mesh size and the increase of the number of computational cells. So the size of the computational region is limited by the CPU power that we can use. This is a practical way of choosing the value of ε and the size of computational region.

Since the temperature at the interface is close to the equilibrium one in the simulation of dendrite growth, we can consider the form of $f(T_i,\nu)$ as

$$f(T_i,\nu) = b\sigma(\nu)(1-T_i)$$

where b is a positive constant. By assuming isotropy ($\sigma(\nu) \equiv 1$) and using Gibbs-Thomson relation, we obtain

$$b\bar{d}_0 = \Delta$$

where $\bar{d}_0$ is a dimensionless capillary length and Δ is a dimensionless supercooling. Then Eq.7 is transformed to

$$\beta(\nu)V = \Delta \cdot (1-T_i) - d_0(\nu)(n-1)\kappa. \tag{8}$$

where

$$\beta(\nu) = \bar{d}_0 a(\nu)/\sigma(\nu) \quad \text{and} \quad d_0(\nu) = \bar{d}_0/\sigma(\nu) \tag{9}$$

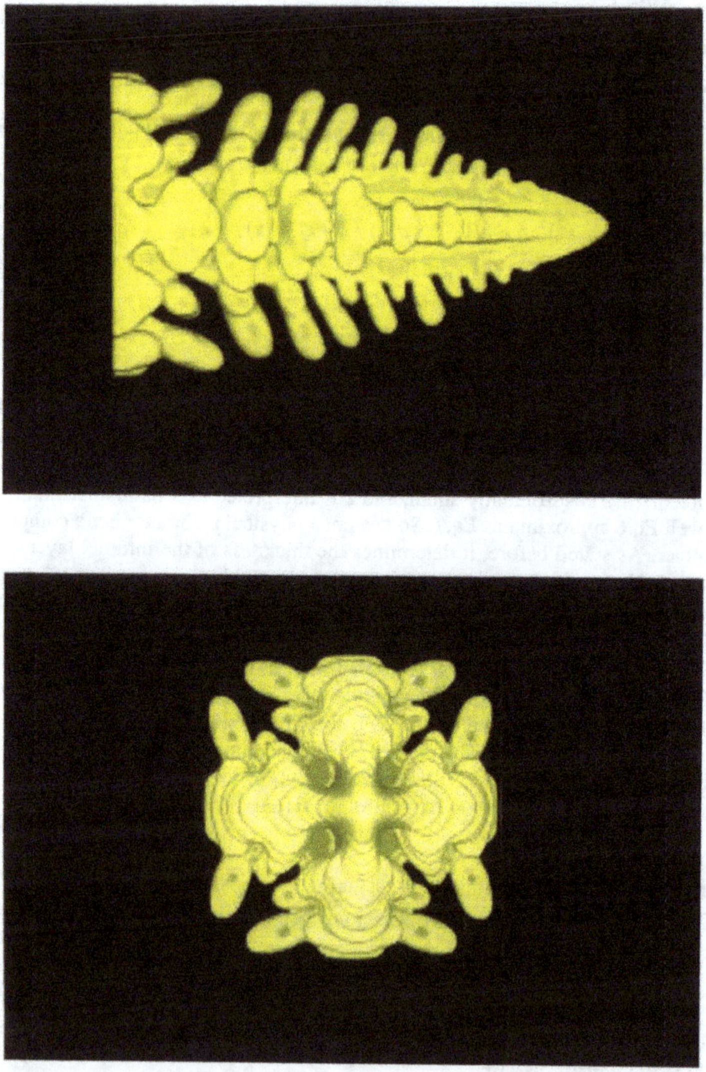

Fig. 1 The rotating perspectives of the computed crystal. The parameters used are the following.
vessel size = $6.0 \times 3.6 \times 3.6$, $\bar{\tau} = 3.0 \times 10^{-4}$, $\varepsilon = 1.0 \times 10^{-2}$, $\alpha = 0.9$, $\bar{\gamma} = 10.0$, $\delta = 0.3$, $K = 4.0$, amplitude of noise = 5.0×10^{-4},
number of the computational cells = $200 \times 60 \times 60$ (corresponds to quater region), time mesh size = 1.0×10^{-4} and the time step = 13000th.

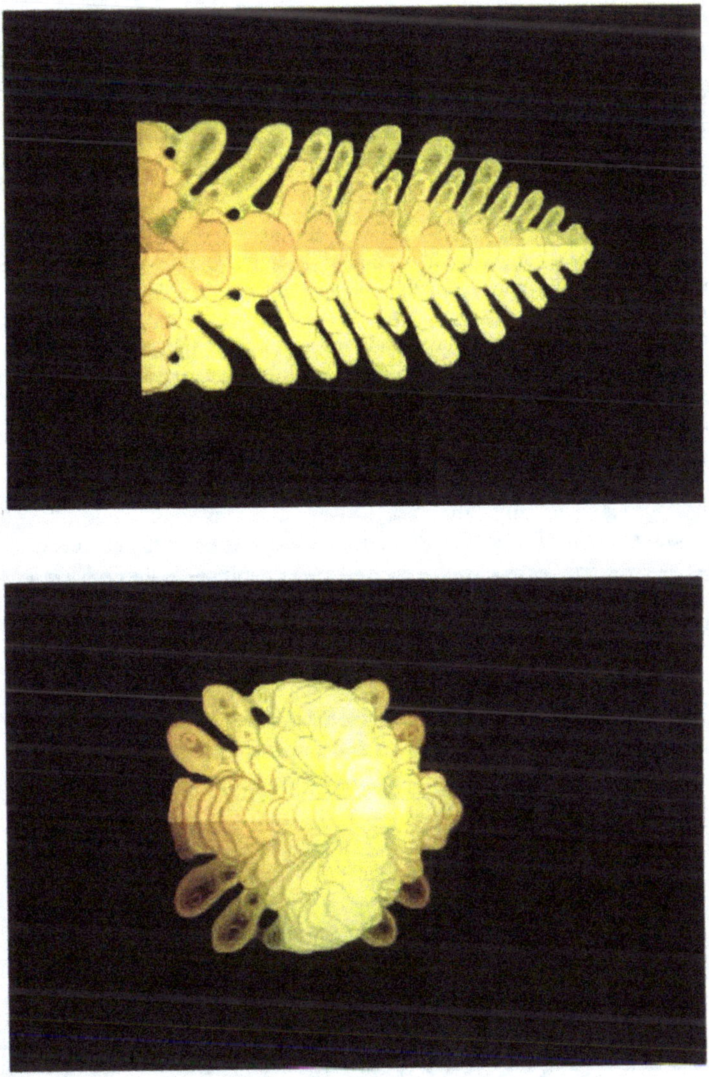

Fig.2 All the parameters are the same as Fig.1 except for $\delta = 0.2$ and the time step = 15000th.

Eq.8 is an interfacial equation that is approximated by the equation Eq.6. When $\beta(\nu)$ and $d_0(\nu)$ are given, we can determine $a(\nu)$ and $\sigma(\nu)$ from Eq.9, except for the critical case such as $d_0(\nu) \leq 0$ holds for some direction ν. Then Eq.8 can be simulated by Eq.6 by using $a(\nu)$ and $\sigma(\nu)$ obtained below. In the present simulation, $\beta(\nu)$ and $d_0(\nu)$ are proportional since $a(\nu)$ is taken as a constant, and $\beta(\nu) \approx 1.9 \times 10^{-3} / \sigma(\nu)$ and $d_0(\nu) \approx 6.3 \times 10^{-4} / \sigma(\nu)$ holds.

3. Simulations

Two simulations were made using the equation system Eq.3 and Eq.2. In the both simulations, the parameters and conditions are the following. The shape of the vessel is a rectangular parallelepiped. Its edges parallel to the direction (1,0,0), (0,1,0) and (0,0,1), which are the easiest directions for the growth in this model. Initially the vessel is filled with uniformly supercooled melt and then the nucleation occurs at the center of the one wall. After nucleation, the crystal grows into the supercooled melt under the adiabatic condition. All through the simulations, small noises are added to the term $m(T,\nu)$ using an uniform random number sequence of small amplitude on each time step, which accelerates the selection of growing side branches[23]. In these simulations, the computation is done in the quarter region assuming the 2 mirror symmetries around the (1,0,0) axis. The only difference between the two simulations is the strength of anisotropy. In the first simulation, anisotropy is stronger than the second, and fig.1 and fig.2 are rotating perspectives of the well grown dendrites in the each simulation[24].

4. Discussions

From fig.3 we can roughly estimate the the velocity V and the curvature radius ρ of the top of the principal branch as $V \approx 5$ and $\rho \approx 0.1$. Then the diffusion length ℓ is 0.2. Therefore the relation $\rho = 2\pi\sqrt{d_0 \ell}$ approximately holds.

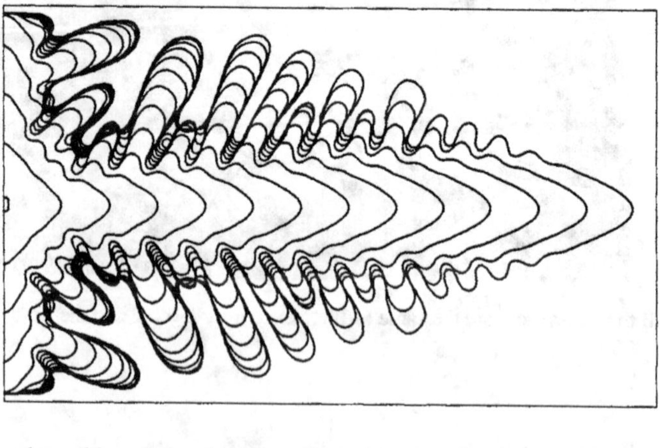

Fig.3 Superimposition of time-lapsed sections of the growing computed crystal. The last section corresponds to fig.1.

This model can simulate the dendritic crystal growth very realistically as shown in fig.1 and fig.2. Fig.1 shows a typical form of the dendritic crystals seen in the melt growth. The two types of destabilization are observed near the top of the principal branch. The one is a four mode fingering around the (1,0,0) axis, and the other is appearance of the side branches. It is observed that the four mode fingering occurs first and then the side branches appear and grow from the four mode fingers. Once the side branch appear, the competition between adjacent branches occurs and then the winner and loser are selected as the crystal grows (fig.3). This is also seen in the experiments. This process is explained by lateral inhibition caused by the release of the latent heat and its diffusion.

In fig.2, the shape of the crystal is rather different from the one of fig.1. The total morphology and especially the shape of the top of the principal branch seem to correspond to the ones of oscillating growth in the two dimensional simulation[23,24]. In fact, the oscillation of the growth velocity of the principal branch is observed in this simulation, while no oscillation occurs in the former simulation.

From the technical view point of numerical simulations the principal advantage of this model or generally the phase field model is that we need no special handling of the interface when the crystal shape becomes complicated or even the topology of the interface changes. Also the spacial dimension is not the problem except for the needed CPU power. It took about 2 or 3 hours as CPU time using CRAY2 for the each simulation. Perhaps it's very hard to simulate the formation of such a complicated shape in the three dimensional space by using another model.

Acknowledgements

I thank to the Geometry Super Computer Project funded by NSF and the university of Minnesota for use of facilities. And CRAY2 time was provided by the Minnesota Supercomputer Institute. Three dimensional visualization of the data is completely owes to Dr. C.Gunn. Also I thank to Prof. Y.Sawada and Dr. A.Tanaka for the fruitful discussions. Finally I thank to Prof. Y.Oono for giving me CRAY2 time for checking the results.

References

1. S.C.Huang and M.E.Glicksman, Acta. Metal. **29** (1981), 701 and 717.
2. J.P.Franck and J.Jung, J. Low Temp. Phys. **64** (1986), 165.
3. J.H.Bilgram, M.Fireman and W.Kanzig, Phys. Rev. **B37** (1988), 685.
4. H.Honjo and Y.Sawada, J. Cryst. Growth **58** (1982), 297.
5. A.Dougherty, P.D.Kaplan and J.P.Gollub, Phys. Rev. Lett. **58** (1987), 1652.
6. J.Maurer, P.Boissou, B.Perrin and P.Tabeling, Euro. Phys. Lett. **6** (1989), 67.
7. R.Brower, D.Kesseler, J.Koplik and H.Levine, Phys. Rev. Lett. **51** (1983), 1111 and Phys. Rev. **A29** (1984), 1335.
8. E.Ben-Jacob, N.D.Goldenfeld, J.S.Langer and G.Schon, Phys. Rev. Lett.**51** (1983), 1930 and Phys. Rev. **A 29** (1984), 330.
9. P.Pelce and Y.Pomeau, Studies in Appl. Math. **74** (1986), 245.
10. B.Caroli, C.Caroli, B.Roulet and J.S.Langer, Phys. Rev. **A33** (1986), 442.
11. A.Barbieri, D.C.Hong and J.S.Langer, Phys. Rev. **A35** (1986), 1802.

12. M.Benamar and Y.Pomeau, Euro. Phys. Lett. **2**(4) (1986), 307.
13. D.Meiron, Phys. Rev. **A33** (1986), 2704.
14. D.Kessler, J.Koplik and H.Levine, Phys. Rev. **A33** (1986), 3352.
15. D.Kessler and H.Levine, Phys. Rev. Lett. **57** (1987), 3069.
16. Y.Saito, G.Goldbeck-Wood and H.Muller-Krumbhaar, Phys. Rev. Lett. **58**(1987), 1541
17. G.Caginalp, Arch. Rat. Mech. Anal. **92** (1986), 205.
18. G.Caginalp and P.C.Fife, SIAM J. Appl. Math. **48** (1988), 506.
19. J.B.Collins and H.Levine, Phys. Rev. **B31** (1985), 6119.
20. G.Fix, in *Free boundary Problems*, edited by A.Fasano and M.Primicero, Reseach Notes in Mathmatics, vol2, Pitman, New York(1983).
21. R.Kobayashi, RIMS Kokyuroku **614** (1987), 39. (in Japanese)
22. M.Mimura, R.Kobayashi and H.Okazaki, videotape (1986).
23. R.Kobayashi, submitted to Physica **D**.
24. R.Kobayashi, *Computing Optimal Geometries,* Proceedings of an AMS Special Session, edited by Jean Taylor, videotapes (1991).

PATTERN DYNAMICS IN AN EXCITABLE REACTION-DIFFUSION SYSTEM

TAKAO OHTA
Department of Physics, Ochanomizu University, Tokyo 112, Japan

ABSTRACT

We show that excited domains arrayed periodically in space in a Bonhoeffer-van der Pol type model system exhibit breathing motion when the time-evolution of the inhibitor is slow compared to that of the activator. This is obtained by the linear stability analysis for the interface (domain boundary) equation of motion. The post-threshold behavior is also studied by deriving a coupled set of equations for the phase and the amplitude of the oscillating domains. We start with an isolated oscillating domain and take into account the interaction among the domains perturbatively. It is shown that there is a parameter regime where the system exhibits bistability such that in-phase and anti-phase oscillations can coexist.

1. Introduction

A non-equilibrium open system provides us with interesting dynamical phenomena which are not seen in thermodynamic equilibrium.[1] One is a limit cycle oscillation which is a characteristic feature of non-variational dissipative systems. Another property is a formation of a stable spatially localized motionless pattern.[2-4] In certain dynamical systems such as a reaction-diffusion equation this also originates from the absence of a Lyapunov functional,[5,6] the same root as a limit cycle oscillation. Recently localized motionless (periodic) patterns are observed experimentally in several systems[7-11] as well as a theoretical analysis.[12]

In this paper, we study pattern dynamics in an excitable reaction-diffusion system described by the Bonhoeffer-van der Pol type equation

$$\varepsilon\tau\partial_t u = \varepsilon^2\partial_x{}^2 u + f(u) - v \tag{1.1a}$$

$$\partial_t v = D\partial_x{}^2 v + \beta u - \gamma v \tag{1.1b}$$

where u(x,t) and v(x,t) are called the activator and the inhibitor, respectively. The constants D, ε, τ, β and γ are all positive. The function f(u) is assumed to have a cubic-like nonlinearity such that

$$f(u) = u(1-u)(u-a) \tag{1.2}$$

where a is a constant. We consider Eqs. (1.1) in an infinitely extended one-dimensional system.

Various phenomena[8,9,13-15] in open systems far from equilibrium are described by the coupled set of equations (1.1). Here we point out another very important feature of Eqs. (1.1). In the limit $\varepsilon\tau \to \infty$ keeping other parameters ordinary magnitude, we may put $\partial_t v = 0$. Solving formally Eq. (1.1b), Eq. (1.1a) becomes a closed equation for u(x,t):

$$\varepsilon\tau\partial_t u = \varepsilon^2\nabla^2 u + f(u) - \beta \int d\vec{r}'\, G(\vec{r}, \vec{r}')\, u(\vec{r}, t) \tag{1.3}$$

where $G(\vec{r}, \vec{r}')$ is defined through the relation

$$(-D\nabla^2 + \gamma)\, G(\vec{r}, \vec{r}') = \delta(\vec{r} - \vec{r}') \tag{1.4}$$

and we have generalized (1.1) to arbitrary spatial dimensions. Notice that Eq. (1.3) can be written as

$$\varepsilon\tau\partial_t u = -\frac{\delta H\{u\}}{\delta u(\vec{r}, t)} \tag{1.5}$$

where

$$H\{u\} = \int d\vec{r} \left[\frac{\varepsilon^2}{2}(\vec{\nabla} u)^2 + W(u) + \frac{\beta}{2}\int d\vec{r}'\, G(\vec{r}, \vec{r}')u(\vec{r})u(\vec{r}')\right] \tag{1.6}$$

with $dW/du = -f(u)$. When β and γ are positive as we have assumed, $H\{u\}$ plays the role of a Lyapunov functional. It is emphasized here that the original set of equations (1.1) is not variational as far as β is *positive*. It is also emphasized that competition of the short range interaction (the first term in (1.6)) and a long range interaction such as the last term in (1.6) is essential to get stable periodic structures. In thermodynamic equilibrium, this has been recognized in the study of mesophase structures.[16-19] Thus applicability of Eqs. (1.1) is not necessarily limited to the systems far from equilibrium but it covers pattern formation problems wider than expected at first sight.

From the above consideration, it is clear that Eqs.(1.1) admit spatially periodic steady solutions. An isolated excited domain is included as a special case of the infinite period. We consider the situation where the boundary between excited and rest domains is sufficiently thin so that dynamics of the system can be well described by the motion of the domain boundary (interface). The interface equation of motion is obtained by applying the singular perturbation method.[20] In one dimension we employ the interfacial approach to investigate the time-evolution and the stability of the periodic pattern. It is shown that when a parameter of the system is changed the solution exhibits a Hopf bifurcation. That is, the motionless periodic domains lose their stability and begin to oscillate. We call it a breathing motion. Furthermore, it is shown that the interaction between domains causes synchronization of the breathing motion. This part has been published in Ref. 21.

Next we study the oscillatory state at post-threshold. We start with an isolated localized breathing domain.[2] We account for the interaction among the domains perturbatively. Our main concern is how the system evolves when the bifurcation parameter is increased slightly away from its threshold value. Since the frequency modulation is expected to be weak compared to the intrinsic frequency at threshold, the phase description of oscillation can be applied.[22,23] We will derive the phase equation coupled with the amplitude based on the simplified model equation for the interfaces.[24]

A coupled oscillator approach[22,23] has been developed for modelling the mutual entrainment of assembly of oscillating units. Almost all the previous studies, however, start with a dynamical system described by an oscillatory reaction equation in each unit. The coupling among the different units is treated by introducing the diffusive or the direct interactions between the units.

It should be emphasized that the present derivation of coupled oscillators is entirely different from those previous ones. We start with the interface equation of motion for localized domains. The interaction among the domains is not simply the diffusive or the direct type. In the present system, it arises mainly from the diffusion of the inhibitor so that the interaction is inherently nonlocal in both space and time. Thus the origin of the interaction has a clearer physical basis compared to the previous modelling.

The organization of this paper is as follows. In the next section, we derive the interface equation of motion. In section 3, the reduction to a coupled oscillator system is carried out. The property of the coupled oscillator is also analyzed. The final section (section 4) gives the concluding remark.

2. Interface Equation of Motion

In this section, we derive the equation of motion for interfaces. Hereafter we put D=1 in (1.1b). In order to carry out the calculation in an exact manner, we utilize here the piecewise linear form of f(u):

$$f(u) = -u \qquad \text{for } u < a$$

$$\text{and} \qquad 1 - u \qquad \text{for } u > a. \tag{2.1}$$

We denote the period and the width of each excited domain by 2d and $2x_0$, respectively. An excited domain is the region where $u > a$. In the limit $\varepsilon \to 0$, the steady solution is readily obtained from (1.1) with (2.1) by the continuity condition of u and its first derivative at $x = \pm d$ and the Neumann condition at $x = \pm x_0$. The result is displayed in Fig. 1.

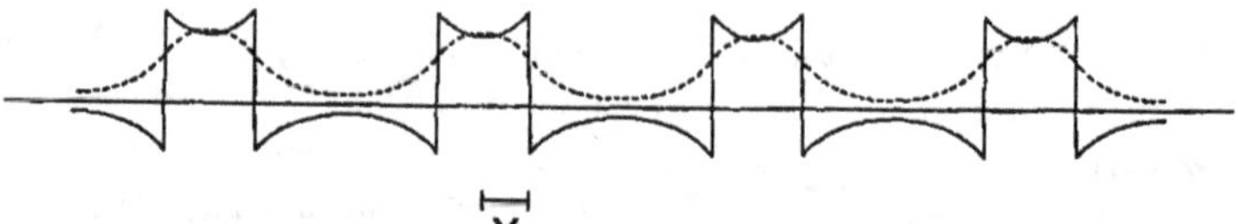

Fig. 1. Spatial variation of u (solid line) and v (dashed line).

Now we consider a slightly deformed periodic pattern. Let us denote the position of the right interface of the i-th excited domain by $X^+_i(t)$ and that of the left interface by $X^-_i(t)$. In order to derive the equation for $X^\pm_i(t)$ we use a singular perturbation method.[20] First, we analyze the outer solution away from the interfaces. In this case, finiteness of the interface width can be ignored. We may solve (1.1a) with (2.1) putting $\varepsilon=0$ as

$$u(x, t) = - v(x, t) + \sum_i \Theta(x - X^-_i(t))\, \Theta(- x + X^+_i(t)) \tag{2.2}$$

where $\Theta(x)$ is a Heaviside step function and we have defined the interface position by the condition $u(X^\pm_i) = a$. Substituting (2.2) into (1.1b), we obtain a closed linear equation for v which is readily solved for given $X^\pm_i(t)$.

Next we solve the inner solution of (1.1a). When we are concerned with the length scale of the order of the interface width $\varepsilon \ll d$, the adjacent interfaces may be regarded infinitely distant and those effect can be ignored. In one dimension, we obtain for the i-th interface

$$\tau V = c(a, v_I) \tag{2.3}$$

where $V = \pm\, dX^\pm_i/dt$. For the piecewise linear form of $f(u)$, $c(a, v_I)$ is given by

$$c(a, v_I) = 2(1/2 - a - v_I)\{(a + v_I)(1 - a - v_I)\}^{-1/2} \tag{2.4}$$

The value v_I of $v(x, t)$ at the interface is defined by $v_I = v(X^\pm_i(t), t)$.

Here we introduce the deviation of the interface positions $y^\pm_n(t)$ as

$$X^\pm_n = 2dn \pm x_0 + y^\pm_n(t) \tag{2.5}$$

Equation for $y^\pm_n$ up to linear order is given from (2.3) and (2.5) by

$$\tau\partial_t\,\vec{y}_n(t) = -2A[\,-\Gamma I\vec{y}_n(t) + \sum_m \int_{-\infty}^{\infty}\frac{dk}{2\pi}\int_0^t ds\; e^{-i2kd(n-m)-(\delta^2+k^2)s}\,\phi(x_0)\,\vec{y}_m(t-s)\,] \tag{2.6}$$

where I is the unit matrix. The vector $\vec{y}_n(t)$, the 2x2 matrix $\phi(x_0)$ and Γ are defined, respectively, by

$$\vec{y}_n = \begin{pmatrix} y_n^+ \\ y_n^- \end{pmatrix}, \qquad \phi(x_0) = \begin{pmatrix} 1 & -e^{-i2kx_0} \\ -e^{+i2kx_0} & 1 \end{pmatrix} \tag{2.7}$$

$$\Gamma = \frac{1}{\delta}\,\frac{\sinh\{(d-x_0)\delta\}}{\sinh \delta d}\,\sinh \delta x_0 \tag{2.8}$$

Equation (2.6) is our basic equation to study the stability of the periodic domain structure. We can see that the domains interact with each other nonlocally in both space and time through the diffusion current of v.

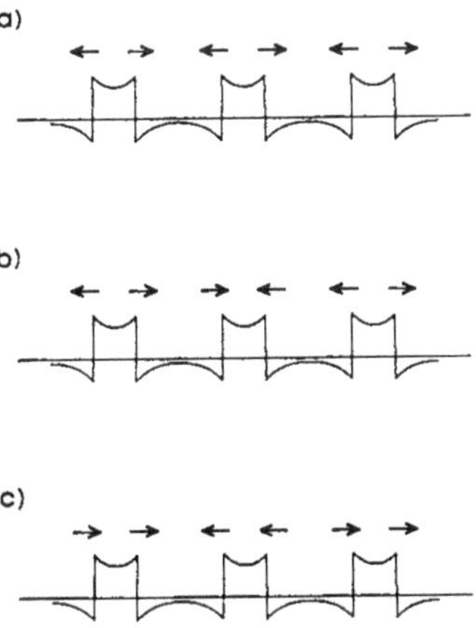

Fig. 2. Domain motions of the OI mode (a), the OA mode (b) and the AA mode (c)

By solving Eq. (2.6) we can determine the bifurcation threshold. Note that there are basically three oscillating modes as shown schematically in Fig. 2. The mode OI is an in-phase oscillation of all the domains. On the other hand, the mode OA is an anti-phase oscillation of the alternative domains. We call the third mode in Fig. 2 an acoustic anti-phase (AA) oscillation. The result of our stability analysis for γ=1/3 and d=2 is given in Fig. 3. The critical value τ_c of τ is plotted as a function of the equilibrium domain width

x_0. The full line indicates the bifurcation line for the OI mode, above which the motionless periodic state is stable. Thus if we decrease the value of τ for given x_0, the domains lose their stability when we cross the line. The dotted line is the line for the OA mode. The critical value of τ for the AA mode is shown by the broken line in Figs. 3. The value of τ_c for the OA mode is slightly larger than that for the OI mode in the region $0.2 < x_0 < 0.35$. For example, $\tau_c = 0.2260$ for the OA mode and $\tau_c = 0.2216$ for the OI mode at $x_0 = 0.3$. We have carefully checked that this is not due to a numerical error but is intrinsic.

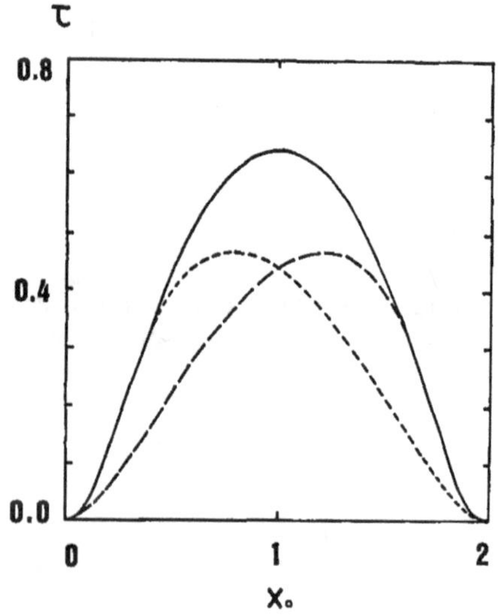

Fig. 3. Bifurcation threshold for the OI mode (solid line), the OA mode (dotted line) and the AA mode (broken line).

3. Reduction to Coupled Oscillators

In order to study the dynamical behavior at post-threshold, we here make a further reduction of degrees of freedom. Our strategy is as follows. First we consider an isolated breathing domain. The interaction among domains is taken into account by a perturbation expansion. Thus we need to derive the interface equation of motion for an isolated domain up to nonlinear terms. We have actually carried out this program and verified that the Hopf bifurcation is supercritical. Thus we here add a simple cubic term $y^{\pm}_n(t)^3$ to Eq. (2.6). This simplification does not alter any essential features of the system. The interaction term (the last term in (2.6)) is treated perturbatively. The second approximation is neglection of any higher harmonics of oscillation modes supposing that we are in the vicinity of the bifurcation threshold. Thus we put

$$y_n{}^{\alpha}(t) = \sigma_n(t)\, A^{\alpha} \cos(\omega_0 t + \phi_n(t)) \tag{3.1}$$

where $A^+ = -A^- = A > 0$ and ω_0 is the frequency at the threshold. The amplitude and phase modulations $\sigma_n(t)$, $\phi_n(t)$ are assumed to be weak over one period $2\pi/\omega_0$. Substituting (3.1) into (2.6) and time-averaging over $2\pi/\omega_0$, we readily obtain

$$\tau \frac{d\phi_n}{dt} = 4 \sum_{m \neq n} \frac{\sigma_m(t)}{\sigma_n(t)} G_{n-m} \sin(\phi_n - \phi_m - \alpha_{n-m}) \tag{3.2}$$

$$\tau \frac{d\sigma_n}{dt} = 3A^2\sigma_n - 6\sigma_n^3 - 4 \sum_{m \neq n} \sigma_m(t) G_{n-m} \cos(\phi_n - \phi_m - \alpha_{n-m}) \tag{3.3}$$

Here we do not write down the expressions of the coupling constant G_n and the phase factor α_n since those are lengthy.[24]

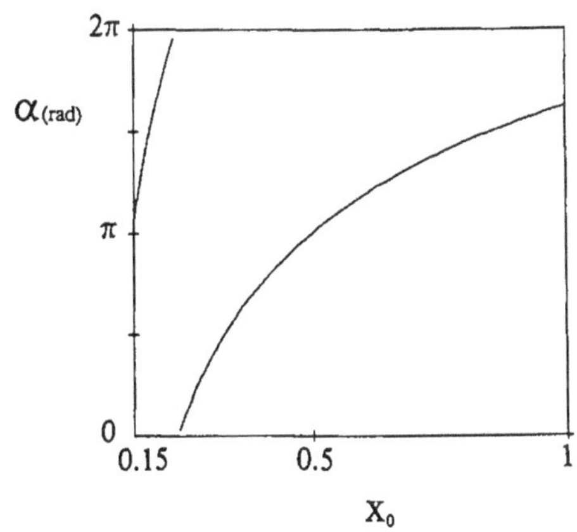

Fig. 4. Phase factor α as a function of x_0.

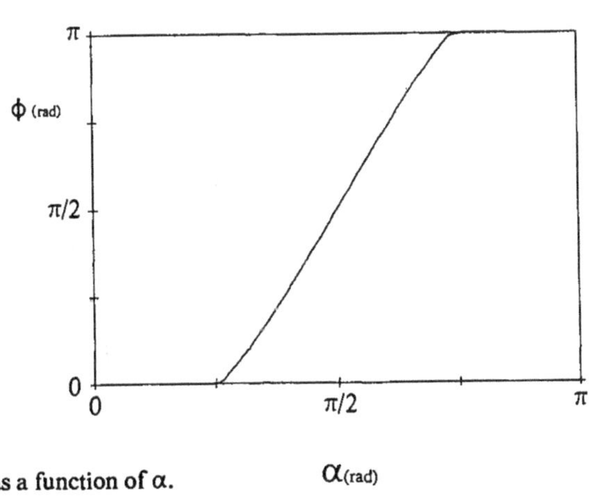

Fig. 5. Marginal value of ϕ as a function of α.

The form of Eqs. (3.2) and (3.3) has a similarity to the one derived by Aronson et al in a different context.[25] They have obtained a coupled oscillator from two limit cycles interacting diffusively or directly. As was emphasized in Introduction, the present system is different from a simple coupled limit cycle. The most essential point is the time-delayed interaction between interfaces. This gives rise to the phase factor α_n. Aronson et al have also shown that a phase factor appears in their model if the diffusive coupling is nonscalar. Thus the physical origin of the phase factor is different from the present system.

We have found numerically that G_n depends on n as $G_n \sim \exp(-n^2)$. This is expected since the interaction arises from the diffusion of v. Therefore we may retain practically only the nearest neighbor coupling. The phase factor $\alpha_n \equiv \alpha$ with $n=\pm 1$ is shown in Fig. 4 as a function of x_0. Here and after, all the numerical computations are carried out for the

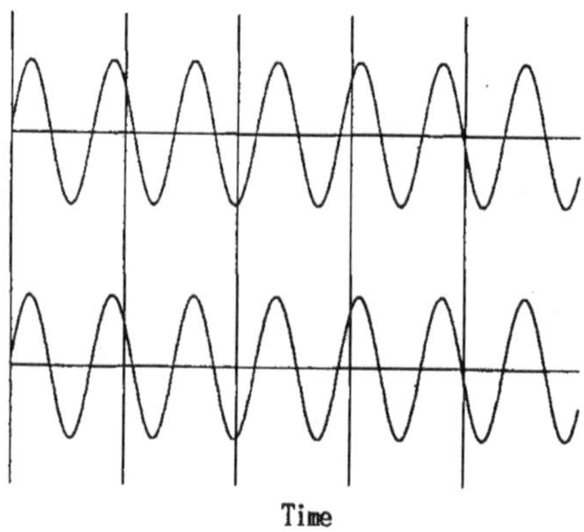

Fig. 6. In-phase oscillation with the initial phase difference $\pi/6.64$.

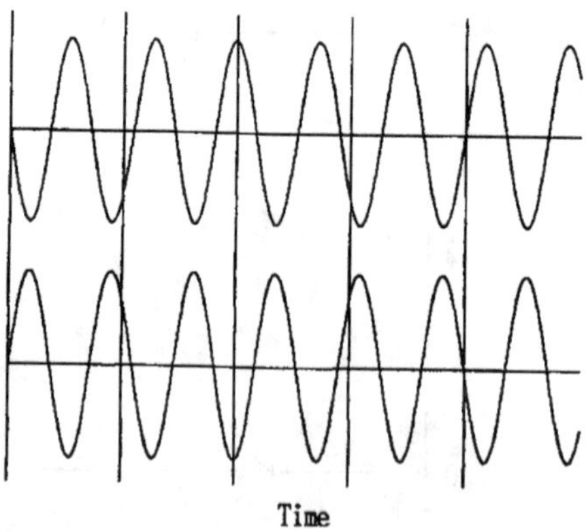

Fig. 7. Anti-phase oscillation with the initial phase difference $\pi/6.63$.

spatial period 2d = 4 and $\gamma = 1/3$. Notice that for $0.22 < x_0 < 0.33$ the phase factor α (mod. 2π) is small, while it exceeds $\pi/2$ when x_0 is increased. This is indeed the region for an anti-phase oscillation obtained by the linear stability analysis in section 2.

It is emphasized that the stability analysis of Eqs. (3.2) and (3.3) shows a bistability of in-phase and anti-phase oscillations. This is confirmed by solving (3.2) and (3.3) numerically for interacting two domains. Figure 5 shows the initial values of ϕ which is marginal between the in-phase and the anti-phase solutions starting with $\sigma_1 = \sigma_2 = 1$ at $t = 0$ and for $\delta = 3A^2/4G_1 = 2$. The actual oscillations for $\alpha = \pi/3$ and $\delta = 2$ are displayed in Figs. 6 and 7. If we start with the initial phase difference $\pi/6.64$, the oscillation is in-phase as in Fig. 6 while if the phase difference is $\pi/6.63$, the oscillation becomes anti-phase as in Fig. 7.

4. Conclusion

Starting from the nonoscillatory reaction-diffusion system, we have derived the coupled set of equations for the phase and the amplitude of oscillating domains. The derivation consists of two steps. One is the reduction to the interface equation of motion.[21] The second is further reduction of degrees of freedom to get the time-evolution equations of the phase and the amplitude. The main conclusion is that both in-phase and anti-phase oscillations can exist stably depending on the parameters. Furthermore it is found that there is a bistable region where these two oscillations can coexist. It would be quite difficult to see these properties if we analyzed directly the original reaction-diffusion equation.

The theory described in section 3 is based on a perturbation expansion such that the interaction between the domains is small. The unperturbed state is an isolated breathing domain. This assumes implicitly that the distance between adjacent domains is sufficiently large, i.e., $x_0 << d$. Hence for larger values of x_0, the validity of the present theory becomes less accurate.

It is mentioned that quite recently the above conclusion that a stable anti-phase oscillation can exist when x_0 is small compared with d has been confirmed by computer simulations of (1.1).[26] Radehaus et al[27] have also observed a stable anti-phase oscillation in computer simulations although their model equation is slightly different from (1.1).

I would like thank to Professor M. Mimura for valuable discussions. Particular thanks are due to Professor T. Ikeda for showing me his result of computer simulations prior to publication. I am also grateful to A. Ito, A. Tetsuka and H. Nakazawa for their collaboration. This work is supported by the Grant-in-Aid of Ministry of Education, Science and Culture of Japan.

6. References

1. G. Nicolis and I. Prigogine, *Self-Organization in Non-Equilibrium Systems* (Wiley, New York, 1981).

2. S. Koga and Y. Kuramoto, Prog. Theor. Phys. **63** (1980) 106.
3. I. S. Aranson, K. A. Gorshkov, A. S. Lomov and M. I. Rabinovich, Physica D **43** (1990) 435.
4. T. Ohta, R. Kobayashi and M. Mimura, Physica D **34** (1989) 115.
5. O. Thual and S. Fauve, J. Phys. **49** (1988) 1829.
6. T. Ohta and M. Mimura, in *Formation, Dynamics and Statistics of Patterns* Vol.1, edited by K. Kawasaki et al. (World Scientific, Singapore, 1990).
7. A. Joets and R. Ribotta, Phys. Rev. Letters, **60** (1988) 2164.
8. K. Toko, M. Nosaka, T. Fujiyoshi, K.Yamafuji and K. Ogata, Bull. Math. Biol. **50** (1988) 225.
9. H. G. Purwins, Ch. Radehaus, T. Dirksmeyer, R. Dohmen, R. Schmeling and H. Willebrand, Phys. Letters **A136** (1989) 480 and the earlier references cited therein.
10. Q. Quyang and H. L. Swinney, Nature **352** (1991) 610.
11. U. Rau, K. M. Mayer, J. Parisi, J. Peinke, W. Clauss and R. P. Huebener, Solid-State Elect. **32** (1989) 1365.
12. E. Schoell and D. Drasdo, Z. Phys. **B81** (1990) 183.
13. R. FitzHugh, Biophys. J. **1** (1961) 445.
J. Nagumo, S. Arimoto and S. Yoshizawa, Proc. IRE **50** (1962) 2061.
14. J. J. Tyson and P. C. Fife, J. Chem . Phys. **73** (1980) 2224.
15. J. J. Tyson and J. P. Keener, Physica **32D** (1988) 327 and the references cited therein.
16. T. Ohta and K. Kawasaki, Macromolecules, **19** (1986) 2621.
17. T. Garel and S. Doniach, Phys. Rev. **B26** (1982) 325.
18. D. Andelman, F. Brochard and J. F. Joanny, J. Chem. Phys. **86** (1987) 3673.
19. M. Hoshino, H. Nakano and H. Kimura, J. Phys. Soc. Jpn. **46** (1979) 1709.
20. P. C. Fife, SIAM-AMS Proc. **10** (1976) 23.
21. T. Ohta, A. Ito and A. Tetsuka, Phys. Rev. A **42** (1990) 3225.
22. A. T. Winfree, " *The Geometry of Biological Time* ", Biomath. 8 (Springer, Berlin, 1980).
23. Y. Kuramoto, *Chemical Oscillations, Waves and Turbulence* (Springer, Berlin, 1984).
24. T. Ohta and H. Nakazawa, Phys. Rev. **A** (to be published).
25. D. G. Aronson, G. B. Ermentrout and N. Kopell, Physica D **41** (1990) 403, which contains an extended list of previous works on this subject.
26. T. Ikeda, unpublished work.
27. C. Radehaus, R. Dohmen, H. Willebrand and F. -J. Niedernostheide, Phys. Rev A **42** (1990) 7426.

PATTERN FORMATION OF THE RIPPLE PHASE IN BILAYER MEMBRANES

KATSUYA HONDA, HATSUO KIMURA and HIROYA KODAMA

Department of Applied Physics, Faculty of Engineering
Nagoya University, Nagoya 464-01, Japan

ABSTRACT

The model free energy is presented to discuss the phase transition of the lipid bilayer membrane with lamellar structure. The ripple phase can be found to occur intermediately between two flat, L_α and L_β, phases due to largely fluctuating order of hydrocarbon chains. The transient phenomena from a unstable flat to the stable periodic state is studied by numerical and theoretical approaches for the nonlinear dynamic equation.

1. Introduction

Recently the phase transition from the flat (L_α or L_β) phase to the ripple (P_β) phase of the lamellar bilayer membranes has occupied many interests.[1] For all theoretical approaches, the problem what mechanism plays a crucial role of constructing the periodic structure has not yet solved. The experiments have indicated unanimously that the transition temperatures are independent of the water amount larger than 40 g%.[2] From this fact, the interaction between neighboring lamellas can be ignored, then only a set of two layers, *i.e.*, a bilayer, surrounded by the water is taken into account. In a previous paper,[1] we have proposed the model free energy for the system investigated within a linear analysis, which will be developed here.

2. Model Free Energy

The spatial variation of the membrane structure is assumed to be only along the x axis. We introduce two kinds of fields, $u_j(x)$ and $\sigma_j(x)$ (j =1,2), which denote the displacement of the head groups along the perpendicular direction to the flat surface and the configuration of hydrocarbon chains, respectively. The suffix j is associated with the upper (j =1) or lower (j =2) layer. The latter depending roughly on the number of the trans bonds in the chains, is controlled by the following Gibbs-type free energy,[3]

$$g(\sigma(x)) = -(a/2)\sigma(x)^2 + (b/4)\sigma(x)^4 - \eta(T)\sigma(x), \tag{1}$$

where the coefficients a and b are positive constants and $\eta(T)$ is assumed to decrease with temperature and change its sign at $T = T_0$. The equilibrium value $\bar{\sigma}$ determined by minimizing Eq.1 decreases with temperature and follows the abrupt change at $T = T_0$, corresponding to the main transition. As far as we consider the physical situation of membranes as the isolated unframed systems, their tension is well known to vanish.[4] Moreover, the width of the membranes, i.e., $u_1(x) - u_2(x)$, can be regarded to be constant and its spatial variation can be neglected.

Now we propose the model free energy for a bilayer in terms of $u_+(x) = u_1(x) + u_2(x)$ and $\sigma_\pm(x) = \sigma_1(x) \pm \sigma_2(x)$;

$$G = \int_0^{L_0} dx A(x), \tag{2}$$

$$A(x) = g(\sigma_1(x)) + g(\sigma_2(x)) + (1/4)C(d^2u_+(x)/dx^2)^2$$
$$+ (1/4)U\{(d^2\sigma_+(x)/dx^2)^2 + (d^2\sigma_-(x)/dx^2)^2\} + (1/2)K\sigma_-(x)(d^2u_+(x)/dx^2), \tag{3}$$

L_0 being the length corresponding to the flat phase. This has been constructed as simply as possible accompanying with indispensable second derivative terms. The key point to lead successfully to the ripple phase is the last term representing a coupling between the upper and lower layers. The relative difference of the order parameters, $\sigma_-(x)$, can be a origin of bending the whole membrane, due to the excluded volume effect by the hydrocarbon chains. Minimum condition of Eq.2 with respect to $u_+(x)$ gives

$$d^2u_+(x)/dx^2 = -(K/C)\sigma_-(x), \tag{4}$$

by symmetry considerations for vanishing integral constants. When $\sigma_-(x)$ =0, the center line of the membrane is horizontal, i.e., $u_+(x)$ =0, while the periodic $\sigma_-(x)$ corresponds to the ripple structure. Substituting Eq.4 into Eq.2 and taking its minimum with respect to $\sigma_\pm(x)$, we get

$$0 = -a\sigma_+(x) + (b/4)\sigma_+(x)\{\sigma_+(x)^2 + 3\sigma_-(x)^2\} - 2\eta(T) + Ud^4\sigma_+(x)/dx^4, \tag{5}$$

$$0 = -(a + K^2/C)\sigma_-(x) + (b/4)\sigma_-(x)\{3\sigma_+(x)^2 + \sigma_-(x)^2\} + Ud^4\sigma_-(x)/dx^4. \tag{6}$$

Equations 5 and 6 always have a uniform solution $\sigma_+(x) = 2\bar{\sigma}$ and $\sigma_-(x)$ =0, corresponding to the flat patterns in the L_α and L_β phases.

By the linear stability analysis around the uniform solution, we can find the uniform $\sigma_+(x) = 2\bar{\sigma}$ and the periodic $\sigma_-(x)$ if $d^2g(\bar{\sigma})/d\bar{\sigma}^2 \leq K^2/C$. Therefore nonlinear effects in the ripple phase are studied only in term of $\sigma_-(x)$. This simplification can be justified near the transition temperature. With use of scaled quantities such as $\sigma_-(x) \propto \psi(\xi)$ and $x \propto \xi$, let us study dynamic properties on the basis of

$$\partial\psi(\xi)/\partial t = -\delta\Delta G/\delta\psi(\xi), \tag{7}$$

$$\Delta G = (1/2)\int d\xi\{-r\psi(\xi)^2 + (1/2)\psi(\xi)^4 + (\partial^2\psi/\partial\xi^2)^2\}, \tag{8}$$

where r is positive in the ripple phase region and the mobility is set to unity without loss of generality. Here it is noted from Eq.4 that the constraint

$$\int d\xi\psi(\xi) = 0 \tag{9}$$

should be required, for the membranes are flat as a whole beyond the mesoscopic scale. We will have the interesting patterns depending sensitively on initial states.

3. Numerical Solutions

Let us perform the numerical calculation of Eq.7 with suitable initial configurations. The periodic boundary condition was used. At first we started from a flat pattern, $\psi(\xi)$ =0, with small amplitude noise. The upper part of Fig.1 displays the result presenting coexistence of the periodic structure and the flat regions taking on the value of $\pm\sqrt{r}$ with small lumps. The constraint mentioned above requires the accompany of the kinks or anti-kink. The amplitude of the former are slightly larger than $\sqrt{r}$. This pattern does not show a stationary solution of Eq.7, though its time dependence becomes very slow after that. As seen from Eq.8, the free energy becomes lower as the flat regions grow. Then the periodic structures like those seen in the ripple phase should be metastable. However, the periodic structure is generated stably from a localized defect, as shown in the lower part of Fig.1. Fronts propagate toward the unstable region, where $\psi(\xi)$ =0. This should correspond to the creation process of the periodic ripple pattern when the system is shifted suddenly from the L_α to the P_β phase, which has been observed by the freeze-fracture electron micrograph.[5]

From the numerical results we have calculated the critical exponents for various quantities near r =0. For example, the amplitude and wavelength of the periodic pattern are proportional to $r^{0.48}$ and $r^{-0.26}$, respectively. From the periodic parts of patterns like the upper figure, the similar exponents, 0.49 and -0.24, respectively, have been obtained. The velocity of the propagating front decreases as $r^{0.73}$ as the transition temperature is approached. They will be confirmed by the following theoretical arguments.

4. Dimensional and Marginal Stability Analyses

Dimensional analysis is a relevant and simple way to get the critical exponents. Assuming that every term in Eq.7 has the identical order of r and substituting $\psi \sim r^\alpha$, $\xi \sim r^\beta$ and $t \sim r^\gamma$ into Eq.7, we can get immediately

$$\alpha = 1/2, \qquad \beta = -1/4, \qquad \text{and} \qquad \gamma = -1, \tag{10}$$

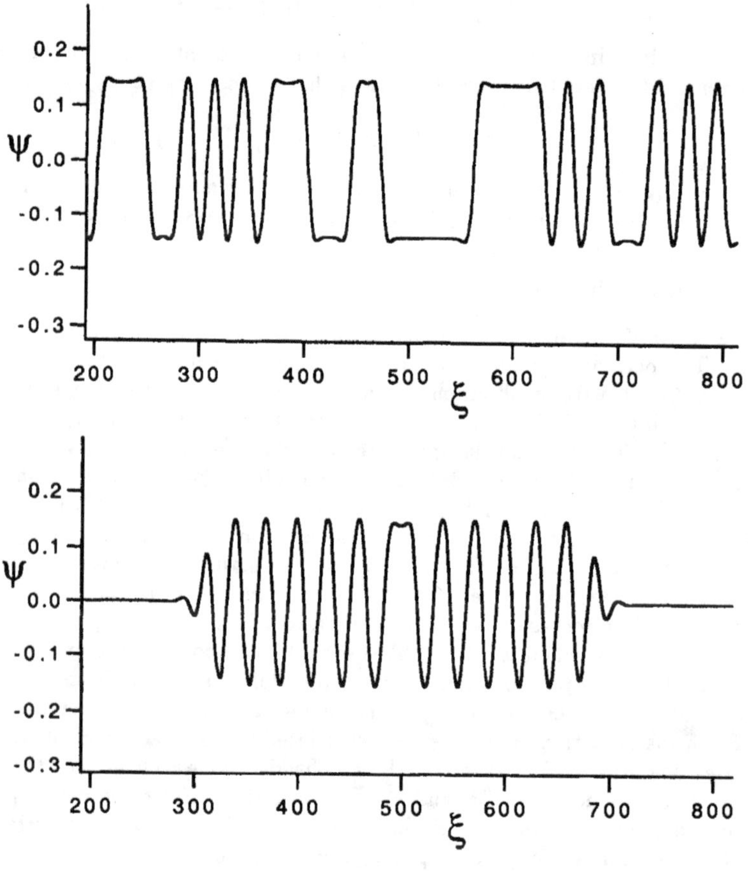

Fig.1 The numerically obtained profiles of $\psi(\xi)$ on the basis of Eq.7 in the case of r =0.02. (1) The upper pattern is generated from the uniform initial state, $\psi(\xi)$ =0, accompanied by the small amplitude perturbation. (2) The lower is a snapshot starting from a localized defect at the center.

which are in good agreement with the numerical results mentioned above. The dimension of the front velocity is given as $[v] = \beta - \gamma = 3/4$.

In the marginal stability theory,[6] we get the most dominant mode q^* from the relations; $\partial \Re\omega(q^*)/\partial \Im q^* = 0$ and $\Re\omega(q^*)/\Re q^* = \Re d\omega/dq\,|_{q=q^*} = v^*$, where v^* is the asymptotic front speed and $\omega(q)$ denotes the dispersion relation given by substituting $\psi \propto e^{\omega t - q\xi}$ into the linearized version of Eq.7, *i.e.*, $\omega(q) = r - q^4$. The conclusions are given as

$$\Re q^* = (r/24)^{1/4}, \qquad \Im q^* = (3r/8)^{1/4}, \tag{11}$$

$$\Re\omega = 3r/4, \qquad \Im\omega = r/\sqrt{3}, \tag{12}$$

and

$$v^* = 2^{11/4} 3^{-3/4} r^{3/4}. \tag{13}$$

We should note that both approaches give the same results.

5. Summary

We have proposed the model free energy to describe the phase transition of the bilayer membranes from the flat to the periodic ripple phase. The coupling between the upper and lower layers has been emphasized to play an essential role of generating the ripple phase. The numerical calculations of the nonlinear dynamic equation have clarified its novel character depending sensitively on the initial states. This has implied that the ripple phase with the periodic structure should be metastable. If we start from the localized perturbation, however, the stable periodic structure can be obtained. The dimensional and marginal stability analyses have successfully confirmed the numerical results. Consult with ref.7 about details.

Acknowledgement

We would express our thanks to D. Andelman, T. Kawakatsu and T. Ohta for stimulating discussions.

References

1. K. Honda and H. Kimura, J. Phys. Soc. Jpn. **60** (1991) 1212, and see references therein.
2. M. Kodama, M. Kuwabara and S. Seki, Biochim. Biophys. Acta **689**(1982) 567.
3. H. Kimura and H. Nakano, J. Phys. Soc. Jpn. **43** (1977) 1477.
4. F. David and S. Leibler, J. Phys. II France 1 (1991) 959.
5. K. Tsuchida, K. Ohki, T. Sekiya, Y. Nozawa and I. Hatta, Biochim. Biophys. Acta **898** (1987) 53.
6. G. T. Dee and J. S. Langer, Phys. Rev. Lett. **50**(1983) 383.
7. K. Honda, H. Kimura and H. Kodama, submitted for publication.

Wave Form Analysis and Computer Simulation in Chemically Driven Nonlinear Waves

Shoichi KAI Toshio MORI Mutsumaru MIKI

Department of Electrical Engineering, Kuyshu Institute of Technology, Tobata, Kitakyushu 804, Japan

Abstract

The experimental and theoretical investigations are presented of chemically driven dynamic instability of an oil-water interface in a cylindrical and annular glass container. The immiscible liquids are water containing a surfactant (TSAC) and nitrobenzene containing iodine. The reaction at the interface leads to complex deformation patterns including rotating solitary waves, multiple wave trains, periodic and non-periodic oscillations, source-to-sink propagation, and intermittent behavior. A pattern diagram is established for the two solute concentrations ranging from 10^{-4} to 10^{-1}M. It shows 5 distinct regions of different dynamic regimes, determined by interaction of at least three mechanisms: dynamic wetting of the container wall, capillary effect, and Marangoni instability of the liquid-liquid interface. The influences of aspect ratio, concentration product, and temperature are investigated and local time traces are derived from electropotential differences. Their spectral analysis reveals details of periodic or irregular motion. Transitions between the dynamic modes of the system during its temporal evolution are recorded. Quantitatively determined profiles of regular waves are analyzed by exponential functions and a simple model for this is proposed. The quantitative wave profiles are experimentally obtained by use of image processing technique and compared with the results of the computer simulation of the model equation based on the diffusion reaction equation with drift terms. The profile obtained from the computer simulation well agrees with the experimental results most of which are described in detail, where soliton-like, standing and traveling waves have been observed. However some aspects cannot be obtained by this model.

1 INTRODUCTION

Spatial and temporal ordering on a macroscopic scale is a well studied phenomenon occurring in nonequilibrium physics and chemistry, such as for example in the

Rayleigh-Benard convection and the Belousov-Zhabotinskii chemical reaction[1-4]. In these cases, the ordering processes are usually governed by highly nonlinear interactions. Patterns arising from hydrodynamic instabilities, for instance, are frequently discussed in terms of mode selection, wave propagation, defect dynamics, or the influence of boundary conditions and container geometry [1,2] and a wide variety of stationary, transient, chaotic or intermittent behavior is known. Much effort has been also devoted to understand similar structures in chemical systems like the Belousov- Zhabotinskii reaction in which multistability, limit cycle behavior, chaos , and periodic spatiotemporal patterns, in particular travelling waves due to reaction-diffusion coupling are observed [3,4]. These phenomena have been treated theoretically in the framework of nonequilibrium thermodynamics [5].

In recent years, several authors [6-11] have investigated complex dynamic deformations of a liquid-liquid interface which occur when chemical reactions are coupled with hydrodynamic instabilities, as often observed in the chemical engineering field [12]. Large-scale chemically driven motion of an interface between an organic and an aqueous solution was first shown in 1971 by Dupeyrat and coworkers [6] who furtheron made salient contributions concerning the hydrodynamical and physicochemical effects leading to the instability [7]. In this system a surfactant is dissolved in the aqueous phase and the interfacial tension is changed by reaction at the interface with a compound supplied from the organic phase. This results in pronounced dynamic interfacial deformations, e.g. several waves rotating along the container wall as shown in Fig. 1, soliton like behavior [13], or irregular nonperiodic motion. No comprehensive theoretical picture for these has been developed yet. A stability theory for surface deformation due to a first order chemical reaction between gas and liquid was achieved by Sakata et al. [14], but the wave form and the motion after occurrence of the instability was not discussed in any detail. Recently, waves were analyzed by approximating convective flow near the interface by a Hele-Shaw cell and introducing an phenomenological hydrodynamic equation for interfacial tension changes, where no chemical reaction term was taken into account [15].

Our studies emphasize the wide variety of phenomena induced by Marangoni instability of the interface due to changes in interfacial tension [16,17,22]. We distinguish among the "capillary effect", interfacial turbulence due to the Marangoni instability, and changes of the contact angle of the wetting behavior at the boundary of the container [18,22]. Furthermore there is an interesting analogy between these chemically driven patterns and the gravitational waves in the fluid mechanically induced vibration; the so-called bell mode instability [19,20]. In this article we describe our study of the dynamic features of interfacial instability experimentally and theoretically. A number of spatial profiles can be quantitatively represented by exponetials. This mostly phenomenological survey is concluded by pointing to various mechanistic aspects with repect to the coupling between reaction and hydrdynamic and diffusive transport. On the bases of these observation we introduce a model by which we make a computer simulation [21].

2 EXPERIMENTAL

2.1 Materials

The immiscible solvents were water and nitrobenzene. When placed in contact with each other the two liquids form a well defined interface. The solutes were the surfactant trimethyl-stearyl- ammonium chloride (TSAC, $C_{18}H_{37}N(CH_3)_3Cl$) and iodine (I_2)[7].

All chemicals were purchased from Tokyo Kasei Co. (best grade) and used without further purification. Solutions were made by dissolving TSAC in water (aqueous phase A) and I_2 in nitrobenzene saturated with potassium iodide (organic phase B). About 120 combinations of concentrations of TSAC, c_A, and of iodine, c_B, were prepared, each concentration ranging from 10^{-4} to 10^{-1} M.

2.2 Containers and Sample Preparation

Interfacial patterns were investigated in two container geometries (Fig. 2): (1) cylindrical containers (beakers) with inner diameter ϕ ;and (2) annular containers consisting of two concentric glass cylinders (diameters ϕ_1,ϕ_2) sealed to a bottom plate and separated by a ring-shaped gap (spacing d). The vessels used in the experiments are listed in Ref.22.

A volume of solution B (typically 2.7 ml for a cylindrical container with $\phi = 19$ mm) is first placed into the vessel, then an equal volume of solution A is poured on top of the organic phase. Due to its higher density ($\rho_B = 1.35$ g/cm^3) solution B always occupies the lower part of the vessel, while solution A ($\rho_A = 1.0$ g/cm^3) stays on top. Upon contact between the solutions, a chemical reaction starts that produces $C_{18}H_{37}(NH_3)_3I$. In the course of time this product becomes weakly visible as a yellow precipitate at the interface. Immediately after the start the contact area is distorted, indicating the onset of interfacial instability. For instance, in a beaker with ϕ =19 mm and $c_A = c_B = 10^{-3}$ M waves with an amplitude of 10mm start to propagate along the wall at a velocity of 13 cm/s, corresponding to an angular velocity of 14 rad/s (similar to the example for an annulus shown in Fig. 1). Samples were prepared for 8 values of ϕ (see Ref.22) with varying aspect ratio $r = \phi/\delta$, where δ is the solution depth between bottom and interface.

For starting an experiment in an annulus (see Ref.22) the liquids were poured into the gap between inner and outer cylinder. In this arrangement the interfacial motion is more distinct and a larger variety of dynamic features is observed as compared to the beakers. A special container was used for electric potential measurements, as specified in Section 2.3. All experiments were performed at 24 degrees unless stated otherwise. The temperature was controlled within ±1 degree.

2.3 Methods

The dynamic behavior of the interface and its temporal evolution was visually observed and classified as a function of several system parameters: mainly the con-

Fig. 1. Photograph of interfacial waves propagating in an annular container (outer diameter ϕ_1 = 50.3 mm; spacing d=2.6mm; height H=50 mm). Equal volumes (7 ml) of solutions A and B with concentration $c_A = 10^{-3}$ M and $c_B = 2 \times 10^{-3}$ M were poured into the space between the two cylindrical columns. The amplitude of the 4 waves is 10 mm.

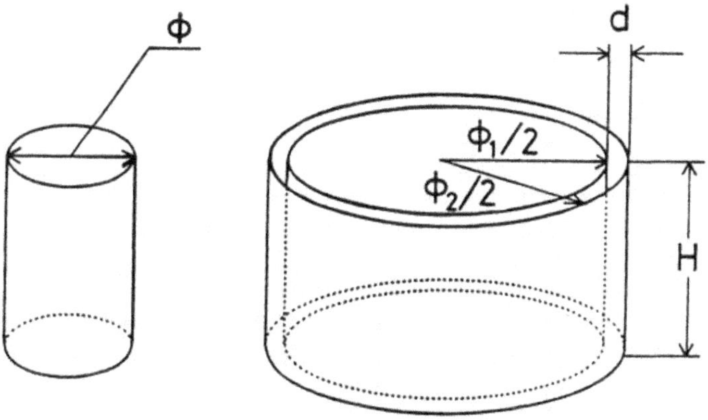

Fig. 2. Schematic drawings of the containers : (a) Cylindrical glass vessel (beaker) with inside diameter . (b) Annular container consisting of two concentric columns sealed to a bottom plate; ϕ_1 is the outer diameter of the smaller, ϕ_2 the inner diameter of the larger column, d the thickness (spacing) of the annulus. H is the height of the containers.

centrations c_A, c_B, the container geometry (aspect ratio r, diameters), and the temperature T. Measured quantities were the number of waves, n, the wave amplitude, h, the propagation velocity, v, (or rotation frequency, ω), and the lifetime, τ, of a dynamical state. Wave forms and profiles were described qualitatively and also recorded with a TV camera (Panasonic VY51) linked to a video tape recorder (JVC BR- 8600) and a video signal processor (ADS Co. PIP-4000) linked to a personal computer (NEC, PC9801) by special interface circuits.

For several cases of waves in an annulus, the video images were further analyzed in two ways: (1) The wave profiles were copied directly from a snapshot displayed on the TV monitor. No correction was made to account for the distortions at the container edges due to the projection of the circular vessel geometry on the screen. (2)In order to observe dynamic changes and motions of waves in whole spatial area along a concentric container simultaneously, we use a cone-shaped mirror with an angle 45 degrees as schematically illustrated in Fig.3. The images are taken from the above by CCD(charge coupled device)- TV camera which is connected to the video tape recorder and the personal computer. A personal computer controls a video tape recorder and then correctly reforms wave profiles to original forms from images on a cone-shaped mirror. An example by this method is shown in Fig.4 (see also Fig. 9 below).

3 RESULTS

3.1 Dynamical Features of Interfacial Motion

3.1.1 Classification of waves, periodic and nonperiodic motion

The interfacial system exhibits a complex variety of structural and dynamical features. As evident from simple visual observation there occur macroscopic dynamic phenomena such as solitary waves of different shapes [13], multiple rotating wave trains (already shown in Fig. 1) with constant or variable number of waves, periodic and non-periodic oscillatory deformations that do not rotate, steady source-to-sink propagation, splitting and recombination of waves [9]. As an additional example we show in Fig. 5 the propagation of a solitary wave as observed on a TV screen. Its profile is of the relaxation oscillation type and remains unchanged during rotation. Figure 5 shows various wave profiles for such waves.

The details of the interfacial dynamics always depend sensitively on the initial and boundary conditions of the experiments, *e.g.* on factors such as temperature control, solution depth, container curvature, or surface properties of the glass wall. In this section we characterize the dynamic behavior found in cylindrical containers with aspect ratio $r = 0.44$ at room temperature. This system shows qualitatively all the features that are characteristic for a wide range of experimental conditions.

The waves moving along the boundary of the container have characteristic profiles [22]. These are always highly asymmetric which points to the nonlinear character of the underlying processes in that the interfacial instability occurs not only

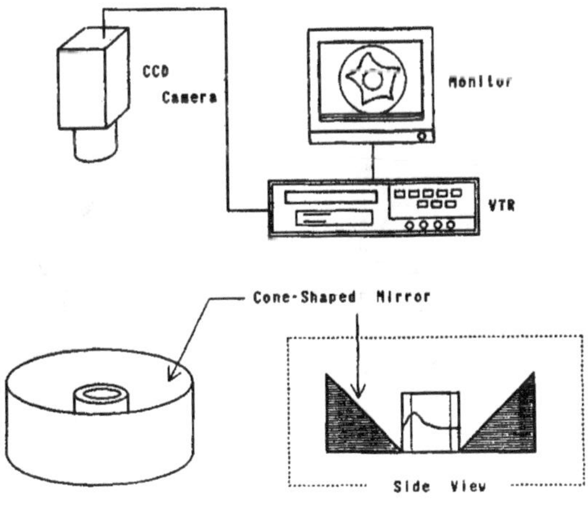

Fig. 3. Observation and analysing set-up for rotating waves.

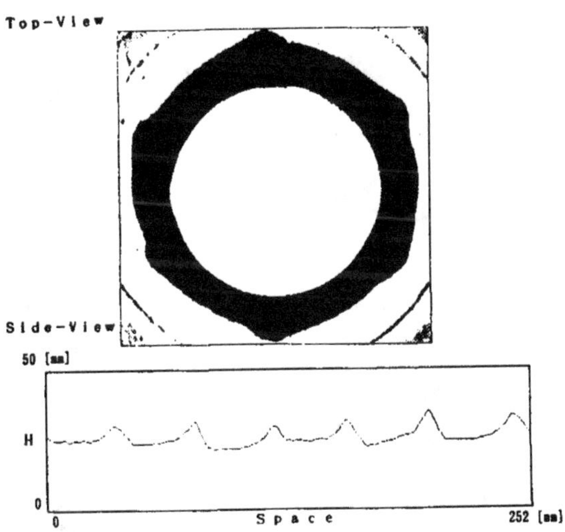

Fig.4. Typical example of observed waves. Top figure: images on a cone-shaped mirror. Bottom figure: reformed waves by a computer after taking an image data.

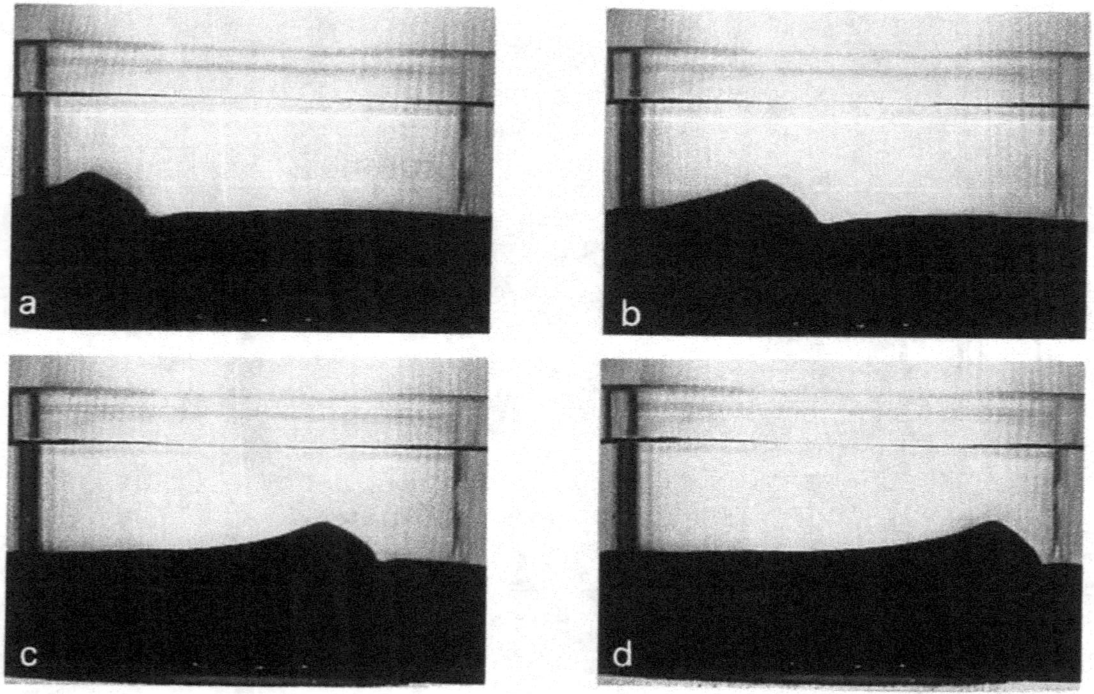

Fig. 5. A solitary wave with relaxation-type profile propagates in the gap of an annular container ($\phi_1 = 85$ mm, $\phi_2 = 80.4$ mm) with $c_A = 10^{-2}$ M and $c_B = 3 \times 10^{-2}$ M at T=26°C, as photographed from a video screen. Time interval between successive snapshots is 1/15 s.

for one elementary frequency but involves a complex spectrum of higher harmonics. Some specific features of the rotating or oscillatory macroscopic deformations, strongly dependent on the precise experimental conditions, are the following [10,22]. The typical phase diagram for the motions and pattern is shown in Fig.6. Characteristic electrochemical potential signals for these corresponding wave profiles have also been given in Ref.22 in detail.

3.1.2 Phase diagram

From a large number of visual observations and potential measurements we established phase diagrams in the (c_A, c_B)-plane for cylindrical containers with varying diameter and aspect ratio. In Fig. 6 such a diagram is shown for ϕ = 19 mm and r= 0.44. The characteristic features of this diagram are the same for all r and 10.5 mm < ϕ < 24 mm, and life time, amplitude, velocity, temporal evolution and temperature dependence of the motions change only little. (Significant differences were observed for ϕ < 5 mm.) For annular containers, too, we find essentially the same behavior.

The qualitative distinctions made in the diagram are based on the most typical, rather stable intermediate regime at the given concentrations, whereas other regimes that may appear during an induction or break-down period are disregarded. Thus, we distinguish between five regions (Fig.6) for which different types of motion are characteristic. In regions I and II most of the steady periodic motions take place including solitons, multiple waves, and oscillations with pronounced amplitude (symbol " o "). Steady non-periodic (chaotic) motion can appear in region II, especially for the highest concentrations. Source-to-sink propagation is mostly observed in an annulus with small spacing in the boundary region between I and II. In the lower part of region I the wave rotation may become somewhat unsteady (symbol "△") or lose its continuity which leads to intermittent bursts of active motion (symbol " ◯ "). In region III we frequently find the isotropic stretching motion (symbol "□"), and in region IV non-periodic motion with relatively small amplitude (symbol "■"). Finally, in regions IV and V there are many cases in which no macroscopic motion can be detected (symbol " • "), but in both regions the formation of a microemulsion at the interface forming small scale, irregular flow patterns can be readily observed under the microscope. One phase boundary almost coincides with the diagonal in the concentration plane. In this context we note an important property of the shape of the interface: after the reaction finishes, the interface must remain concave with respect to the organic phase if $c_A < c_B$ and concave with respect to the aqueous phase if $c_A > c_B$. This follows from the numerical values for the concentration dependence of γ_A and γ_B (see Ref.22).

Some general differences between regions I and II may be related to the difference between an oil-water and an water-oil emulsion. This is suggested by the fact that in I the organic liquid is pulled down along the container wall when a wave passes, while in II it is lifted up into the aqueous phase.

Motions with relatively long lifetimes are found along the diagonal, those with the longest (τ > 15 min) for concentrations $c_A = c_B$ ranging between 5×10^{-3}

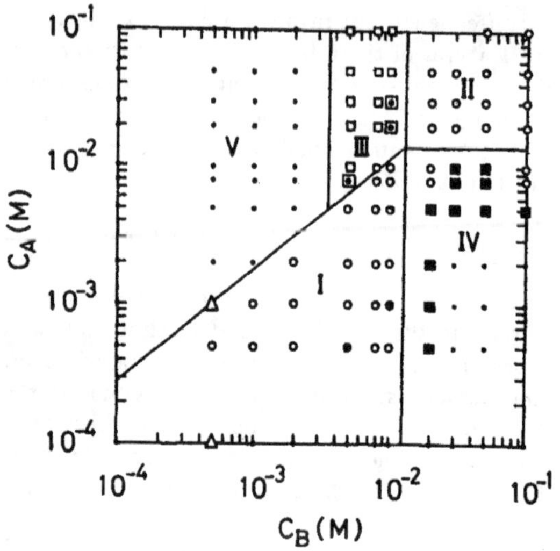

Fig. 6. Phase diagram of interfacial motions in the concentration plane (c_A: surfactant concentration in water, c_B: iodine concentration in nitrobenzene) for a cylindrical vessel (ϕ = 19 mm, r = 0.44, T = 24°C). Details are given in the text.

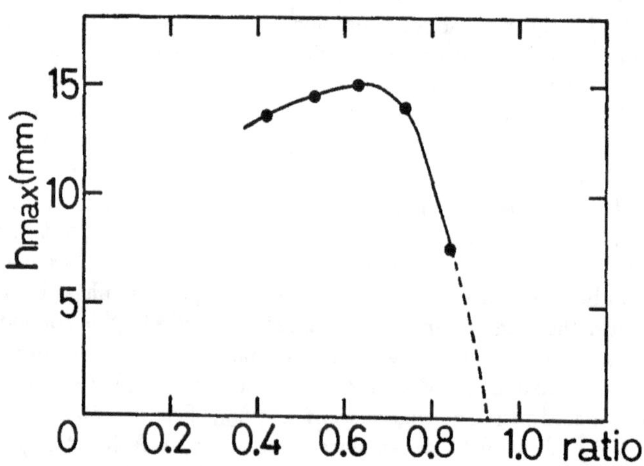

Fig.7. Aspect ratio dependence of wave amplitude h_{max} and life time. ($C_A = C_B = 1 \times 10^{-2}$ M, T = 24°C, ϕ = 19 mm)

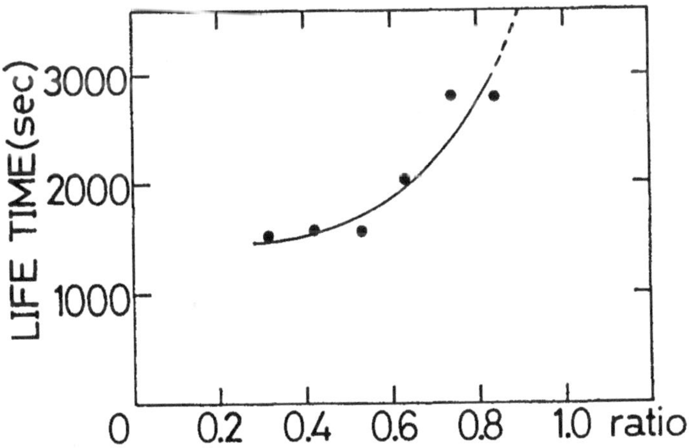

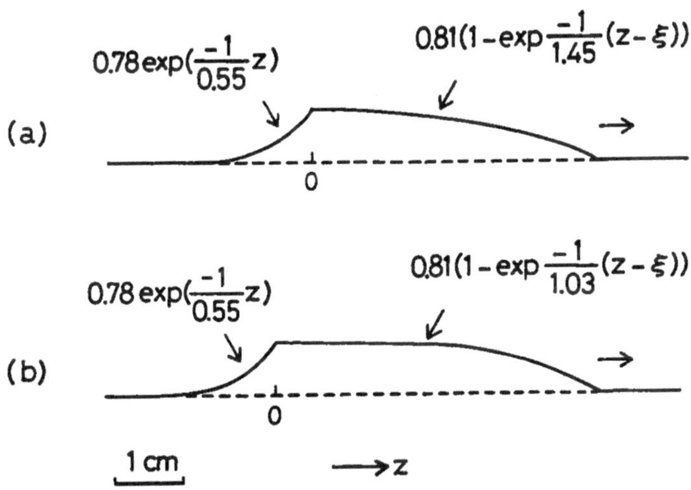

Fig. 8. Wave profiles in an annular container (ϕ_1 =80 mm, ϕ_2 = 75 mm) with $c_A = c_B = 5 \times 10^{-3}$ M at T=20 C. Time after start of experiment: (a) 5 min and (b) 25 min. Wave velocities: (a) 2.6 cm/s, (b) 3.5 cm/s.

and 9×10^{-2} M. Furthermore, we note that the dependence of the lifetime on the temperature is opposite in regions I and II: in I, increases linearly, in II, decreases approximately exponentially with T.

The motion and the life time also depend strongly on the aspect ratio of containers. In Fig.7, as an example, the maximum amplitude h_{max} of a wave and the life time of a wave are shown versus aspect ratio of containers. The life time simply increases very rapidly as r is increased. On the other hand the amplitude shows the maximum then decreases. In this combination for $r = 1.0$ there is no wave amplitude.

3.2 Temporal Changes of Interfacial Motion

Due to closed conditions the 0 gradual depletion of the reacting compounds necessarily leads to changes of motion which can be roughly devided into an induction, an intermediate, and a break-down phase. In one example (see [9]) first oscillations built up at the wall of a beaker, then about 3 waves rotated along the wall with constant velocity. Later this number decreased stepwise down to one, showing at each step a short period of transient behavior with fluctuating number of waves.. This decrease was first accompanied by an increase then by a decrease of the amplitude, until the wave motion vanished completely. As a general trend the number of waves decreases towards the end of an experiment and the profiles become more symmetric and smoother in time. The details of these trends depend, of course, on the particular region in the phase diagram.

These qualitative features were confirmed in an experiment with similar wave shape that was reconstructed quantitatively from a time series. In Fig. 8 the profiles $h(z)$ in a moving frame (variable z) could be fitted by exponential functions. For the lower of the two shown cases these are:

$$h(z) = 0.81(1 - \exp(-0.97(z - \xi))), \tag{1}$$

for the leading slope and

$$h(z) = 0.78 \exp(-1.82z), \tag{2}$$

for the decaying tail, where $z = x - vt$ with x denoting the position and v the velocity of the waves and $\xi = 4.7$ cm. For the upper profile, which was obtained at an earlier time, the decay function is the same but the leading slope rises faster and is shorter ($\xi = 4.1$ cm). Thus, the analyzed example with exponential profiles suggests that the phenomenon is related to the relaxation type oscillations as known from other chemical oscillators [3].

Figure 9 shows various wave profiles obtained from real experiments for various concentration initially prepared using a cone-shaped mirror in Fig.4.

Figure 10 depicts a scenario of wave collision recorded in a system with $c_A = c_B = 5 \times 10^{-3}$ M. Within 1 s two waves pass through each other and emerge in only slightly modified form. In this collision process the waves behave like common solitons. Two other types of wave interaction have been observed: waves with large

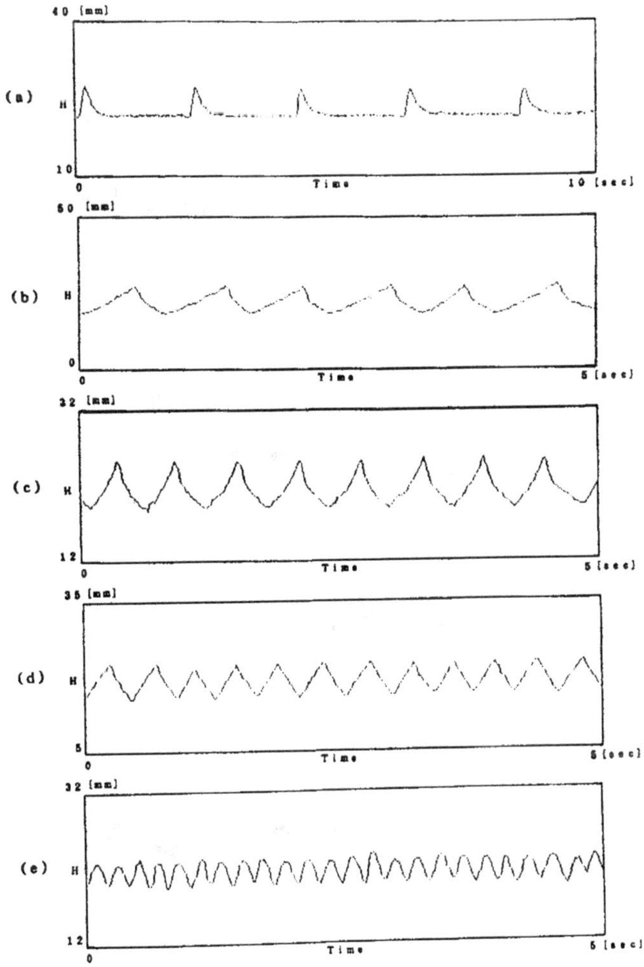

Fig.9. Various waves observed in the present system. (a),(b) relaxation oscillation type (c)cusp-like (d)triangle (d)sinusoidal-like waves.

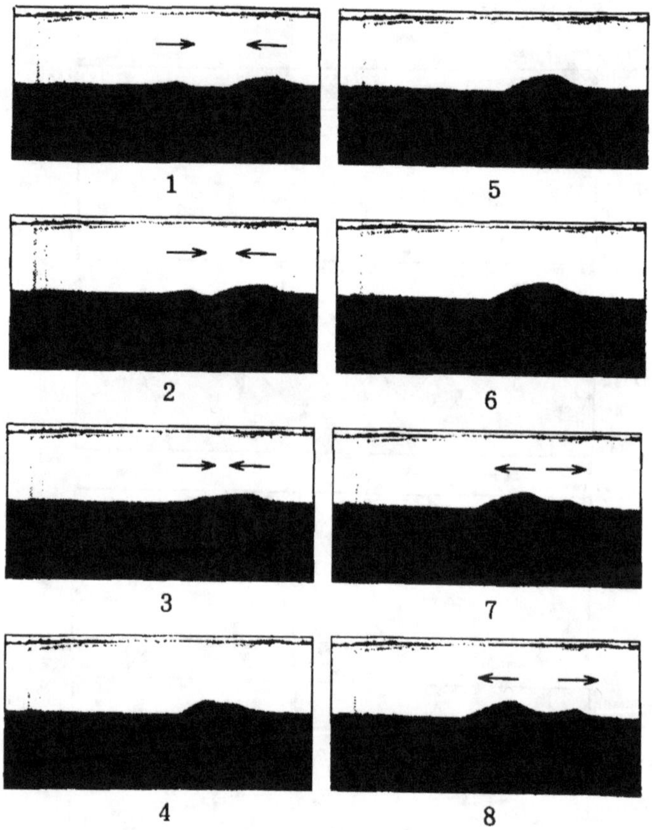

Fig.10. Soliton like motion showing "pass-through each other".

amplitude absorb waves with small amplitude upon collision; two colliding waves with equal amplitude may annihilate each other.

4 DISCUSSION

4.1 Summarizing Comments

A large variety of spatiotemporal structures occurring at a liquid-liquid interface has been described in terms of visual observation, electropotential measurements, construction of a phase diagram, and measurement of wave profiles. The phase diagram can be divided into 5 regions: (1) regions I and II where basically four different modes of macroscopic steady motion are observed, *i.e.* solitons, rotating multiple waves, oscillations, and chaotic motion; (2) region III where a periodic stretching motion appears; (3) region IV where the motion is unsteady and intermittent; and (4) the remaining regions where one observes no macroscopic motion but only formation of a microemulsion with microscopic irregular dynamics. The lifetime and the activity of motion in this closed system are largest, if the initial solute concentrations are equal, and increase over a wide range with the concentration product. During evolution sharp transitions between different dynamic regimes occur. The local time course of the motion can be described more quantitatively by the difference of electrochemical potential. The dynamical behavior is not only caused by changes in interfacial tension but changes in the electrical conditions (ionic concentrations) in the bulk and its influence on the interface.

4.2 Mechanistic Aspects

As mentioned previously, surface waves on a water column in a beaker may show properties that are similar to those of the currently investigated system [19,20]. This so-called bell mode instability, which results in macroscopic elastic wavelike surface deformation, is caused by nonlinear interaction between the modes of water waves and an externally applied periodic force with a much higher frequency than that of the waves. In this case, when the frequency of the external driving force deviates from a certain resonance frequency, wave propagation along the glass wall occurs. This phenomenon has been studied in several theoretical papers. The basic mechanism of the interfacial system under consideration would be almost the same, if the chemical reaction processes were regarded as equivalent to the external periodic driving force. However, no explanation in terms of such a direct analogy has been put forward yet.

Since the phenomenon under consideration is driven by the chemical reaction through the interface (adsorption and desorption), the dynamic features must be largely influenced by the reaction kinetics, by transport processes such as diffusion and convection, as well as by the influence of concentration variations on the physical boundary conditions, in particular on the forces acting at the interfaces.

From the phase diagram in Fig.6 and spectral analysis of the electrochemical potential we learn that in region I of the diagram periodic rotation prevails, but that in region II a noisy component is superimposed. For the highest concentrations in region II the nonperiodic irregular motion even dominates and interfacial turbulence can be clearly observed. Accordingly, different mechanisms appear to contribute to the wavelike displacements, all of which are based on the variation of interfacial tensions: The interfacial turbulence is caused by the Marangoni instability [16,17]. It leads to hydrodynamic transport of the ions from the bulk to the interface and usually remains on a small scale. The major reason for the large scale displacements (10 mm) is the capillary effect: the interface is lifted up or pulled down by the changes of interfacial tension between oil and water. This depends largely on the difference and does not involve convective flow. In addition, the dynamical wetting induces changes of the contact angle and thus may contribute to the large scale displacements. It is, however, restricted to areas near the solid boundary. While capillary effect and wetting dominate in region I, the interfacial random convection becomes noticeable only in region II. For the highest concentrations used it has considerable influence and the wavelike motion is then strongly coupled with pronounced interfacial turbulence.

Evidence has been presented about the important role of the concentration product as a driving force of the phenomenon: the larger $c_A c_B$ the faster and more pronounced the motion [22]. The fact that the maximum rotation frequency is a linear function of the logarithm of $c_A c_B$ suggests that the rotation velocity is closely connected to the chemical reaction kinetics. By contrast, the maximum amplitude is essentially independent of this product, because it is mainly determined by differences in interfacial tension, thus by the magnitude of the concentration change during the reaction, rather than by the reaction velocity. It should depend on the resulting concentration difference which we assume to be determined by threshold values of $c_A c_B$ at which the reaction is turned on and off see below).

Among the observed waves we consider here those having relaxation type profiles. The exponentioal shapes shown in Fig.8 and the arguments for the role of reaction, diffusion and driving force present a basis for preliminary modelling. The main processes for the formation of the waveform are controlled by reaction for the rising slope and by diffusion for the decaying slope. Apparently the chemical processes at the interface lead often to such relaxation type oscillations as they occur in other chemical nonequilibrium systems. As a first step, the phenomenon of wave propagation could be interpreted as a change of phase across a set of spatially distributed relaxation type oscillators. But this picture does not lead very far as to understand the nature of the spatial coupling of neighboring volume elements.

4.3 A Simple Model

4.3.1 Qualitative description

Figure 11 shows schematically a possible explanation for the formation of oscillatory behavior at the interface. The role of the capillary effect is emphasized, any change of the contact angle is neglected ($\alpha = \pi/2$) and convection is disregarded: Suppose

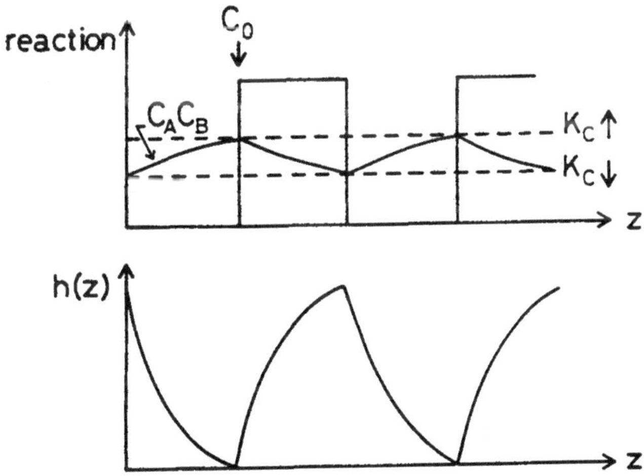

Fig.11. Schematic drawing of a relaxation-type oscillation, based on a phenomenological model (see text). $h(z)$: wave amplitude with $z = \omega t$ (ω=rotation frequency), $K_c \uparrow$:threshold for start of reaction, $K_c \downarrow$: threshold for end of reaction.

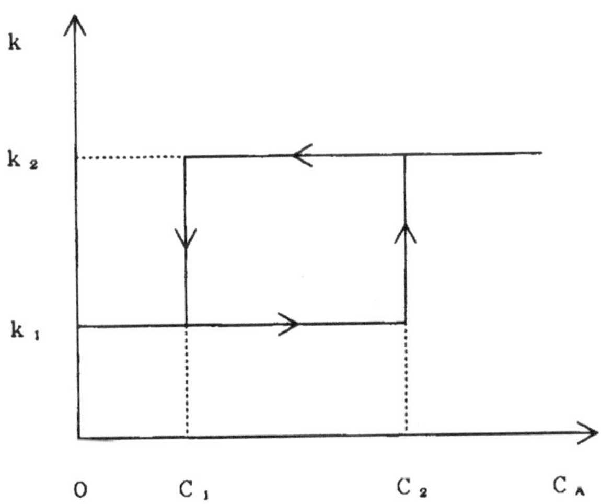

Fig.12. Assumption for hysteresis of reaction rate k due to concentration C_A

that the chemical reaction starts and stops at two different threshold values of the concentration product $K(= c_A c_B)$, $K_c \uparrow$ and $K_c \downarrow$ ($K_c \uparrow > K_c \downarrow$), respectively, and that the state of the system is metastable at any value between these thresholds. We start from the top of a wave. When ions are transported to the interface by diffusion, the interfacial tension decreases, because more surfactant molecules become available to form a monolayer. As a consequence of the capillary effect the interface sinks down in the gravitational field. The increase of ions leads to the reaction start at $K > K_c \uparrow$, ions at the interface are consumed so that the difference of interfacial tensions increases leading to an increase of the wave amplitude. The reaction continues until $K < K_c \downarrow$ with a time scale faster than that of diffusion. After the reaction has stopped due to depletion of available reactant, diffusion comes into play again, thus allowing the ion concentration to be newly built up at the interface so that repitition of the sequence can occur. In Fig. 11 the time course of this repetitive competition between the reaction and the diffusion controlled steps is depicted. Wave propagation could be introduced simply by a propagating phase shift between local oscillators, but this includes no assumption about the nature of the spatial coupling [4]. Evidently, a more sophisticated mechanism for spatial coupling leading to propagation phenomena has to be implemented.

Also the bell mode instability, caused by the nonlinear coupling between fast modes (the reaction as the external driving force) and slow modes (the macroscopic surface wave), may play an important role for a complete mechanism. When interfacial turbulence due to the Marangoni effect is only weakly developed, the chemical reaction probably enhances more and more the deformations and nonlinear coupling both to the amplitude and to the turbulent flow component takes place.

4.3.2 Theoretical model

According to the chemical reaction and diffusion as described in the above, we consider the simplist reaction diffusion mechanism i.e. we introduce a simple switching mechanism in Fig.12. The model equation is following.

$$\partial C_A/\partial t = -kC_AC_B + D'_A(C_0 - C_A) + D_A\partial^2 C_A/\partial x^2 + v_0\partial C_A/\partial x + v_1(\partial C_A/\partial x)^2 \quad (3)$$

Here C_A, C_B and C_0 are the concentration of surfactant ion, the concentartion of I^- ion and the initial concentration of surfactant ion at the interface respectively. k is the reaction rate constant. D_A and D'_A are the lateral diffusion constant in the interface and the effective diffusion constant from a bulk containing the boundary layer thickness. v_0 and v_1 express the drift terms coming from hydrodynamic flow effects and all hydrodynamic effects are included in these terms. The coordinate x indicates the space along the wall of a curclar container in one dimension. Equation (3) is nondimensionalized as,

$$\partial C'_A/\partial t' = A - BC'_AC_0 + \alpha\partial^2 C'_A/\partial x'^2 + \beta\partial C'_A/\partial x' + \gamma(\partial C'_A/\partial x')^2, \quad (4)$$

by $A = (l^2/D_A)D'_A$, $B = (l^2/D_A)(kC_B + D'_A)$, $\alpha = const.$, $\beta = \beta'(v_0 l/D_A)$ and $\gamma = \gamma'(v_1 C_0/D_A)$. Here $C'_A = C_A/C_0$, $t' = tD_A/l^2$ and $x' = x/l$. Hereafter we neglect a superscript " ' " of nondimension variables. α, β' and γ' are coefficients

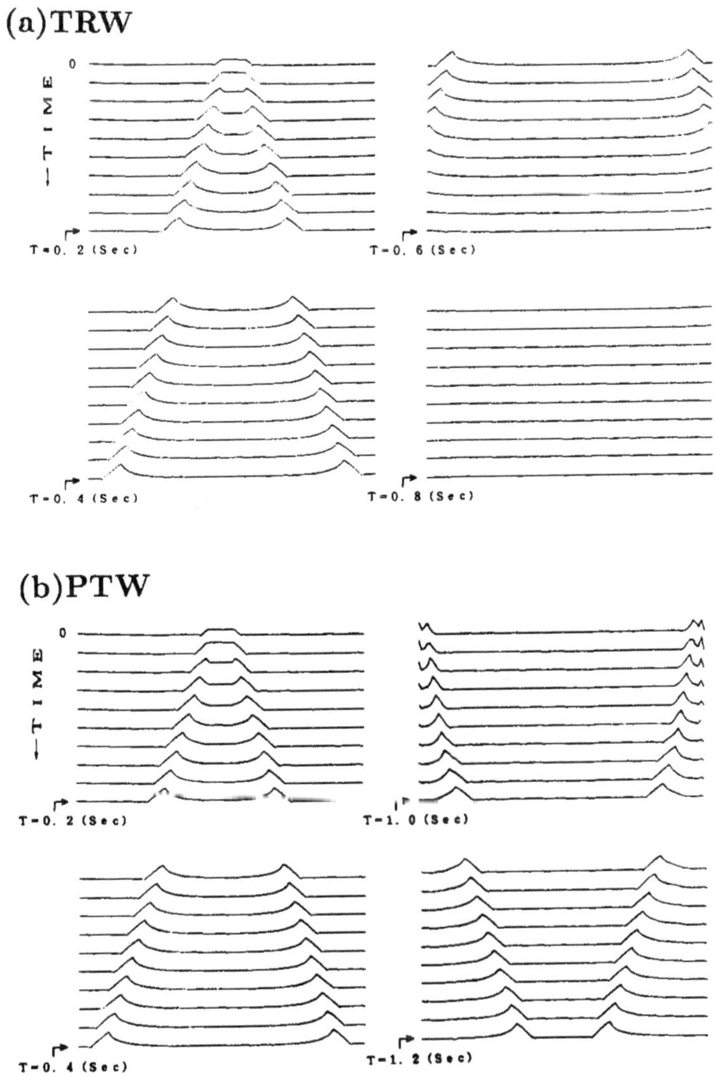

Fig.13. Example of computer simulation. (a)TRW (b)PTW

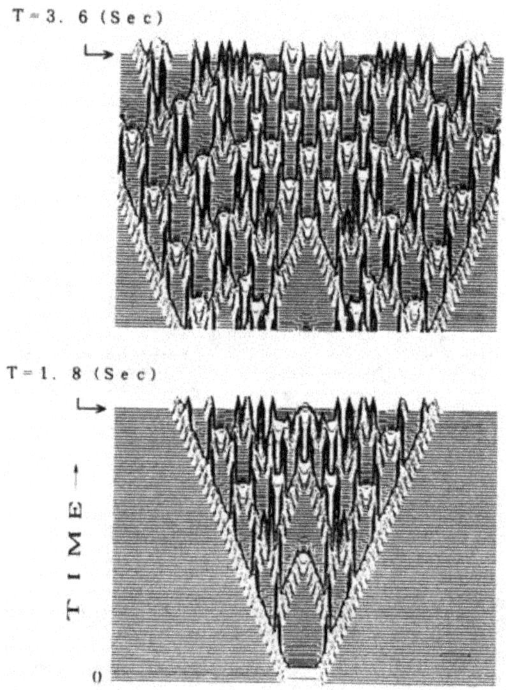

Fig.14. Examples of BUNKI (bifurcation) wave (BW) ($\beta = 0$ and $\gamma \neq 0$). TW spontaneously bifurcate and annihilate with collision (TW to RAND).

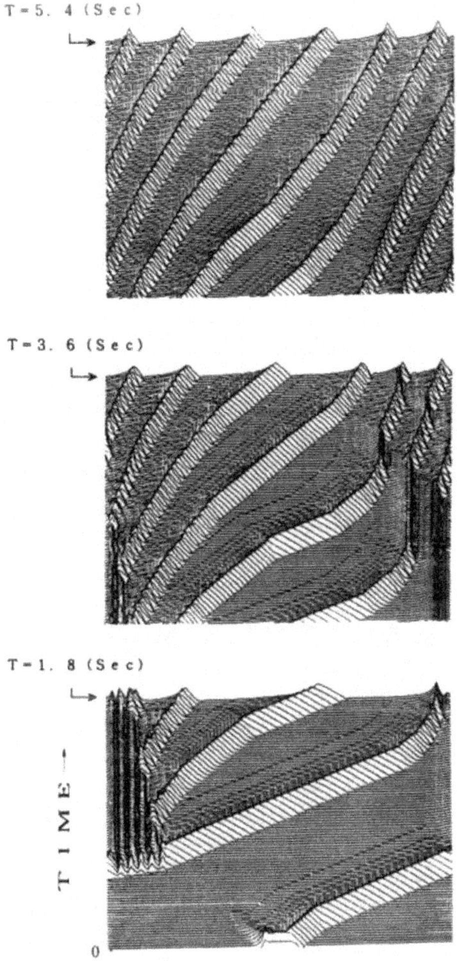

Fig.15. Asymmetric motion (ATW: travel to one direction) to standing wave ($\beta \neq 0$ and $\gamma = 0$)(SW-A) No collision occurs.

indicating contribution to the phenomenon. The surface tension Π (dyn/cm) due to the concentration difference of a surfactant TSAC is given by equation,

$$\Pi = 9.3/(\log C_A/2 + 1) + 28, \tag{5}$$

obtained experimentally [9,10]. By the capirally effect caused by this surface tension ,therefore, the height Δh (cm) of waves can be estimated as an order of centimeter from the relation,

$$\Delta h = (4\cos\theta/\Delta\rho g l) \times \Pi. \tag{6}$$

Here $\Delta\rho$, g and l are the density different (~ 0.3) between two fluids, the gravitational accelaration (980 cm/s^2) and the space width (1 to 5 mm) between container walls, respectively. θ is a contact angle (assumed to be ~ 0 when the surfactant exists.). The simulation has been done by use of difference equation form of Eq.(4). The space x is divided into 100 and each one is dominated by Eq.(4). Therefore we have 100 linearly-coupled oscillators in space in one dimension.

The initial conditions are set as follows. (1)The ion of RAC is uniformly distributed in space. (2)The initial perturbation is given only at one oscillator. This corresponds to triggering. (3)The boundary condition is periodic. (4)$\Delta t = 5 \times 10^{-5}$. (5)The other parameters are used from actual experimental values:$C_0 = 1 \times 10^{-3}$ M, $C_B = 8 \times 10^{-3}$ M, $k_1 = 1$ (Ms)$^{-1}$, $k_2 = 5000$ (Ms)$^{-1}$, D'_A=1s^{-1}, D_A=0.05 cm^2/s and l=10 cm. The calculation is done by a personal computer (NEC PC-9801VX).

4.3.3 Simulation

Figure 13 shows the typical wave profile and the temporal sequencies of a wave motion obtained from the simulation. Here all temporal interval for display are chosen at 0.02 s. The initial perturbation is given at the center.

In Fig.13a, two waves which behave symmetrically are formed, travel to the opposite direction each other (noting to be a periodic boundary condition) and annihilate by collision. It shows typical property of dissipative waves. After collision no wave forms any more. This property is basically due to the coefficient γ of the second order drift term. We call therefore this wave the trigger wave (TRW). In Fig.13b, two waves nucleated by an initial trigering travel to the opposite direction each other and without annihilation after their collision they pass through. This is repeated forever as long as chemicals are left. This behavior is well-known as one of a typical soliton, which is rather due to β of the first order term for drift. We call this the passing-through wave (PTW). One can note that the wave profile obtained from simulation is quite similar to the experimentally observed one seen in Figs 4,5 and 9. Since parameter regions for various motions are given in Table 1, we will not give detailed parameter values in each figure caption.

Figure 14 shows that TRW spontaneously bifurcates in a cascade way and annihilates by collision which make more complex state at $\beta = 0$. This is called the random wave (RAND). The larger the γ, in this case, the faster the traveling velocity of waves. When γ=0 and $\beta \neq 0$, a motion is not symetric and one wave is created by an initial trigger which travels to one direction as shown in Fig.15 (A(O)TW). β indicates the traveling velocity.

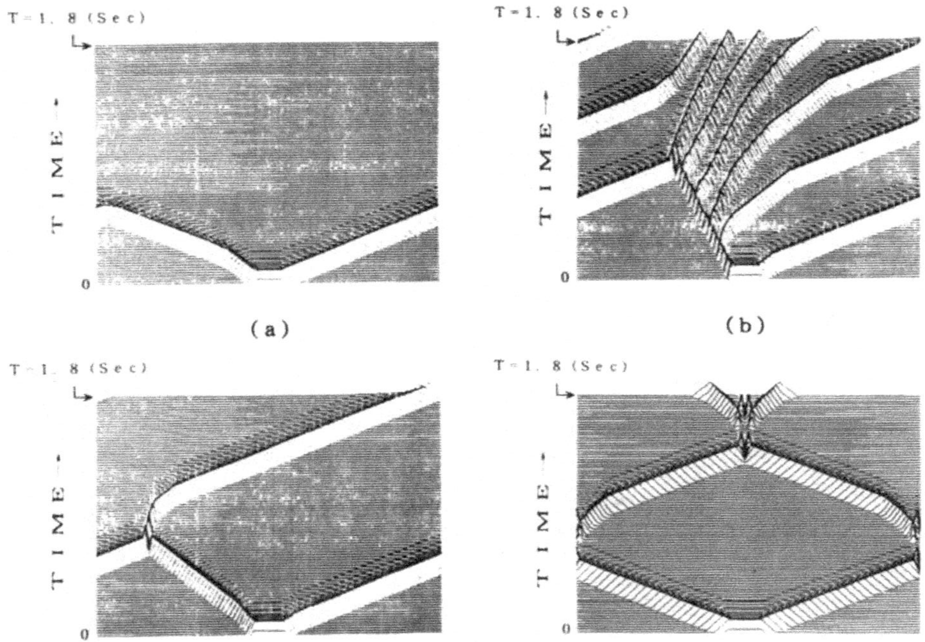

Fig.16. Asymmetric motion and collision ($\beta, \gamma \neq 0$). (a) After collision, the velocity is changed. (b) Small additional waves can be seen.

Increasing β, the waves self-bifurcate but never collide unlike a previous case. They change the velocities and become perfect standing waves from traveling waves. This is called the bifurcated wave (BW) and the standing wave (SW), and expresses by "BW and SW" in Table 1.

When both β and γ are not zero, various non-symmetric behaviour can be observed. A typical aspect is shown in Fig.16. One notes here that due to non-symmetric property the waves do not disapper by collision. Instead imediate after collision the velocity is changed. However later the velocity recovers to the initial value. When increasing β, waves often bifurcate and tend to be SW. We indicate this "SW-A" (asymmetric wave) in Table 1. Further we could find a region where waves are intermittently formed (○ in Table 1). For very large β and γ our simulation could not be possible because of too fast travels of waves.

In actual experiments, SW has never observed. On the other hand, while we actually observe for example both a cusp-like and triangle wave forms as shown in Fig.9, no such profiles are available in our simulation (see Table 2).

5 CONCLUSIONS

The investigated temporal and spatial structures experimentally forming at an oil-water interface are driven by a chemical reaction coupled with transport processes and hydrodynamic convection. The macroscopic motions are highly nonlinear and occur under far from equilibrium conditions. There are mainly two mechanisms involved: one is convection due to the Marangoni instability, that is hydrodynamic transport arises from gradients in interfacial tension and induces turbulent behavior, the other is the capillary effect, meaning that interfacial tension differences lead to vertical displacements of the interface. Furthermore, changes of the contact angle have to be considered. Which of these plays a more important role depends on the conditions of the experiment. The first mechanism probably dominates for relatively high, the second for relatively low solute concentrations.

It may be necessary to include an additional mode of deformation, namely the elastic one, *e.g.* for thoroughly understanding periodic isotropic stretching motion which resembles the typical bell mode with the wavenumber $k = 0$. Because of the phenomenological similarities between our reactive system and the bell mode instability it will be useful to consider such an elastic hydrodynamic instability in order to obtain further ideas for the appropriate mechanism.

The computer simulation of the present model, using reaction diffusion equation containing drift terms and assumption of switching mechanism of reaction rate dependent on concentration, has well described the wave profile of relaxation-oscillation type and motions. Reasonable parameter values are used in calculation obtained from experimental facts ,and the resulted value of the height and the form of waves show the corresponging value and profile observed in experiments For instance, the wave form seen in Fig.13 well agrees with the experimentally obtained one in Figs.4 and 5. However it cannot describe all experimental facts. The results are summarized in Table 2. The basic feature of the phenomenon can be explained

Table 1. Classification of waves in α - β - γ paramctor regions, (a) $\beta = 0$, (b) $\gamma = 0$, (c) $\alpha = 0.01$. A(O)TW:asymmetrically (one direction) traveling wave ATRW:asmmetrically taveling triger wave TRW: triger wave SW: standing wave BW:bifurcation wave RAND:random wave N:no available to simulate × : no wave

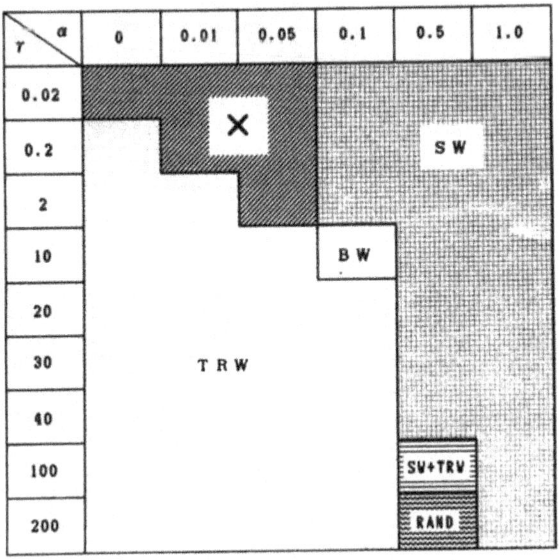

(a)

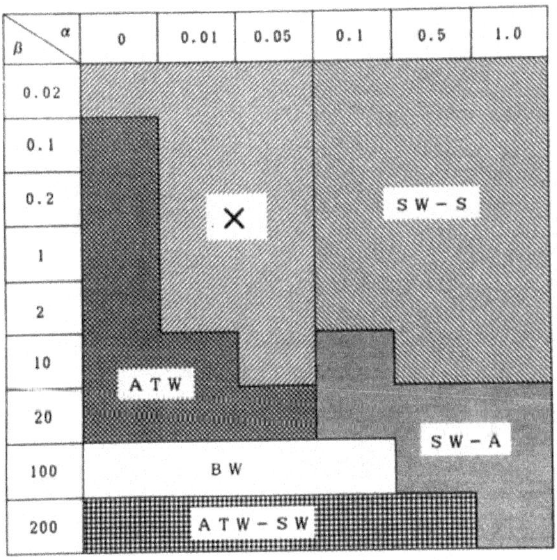

(b)

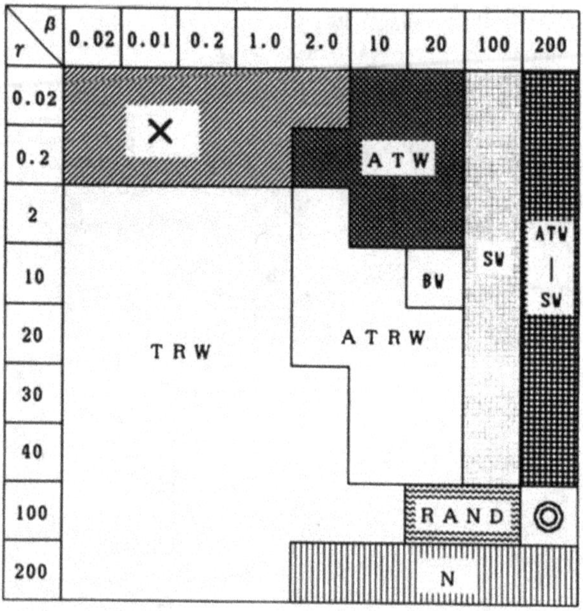

(c)

Table 2. Comparison with simulation and experimental results. ○ :observed × : no observation

Wave Motion	Simulation	Experiment
ATW	○	○
ATRW	○	○
TRW	○	○
RAND	○	○
PTW	○	×
SW	○	×

by the present equation for an interface boundary layer.

More complete equation may be desired. Especially analytical introduction of drift terms from basic equations of this phenomenon will be needed, which we have automatically introduced in our equation regarding as the contribution from hydrodynamic effects. Most will be future subjects.

This work is in part suported by the Grant-in-Aid for Scientific Research (No.02452047) from the Ministry of Education, Science and Culture of Japan.

References

[1] S. Chandrasekhar, *Hydrodynamic and Hydromagnetic Stability* (Clarendon Press, Oxford, 1961).

[2] H. L. Swinney and J. P. Gollub (Eds.), *Hydrodynamic Instabilities and the Transition to Turbulence*, 2nd edition (Springer-Verlag, Berlin Heidelberg, 1986).

[3] R. J. Field and M. Burger (Eds.), *Oscillations and Traveling Waves in Chemical Systems* (Wiley Interscience, New York, 1985).

[4] J. Ross, S. C. Muller and C. Vidal, *Chemical waves*, Science 240 (1988) 460-465.

[5] P. Glansdorff and I. Prigogine, *Thermodynamic Theory of Structure, Stability and Fluctuations* (Wiley Interscience, New York, 1971).

[6] M. Dupeyrat and J. Michel, *Instability observed at a liquid-liquid interface containing surface-active compounds*, J. Exp. Suppl. 18 (1971) 269-273.

[7] E. Nakache, M. Dupeyrat and M. Vignes-Adler, *Experimental and theoretical study of an interfacial instability at some oil-water interfaces involving a surface-active agent*, J. Colloid Interf. Sci. 94 (1983) 187 ; *The contribution of chemistry to new Marangoni mass-transfer instabilities at the oil/water interface*, Faraday Discuss. Chem. Soc. 77 (1984) 189-196.

[8] K. Yoshikawa, S. Maeda and H. Kawakami, *Various oscillatory regimes and bifurcations in a dynamical chemical system at a interface*, Ferroelectrics 77 (1988) 281-98.

[9] S. Kai and S.C. Muller,*Spatial and temporal macroscopic structures in chemical reaction systems*, Sci. Form 1 (1985) 9-39.

[10] S. Kai, E. Ooishi and M. Imasaki, *Experimental study of nonlinear waves on interface between two liquid phases with chemical reaction*, J. Phys. Soc. Jpn. 54 (1985) 1274-81.

[11] E. Sackmann, in *"Temporal Order"*, eds. L. Rensing and N. I. Jaeger (Springer-Verlag, Berlin, 1984) 153.

[12] C. V. Sternling and L. E. Scriven, *Interfacial turbulence: Hydrodynamic instability at some oil-water interface and the Marangoni effects*, A. I. Ch. E. J. 5 (1959) 514-523.

[13] A. C. Scott, *Sine-Gordon breather dynamics*, Phys. Scripta 20 (1979) 509-513.

[14] T. Fukuda and M.Kotani, *A numerical study of nonlinear diffusion equation governing surface deformation in the Marangoni convection*, J. Phys. Soc. Jpn. 55 (1986) 3857-3862.

[15] R. Takaki and K. Sannomiya, *Self-excited wave on an interface in a chemical reaction*, Forma 4 (1989) 35-47.

[16] L. E. Scriven and C. V. Sternling, *The Marangoni Effects*, Nature 187, (1960) 185-188.

[17] A. Sanfeld and A. Steinchen, *Motion induced by surface-chemical and electrochemical kinetics*, Farad. Discuss. Chem. Soc. 77 (1984) 1-11.

[18] G. H. Findenegg and R. Loring, *Fluid absorption up to the critical point. Experimental study of a wetting fluid/solid interface*, J.Chem. Phys. 81 (1984) 3270-6.

[19] J. J. Mahony and R. Smith, *On a model representation for certain spatial resonance phenomena*, J. Fluid Mech. 53 (1972) 193-207.

[20] I. Huntley, *Spatial resonance of a liquid-filled vibrating behavior*, J. Fluid Mech. 80 (1977) 81-97.

[21] S. Kai, T.Mori and M.Miki, in *"Research of Pattern Formation" ed. by R.Takagi*, KTK Scientific Publishers, Tokyo, 1991.

[22] S.Kai, S.C. Muller, T. Mori and M. Miki *Chemically driven dynamic patterns at an oil-interface*, Physica D 50(1991) 412-428.

ON MATHEMATICAL MODELS OF REACTION-DIFFUSION PROCESSES ON CATALYTIC SURFACES

Makoto KATORI
Department of Physics, Faculty of Science, University of Tokyo, Bunkyo-ku, Tokyo 113, Japan

and

Norio KONNO
Department of General Education, Muroran Institute of Technology, Mizumoto-cho, Muroran, Hokkaido 050, Japan

Stochastic Ising models for phase transitions in magnetic systems satisfy the condition of detailed balance and thus they have the Gibbs states as stationary states. On the other hand, some spin systems which do not satisfy the condition of detailed balance can have nontrivial stationary states as a result of nonequilibrium phase transitions. In the present paper, the reaction-diffusion processes of Dickman are studied as typical examples of this class.

1. Introduction

Recently Dickman[1,2] proposed some reaction-diffusion processes as mathematical models for chemical reactions on catalytic surfaces. His models consist of the following three processes.

(i) Adsorption of atoms on the surface due to attractive interactions is represented by creation of particles on the surface by contact processes.

(ii) Neighboring particles on the surface react to make a compound molecule, which desorbs with a constant rate leaving empty sites on the surfaces. This process is represented by multi-particle annihilation.

(iii) Jumping processes of particles are also introduced to represent diffusion due to thermal fluctuations.

Dickman gave some solutions of mean-field-type approximations and performed computer simulations to show the existence of phase transitions from an extinction state (devoid of particles) to an active steady state (with particles surviving even in the limit $t \to \infty$).[1,2] The extinction state may correspond to a poisoning state of catalysts.[3] It is important to determine phase diagrams for testing the validity of these models to describe the poisoning of catalysts. Especially competition between diffusion and multi-particle annihilation should be studied.

From the view point of statistical physics and study of interacting particle systems, Dickman's models shall attract much attention, because they are related to important interacting particle systems which have been well studied by mathematicians.[4–6] They can be viewed as the combinations of the exclusion processes and some modified versions of the contact process (CP) of Harris.[7] The active (survival) states of his models may be related to the nontrivial upper invariant measure of the CP, which does not satisfy the condition of the detailed balance. The multi-particle annihilation makes the processes fail to have a certain monotonicity property known as *attractiveness.* However the pair annihilation model of Dickman is closely related to the branching annihilating random walk (BARW) studied by Bramson and Gray.[8]

2. Models

The reaction-diffusion models of Dickman are continuous time Markov processes whose state space is $X = \{0,1\}^{\mathbf{Z}^d}$: at each site $x \in \mathbf{Z}^d$ a variable $\eta_t(x)$ takes 0 and 1 representing a vacant and a particle on site, respectively. The formal generator of the processes Ω consists of three parts:

$$\Omega = \Omega_{cre} + \Omega_{an} + \Omega_{dif}. \tag{2.1}$$

Let $C(X)$ be a set of the continuous functions on X and Y be a collection of all finite subsets of $\mathbf{Z}^d$. We define $\eta^A \in X$ for $A \in Y$ as

$$\eta^A(u) = \begin{cases} 1 - \eta(u), & \text{if } u \in A, \\ \eta(u), & \text{otherwise.} \end{cases}$$

The first term of (2.1) Ω_{cre} denotes the creation process with a rate λ by a nearest-neighbor interaction (the contact process); for $f(\eta) \in C(X)$,

$$\Omega_{cre} f(\eta) = \lambda \sum_{x \in \mathbf{Z}^d} \sum_{y:|y-x|=1} (1 - \eta(x))\eta(y)[f(\eta^{\{x\}}) - f(\eta)]. \tag{2.2}$$

The diffusion term Ω_{dif} is a exclusion process with jumping to neighboring sites with a rate D; for $f(\eta) \in C(X)$,

$$\Omega_{dif} f(\eta) = D \sum_{x \in \mathbf{Z}^d} \sum_{y:|y-x|=1} \eta(x)(1 - \eta(y))[f(\eta^{\{x,y\}}) - f(\eta)]. \tag{2.3}$$

For each model the annihilation term Ω_{an} is given as following for $f(\eta) \in C(X)$.

Single annihilation model (SAM);

$$\Omega_{an} f(\eta) = \sum_{x \in \mathbf{Z}^d} \eta(x)[f(\eta^{\{x\}}) - f(\eta)]. \tag{2.4a}$$

Pair annihilation model (PAM);

$$\Omega_{an} f(\eta) = \sum_{x \in \mathbf{Z}^d} \sum_{i=1}^{d} \eta(x)\eta(x + e_i)[f(\eta^{\{x,x+e_i\}}) - f(\eta)]. \tag{2.4b}$$

Here e_i's are the unit vectors of $\mathbf{Z}^d$.

There is a unique Markov process η_t on X corresponding to the Markov semigroup $S(t)$ generated by Ω.[4]

3. On the Extinction Region of the SAM

The SAM consists of the CP and the exclusion process. Because both of the processes are self-dual with their coalescing dual processes,[4] so is the SAM as shown below. Let $A_t = \{x : \eta_t(x) = 1\}$, then A_t is a Markov process on Y and is a coalescing dual of η_t. It is easy to find the following duality for any $\eta \in X$ and $A \in Y$,

$$P^{\eta}(\eta_t(x) = 0, ^{\forall} x \in A) = P^{A}(\eta(x) = 0, ^{\forall} x \in A_t). \tag{3.1}$$

Consider the case that the initial state is δ_1 (i.e. $\eta \equiv 1$) and take the limit $t \to \infty$ in (3.1). Then we obtain,

$$1 - E_{\bar{\mu}}[\prod_{x \in A}(1 - \eta(x))] = \sigma(A). \tag{3.2}$$

Here $\sigma(A)$ is a survival probability of the process with the initial state $\{x : \eta_0(x) = 1\} = A$;

$$\sigma(A) = \lim_{t \to \infty} P^A[A_t \neq \emptyset]. \tag{3.3}$$

By using the survival probability, the extinction of process is characterized as,

$$\text{the process is extinct} \iff \sigma(A) = 0, \quad ^{\forall} A \in Y. \tag{3.4}$$

We define $H(\eta, A) \equiv \prod_{x \in A}(1 - \eta(x))$. Because $H(\eta, A) + H(\eta, B) \leq H(\eta, A \cup B) + H(\eta, A \cap B)$ for any $A, B \in Y$, we obtain the following inequality (*submodularity*[4]) by using the relation (3.2),

$$\sigma(A \cup B) + \sigma(A \cap B) \leq \sigma(A) + \sigma(B). \tag{3.5}$$

By this inequality, the extinction of the process (3.4) will follow once we have shown that $\sigma(\{x\}) = 0$ for any singleton $\{x\}$. That is, it is sufficient to show that $E_{\bar{\mu}}[\eta(x)] = 0$ to conclude the extinction of the SAM.

Now we give a theorem on the extinction of process for the SAM.

Theorem 3.1. In the parameter region $D < \{1-(2d-1)\lambda\}/\{(2d-1)(2d\lambda-1)\}$, the process of the SAM becomes extinct.

This theorem is proved[9] by using the method which has been used to show ergodicity for the basic CP in refs. 10 and 11. For the basic CP, see also ref.12.

4. PAM in Low Dimensions

In the PAM, the simple duality as (3.1) does not hold. Then we have to analyze the original process η_t to know the conditions of the survival and extinction of the process. It should be noticed that the present process is not *attractive*. Here an attractive spin system is one in which the addition of an extra occupied site to the state of the process does not decrease the rate at which any of the vacant site become occupied, nor does it increase the rate at which the other occupied site are vacated. In the PAM, the addition of a new particle at a vacant site neighboring particles may reduce rather than increase the total number of particle.

Bramson and Gray[8] introduced a Markov process which they called the branching annihilation random walk (BARW). In this process, a particle gives birth to a new particle at one of the neighboring sites with a rate $1/(2d)$ (*branching*) and a particle jumps to one of the neighboring sites with a rate $\rho/(2d)$ (*random walk*). And if a particle lands on a site which is already occupied, either by jumping there or as the result of branching, then both particles disappear (are *annihilated*). This process is closely related to the PAM and we find the argument which was given for the BARW is also applicable to the present problem. In order to clarify the relationship between these two processes, we introduced in ref.13 a generalized process called *Restricted Branching Random Walk* (RBRW) which includes both of them as special cases.

Definition 4.1. A restricted branching random walk with parameters $(b_1, r_1; b_2, r_2)$ (RBRW$(b_1, r_1; b_2, r_2)$) is a Markov process generated by the following generator; for $f(\eta) \in C(X)$,

$$\begin{aligned}
&\Omega_{b_1,r_1;b_2,r_2} f(\eta) \\
&= \sum_{x\in \mathbf{Z}^d} \sum_{y:|y-x|=1} \eta(x)\{b_1(1-\eta(y)) + b_2\eta(y)\}[f(\eta^{\{y\}}) - f(\eta)] \\
&+ \sum_{x\in \mathbf{Z}^d} \sum_{y:|y-x|=1} \eta(x)\{r_1(1-\eta(y)) + r_2\eta(y)\}[f(\eta^{\{x,y\}}) - f(\eta)].
\end{aligned} \tag{4.1}$$

Remark 4.1. It is easy to find that the BARW of Bramson and Gray is the RBRW$(1/(2d), \rho/(2d); 1/(2d), \rho/(2d))$ and that we obtain the PAM by letting $b_1 = \lambda, r_1 = D, b_2 = 0$ and $r_2 = 1/2$ in the RBRW.

Our results are following.[13]

Theorem 4.1. Consider the one-dimensional RBRW$(b_1, r_1; b_2, r_2)$. If the ratios r_1/b_1 and r_2/b_1 are both sufficiently small, then for all non-empty initial states,

$$P(\eta_t \neq \emptyset \quad \text{for all } t \geq 0) > 0, \tag{4.2}$$

i.e. the process survives with a positive probability.

Theorem 4.2. Consider the RBRW$(b_1, r_1; b_2, r_2)$ and suppose that the dimensionality $d \leq 2$. If the ratio r_1/b_1 is sufficiently large, then for any initial state η_0

with finite particles,

$$P(\eta_t = \emptyset \quad \text{for some } t \geq 0) = 1, \tag{4.3}$$

i.e. the process becomes extinct with probability one.

Noticing Remark 4.1 we obtain the following corollary from the above two theorems.

Corollary 4.1. Consider the PAM of Dickman. Suppose the rate of the diffusion (exclusion process) is positive and finite; $0 < D < \infty$.

i) For sufficiently large λ and all non-empty initial states, the process survives with a positive probability when d=1.

ii) For sufficiently small $\lambda > 0$ and any initial state with finite particles, the process becomes extinct with probability one when $d \leq 2$.

It should be remarked that Corollary 4.1 implies that there should be at least one transition from the extinction state to the survival state when we increase the rate λ for any finite value of D in the one-dimensional case. As mentioned in refs.1 and 9, the pair level mean-field approximation suggests for the PAM that the value of D at the phase boundary between the extinction state and the survival state increases monotonically as λ decreases and terminates at a finite value $D^* = 1/\{2(2d-1)\}$ for $\lambda = 0$ and that the process can survive for any $\lambda > 0$ if $D > D^*$. The above corollary implies that in the low dimensional case ($d \leq 2$), the true phase boundary is qualitatively different from this mean-field prediction. That is, we will find a finite region of λ where the process becomes extinct for any finite D. In other words, $D^* = \infty$ for $d \leq 2$. We expect that D^* becomes finite for high dimensions $d \geq d_0$ with some d_0 (which may be 3).

References

1. R.Dickman, *Phys.Rev.* **B40** (1989) 7005.
2. R.Dickman, *Phys.Rev.* **A42** (1990) 6985.
3. E.R.Grannan and G.Swindle, *J.Stat.Phys.* **61** (1990) 1085.
4. T.M.Liggett, *Interacting Particle Systems*, (Springer-Verlag, New York, 1985).
5. D.Griffeath, *Additive and Cancellative Interacting Particle Systems*, *Springer Lecture Notes in Mathematics*, Vol.724 (Springer-Verlag, New York, 1979).
6. R.Durrett, *Lecture Notes on Particle Systems and Percolation*, (Wadsworth, Pacific Grove, California, 1988).
7. T.E.Harris, *Ann.Probab.***2** (1974) 969.
8. M.Bramson and L.Gray, *Z. Wahrsch.verw.Gebiete* **68** (1985) 447.
9. M.Katori and N.Konno, submitted to *Physica* **A**
10. D.Griffeath, *Adv.Appl.Prob.***7** (1975) 179.
11. M.Katori and N.Konno, *J.Phys.Soc.Jpn* **59** (1990) 877.
12. M.Katori and N.Konno, *J.Stat.Phys.* **63** (1991) 115.
13. M.Katori and N.Konno, submitted to *J.Stat.Phys.*

RIGID ROTATION AND CORE MEANDERING OF SPIRAL WAVES IN REACTION-DIFFUSION SYSTEMS

Shinji Koga

Osaka Kyoiku University, Osaka 543, Japan

Spiral wave pattern formations including 3D scroll waves in reaction-diffusion systems can be maintained in chemical and biological media [1]-[2]. This wave pattern together with the interaction of the spirals [3]-[5] have been focused much attention recently. In the media in the absense of any obstacles, it is possible to classify the spiral waves in 2D media into three proto-types;(1) rigid rotations (2) core meanderings and (3) spiral wave turbulence, when we consider the equation

$$\frac{\partial}{\partial t}W = a + W - (1 + ic_2)|W|^2W + \Delta W, \tag{1}$$

where W is the complex field, c_2 is the real parameter, and a is introduced to distort the cicle limit cycle $|W| = 1$. The parameter a is not very significant physically, that is, the above three types can be realized even in the case a=0. However, this additional term is here introduced to reflect the relaxation-oscillation character slightly, as long as a is small.

In this brief note, we first outline the theory for obtaining the generalized amplitude and the generalized phase function [6] corresponding to those of the $\lambda - \omega$ system, and then we mention the core meandering [7] which is the spiral wave turbulence [8] supressed by the boundary effect.

The rigidly rotating spiral wave rotates around the core as the pivotal point. In order to analyze the behavior at the core, we presented the theory based on the assumption

$$\mathbf{X}(r, \theta, t) = \mathbf{X}(r, \omega t + m\theta - S(r)). \tag{2}$$

The second argument is called the phase, and ω, m, r and θ are the frequency, the winding number, and the polar coordinates. By substituting Eq.(2) into the reaction-diffusion equation, we have derived the generalized amplitude and the generalized phase function. These two quantities can be calculated in concrete models by transforming the formulas further. The details of this transformation will be discussed elsewhere. As a result, we have obtained these quantities in Eq.(1) which are quite similar to those of the $\lambda - \omega$ system.

We next find the core meandering phenomenon in Eq.(1). This is primarily originated by the effect due to the boundary, together with the instability of the spiral. Namely the system size is small enough to supress the spiral wave turbulence. Here we adopt the following definition of the core moving in time;

$$\frac{\partial}{\partial t} W = 0. \tag{3}$$

When we choose the parameters as; $c_2 = 2$, the system size is 16, and $0 \leq a \leq 0.15$, the core moves around the medium almost parallel to the boundary.

Acknowledgements

We thank Dr. S. Sasa and Mr. T. Mizuguchi for helpfull discussions.

References

[1] A.T.Winfree, The Geometry of Biological Time (Springer-Verlag, New York, 1980)

[2] see for the complihensive studies

Waves and Patterns in Chemical and Biological Media (H.L. Swinney and V.I. Krinsky eds.) Physica D **49** (1991)

[3] S. Rica and E. Tirapegui, Physica D **48** (1991) 396

[4] P. Pelcé and J. Sun, Physica D **48** (1991) 353

[5] H. Sakaguchi, Prog. Theor. Phys. **82** (1989) 7

[6] S. Koga, Prog. Theor. Phys. **67** (1982) 164 and 454

[7] G.S. Skinner and H.L Swinney, Physca D **48** (1991) 1

[8] Y. Kuramoto and S. Koga, Prog. Theor. Phys. **66** (1981) 1081

LAYER DYNAMICS IN LATERAL INHIBITION MODELS

MASAYASU MIMURA
Department of Mathematics, Hiroshima University,
Higashi-Hiroshima 724, Japan

ABSTRACT

Despite their apparently simple structure, reaction-diffusion equations exhibit a surprising variety of spatio and temporal patterns. Typical examples are stationary and breathing pulses, traveling waves and chaos. Attempts of theoretical understanding of pattern formation in reaction-diffusion systems are made in two ways. One is numerical simulations of pattern evolution, and the other is via phase and interface-dynamics which allow approximate but reasonable formation of dynamical theory of patterns. Pattern formation in lateral inhibition model equations is discussed mainly in the framework of activator-inhibitor interaction which often arises in chemistry, developmental biology, population dynamics and in other fields.

1. Introduction

During the past decade, various interesting phenomena of patterns, waves and interfaces have been observed in natural sciences such as population dynamics[17] , morphogenesis[13] , chemical and biochemical reaction processes[23], combustion[4] , solidification[22]. Interfaces or layers are present as sharp gradient with narrow width between two qualitatively different states such as liquid and solid or two quantitatively different chemical or physical states. Typical examples are spiral waves in the Belousov-Zhabotinsky reactions, flame front in combustion and dendritic crystal growth in supercooling liquid. The dynamics of interfaces and layers can be often described by reaction-diffusion model equations with layer parameter, say ε, which is sufficiently small. The simplest but suggestive equation describing the evolution of internal layers is

$$u_t = \varepsilon \Delta u + \frac{1}{\varepsilon} f(u), \tag{1.1}$$

where Δ is the Laplace operator in $\mathbf{R}^N$ and $f(u)$ has a cubic-like nonlinearity. A typical example is $f(u) = u(1-u)(u-a)$ with the constant $a \in (0,1)$. Here $u(t,x)$ means the phase variable at time t and position x. It is obvious that two equilibria

$u \equiv 1$ and $u \equiv 0$ are both stable, while $u \equiv a$ between them is unstable. For this reason, (1.1) is called a bistable equation. If ε is sufficiently small, one could expect that the dynamics of the solution $u(t,x)$ of (1.1) consists of two stages : In the first stage, $u(t,x)$ tends to either 1 or 0, by which the region is separated into two different subregions and there occur internal layers with width $O(\varepsilon)$ between these subregions. In the second stage, these internal layers propagate. Recently, the study of the dynamics of layers generated in (1.1) has been well developed. One approach is to regard such layers as (diffusive) interfaces in the limit $\varepsilon \searrow 0$ and to consider the dynamics of these interfaces. Let $\Gamma(t)$ be an interface in $\mathbf{R}^N$, which is the boundary between two regions where u takes either 1 or 0. Then the equation for the interface $\Gamma(t)$ is approximately described by

$$\frac{d}{dt}\Gamma(t) = \Big[c - \varepsilon(N-1)\kappa\Big]\nu, \tag{1.2}$$

where ν and κ are respectively the unit normal vector and the mean curvature at Γ[14]. Here c is the velocity of the traveling front solution of the one dimensional version of (1.1)

$$u_t = u_{xx} + f(u) \quad t > 0, \quad x \in \mathbf{R}$$

connecting with two stable equilibria at $x = \pm\infty$. It is known that the velocity is uniquely determined and the corresponding traveling front solution is stable. (1.2) clearly shows that the dynamics of the interface is obeyed by the bistable kinetic effect and the geometrical one. Let us show an application of (1.2) to the understanding of chemical patterns . Foerster et al.[5] consider the collision of expanding circular chemical waves arising in the Belousov-Zhabotinsky (B-Z) reaction. They measured the temporal evolution of cusp like structures which appear after the collision of two circular waves by using a computerized spectrophotometric video technique . From a modelling view point, the spatio-temporal dynamics of the B-Z reaction can be described by the simplified Oregonator with diffusion :

$$\begin{cases} \dfrac{\partial u}{\partial t} - D\Delta u &= \dfrac{1}{\varepsilon}(u^2 - hv\dfrac{u-q}{u+q}) &\equiv \dfrac{1}{\varepsilon}f(u,v) \\ \dfrac{\partial v}{\partial t} - D\Delta v &= u - v &\equiv g(u,v) \end{cases} \tag{1.3}$$

with some constants h, q and ε which are experimentally given by $h = 3, q = 2 \times 10^{-4}$ and $\varepsilon = 10^{-2}$. The nulclines of f and g are drawn in Fig. 1 where there is only one constant equilibrium $(\overline{u}, \overline{v})$ which is asymptotically stable. Singular perturbation arguments show that v approximately takes the value of $\overline{v}$ in the vicinity of the front layer of waves so that u satisfies approximately the scalar equation

$$\frac{\partial u}{\partial t} = D\Delta u + \frac{1}{\varepsilon}f(u, \overline{v}), \tag{1.4}$$

so that $f(u, \overline{v})$ is exactly a cubic-like function of u. Thus, one finds that the interface equation (1.2) can be derived from (1.4) in the limit $\varepsilon \searrow 0$ when D is of the order $O(\varepsilon)$. Fig. 2 shows the numerical calculation of (1.2) which is in good agreement with the experimental observation. This result clearly shows that the interface equation (1.2) is a nice approximation to understand the dynamics of chemical waves of the B-Z reaction.

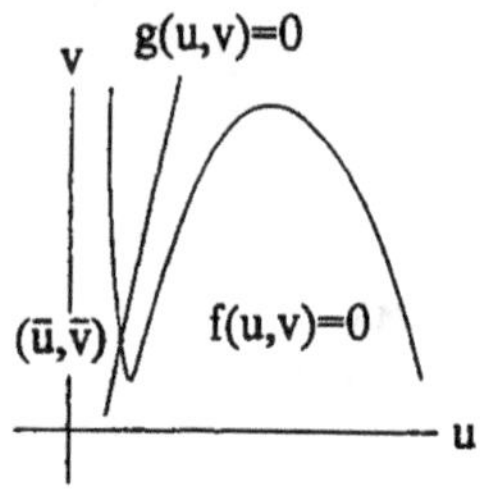

Fig.1 Nulclines of f and g in Oregonator

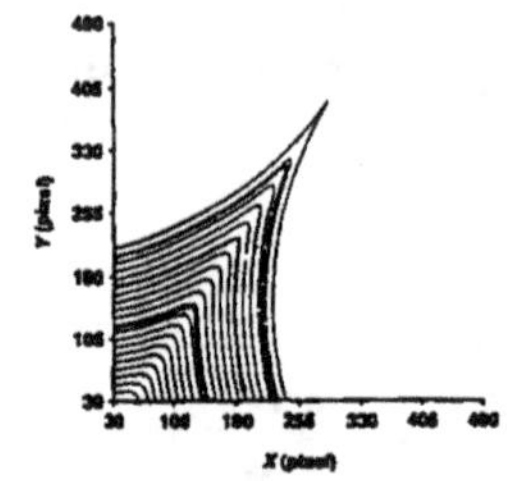
Fig.2 Superposition of the numerical contour and experimentally determined contour(dots) map

The approach of interfacial dynamics can be applied to more general systems of equations of the form

$$\frac{\partial u}{\partial t} = \varepsilon D\Delta u + \frac{1}{\varepsilon}F(u), \tag{1.5}$$

where $u = (u_1, u_2, \cdots, u_n)$, D is the diagonal matrix with elements $d_i > 0$. Here we assume that $F(u)$ is of bistable nonlinearity, that is, $F(u) = 0$ takes exactly three solutions $\underline{u}$, u_0 and $\overline{u}$ where $\underline{u}$ and $\overline{u}$ are both stable equilibria of (1.5).

Let us show one example which arises in population ecology.

$$\begin{cases} \dfrac{\partial u_1}{\partial t} &= d_1\Delta u_1 + a_1(1-\alpha u_1 - u_2)u_1 \\ \dfrac{\partial u_2}{\partial t} &= d_2\Delta u_2 + a_2(1-u_1 - \beta u_2)u_2 \end{cases} \tag{1.6}$$

where u_1 and u_2 are the population densities of two competing species, d_i is the diffusion rate and a_i is the intrinsic growth rate, and α, β are intraspecific competition rates of u_i $(i = 1, 2)$. Suppose that α and β are chosen to satisfy

$$\alpha < 1 < \frac{1}{\beta},$$

which means ecologically that the intraspecific competition is weaker than the inter-specific one. Simple calculations show that $(a_1/b_1, 0)$ and $(0, a_2/b_2)$ are both stable, while $\left(\dfrac{1-\beta}{1-\alpha\beta}, \dfrac{1-\alpha}{1-\alpha\beta}\right)$ is unstable, that is, the kinetics in (1.6) is of bistable non-linearity.

For (1.5), we assume that there is a stable 1-dimensional traveling front solution $u(z)$ $(z = (x-ct)/\varepsilon)$ with the velocity c, connecting with $\overline{u}$ and $\underline{u}$ at $z = \pm\infty$. That is, $u(z)$ satisfies

$$(1.7) \qquad \begin{cases} Du'' + cIu' + F(u) = 0, & -\infty < z < \infty \\ u(-\infty) = \underline{u}, \quad u(\infty) = \overline{u}, \end{cases}$$

where $' = \dfrac{d}{dz}$. Unfortunately, the existence and stability of the velocity c have not yet been proved except for special cases. For the specific system (1.6), for instance, the existence of traveling front solutions is shown[3], but the uniqueness of solutions have been still unsolved. However, numerical simulations confirm that (1.6) has the only one velocity and the corresponding traveling front solution is stable. This fact seems to be plausible but it does not valid in general. In fact, for suitable nonlinear functions $F(u)$, there possibly coexist stable and unstable traveling front solutions of (1.5) even if it is a two coupled system[9]. This is the reason why we have to assume the existance and stability of traveling front solutions for (1.7)

Let $\mathcal{L}$ be the linearized operator of (1.7) around the traveling front solution $u(z)$ with the velocity c,

$$\mathcal{L} = D\frac{d^2}{dz^2} + cI\frac{d}{dz} + F'(u(z)),$$

where F' is the Jacobian at $u = u(z)$, and let $\mathcal{L}^*$ be the adjoint operator of $\mathcal{L}$. We note that both $\mathcal{L}$ and $\mathcal{L}^*$ have the zero eigenvalue $\lambda_0 = 0$ since any spatial translation of $u(z)$ is also a solution of (1.6). Let ξ_0 and ξ_0^* be the eigenfunction of $\mathcal{L}$ and $\mathcal{L}^*$ associated with λ_0, respectively, such that $<\xi_0^*, \xi_0>= 1$ holds. With $\alpha =<\xi_0^*, D\xi_0>$, the interface equation of (1.5) in the limit $\varepsilon \searrow 0$ is described by

$$(1.8) \qquad \frac{d}{dt}\Gamma(t) = \Big[c - \varepsilon(N-1)\alpha\kappa\Big]\nu.$$

When $D = dI$, it is easy to see $\alpha = d$, so that (1.8) is essentially equivalent to (1.2)[18].

Recently, the qualitative study of (1.2) or (1.8) has been investigated by numerous authors, from theoretical and numerical view points. One global property of interfaces in the plane $(N = 2)$ is that if the initial interface $\Gamma(0)$ is given by a smooth, embedded closed curve, $\Gamma(t)$ expands $(c > 0)$, or shrinks to a point $(c \leq 0)$, becoming round[7,21]. Fig. 3 demonstrates the numerical curves of $\Gamma(t)$ $(N = 2)$ for different values of t . These simulations clearly show that even if the initial curve is of complicated shape, it becomes smooth and round . On the other hand, for higher dimensions $(N \geq 3)$, the situation drastically changes, that is, even if the initial curve is of so smooth and simple shape, singularities may develop in curves.

C > 0 (expanding) Fig.3 Dynamics of interfaces in the plane C < 0(shrinking)

2. Activator-inhibitor systems

In order to understand spatio-temporal patterns arising in the fields of chemistry[10], developmental biology[15] and population dynamics[17], various reaction-diffusion systems have been proposed, as several extensions of (1.5). Among them, we meet the following reaction-diffusion equations coupled with another single species v :

$$(2.1) \qquad \begin{cases} \tau \dfrac{\partial u}{\partial t} = \varepsilon D \Delta u + \dfrac{1}{\varepsilon} F(u, v), & t > 0, \ x \in \Omega, \\ \dfrac{\partial v}{\partial t} = d \Delta v + g(u, v), & t > 0, \ x \in \Omega, \end{cases}$$

where $u = (u_1, u_2, \cdots, u_n)$ and v is a scalar function with the diffusion rate d.

One of the features of (2.1) is that the diffusing species u reacts so fast compared with the other diffusing species v, since ε is sufficiently small. We assume that F and g have the following characteristics :

(A.1) $F(u, v) = 0$ has only one stable solution $u = \underline{u}(v)$ for large v and $u = \overline{u}(v)$ for small v in the sense of the first system of (2.1) for fixed v

$$(2.2) \qquad \frac{\partial u}{\partial t} = \varepsilon D \Delta u + \frac{1}{\varepsilon} F(u, v).$$

On the other hand $F(u, v) = 0$ has three solutions $u = \underline{u}(v), u_0(v), \overline{u}(v)$ for intermediate v and two of them, $\underline{u}(v), \overline{u}(v)$ are still stable while $u_0(v)$ is unstable in (2.2). In biological terms, we call the $\underline{u}(v)$-branch the rest state and the $\overline{u}(v)$-branch the active state in (u, v)-space. Then (A.1) indicates that the activation of u-species is inhibited by the presence of v-species. For this reason, u and v are respectively named activators and their inhibitors.

(A.2) $g(u, v)$ is increasing with u and decreasing with v, that is

$$\frac{\partial g}{\partial u_i} > 0 \quad \text{(production)} \quad (i = 1, 2, \cdots, n) \quad \text{and} \quad \frac{\partial g}{\partial v} > 0 \quad \text{(degradation)}.$$

Let us show some examples of F and g.

Example 1. ($n = 1$, Bonhoeffer–van der Pol model)

$$\begin{cases} F(u, v) &= u(1-u)(u-a) - v \\ g(u, v) &= u - \gamma v + I \end{cases} \tag{2.3}$$

with positive constants a, γ and I. It is obvious to see that (2.3) satisfies (A.1) and (A.2) when $0 < a < 1$. The nonlinearity in (2.3) is essentially same as the one in (1.5).

Example 2. ($n = 2$, 2 prey–1 predator population model)

$$\begin{aligned} F(u_1, u_2, v) &= \begin{pmatrix} a_1(1 - \alpha u_1 - u_2 - kv)u_1 \\ a_2(1 - u_1 - \beta u_2 - v)u_2 \end{pmatrix} \\ g(u_1, u_2, v) &= (-r + k\gamma u_1 + u_2)v, \end{aligned}$$

where u_1, u_2 and v are respectively the population densities of two competing species and their predator. a_1, a_2, α and β are positive constants used in (1.6). k is the ratio of predation rates, γ is the transformation rate of predation and r is the death rate of v. This population model is often used to study the coexistance problem of two competing species mediated by a predator[8]. We note that $F(u_1, u_2, 0)$ is exactly the same form as the one in (1.6). We now assume that α, β and k satisfy $\alpha < 1 < \beta$ and $\alpha\beta < 1$ and $k > 1$, under which the lines of $(1 - kv) - \alpha u_1 - u_2 = 0$ and $(1 - v) - u_1 - \beta u_2 = 0$ are classified into three cases as in Fig. 4. For fixed v, the analysis of

$$\begin{cases} \dfrac{\partial u_1}{\partial t} &= d_1 \Delta u_1 + a_1(1 - \alpha u_1 - u_2 - kv)u_1 \\ \dfrac{\partial u_2}{\partial t} &= d_2 \Delta u_2 + a_2(1 - u_1 - \beta u_2 - v)u_2 \end{cases}$$

shows that for $\alpha < \dfrac{1}{\beta} < \dfrac{1-kv}{1-v}$ (small v), $(u_1, u_2) = \left(\dfrac{1}{\alpha}, 0\right)$ stable, for $\alpha < \dfrac{1-kv}{1-v} < \dfrac{1}{\beta}$ (intermediate v), $\left(\dfrac{1}{\alpha}, 0\right)$ and $\left(0, \dfrac{1}{\beta}\right)$ are both locally stable and $\dfrac{1-kv}{1-v} < \alpha < \dfrac{1}{\beta}$

(large v), $\left(0, \dfrac{1}{\beta}\right)$ is stable. That is, it turns out that the nonlinearity of $F(u_1, u_2, v)$ is of cubic-like in (u_1, u_2, v) space.

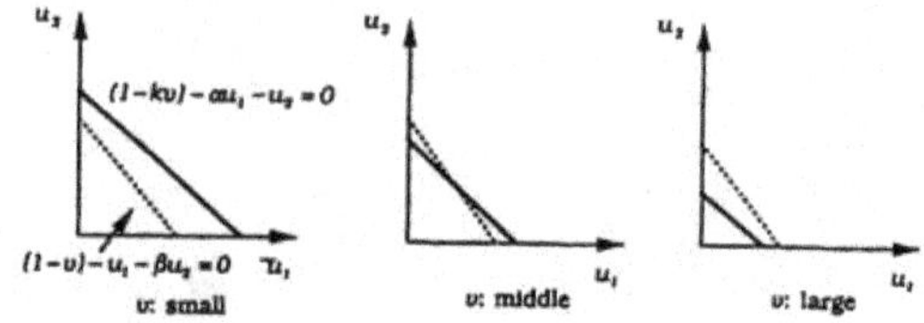

Fig.4 Lines of $(1\text{-}kv)\text{-}\alpha u_1\text{-}u_2=0$ and $(1\text{-}v)u_1\text{-}\beta u_2=0$

Recently it has been reported that systems (2.1) satisfying (A.1) and (A.2) exhibit a surprising variety of spatio-temporal patterns. So far almost all studies of (2.1) are concentrated on the case when d is very small or at most of the order ε. The typical examples are the Fitzhugh–Nagumo equations describing an exciting pulse propagating along the nerve axon ($d = 0$) and the Oregonator (1.3) describing the B-Z reaction. In this case, (2.1) exhibits a traveling pulse in 1 dimension and circular and spiral waves in higher dimensions in the situation where the kinetics of (F, g) is of monostable. On the other hand, the opposite case where $\varepsilon << d$, that is, the inhibitor v diffuses so fast compared with the activator u has been considered. Because of big difference between the diffusion rates, the behavior of the solution drastically changes. Suppose that the activator is initially distributed in some finite area. The inhibitor is, then, produced by the reaction but diffuses so fast that it flows out of the area. Because of the efffect of lateral inhibition, the localized pattern may possibly neither shrink or expand. This suggests us the localization of the activator . Fig. 5 shows two pulses propagating to the opposite two directions when d is small, while Fig. 6 shows one localized pulse when d is large compared with ε. What we would like to be considered with in this paper is the formation of localized patterns due to lateral inhibition. By the above intuitive explanation and numerical simulations, one could think that such localized patterns with internal layers would be stable. However, this is not necessarily true. In fact, for small τ, the localized pulse undergoes sustained oscillation like a breathing motion as in Fig. 7[12,16].

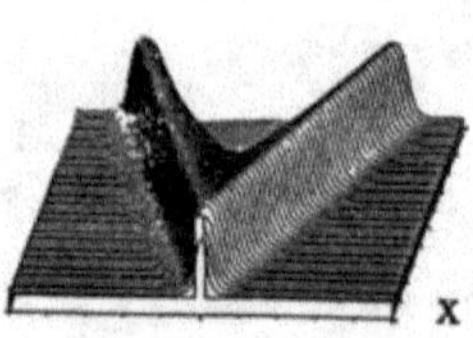

Fig.5 Propagating pulses

Fig.6 Localized pulse

Fig.7 Oscillating pulse

In order to study the dynamics of layers arising in (2.1), we derive the interface equation from (2.1). Using the discussion on the scalar equation (1.1), we know that the region is separated into two subregions, as ε tends to zero : one is $\overline{\Omega}(t)$ where (u,v) satisfies $u=\overline{H}(v)$ and the other is $\underline{\Omega}(t)$ where (u,v) satisfies $u=\underline{H}(v)$. Here $u=\overline{H}(v)$ and $\underline{H}(v)$ define respectively two branches of u as functions of v as in Fig. 8. Let $\Gamma(t)$ be the interface which divides these two subregions. Then v satisfies

$$\frac{\partial v}{\partial t}=\begin{cases} d\Delta v+g(\overline{H}(v),v) & \text{in } \overline{\Omega}(t)\\ d\Delta v+g(\underline{H}(v),v) & \text{in } \underline{\Omega}(t), \end{cases} \tag{2.4}$$

with $v\in C^1(\mathbf{R}^N)$ as the continuity condition [2]. Here we note that the value of v on the interface $\Gamma(t)$, say v_I, is not known *a priori*. By (1.8), the interface equation is given by

$$\tau\frac{d\Gamma}{dt}=\Big[c(v_I)-\varepsilon(N-1)\alpha(v_I)\kappa\Big]\nu, \tag{2.5}$$

where $c(v_I)$ is the velocity of the 1-dimensional traveling front solution of the system for u

$$\frac{\partial u}{\partial t} = Du_{xx}+F(u,v_I) \quad t>0,\ x\in\mathbf{R},$$

with the boundary conditions

$$u(t,-\infty)=\overline{H}(v_I) \quad t>0$$

$$u(t,+\infty)=\underline{H}(v_I) \quad t>0,$$

and $\alpha(v_I)$ is defined in a similar way to α in (1.8). For the models in Examples 1 and 2, it is numerically known that $c(v)$ is a monotone decreasing function of v and that there is v_* such that with $c(v_*)=0$. We thus found that the dynamics of interfaces is described by (2.4), (2.5).

$$\begin{cases} \tau\dfrac{d\Gamma}{dt} = \Big[c(v_I)-\varepsilon(N-1)\alpha(v_I)\kappa\Big]\nu \quad t>0\\ \dfrac{\partial v}{\partial t} = \begin{cases} d\Delta v+g(\overline{H}(v),v) & \text{in } \overline{\Omega}(t)\\ d\Delta v+g(\underline{H}(v),v) & \text{in } \underline{\Omega}(t), \end{cases} \end{cases} \tag{2.6}$$

The essentially different point of (2.5) from (1.8) is that c as well as ν are not constants but functions of v_I, that is, the interface $\Gamma(t)$ is truely coupled with the inhibitor v . Fig. 9 numerically demonstrates the dynamics of interfaces of (2.6) with the dynamics given in Example 1. It thus turns out that even if $\Gamma(0)$ is smooth in

$\mathbb{R}^2$, $\Gamma(t)$ exhibits complicated patterns[27]. This fact is totally different from the one in the scalar version (1.1) which does not interact with the inhibitor.

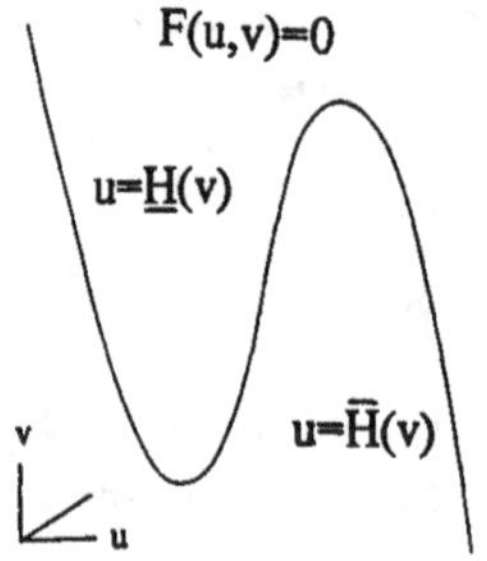

Fig.8 Stable branches $u = \bar{H}(v), u = \underline{H}(v)$ of $F(u,v) = 0$

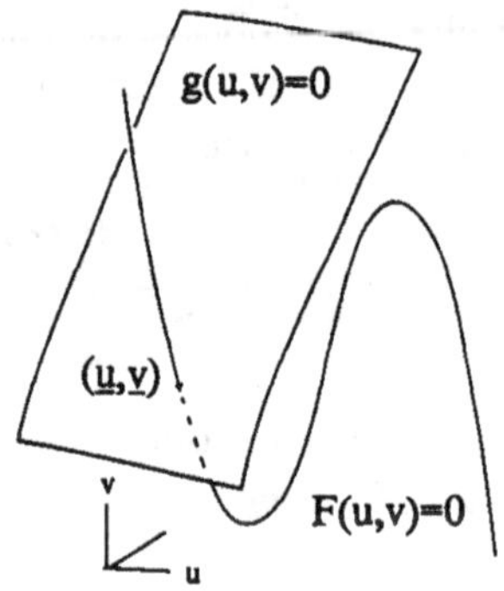

Fig.10 Monostable state $(\underline{u},\underline{v})$ of $(F,g) = 0$

Fig.9 Instability of almost round shape to five cells pattern

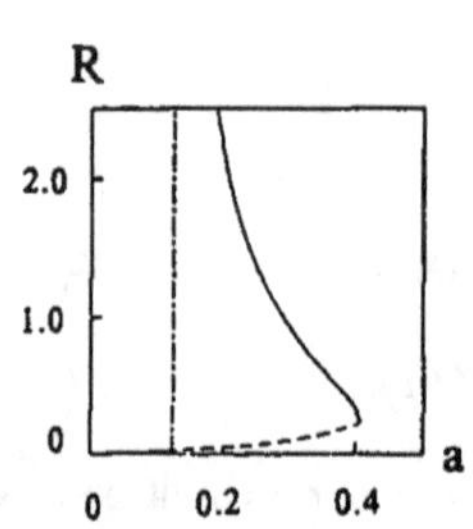

Fig.11 Solution branch of (2.6) with $N = 2$. The dashed line indicates the unstable branch given by solving the stationary problems (2.1) with (2.8)

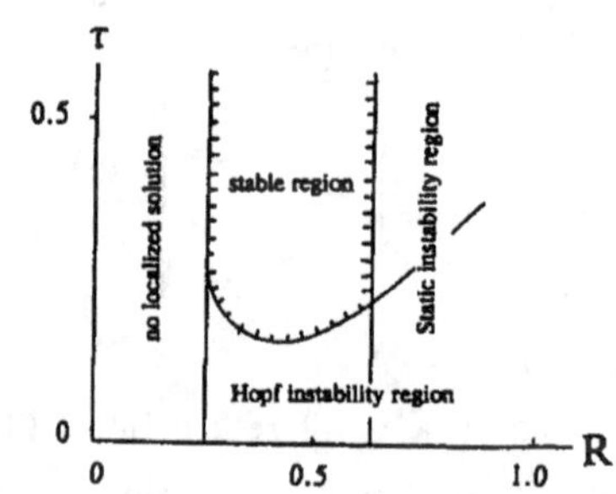

Fig.12 Stability diagram of a radially symmetric solution ($N = 2$)

Recently, the existence of global solutions to (2.4), (2.5) has been proved in weak formulation, although the uniqueness problem is still unsolved except for one dimension[6]. On the other hand, the study of dynamics of interfaces is only just beginning for higher dimensions. At present , we only show the existence and stability of radially symmetric stationary solutions of (2.6) in $\mathbf{R}^N$. We assume that $F = 0$ and $g = 0$ intersects at one point $(\underline{u}, \underline{v})$ as in Fig. 10. With the variable r, two unknowns $(v(r), R)$ satisfy

$$(2.7) \qquad \begin{cases} 0 = \left[\dfrac{d^2}{dr^2} + \dfrac{N-1}{r}\dfrac{d}{dr}\right] v + g(\overline{H}(v), v) & \text{in } 0 < r < R \\ 0 = \left[\dfrac{d^2}{dr^2} + \dfrac{N-1}{r}\dfrac{d}{dr}\right] v + g(\underline{H}(v), v) & \text{in } R < r < +\infty \\ 0 = -\dfrac{\varepsilon\alpha(v(R))(N-1)}{R} + c(v(R)), \end{cases}$$

and the boundary conditions

$$\begin{cases} \dfrac{du}{dr} = 0 = \dfrac{dv}{dr} & \text{at } r = 0 \\ v \in C^1 & \text{at } r = R \\ \lim\limits_{r\to\infty} (u, v) = (\underline{u}, \underline{v}). \end{cases}$$

If $N = 2$, R means the radius of a circle of the interface. For a simple example, we consider two-components system (u, v) where F and g are specified as

$$(2.8) \qquad \begin{cases} F(u, v) = H(u - a) - u - v \\ g(u, v) = u - \gamma v \end{cases}$$

where $H(\xi)$ is the Heaviside step function, a and γ are constants satisfying $0 < a < 1$ and $\gamma > 0$. Because of this specified F, the solution of (2.7) can be explicitly solved. Fig. 11 demonstrates the solution branch ($N = 2$) with $\gamma = 3$ when a is varied, which shows that the radius R is uniquely determined, depending on the value of a. For the stability of this radially symmetric solution, we can show that (i) for large values of τ, it is stable when R is small, while it is unstable and radially asymmetric stationary solutions appear when R is large (ii) even if R is small, the solution becomes unstable and it undergoes a sustained (breathing) oscillation. Fig. 12 is the stability diagram when τ and R are used as the bifurcation parameters. A remarkable fact is the occurrence of codimension 2 bifurcation for suitable values of τ and R. This suggests us the possibility that interfaces exhibit complicated curves. Although the conclusions above are proved for specific nonlinearities (2.8), numerical simulations confirm that these phenomena possibly occur for more general nonlinearities[20,27].

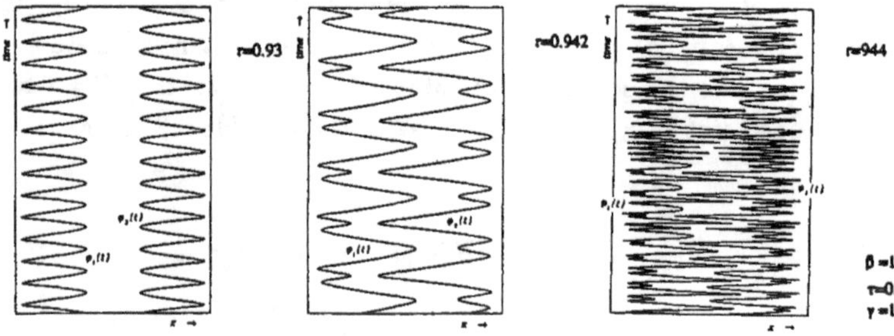

Fig. 13

$\beta=1.01$, $d=1.0$, $D=0.2$, $\tau=0.36$ $k=1.35$, $\alpha=0.818$, $\gamma=1.0$, $a_1=a_2=1.0$

Our discussion above is restricted to the linear stability of radially symmetric stationary solutions. Quite recently, we have observed complicated dynamics of interfaces in 1 dimensional finite interval with the Neumann boundary conditions when τ is sufficiently small. The nonlinearities used there are the ones in Example 2. Fig. 13 clearly shows the appearance of complex oscillations interfaces through period doubling cascades[8]. Unfortunately, such phenomena are unclear even in 1 dimensional space, because these are beyond the scope of the linear stability analysis.

3. Discussion

The characteristic property of the interface equation is that it truely coupled with the fast diffusing but slowly varying inhibitor v. This results in the nonlocal and time-delayed interaction in the dynamics of interfaces. In this sense, the present problem has a sort of similarity, to some extent, to the phase-field approach to crystal growth[1,11], where the local temperature plays the role of the inhibitor[19].

As a real experiment closely related to the present work, we should mention the formation of chemical patterns in the chlorite-iodide-malonic acid reaction , which has been quite recently observed by de Kepper and his collegues[10].

Acknowledgements

The author thanks R. Kobayashi and T. Ikeda and many colleagues in Applied analysis group in the Department of Mathematics, Hiroshima University for nice collaboration with him on numerical simulations and visualizations into video tapes related to the present work.

References

1. G. Caginalp, Arch. Rat. Mech. Anal., 92 (1986), 205.
2. X. Y. Chen, Hiroshima Math. J., 21 (1991), 47.
3. C. Conley and R. Gardner, Indiana Univ. Math. J., 23 (1984), 321.
4. P. C. Fife, *Dynamics of internal layers and diffusive interfaces*, CBMS–NSF. Regional Conf. Series in Appl. Math. 53, 1988.
5. P. Foerster, S. C. Müller and B. Hess, Science, 241 (1988), 685.
6. Y. Giga and S. Goto, *Motion of hypersurfaces and geometric equations*, preprint.
7. M. A. Grayson, J. Differential Geometry, 26 (1987), 285.
8. T. Ikeda and M. Mimura, *An interface approach to regional segregation of two competing species mediated by a predator*, to appear in J. Math. Biol.
9. H. Ikeda, M. Mimura and Y. Nishiura, Nonlinear Analysis, 13 (1989), 507.
10. P. De Kepper, V. Castets, E. Dulos and J. Boissonade, Physica D, 49 (1991), 161.
11. R. Kobayashi, *Modeling and numerical simulations of dendritic crystal growth*, to appear in Physica D.
12. S. Koga and Y. Kuramoto, Prog. Theor. Phys., 63 (1980), 106.
13. H. Meinhardt, *Models of Biological Pattern formation*, Academic Press, New York, 1982.
14. P. de Mottoni and M. Schatzman, C.R. Acad. Paris, 309 (1989), 453.
15. J. D. Murray, *Mathematical Biology*, Springer, Berlin, 1989.
16. Y. Nishiura and M. Mimura, SIAM J. Appl. Math., 49 (1989), 481.
17. A. Okubo, *Diffusion and Ecological Problems : Mathematical Models*, Springer, Berlin, 1980.
18. T. Ohta, Prog. Theor. Phys., 73 (1985), 1377.
19. T. Ohta and M. Mimura, Formation Dynamics and Statistics of Patterns, K. Kawasaki et al. ed., World Scientific Publishing Co., 1980, 55.
20. T. Ohta, M. Mimura and R. Kobayashi, Physica D, 34 (1989), 115.
21. S. Osher and J. A. Sethian, J. Comput. Phys., 79 (1988), 12.

22. P. Pelcé (ed.) *Dynamics of Curved Fronts, Perspectives in Physics*, Academic Press, New York, 1988.

23. A. T. Winfree, *The Geometry of Biological Times*, Springer, Berlin, 1980.

Video tapes

24. M. Mimura, R. Kobayashi, H. Okazaki and H. Ikeda, *Pattern formation in excitable media*, (1986) (Instabilities of traveling pulses, targets, spirals, breathing motions).

25. M. Mimura, R. Kobayashi, H. Okazaki and T. Yamanoue, *Dynamics of interfaces in systems of reaction-diffusion equations*, (1988) (Solidifications, targets, spirals).

26. R. Kobayashi and M. Mimura, *Pattern dynamics in reaction-diffusion systems far from equilibrium* (1989) (Phase separation and coarsening).

27. T. Ikeda and R. Kobayashi, *Numerical simulations to interfacial dynamics*, (1989).

HYDRODYNAMIC EFFECTS INDUCED BY CHEMICAL WAVE PROPAGATION

Hidetoshi MIIKE, Hideaki YAMAMOTO, Masahiro MOMOTA*
Department of Electrical and Electronic Engineering,
Yamaguchi University, Tokiwadai 2557, Ube, 755 Japan

and

Hajime HASHIMOTO
Department of Information Science, Ohshima National College of
Maritime Technology, Komatsu 1091-1, Ohshima-cho, 742-21, Japan

Abstract

A chemical wave propagation accompanied with an explosive hydrodynamic flow is observed in a thin layer of the Belousov-Zhabotinsky solution. Dynamic behaviors of chemical activity and hydrodynamic flow are measured quantitatively by microscope sequential image processing system. The underlying mechanisms of the wave propagation are discussed briefly.

1. Introduction

Belousov-Zhabotinsky (BZ) solution is known as a typical example of the oscillating chemical reaction[1,2]. In a stirred flow reactor, a variety of temporal patterns such as regular oscillation, chaotic behavior and intermittency[3-5] have been reported. On the other hand, in an unstirred batch reactor of the BZ solution, spontaneous pattern formation of spatio-temporal structures are observed, e.g., target patterns and spiral-shaped chemical waves [6-8]. They are recognized as excellent examples of the dissipative structure[9], and are treated theoretically by modeling the nonlinear coupling of reaction and diffusion[1,2].

Pattern dynamics due to the coupling between reaction-diffusion and hydrodynamic flow has attracted increasing attention. Convective effects accompanied with the chemical wave propagation in covered and uncovered horizontal Petri dishes have been studied by several authors[6,10-12]. Recently, one of our author and his collaborator detected new hydrodynamic phenomena on chemical waves such as a traveling convective flow with a circular wave propagation[13] and an oscillatory flow induced by periodic passage of spiral wave trains even under covered liquid/gas interface[14,15]. In Fig.1, a typical example of well controlled spiral waves (Fig.1a) and induced hydrodynamic oscillation (Fig.1b) are shown. However, the mechanism of these hydrodynamic instability had not been well established yet.

* Permanent address: Department of Information and Electronic Engineering, Tokuyama College of Technology, Tokuyama, 745 Japan

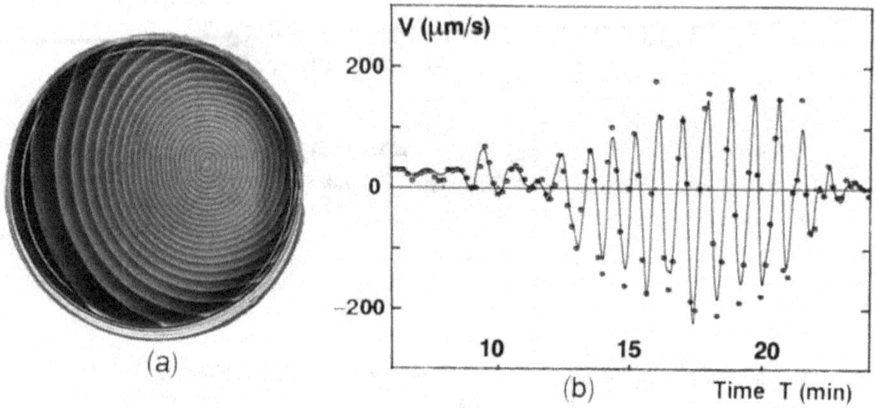

Fig. 1 Typical spiral patterns (a), and induced oscillatory hydrodynamic flow (b).

In this report, experiments of propagating one trigger wave are carried out in an excitable BZ–solution under well controlled condition. Utilizing microscope image processing technique[14)], the chemical wave propagation and induced hydrodynamic flow are observed quantitatively. Direct correlation between wave propagation velocity and velocity of the hydrodynamic flow is investigated. We report on a big wave having rapid and accelerating propagation velocity, and the wave brings an explosive hydrodynamic flow in the BZ–solution layer.

2. Materials and Methods

2.1 Sample preparation

An excitable solution of the BZ–reaction was obtained by preparing a mixture of 48 mM NaBr, 340 mM $NaBrO_3$, 95 mM $CH_2(COOH)_2$, and 378 mM H_2SO_4. About 5 min after mixing, the color of the solution changes from yellowish brown (Br^{2+}) to colorless (clear), then the catalyst and indicator ferroin (3.5 mM) was added. A volume of the mixture resulting about 0.85 mm layer depth was poured into a dust free Petri dish of 7 cm diameter. For measuring hydrodynamic flow, polystyrene particles (diameter 0.48 μm) serving as scattering centers were mixed into the solution. After mixing the catalyst, one circular wave was triggered by a silver wire (0.2 mmϕ). An air gap of about 12 mm was left between the layer surface and a glass cover. All experiments were carried out under the covered condition. Fresh samples were prepared for every new experiments. The room temperature was controlled within 25 ± 1 °C.

2.2 Measuring system

For the observations of chemical activity and hydrodynamic flow a microscope sequential image processing system is introduced. The system comprises of a He–Ne laser (Nihon Kagaku Eng. NE0–5M : 632.8 nm), an inverse microscope (Nikkon TMD), a TV–camera

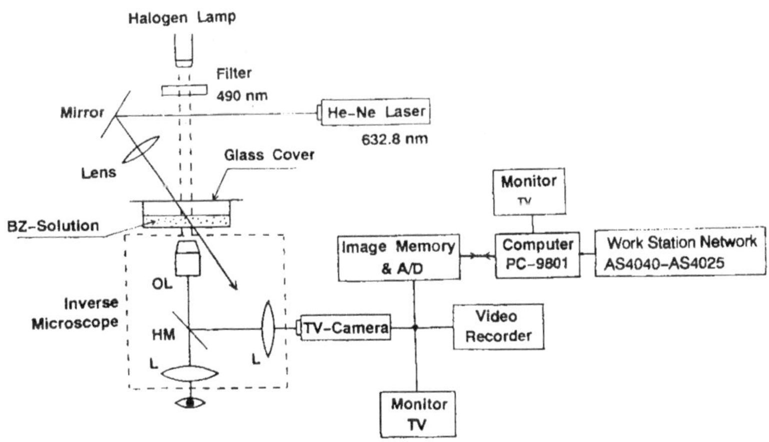

Fig.2 An experimental system block diagram.

(National WV-1850), a video recorder (Mitubishi HV-V36), a personal computer (NEC 9801RA) with image acquisition board (Microtechnica MT98FMM), and work station networks (Toshiba-SUN AS4040 & AS4025). A schematic is shown in Fig.2. For visualizing the motion of polystyrene particles the He-Ne laser light is illuminated at a slightly tilted angle (about 30 deg). A homogeneous transmitted monochromatic light (490 nm) allows us to detect propagating waves of chemical activity. The velocity of the hydrodynamic flow and that of propagating chemical waves are estimated by sequential image processing techniques that utilize spatial filtering velocimetry[16] and spatio-temporal mutual correlation analysis[17-19]. In this study, an original development is proposed in the spatial filtering velocimetry based on a sequential image processing for long video data analysis. Outline of the velocimetry is introduced briefly in the following section.

2.3 Adaptive spatial-filtering velocimetry by image processing

As illustrated in Fig.3, when we superpose a sinusoidal spatial filter on a sequential image, a reduced signal $A(t)$ is defined as

$$A(t) = \sum_x \sum_y S(x,y,t)\ \sin\{K\cdot(r-V_s t)\} \qquad (1)$$

where $S(x,y,t)$ is a brightness function of the sequential image, K is a wave vector of the spatial filter ($K = 2\pi / \lambda$), and V_s is a translational velocity of the sinusoidal filter. As is well known, resolution and dynamic range of the velocimetry depend on the wave length λ and the translational velocity V_s of the spatial filter. In the image processing, contrast to the hardware technique of the spatial filtering[20], we can optimize the spatial filter according to the velocity and the size of moving object. In the actual analysis, a parallel processing of the sequential image acquisition and a projection accumulation is realized in our system. When we want to

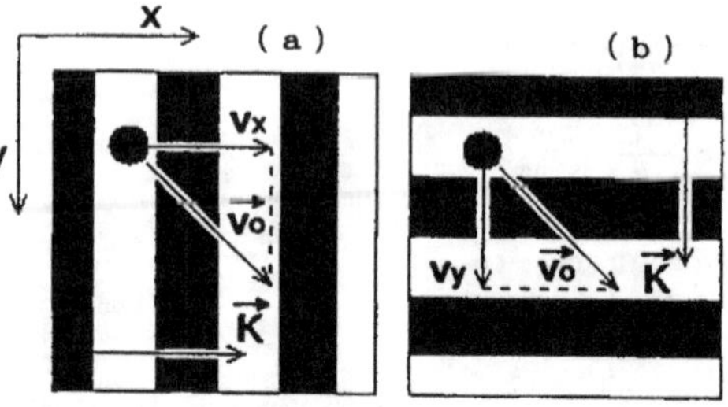

Fig. 3 A particle motion in sinusoidal spatial filters. The wave vectors: (a) $K=K_x$, (b) $K=K_y$.

measure x-component of the velocity, the projection accumulation to the y-axis and subtraction of background level (*BG*) are carried out with data acquisition, simultaneously:

$$B(x,t) = \sum_{y} \{S(x,y,t)-BG\} \quad (2)$$

Now, we obtain a long sequence of one dimensional image $B(x,t)$ as stored data. Then, velocity analysis are continued after the acquisition. The procedures are 1)spatial filtering,

$$A_x(t) = \sum_{x} B(x,t)\, sin\{K_x(x - V_{sx}\, t)\} \quad (3)$$

and 2)power spectrum analysis by maximum entropy method (MEM)[21] (See details of the analysis in ref. 16). The spectrum analysis is carried out every one second utilizing the image data of 60 frames. In the analysis, the wave length λ and the translational speed V_s are changed according to the velocity V_0 of moving object under the conditions:

$$V_s / D = C(constant) \quad (4)$$

$$V_s = \gamma\, V_0 \quad (5)$$

where the constants are selected as C=1/4 (1/frame) and γ=2 considering the oscillatory nature of the hydrodynamic flow. On the analogy of the laser Doppler velocimetry, the observed signal frequency f is

$$f = \{K\cdot(V_s \pm V_0)\} / 2\pi$$

$$= f_s \pm f_0, \quad (6)$$

where f_s is the shift frequency by the translating filter with velocity V_s, and f_0 is the signal frequency from moving object with velocity V_0. Real time analysis is impossible in our method, however, we can adopt a dynamic optimizing technique by soft ware processing. Hence, a computerized reliable velocimetry is established for the long video data processing. Example of the velocity analysis is shown in Fig.1b. The solid curve in the figure corresponds to the obtained velocities every one second by the proposed image processing. The open circles represent manual analysis every 5 seconds. The result seems to be satisfactory.

3. Results and Discussion

3.1 A big second wave

Experiments are carried out by triggering one circular wave at an edge of the Petri dish. During the wave propagation initiation of other waves is suppressed. In Fig.4, a sequence of temporal evolution is shown. The wave is triggered at time t=0, then it propagates with almost a constant velocity (about 170 μm/s) in the excitable medium. Then, the wave reaches at the opposite side of the dish (at about t=5 min), and vanish–

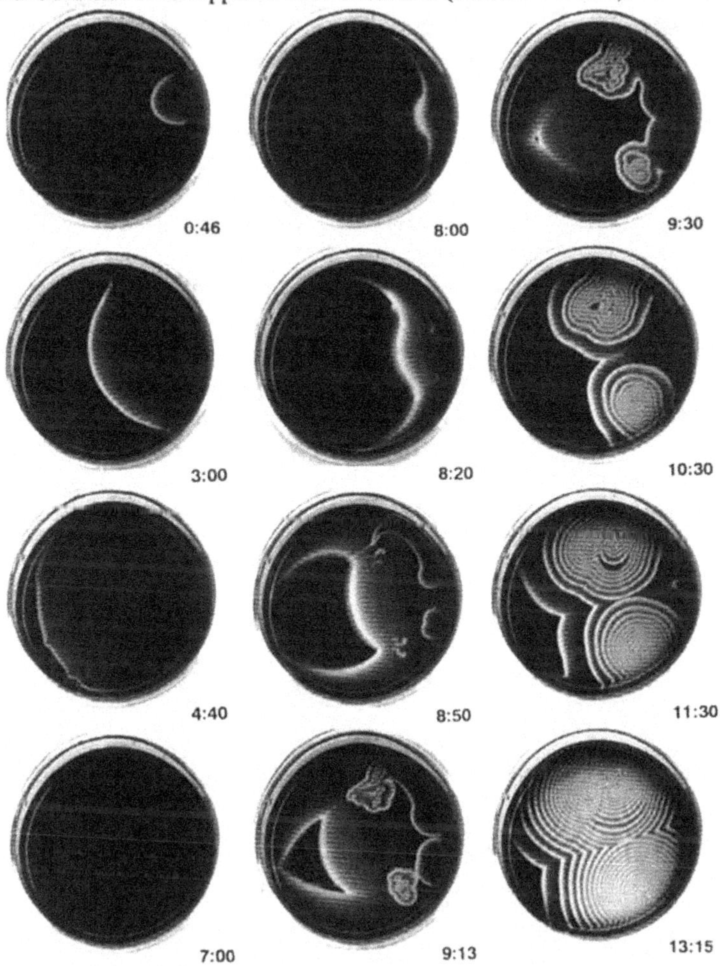

Fig.4 Pattern formation induced by a trigger wave. A circular wave is triggered at t=0. It propagates with normal velocity (3:00–7:00). At about 8 min, the second wave appears spontaneously, and propagates with rapid and accelerating speed (8:00–9:13). Finally, the second wave decomposes and spiral patterns are established autonormously (9:30–13:15).

es away. Until about t=8 min, the solution stays in quiescent state (the reduced state), then the second wave starts spontaneously from almost the same position of the first triggered wave. In some case, as shown in Fig.4, the second wave has very different characteristics comparing with the first one: 1) rapid and accelerating velocity of propagation, 2) big and broad wave front (oxidized state), 3) decomposition and chaotic behavior of the wave front. Finally, the decomposed wave fronts induce spiral waves autonomously (see 9:30–13:15 in Fig.4). Difference of the propagation velocity between the first and the second waves is shown in Fig.5 as a temporal trace. The velocity of the second wave develops to almost three times than that of the first wave.

The characteristics of the second wave resemble to that of the phase diffusion wave[22]. The BZ–solution prepared in this study is a quiescent but excitable medium, however, it oscillates spontaneously with period about 10 min in the Petri dish. Because of the long oscillation period, we can investigate a detailed pathway from triggered wave to phase wave. When we trigger the first wave, the solution is homogeneous. Namely, the phase difference in the solution layer is negligible. On the other hand, it takes about 5 min to reach the first wave to the opposite end of the dish. Then, after the wave propagation, a spatio–temporal phase gradient is established in the oscillatory BZ–solution layer. Spontaneous excitation of the second wave can be regarded as a memory effect of this phase difference. The second wave propagates in the phase field, then the phase diffusion wave might be induced. Rapid and accelerating nature of the second wave can be explained by the phase diffusion wave which resulting the big and broad wave front[22]. However, decomposition and chaotic behavior of the wave front can not be explained within the scheme of the phase diffusion wave. We have to investigate the hydrodynamic effects in the solution layer.

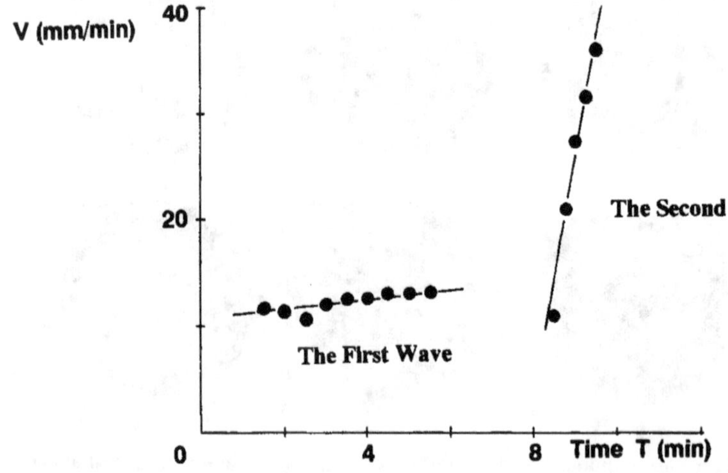

Fig.5 Temporal trace of propagation velocity of the chemical wave. The characteristics of the wave seems to change from normal trigger wave to "a big wave" (a phase diffusion wave accompanied with explosive hydrodynamic flow).

3.2 An explosive hydrodynamic flow induced by the big wave

Measurements of the hydrodynamic flow are carried out at the center of the dish during the wave propagation. Typical examples of temporal trace of the hydrodynamic flow are shown in Fig.6. The lower curve represents a case of normal trigger wave. As reported previously[13], a convective flow traveling with chemical wave propagation is confirmed. However, hydrodynamic effect induced is rather weak : the velocity of the flow is normally below than that of the chemical waves (about 70–120 μm/s). On the other hand, under a certain condition, we can observe an explosive hydrodynamic flow induced by a chemical wave propagation. Especially, in the second wave, the maximum flow velocity reaches sometimes to about 3000 μm/s (see the upper curve in Fig.6). The velocity of the flow is almost 20 times as that of the trigger wave. It can be easily recognized that the explosive flow brings decomposition and/or chaotic behavior of the wave front. We tentatively call this type of wave as "a big wave".

In Fig.7, another example exhibiting the difference between the normal trigger wave (Fig.7a) and the big wave (Fig7.b) is shown by pseudo color image. We confirm again the remarkable characteristics of the big wave as discussed in the preceding section, however, the mechanisms establishing the big wave are not known. In considering the underlying mechanisms for rapid and accelerating propagation of the big wave, we have to taken into account the following effects: 1)the phase diffusion wave, 2)hydrodynamics induced by the wave, 3)deformation of the wave front by the hydrodynamics, and 4)reformation of the deformed front by the curvature effect. The velocity v of the phase diffusion wave[22] is defined as the ratio of the variation of the phase Φ with time divided by the variation of the phase in space

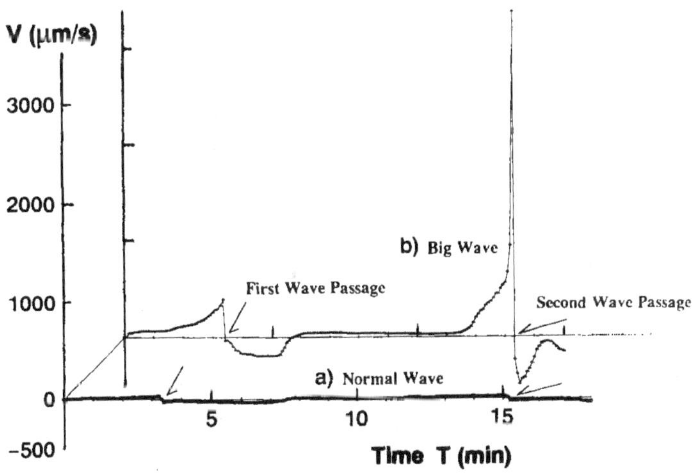

Fig.6 Typical examples of hydrodynamic flows induced by the trigger wave (a), and the big wave (b). Note that the maximum velocity of the second wave reaches to about 3000 μ m/s.

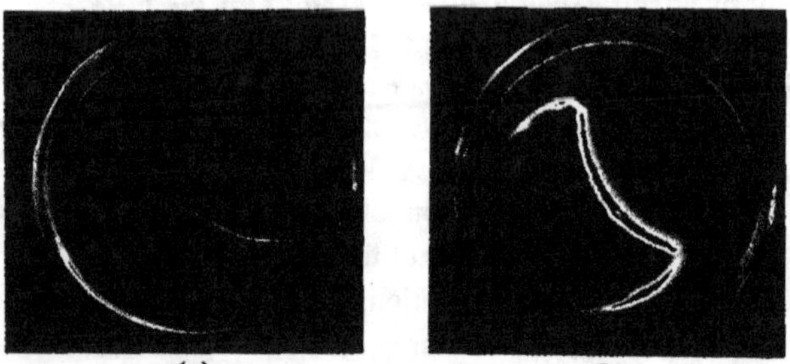

(a) (b)

Fig.7 Pseudo color representations of the normal trigger wave (a), and the big wave (b). The chemical activity was measured by TV–camera (CCD) under 490 nm interference filter.

$$v = -(\delta\Phi/\delta t) / (\delta\Phi/\delta r) \qquad (7)$$

Then, as the phase difference becomes smaller the velocity of the wave increases. Another key mechanism is hydrodynamic instability induced by the propagating chemical wave. Two possibilities are proposed. They are 1)Marangoni convection[23] caused by inhomogeneity of surface tension (σ), and 2)natural or free convection caused by density(ρ) gradient[24]. In the chemical wave, we have to consider both solute and thermal gradients both in density and surface tension. A combined Rayleigh number of the simple convection is

$$R_a = (r^4 g / \mu) \{ (d\rho_\tau / \kappa dz) + (d\rho_c / D dz) \} \qquad (8)$$

where μ=0.0089 g/cm.s (kinematic viscosity for water), g=980 cm/s^2, D is the average diffusion coefficient of the solute (10^{-5} cm^2/s), κ is thermal diffusivity (10^{-3} cm^2/s), r is radius of a vertical cylinder. Pojman and Epstein[24] suggested a possible mechanism of convection by chemical waves in horizontal tube: the difference in partial molar volume of the oxidized and reduced forms of the metal catalyst causes a horizontal density gradient ($d\rho_c/dz$) which is always unstable. On the other hand, an unbalance in surface tension produces a motion on the surface which brings about a flow in the bulk of the fluid due to the viscosity of the medium[12] (Marangoni convection). According to the formulation in eq.(8), a combined Marangoni number M_a can be written as

$$M_a = -(r / \mu) \{ (d\sigma_\tau / \kappa dz) + (d\sigma_c / D dz) \}, \qquad (9)$$

where r represents a layer depth of the solution. A horizontal gradient of surface tension at the wave front may bring the convective flow without threshold. The other important mechanism is the curvature effect[25] on the chemical wave. The normal velocity N of the wave is

$$N = C - DK, \qquad (10)$$

where C is the velocity of plane wave, D is the diffusion coefficient of the active chemical species, and K is the curvature of the wave front. The wave front having negative curvature propagates quickly, and vice versa. Then, deformed wave front can be reformed by the curvature effect.

4.Concluding Remarks

As discussed above, underlying mechanisms for the big wave are rather complicated. We are not sure that the other new mechanisms are necessary or not to explain the dynamic behavior of the big wave. Especially, our main interest is focused on a question how hydrodynamic flow feedback to the acceleration of the chemical wave propagation. The question seems to be also a key mechanism to understand the oscillatory hydrodynamic flow (see Fig.1b), which is induced by the spiral wave trains (see Fig.1 a).

In summary remarkable findings are the followings:

1)The excitable solution prepared in this study can be regarded as an oscillatory medium having very long oscillation period T_0 (about 10 min at 25°C),

2)The propagation velocity of the normal wave is almost constant, and amplitude of the induced hydrodynamic flow is small (normally below than the wave speed),

3)A big wave having very rapid and accelerating propagation velocity is observed under a certain experimental condition. Remarkable characteristics of the big wave are:

a)having a broad and obscure wave front,

b)accompanied with an explosive hydrodynamic flow,

c)the developed flow brings decomposition of the wave front, thus induces spontaneous pattern formation of spiral waves.

Numerical simulations for the hydrodynamic instability and the wave propagation are now in progress.

References

1)C. Vidal and P. Hanusse, Int. Rev. Phys. Chem., 5 (1986) 1.

2)R.J.Field and M.Burger, *Oscillations and Traveling Waves in Chemical Systems*,(Wiley, New York, 1985).

3)Y.Pomeau, J.C.Roux, A.Rossi, S.Bachelart and C.Vidal, J. Phys. Lett., 42 (1981) L271.

4)R.H.Simoyi, A.Wolf and H.L.Swinney, Phys. Rev. Lett., 49 (1982) 245.

5)J.C.Roux, Physica, 7D (1983) 57.

6)A.M.Zhabotinsky, A.N.Zaikin, J. Theor. Biol., 40 (1973) 45.

7)A.T.Winfree and S.H.Strogatz, Nature, 311 (1984) 611.

8)K.I.Agladze and V.I.Krinsky, Nature, 296 (1982) 424.

9)G.Nicolis and I.Prigogine, *Thermodynamics of Structure, Stability and Fluctuations*, (Wiley, New York, 1971).

10)M.Orban, J. Am. Chem. Soc., 102 (1980) 4311.
11)K.I.Agladze, V.I.Krinsky and A.M.Pertsov, Nature, 308 (1984) 834.
12)S.C.Müller, Th.Plesser and B.Hess, Ber. Bunsenges. Phys. Chem. 89 (1985) 654.
13)H.Miike, S.C.Müller and B.Hess, Physics Letters A, 141 (1989) 25.
14)H.Miike, S.C.Müller and B.Hess, Chem. Phys. Lett., 144 (1988) 515.
15)H.Miike, S.C.Müller and B.Hess, Phys. Rev. Lett., 61 (1988) 2109.
16)H.Miike, K.Koga, M.Momota and H.Hashimoto, Jpn.J.Appl.Phys., 26 (1987) L1431.
17)H.Miike, Y.Kurihara, K.Koga and H.Hashimoto, Jpn.J.Appl. Phys., 25 (1986) L409.
18)K.Koga, H.Miike and M.Momota, Trans. IECE Japan, E70 (1987) 719.
19)K.Koga and H.Miike, Proceedings of ISSSE'89 (Erlangen) 517.
20)Y.Aizu and T.Asakura, Appl. Phys. B, 43 (1987) 209.
21)S.Haykin, *Nonlinear Method of Spectral Analysis*, (Springer Verlag, 1979).
22)J.M.Bodet, J.Ross and C.Vidal, J. Chem. Phys., 86 (1987) 4418.
23)H.Linde, in Convective Transport and Instability Phenomena, eds. Zierep and H.Oertel,Jr., (Karlsruhe, Braun, 1982), p. 177.
24)J.A.Pojman and I.R.Epstein, J.Phys.Chem., 94 (1990) 4966.
25)J.P.Keener, SIAM J.Appl. Math., 46 (1986) 307.

Numerical Studies of Target Pattern Formation in a Chaotic Medium

Hiroyuki NAGASHIMA
Department of Physics, Faculty of Liberal Arts,
Shizuoka University, Ohya, Shizuoka,422 Japan

The Rössler oscillators coupled by diffusion were numerically studied. It is found that a target pattern of incoming waves emerges from a pacemaker whose frequency is lower than that of the surrounding oscillators. As a whole the patterns are stable against small perturbations, while the original instabilities still exist in the system.

From the practical and theoretical points of view, it is worth-while investigating whether macroscopic patterns such as target patterns and spiral patterns emerge in chaotic reaction-diffusion systems.

Agladze, Krinsky and Pertsov[1] observed distorted target patterns in a chaotic Belousov-Zhabotinsky reaction. We studied numerically the target pattern formation in a chaotic reaction-diffusion medium[2] which is expressed in the form

$$A_t = F(A) + D_M \nabla^2 A + g(A,x), \qquad (1)$$

where the vector A has three components X, Y and Z, D_M is the diffusion matrix, and g(A,x) denotes the inhomogeneous term.

Here, the oscillator dA/dt = F(A) is the Rössler model[3] which is given by

$$F(A) = \begin{pmatrix} -(X+Y) \\ X + 0.2\,Y \\ 0.2 + Z(X-\mu) \end{pmatrix}, \qquad (2)$$

where is μ the control parameter of the oscillatory state.

Here the Laplacian takes the form

$$\nabla^2 = \partial^2/\partial x^2 + \partial^2/\partial y^2 ,$$

in the two-dimensional Cartesian coordinates, or it is sometimes reduced to a one-dimensional Laplacian in the polar coordinates

$$\nabla^2 = 1/r\ \partial/\partial r + \partial^2/\partial r^2 ,$$

The inhomogeneous term $g(A,x)$ acts as the pacemakers of the target patterns and in the present case these are circular and expressed as

$$g(A,x) = \Sigma\, g_i(A,x_i),$$
$$g_i(A,x_i) = \begin{cases} f_i^r - 1 & |x - x_i| < R_i \\ 0 & |x - x_i| > R_i , \end{cases} \qquad (3)$$

where f_i^r is the ratio of the frequency of the pacemaker with the radius R_i and the center x_i to that of the outside medium.

Figure 1 shows typical results of target pattern formation in the two-dimensionally arrayed Rössler oscillators. In the present cases, the diffusion matrix D_M is reduced to a scalar coefficient D. In a chaotic

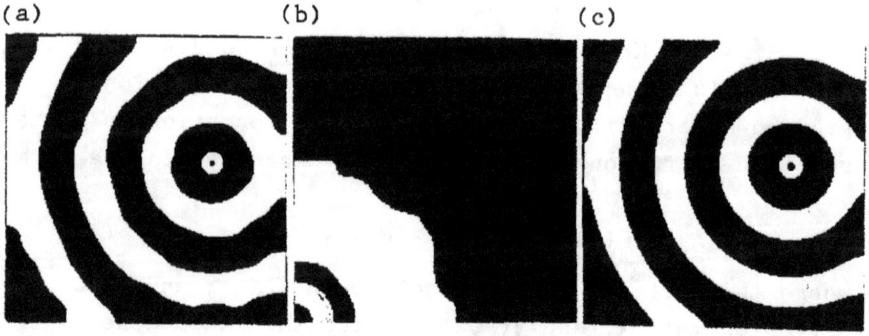

Fig.1 Two-dimensional patterns in the diffusion coupled Rössler oscillators. The black and white parts correspond to Y<0 and Y>0, respectively. The number of oscillators is 200 x 200. (a) chaotic region μ=5.7, f_r=0.5 and the radius of the pacemaker, R=7a. (b) chaotic region, μ=5.7, f_r=2.0 and R=20a. (c) regular region μ=2.7, f_r=2.0 and R=20a.

oscillatory state of the oscillator, μ = 5.7, target pattern is formed from a pacemaker. For f_r > 0, the pattern is stable as a whole and has small and wavy distortion of the nodal lines (Fig. 1(a)). For f_r > 2, more diffusive pattern is observed as shown in Fig. 1 (b). A target pattern emerged in the regular oscillatory state is shown in Fig. 1(c) for reference.

These patterns are accompanied by incoming waves, which are unusual properties of the target pattern formation and due to the artificiality of Eq. 2. As shown in Fig. 1, in order to examine the stability of patterns, we imposed

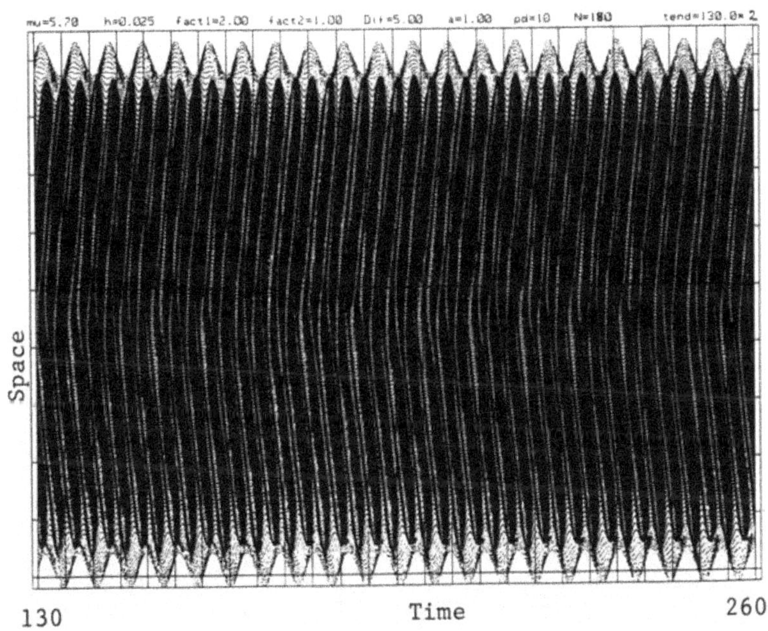

Fig.2 Collision of waves in a chaotic state of the system. The array of oscillator is one-dimensional. The waves emerge from the pacemaker placed at the both ends of the system. The number of oscillator is 180. f_r= 0.5. μ = 5.7.

nonsymmetrical boundary conditions of no flux by putting the pacemakers apart from the center of the computational square region. These conditions act as spatial perturbations to a circular target pattern; the small change caused by the perturbation at boundaries enhanced by the original instabilities of the oscillators.

Figure 2 shows the collision between two patterns and the results show that each pattern is stable as if the collision took place in a regular reaction-diffusion system.

In order to investigate how the original chaotic properties of the Rössler oscillator are reflected to the behavior of the target pattern, a small perturbation at a certain time imposed on the pattern.

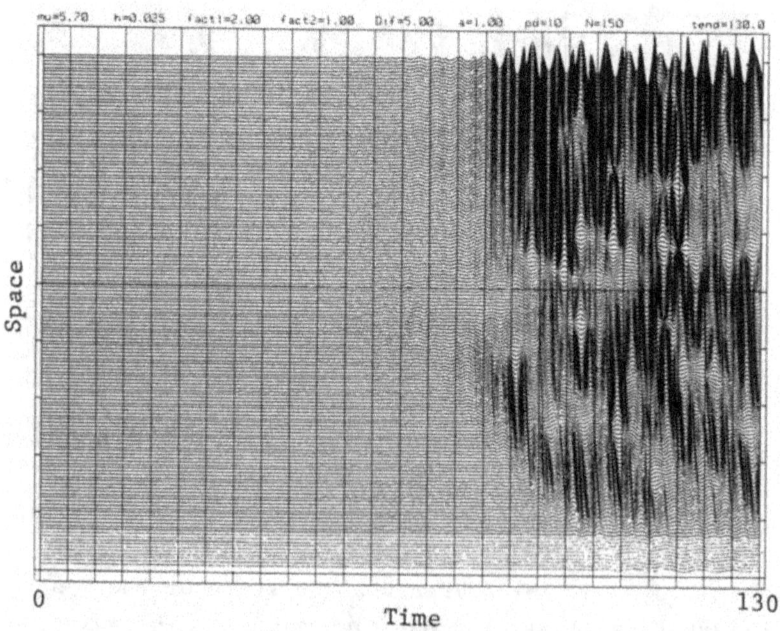

Fig.3 Development of a perturbation at t = 30 on the one-dimendional system. The perturbation changed X into X + 0.001. The number of oscillators is 150. f_r = 0.5. μ = 5.7.

Figure 3 shows the time developement of patterns in one-dimensionally arrayed oscillators. The perturbation $\Delta X = 0.001$ was imposed at $t_0 = 30$ in Fig. 3(b). The development of the difference between unperturbed and perturbed waves is shown in Fig. 3(c). The small difference becomes gradually greater due to the original instabilities of the oscillators and grows up to the magnitude of the order of unity around $t_1 = 100$. This gives the growth rate

$$g = \ln(1/\Delta X)/(t_1 - t_0) \sim 0.1$$

which is consistent with the positive Lyapunov number of the Rössler model, $\lambda_1 = 0.072$.

Therefore in this system the chaotic properties of the oscillators coexist with the pattern formation processes.

References

1. K. I. Agladze, V. I. Krinsky and A. M. Petrosov, Nature 308 (1984) 834.
2. H. Nagashima, J. Phys. Soc. Jpn., 60 (1991) 2485.
3. O. E. Rössler, Phys. Lett. A57 (1976) 397.

ANALYSIS OF THE FRACTAL STRUCTURE OF EARTHQUAKE SIZE DISTRIBUTIONS

Toru Ouchi
Faculty of Sciences
Kobe University
Nada, Kobe 657, Japan.

ABSTARCT

The earthquake size distribution is widely considered to show the fractal structure. We have investigated deviation of the earthquake size distribution from the fractal structure. The concept of the *b* value spectrum is introduced based on the generalized information measures to address this problem. Analysis shows that actual size distributions of earthquakes show large variations from case to case and in some cases deviate significantly from the fractal structure.

1.Introduction

The magnitude frequency distribution (MFD) of earthquakes is considered to obey the following Gutenberg-Richter (GR) formula

$$\log n = a - bM, \tag{1}$$

where n denotes the number of earthquakes with magnitude $M \sim M+\Delta M$. This shows the power law with respect to the energy distribution, since an empirical relation $M \propto \log E$ holds. "a" and "b" are parameters and this "b" has been considered to show one of the most important features of the seismic activity. Recent studies, however, have shown that MFDs of earthquakes do not always obey this formula. Sometimes they deviate significantly from this law and some modified formulae have bee also presented (e.g., Utsu, 1978). This problem also give rise to dependency of the estimation of "b" values on the methods. Conventionally, "b" values are estimated by the least square method or the maximum likelihood method and both results sometimes differ significantly. Also, deviation of MFDs from the GR law has been investigated (Utsu, 1978; Okuda et al., 1991). However, it is not desirable to introduce complicated models to address this situation since we must usually deal with rather limited amount of data. Thus we introduce the concept of the "b value spectrum" based on the generalized information measures to describe characteristic feature of MFDs of earthquakes. In this study, we assume the GR law basically for MFDs of earthquakes and investigate how and how much degree actual MFDs deviate from this law.

2.Estimation of "b" Values

First, we consider the method of estimating "b" values using Kullbuck-Leibler's mean information (hereafter we call this the KL information) which is represented as

$$K(\mathbf{p},\mathbf{q})=\sum_{i=1}^{m} p_i \log(p_i/q_i). \tag{2}$$

Here, $\mathbf{p}=\{p_i\}$ and $\mathbf{q}=\{q_i\}$ denote real (a posteriori) and model (a priori) distributions. This measure represents a divergence between two probability distributions. Now we consider a family of earthquakes and investigate the the characteristic feature of the MFD of these earthquakes. Let p_i and q_i be the probability observed event has a magnitude of $M_i \sim M_i+\Delta M_i$ $(i=1,2,\cdots m)$ and that derived from the GR formula, respectively. Actually they are defined as

$$p_i = n_i/N ,$$

and

$$q_i = B \exp(-BX_i)\ \Delta M,$$

where $X_i=M_i-M_o$ and $B=b\ln 10$. N and n_i are a total number of earthquakes and the number of events with magnitude of $M_i \sim M_i+\Delta M_i$. M_o is the minimum magnitude of earthquakes. When the numbers of events $n_1, n_2, n_3, \cdots n_m$ are given, the parameter value "b" for this earthquake population is estimated so that it maximizes the foregoing KL information (2). This is identical to that of the maximum likelihood estimation (Aki, 1965;Utsu, 1965).

3.Generalized Information Measures

Now we introduce the generalized information measures and use them to estimate model parameters. The following information measures have been presented which are the extended ones of the KL information.

Now let $\{p_i\}$ and $\{q_i\}$ again denote real (a posteriori) and model (a priori) distributions.

(a) Kawata (1987) considered

$$I^{\lambda}=[\sum_{i=1}^{m} p_i(p_i/q_i)^{\lambda}-1]/\lambda \qquad (-1/2\leq\lambda<\infty). \tag{3}$$

(b) Rényi (1961) has presented

$$R^{\alpha}=[\log\sum_{i=1}^{m} p_i(p_i/q_i)^{\alpha-1}]/(\alpha-1). \tag{4}$$

Both measures have a simple relation as

$$\lambda R^{\alpha}= \log\ (\lambda I^{\lambda}+1)$$

and

$$\lambda=\alpha-1. \tag{5}$$

These measures R^{α} and I^{λ} give the KL information in the limit of $\alpha \to 0$ and $\lambda \to 0$.

4.The "b" Value Spectrum

Conventionally, the model parameters are generally estimated by minimizing $K(\boldsymbol{p},\boldsymbol{q})$ which gives the maximum entropy estimation. Model parameters are chosen so that $\{p_i\}$ and $\{q_i\}$ be as close as possible. However, this method has serious difficulties unless the model is valid enough and sufficient amount of data are available. First, when the model is inappropriate, it is not easy to evaluate the results obtained by this scheme. In general, it is not a simple matter to judge the validity of models.

Second, we assume generally $p_i=n_i/N$, but this is not always appropriate since this holds precisely in the limit $N \to \infty$. We must treat usually quite limited or few amount of data like earthquake data. Particularly, for small n_i which is the case for large earthquakes, the estimate of p_i is less reliable. In what follows, we must evaluate p_i, more or less, subjectively when we deal with actual data. Consequently, estimation of b values may differ seriously depending on how we evaluate or namely how much degree we place a weight on large earthquakes that are usually very few and scatter significantly from the GR law.

Estimation of model parameters, in general, highly depend on the evaluation of the data that deviate from assumed models. This is true when we analyze exponential or power law distributions that appear in various natural phenomena. Conversely, however, we can take advantage of this situation to see the actual feature of MFDs of earthquakes and investigate how and to what extent the real distributions deviate from the assumed models.

For this purpose, we use the generalized information measures R^{α} or I^{λ} instead of the KL measure $K(\mathbf{p},\mathbf{q})$. Basic idea of our scheme is to estimate the parameter values of models so as to minimize R^{α} or I^{λ}. Then we investigate the variation of parameter values with respect to α or λ. In this procedure, it does not matter whether we use R^{α} or I^{λ} since both measures are identical as long as we use them for this minimization program. In reality, we minimize the following quantity,

$$\varkappa^{\lambda}=\sum_{i=1}^{m}p_i(p_i/q_i)^{\lambda} \tag{6}$$

in both measures.

Suppose now the probability distribution $\{q_i\}$ be defined through model parameters $\theta=(s_1,s_2,\cdots s_r)$ as

$$\mathbf{q}(s_1,s_2,\cdots s_r) = (q_1(s_1,s_2,\cdots s_r),\cdots \quad ,q_m(s_1,s_2,\cdots s_r)).$$

Let true distribution $\{q_i\}$ be given as

$$\mathbf{q}^0=(q_1{}^0,q_2{}^0,\cdots,q_m{}^0),\ \mathbf{q}^0=\mathbf{q}(s_1{}^0,s_2{}^0,\cdots s_r{}^0)$$

and

$$\theta^0=(s_1{}^0,s_2{}^0,\cdots s_r{}^0).$$

If we obtain a distribution $\{p_i\}$ where $p_i=n_i/N$ $(i=1,2,\cdots m)$, then we can consider a problem of estimation of parameters $\theta^0=(s_1{}^0,s_2{}^0,\cdots s_r{}^0)$. Our scheme is to find parameter values θ that minimize $R^{\alpha}=R^{\alpha}(\theta)$ or $I^{\lambda}=I^{\lambda}(\theta)$. Actually, we aim to get parameters which minimize

$$\varkappa^{\lambda}(\theta)=\sum_{i=1}^{m}p_i(p_i/q_i(\theta))^{\lambda}. \tag{7}$$

This method is identical with the χ^2 method when $\lambda=1$ and the maximum likelihood method when $\lambda=0$, respectively. Also, we can evaluate confidence limits for the parameter values using $I^{\lambda}(\theta)$ (Kawata, 1987). It should be stressed here that the parameter values thus obtained depend highly on α and λ.

Now we apply this scheme of the model parameter estimation to the *b* values of the GR law. Following the procedure shown in section 2, we estimate the *b* values of the GR formula. This time, however, we seek *b* values that minimize not the KL information (2) but the generalized measures (3), (4) and (6). Then we can obtain *b* values as a function of α or λ. This should be called the "*b* value spectrum" and will be used to investigate characteristic feature of MFDs of earthquakes.

5.Analysis of Earthquake Data

We have applied this idea to various earthquake populations in Japan. Fig.1a shows the three earthquake families. Tsuboi (1952) divided earthquakes occurred in and around Japan into three groups A, B and C and calculated *b* values for them by the least square method. He obtained quite different *b* values 1.06 (A), 0.72 (B) and 0.66 (C) for three regions. Utsu (1967) reanalyzed these data by the maximum likelihood method and got almost same b values 0.97 (A), 0.94 (B) and 0.95 (C), respectively.

The b value spectra for these three data sets are shown in Fig.1b. The number of earthquakes of A is large and the variation of b value is small, while the number of events of C is small and the variation of b value is large. Feature of the b value spectrum of B shows just intermediate nature. The case C seems to indicate that the MFD might deviate from the GR law. It is noted that the results by the maximum likelihood method corresponds to the case of $\alpha=1$ ($\lambda=0$) and those by the least square method appear to represent situations for larger part of α or λ. These b value spectra seem to represents characteristic feature of the MFDs of earthquakes fairly well and reveal differences clearly.

We have analyzed various MFDs of earthquakes in Japan and found b value spectra show large variations. Indeed, actual MFDs of earthquakes differ significantly from case to case and in many cases deviate from the GR law or the fractal structure.

6. Discussion

If the GR model is complete and sufficient data both in quality and quantity are available, the parameter b would take similar values independent of α or λ. Then a single b value can represent the situation of the MFD of earthquakes fairly well. Otherwise, it may not be appropriate to describe the feature of the MFDs of earthquakes by a single b value only. Besides, more complicated models are not statistically desirable although they may fit better to certain data sets. This is the reason why we have introduced the concept of the b value spectrum.

Through this b value spectrum, we can analyze MFDs more efficiently, namely, we can see how and to what extent actual MFDs deviate from the GR law. In general, the b value decreases monotonously with α or λ. The effect of α or λ on the b value estimate is not so simple, but when α or λ is large, frequency of larger magnitude is kept less than that of expected ones from the model. Conversely, when α or λ is small, the b value is chosen so that earthquakes with smaller magnitude fit the better to the model. Consequently, events with smaller magnitude are emphasized when α or λ is small while contribution of larger magnitude events becomes large when α or λ is large. When the actual MFD fits well to the GR law, the variation of the b value spectrum will be small but the larger deviation from the GR law produces the steeper slope in the b value diagram. Thus we can see the degree of deviation of MFDs from the GR law from the shape of the b value spectrum.

The actual b value spectra show wide varieties and thus in many cases the GR's fractal law that has been a priori assumed and widely accepted may not always hold for real earthquake populations.

7. Conclusions

We have introduced the concept of the b value spectrum based on the generalize information measures. This enables us detailed analysis of MFDs of earthquakes. Application of this approach to various earthquake populations in Japan shows the b value spectra differ remarkably from data to data, which indicates actual MFDs may deviate from the GR law or the fractal structure. Hence, a simple fractal model like the GR formula may not be sufficient to describe the feature of MFDs of earthquakes. This idea of the "*b* value spectrum" will be used for more general analysis of the fractal structure that appear in various natural phenomena.

References

1.K.Aki, Bull.Earthq.Res.Inst., 43 (1965), 237.
2.Y.Kawata, Proc.Inst.Stat.Math., 35 (1961), 1, (in Japanese).
3.S.Okuda, T.Ouchi,and T.Terashima, submitted to Phys. Earth. Planet. Int., (1991).
4.A.Rényi,A., Proc.Fourth Berkley Symp.Math.Statist.Prob.,1 (1961), 47.
5.C.Tsuboi, J.Phys.Earth., 1 (1952), 47.
6.T.Utsu, Geophys.Bull.Hokkaido Univ.,13 (1965), 99, (in Japanese).
7.T.Utsu, Geophys.Bull.Hokkaido Univ.,17 (1967), 85, (in Japanese).
8.T.Utsu, Zisin 2, 31 (1978), 367, (in Japanese).

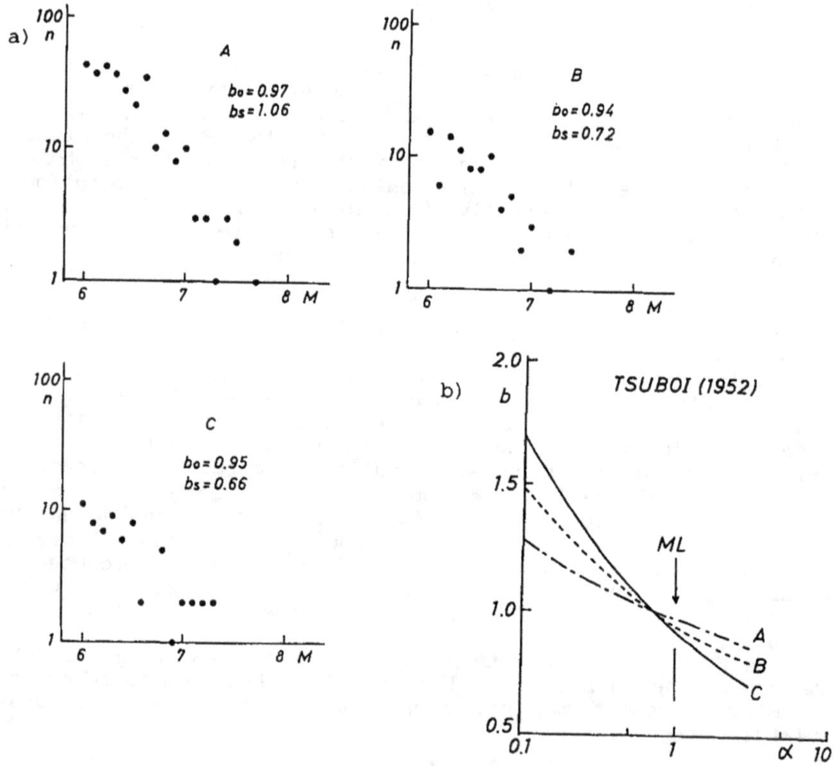

Fig.1 a) Magnitude frequency distributions (MFDs) of earthquakes of regions A, B and C in Japan. A: east Japan of Pacific side; B: west Japan of Pacific side and C: Japan sea side. Numbers of b_o and b_s show estimates by the maximum likelihood and least square methods, respectively. b) The b value spectra for earthquakes in regions A, B and C.

SELF-INDUCED VIBRATION OF EVAPORATING LIQUID DROP

Naoko TOKUGAWA
Tokyo University of Agriculture and Technology
Fuchu, Tokyo 183, Japan
and
Ryuji TAKAKI
Tokyo University of Agriculture and Technology
Fuchu, Tokyo 183, Japan

Abstract

Mechanism of self-induced vibration of an evaporating liquid drop is investigated experimentally and theoretically. In the experiment, normal mode numbers of vertices $n = 2 \sim 8, 10$ were observed for the distilled water and those with $n = 2, 3$ for the ethanol. It was made confirmed there was a temperature difference between the upper and the lower surfaces of a drop of distilled water drop. In the theory, equations are analyzed using weak-nonlinear approximation, and a model is constructed baesd on the nonuniformity of surface tension. A stable steady vibration solution is obtained for each mode $n = 2 \sim 6$ for distilled water. The results for $n = 4 \sim 6$ agree with the experiment.

1 Introduction

When a liquid drop is put on a hot plate, the liquid evaporates rapidly and the drop is observed to float above the plate due to a thin vapor sheet under the drop. Since the heat conductivity of vapor sheet is small, the drop doesn't evaporate instantaneously. Moreover the liquid drop shows a charactaristic vibration with a polygonal plane view. (see Fig.1)

This phenomenon has been observed for a water drop by Holter [1], for liquefied gases by Adachi & Takaki [2]. It is studied theoretically [3] and numerically [4]. But the mechanism has not been understood yet. The purpose of the present work is the mechanism of sustaining the vibration in the presence of viscous resistance.

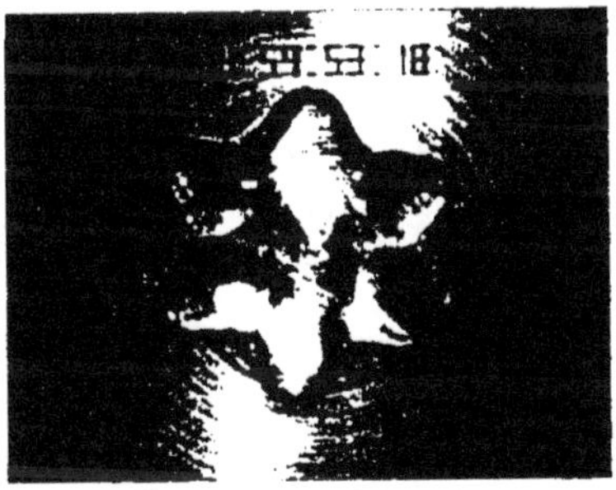

Fig. 1: Vibration of distilled water(n = 6)

2 Experiment

Liquid drops of distilled water and ethanol were put on a hot plate, whose temperature was controled at a constant value of $200 \sim 400°C$.

The vibration was recorded by CCD video camera, and the mode number n was determined by naked eyes. Appearances of vibration modes are shown in Figs.2 and

3. Modes with $n = 2 \sim 8, 10$ were observed for the distilled water, and those with $n = 2 \sim 4, 6$ appeared often. For ethanol, $n = 2, 3$ occured. There seemed to be no definite rule for mode selection of water drop. But mode of ethanol depended only on radius of plane view, and not on the temperature of hot plate.

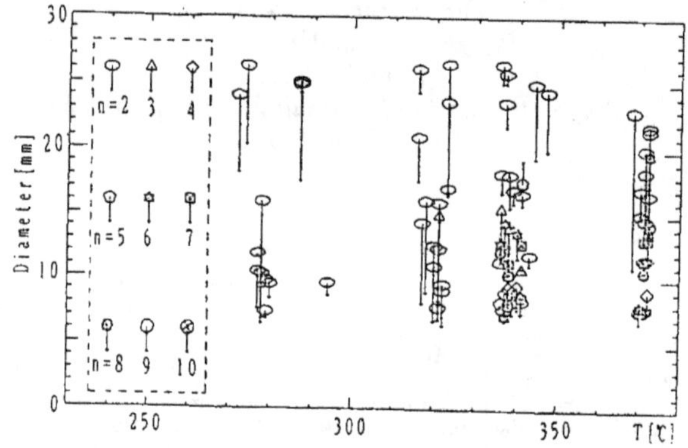

Fig. 2: Mode selection (distilled water)

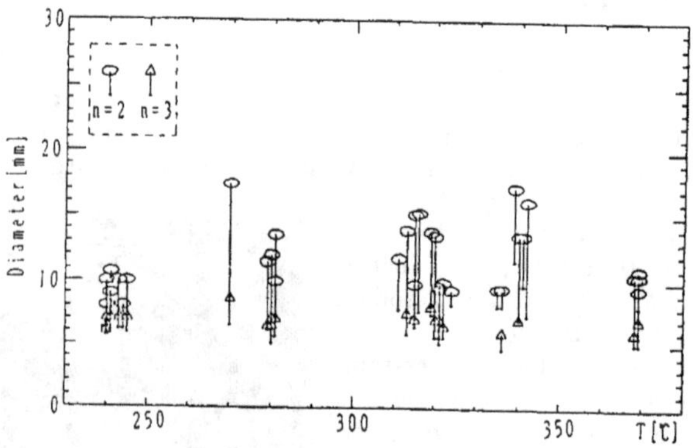

Fig. 3: Mode selection (ethanol)

The temperature distribution of a water drop on the hot plate at about 320°C was measured by chromel-alumel thermocouple with wire diameter 0.1mm. When the thermocouple was within the drop, the vibration tended to stop. Temperature inside the drop was almost constant at $95 \sim 90°C$, and was about 100°C at the lower surface (see Fig. 4). The temperature of upper surface was measured by changing the height of the thermocouple through the upper surface slowly. The temperature at the upper surface was estimated at $90 \sim 80°C$.

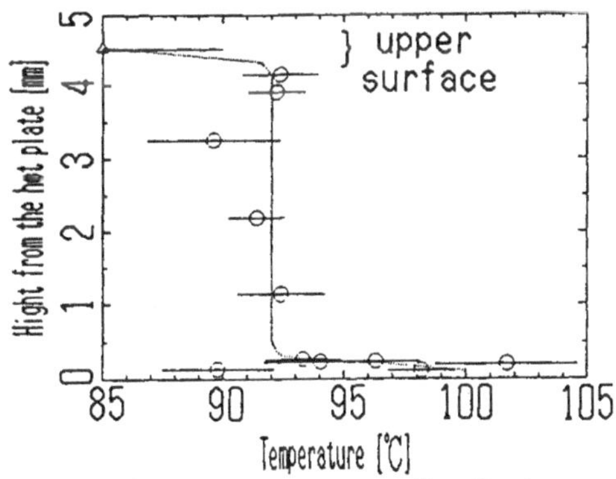

Fig. 4: Temperature distribution

3 Theoretical Analysis

For the sake of simplicity we put the following assumptions:

1. The thin vapor flow under the drop is looked as the Hele-Shaw flow.
2. The shallow water theory is applied for the motion of drop.
3. The flow inside the drop is assumed incompressible and inviscid.
4. Material parameters expect the surface tension are constant during vibration.

Then the governing equations are the formulas for the vapor flow

$$\bar{u} = -\frac{D_0^2}{12\mu}\nabla(\rho_L g h - \sigma\nabla^2 h + p_{atm}) + \frac{1}{2}\nabla\phi, \tag{1}$$

the equation of continuity of the drop

$$\frac{\partial h}{\partial t} + \nabla\cdot(h\nabla\phi) = 0, \tag{2}$$

and the Euler equation for the drop with the friction with the vapor sheet

$$\rho_L h\nabla\frac{\partial\phi}{\partial t} + \rho_L h\nabla_i\phi\nabla_i\cdot\nabla\phi = -\nabla(\frac{1}{2}\rho_L g h^2 - \sigma h\nabla^2 h) + \frac{6\mu}{D_0}(\bar{u} - \frac{2}{3}\nabla\phi) \tag{3}$$

where $\bar{u}$, D_0, μ are the mean velocity, the mean depth, and the viscosity of the vapor sheet, ρ_L, σ, h, ϕ are the density, the surface tension, the depth, rhe velocity of the liquid drop, and p_{atm} is the atmospheric pressure, respectively.

The boundary conditions posed at the peripheral part of the drop are the balance of the surface tension and pressure

$$\left(\frac{1}{2}\rho_L g h^2 - \sigma h\nabla^2 h = h\sigma(\frac{1}{R_1} + \frac{1}{R_2})\right)|_{r=R}, \tag{4}$$

where R_1 and R_2 are radius of curvature, and the kinematic condition

$$\left(\frac{\partial\phi}{\partial r} = \frac{1}{r}\frac{\partial\phi}{\partial\theta}\frac{1}{r}\frac{\partial R}{\partial\theta} + \frac{\partial R}{\partial t}\right)|_{r=R}. \tag{5}$$

The variables are normalized by the avarage radius R_0, the depth h_0, the characteristic surface tension σ_0, and the time scale $t_0 = \sqrt{\frac{\rho_L R_0{}^3}{\sigma_0}}$, but the some notations are used for nondimensional quantities. Equations are linearized with respect to purturbations, expect that the second order of surface tension $\tilde{\sigma}$ is retained. In analogy with linear analysis [3] , the solution is expressed by product of Bessel function $J_n(\mathrm{w}r)$, and $\exp(in\theta)$, with amplitude function $a(t)$, i.e.

$$(\phi, \tilde{h}, \tilde{R})|_{r=R} = \left(\dot{a}(t), a(t)\mathrm{w}^2, a(t)[n - \frac{\mathrm{w}^2}{2(n+1)}]\right) J_n(\mathrm{w})e^{in\theta}, \tag{6}$$

where eqs. (2) , (5) are used. From eqs. (1), (3), (4), the following differential equation is derived :

$$\begin{aligned}\ddot{a} + \frac{A''}{6}\dot{a} + (h_0 + \frac{D_0}{2})\frac{1}{h_0}(1+\tilde{\sigma})\{\frac{\mathrm{w}^2}{(1+\beta H)} - \frac{(1-n^2)}{(1+\beta H)^2}(n - \frac{\mathrm{w}^2}{2(n+1)} + \beta H \mathrm{w}^2)\}a \\ = -(h_0 + \frac{D_0}{2})\frac{1}{h_0}(\frac{2}{H} + \frac{1}{1+\beta H})[J_n(\mathrm{w})e^{in\theta}]^{-1}\tilde{\sigma},\end{aligned} \tag{7}$$

where $\beta = \frac{1}{2} - \frac{\pi}{8}$ is a topological constant and

$$\begin{aligned} G &= \frac{\rho_L g R_0{}^2}{\sigma_0}, \quad H = \frac{h_0}{R_0} \\ \mathrm{w} &= [\{h_0 + \frac{D_0}{2}\}^{-1}\frac{R_0}{G}\frac{n(n^2-1)}{\{1 + 2\beta H + \frac{n-3}{2GH}\}}]^{1/2}. \end{aligned}$$

To include the temperature dependence of surface tension, the temperature outwards at the periphery is considered to change as follows. When the periphery is going outwards (i.e. $\dot{a} > 0$) , the inside cool fluid comes to the surface. On the other hand, as the periphery is going back (i.e. $\dot{a} < 0$) , the hot lower surface is raised and after a short time occupies the periphery. As the simplifest model the drop surface is separated into two regions with high or low temperature. Hence, surface tension is assumed to vary

$$\sigma = \begin{cases} \sigma_{high} = \sigma_0 & (\dot{a} \le 0 \text{ , and } | a - a_{max} | \ge Da) \\ \sigma_{low} = \sigma_0(1+\tilde{\sigma}) & (\text{else}), \quad (\tilde{\sigma} > 0) \end{cases}$$

where a_{max} is the amplitude of vibration of $a(t)$ and Da correspond to the length from the end of lower surface to the tip of periphery ,defined as

$$Da = \frac{h_0}{2}\{[n - \frac{\mathrm{w}^2}{2(n+1)}]J_n(\mathrm{w})e^{in\theta}\}^{-1},$$

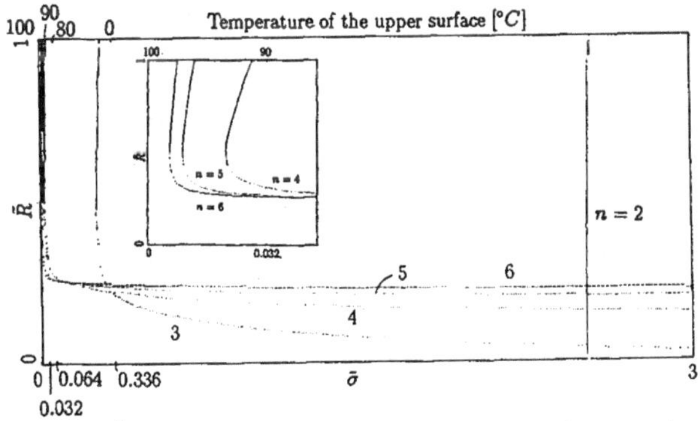

Fig. 5: Amplitude $\tilde{R}$ of stable vibration vs. assumed surface tension of upper surface

and $\bar{\sigma}$ is looked upon as a free parameter. Amplitude $\tilde{R}$ of stable steady vibration is calculated numerically based on eq.(7) with material parameters for distilled water. Dependence of the amplitude on $\bar{\sigma}$ for each mode $n = 2 \sim 6$ is shown in Fig. 5.

It is also calcurated Dependences on Da and w are also examined , but there are no appreciable influences of them.

4 Discussion

The steady vibrations obtained theoretically for water with $n = 4 \sim 6$ agree with the present experiment, because the required lower temperature (i.e. $\simeq 90 \sim 80°C$) agree with mesurment , but fail to agree for $n = 2, 3$. So, the present theory is reasonable for $n \geq 4$. To improve the theory we should consider the real shape of the drop and the effects of vapor flow under the drop. It is also necessary to consider other nonlinear terms.

Other phenomena, such as the mode selection rule observed experimentally can not be yet understood in the present framework and left for future work.

References

[1] N. J. Holter and W. R. Glasscock : J. Acoust. Soc. Am. **24** (1952) 682

[2] K. Adachi and R. Takaki : J. Phys. Soc. Jpn. **53** (1984) 4184

[3] R. Takaki and K. Adachi : J. Phys. Soc. Jpn. **54** (1985) 2462

[4] R. Takaki, A. Katsu, Y. Arai, and K. Adachi : J. Phys. Soc. Jpn. **58** (1989) 129

CHEMICAL REACTION ON SIERPINSKI GASKET

A.Yu. Tretyakov and H. Takayasu
Department of Earth Sciences, Faculty of Science,
Kobe University, Kobe 657, Japan

We study numerically two chemical reaction models on a Sierpinski gasket in a view to find out if any unique features will appear in comparison with the same models in 2 dimensions. We find that while in the case of monomer-dimer reaction model there is no significant difference, the branching annihilating random walk model on Sierpinski gasket leads to completely different behavior.

1. Introduction.

One of the most exciting features of second order phase transitions is their universality. For dynamical phase transitions, occurring on a lattice, it can be formulated as follows: critical indexes for a given process do not depend on the details of a lattice structure and are determined solely by its dimensionality. With the emergence of the physics of fractals a question arises: What if the lattice is fractal? Are the critical indexes going to assume some new value, or they are going to be equal to the critical indexes for the same process in the embedding dimension?

In order to through some light on this problem we have studied two processes on Sierpinski gasket, a lattice with fractal dimension between 1 and 2 (see Fig. 1).

Fig.1. Sierpinski gasket (iteration 3).

The processes we studied can be interpreted as catalytic reactions in which particles appear and react on sites of a lattice, leading in the infinite time limit, depending on the values of control parameters, either to a degenerate state when no more reaction is possible (the catalyst is poisoned) or to a steady state, in which reaction is going on indefinitely. A second order phase transition occurs between the poisoned state and the steady state. The principal aim of our study was to find out if the critical indexes for the steady state concentration of particles in the case of reaction on a Sierpinski gasket differ from those for the same processes in two dimensions.

2. Monomer-dimer catalysis model.

The monomer-dimer catalysis model was introduced by Ziff et. al.[1] and later generalized by D. ben-Avraham et. al.[2] as a model of carbon dioxide oxidation on a metal catalyst. In the model, the catalysis involves two elementary steps: adsorption and surface reaction. In the adsorption step, which takes place with probability 1-r, a deposition attempt is made either by an oxygen dimer (A_2), or a carbon monoxide monomer (B) with relative probability p or q=1-p, respectively. In the case of A_2 deposition attempt a pair of nearest neighboring sites is chosen at random and if both of them are free they become occupied by (immobile) surface atomic oxygens, which we denote by A_s. In the case of the monomer deposition attempt, a site is chosen at random and if it is free it becomes occupied by surface-bound reactant B_s. In the reaction step, taking place with probability r, a pair of nearest neighboring sites is chosen at random, and if it is occupied by A_s and B_s, these two reactants bond to form a carbon dioxide molecule which desorbs immediately. This leaves behind two vacant sites which can then accommodate additional particles.

The most interesting feature of the model is the existence of a well defined region of r and p values where the system can reach a steady state corresponding to continuous reaction[1,2]. Outside of this region the catalyst surface eventually becomes fully covered with one of the species (A_s or B_s) and no more reaction is possible. In the case of transition from A saturated surface to the steady state region concentrations on the surface are changing continuously (second order transition), in the case of transition from B saturated surface to the steady state region concentrations are changing in a discontinuous manner (first order transition).

The results of our simulations (see Tretyakov and Takayasu[3] for details) are shown in Fig.2 and Fig.3. We were not able to locate steady state for the values of r less than 0.6. For r > 0.6 the steady state region appears for a small interval of p values, the interval becomes wider as we are closer to the adsorption controlled limit, but still it is much thinner than in the case of the square lattice[4]. At r=1 we have the first order transition at p_1=0.509 and the second order transition at p_2=0.534 (for the square lattice corresponding values are[5] p_1=0.4723 and p_2=0.6125).

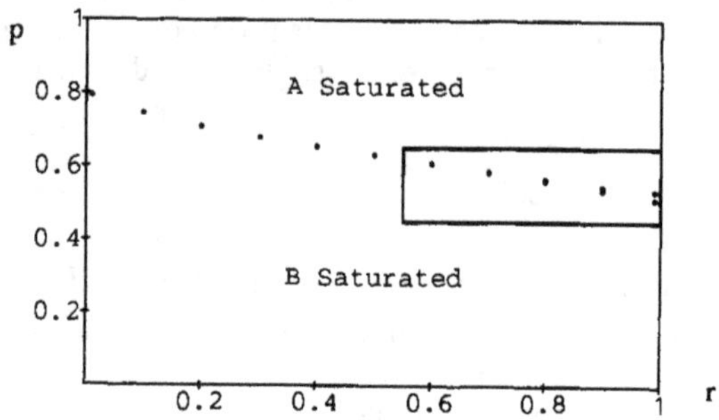

Fig.2. Phase diagram for the monomer-dimer catalysis model on Sierpinski gasket. The region where the steady state has been located is highlighted with a box.

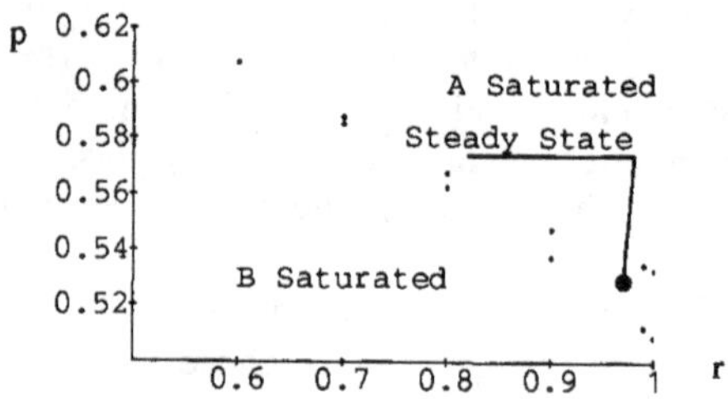

Fig.3. Enlarged phase diagram for the steady state region in Fig.2.

According to Meakin[5] critical exponents (at r=1) for the concentrations of A_s and B_s near the second order transition are β_A=0.61 and β_B=0.69, respectively, and are the same for the square lattice and for hexagonal lattice in two dimensions. Our data (Fig.4) indicate that $\beta_A=\beta_B=0.7\pm0.1$ for the Sierpinski gasket. It looks probable that the Sierpinski gasket and two dimensions fall in the same universality class.

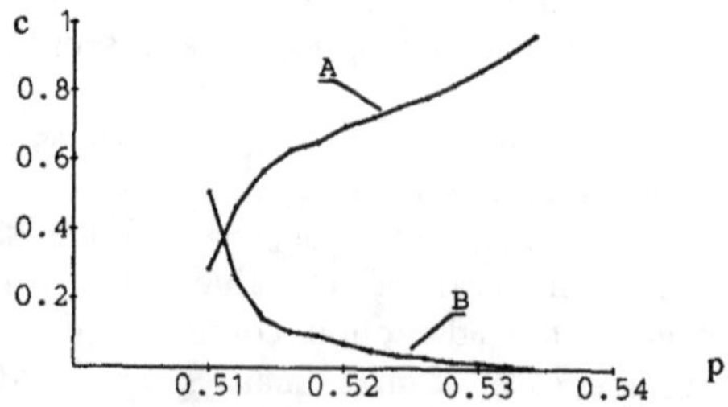

Fig.4. Concentrations of dimer (A) and monomer (B) covered sites of the catalyst surface in the steady state region in the adsorption controlled limit (r=1).

3. Branching annihilating random walk.

Branching annihilating random walk[6,7] (BAW) of particles, occupying sites of a lattice, is a process comprising of two steps, taking place with probabilities ξ and $1-\xi$ respectively: 1. A particle is chosen at random and moves to a randomly chosen nearest neighboring site. If this site is already occupied, the two particles annihilate leaving an empty site. 2. A particle, chosen at random, produces a fixed number of offsprings which are placed on neighboring sites. If a newborn offspring tries to occupy an already occupied site, it annihilates with the occupying particle leaving an empty site.

In the course of BAW particles can eventually disappear (extinction) or a nonzero concentration of particles can exist indefinitely (survival).

The results of our simulations (see Takayasu and Tretyakov[8] for details) are presented in Fig.5 and Fig.6 and in Tab. 1.

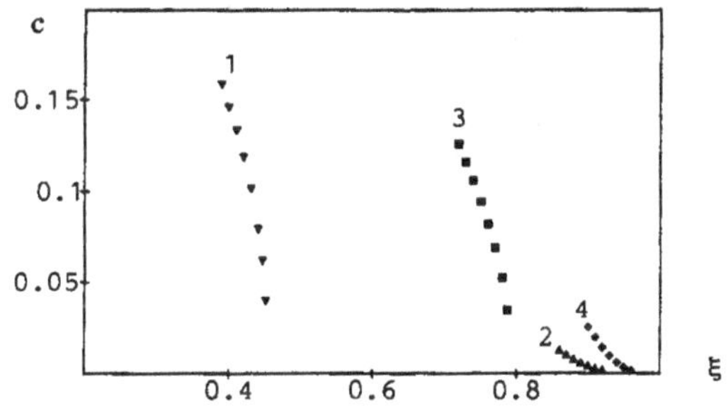

Fig.5. Sierpinski gasket: Steady state concentrations of particles near the threshold (BAW model). Numbers indicate the number of offsprings.

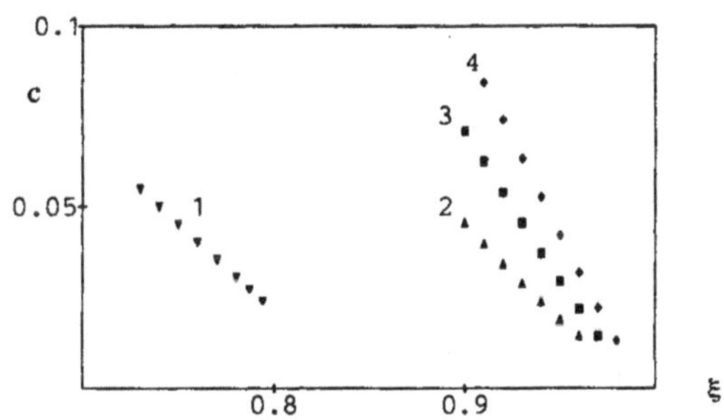

Fig.6. 2-dimension, square lattice: Steady state concentrations of particles near the threshold (BAW model).

Tab.1. Critical probabilities (ξ^*) and critical exponents (β) for BAW model.

Number of offsprings	Sierpinski gasket		2d	
	ξ^*	β	ξ^*	β
1	0.45±0.01	0.5±0.1	0.85±0.01	1.0±0.1
2	1.00±0.01	3.0±0.3	1.00±0.01	1.25±0.02
3	0.79±0.01	0.5±0.1	1.00±0.01	1.26±0.02
4	1.00±0.01	3.0±0.3	1.00±0.01	1.25±0.02

4. Discussion.

As we could see, in the case of monomer-dimer reaction model the critical indexes on Sierpinski gasket, both for monomer and dimer stationary concentrations, are, within the calculation error, equal to the ones estimated by Meakin[5] for 2-dimension. In the case of branching annihilating random walk we have a completely different picture: not only the critical indexes on Sierpinski gasket are definitely different from the ones in 2-dimension (Tab. 1), but also the shape of the stationary concentration dependences on the control parameter (see Fig.5 and Fig.6) clearly indicates that the behavior of the model on Sierpinski gasket differs from the behavior in 2-dimension.

Although our study is incomplete and it would be desirable to have the results for fractal lattices other than Sierpinski gasket, we think that our results indicate on a possibility of a new kind of classification of second order phase transitions in lattice models: those which have the same critical exponents on fractal lattices as in embedding dimension and those which are not universal in this respect.

One of the authors (A.Yu.T.) is indebted to Japanese Society for Promotion of Science for financial support.

1. R. Ziff, E. Gulari and Y. Barshad, Phys. Rev. Lett. 56 (1986) 2553.
2. D. ben-Avraham, D. Considine, P. Meakin, S. Redner and H. Takayasu, J. Phys. A 23 (1990) 4297.
3. A.Yu. Tretyakov and H. Takayasu, Phys. Rev. A 44 (1991) 8388.
4. D. Considine, H. Takayasu and S. Redner, J. Phys. A 23 (1990) L1181.
5. P. Meakin and D. J. Scalapino, J. Chem. Phys. 87 (1987) 731.
6. M. Bramson and L. Gray, Z. Wahrsch. verw. Gebiete 68 (1985) 447.
7. A. Sudbury, Ann. Probab. 18 (1990) 581.
8. H. Takayasu and A.Yu. Tretyakov, submitted for publication to Phys. Rev. Lett..

DYNAMICS OF WAVE FRONTS IN EXCITABLE MEDIA

YAMADA, Hiroyasu and NOZAKI, Kazuhiro
Department of Physics, Nagoya University, Nagoya 464–01, Japan

The dynamics of wave fronts in two dimensional excitable media is analytically studied. By means of the asymptotic perturbation method for wave fronts in reaction-diffusion systems, we derive a space-derivative Burgers' equation which governs the shape of boundary layers and their propagation. Application of Lie group analysis of differential equations yields group-invariant solutions of this equation and their classification. Some classes of the group-invariant solutions determine the remarkable patterns: expanding circle waves, rotating spiral waves, etc.

The theory of wave propagation in excitable media is usually based on the following reaction-diffusion equations: [1]

$$\begin{aligned}\partial_t u &= D\Delta u + f(u, v),\\ \partial_t v &= \delta\Delta v + g(u, v),\end{aligned} \tag{1}$$

where Δ is the Laplacian operator. The variables u and v are known as the activator and inhibitor, respectively. D and δ (≈ 0) are their diffusion coefficients. The nonlinear functions $f(u, v)$ and $g(u, v)$ describe the reaction kinetics of the system. The u-nullcline $f(u, v) = 0$ is N-shaped, the v-nullcline $g(u, v) = 0$ is monotonous, and they have only one intersection at $(u, v) = (U^S, V^S)$. The unique homogeneous steady state (U^S, V^S) is locally stable but excitable: while small perturbations from the steady state are rapidly damped, the suprathreshold stimulus evoke an excitation cycle governed by the local reaction kinetics, before the system returns to the steady state.

In one spatial dimension, an excitation cycle produces a solitary pulse with definite velocity. A wave front is a boundary layer between the rest state and excited state of the activator u in the non-recovering region $v = V^S$, and it can be observed as a trigger wave propagating to resting medium. If width of such boundary layer is sufficiently small, the system in non-recovering region can be reduced to one component system: [1]

$$\partial_t u = D\Delta u + f(u; V^S). \tag{2}$$

Travelling wave-front solutions of Eq. 2 are heteroclinic orbits for the ordinary differential equation in the moving frame, $\xi = x - ct$:

$$Du_{\xi\xi} + cu_{\xi} + f(u;V^S) = 0, \quad u(\xi \to +\infty) = U^S, \quad u(\xi \to -\infty) = U^E, \tag{3}$$

where U^S and U^E are the rest and excited state of activator u respectively. Equation. 3 is a nonlinear eigenvalue problem for wave-front velocity c parametrized by V^S. That is, the level of inhibitor in a wave front determines the wave velocity $c = c(V^S)$ and wave form $u = u(\xi; V^S)$.

In this paper, we assume that one-dimensional solitary wave fronts are known, i.e. Eq. 3 is solved, and use them to construct wave fronts in two-dimensional media. Using a perturbation method and geometrical relations, we derive a space-derivative Burgers' equation, which governs the configuration of boundary layers in two-dimensional space and their propagation. Our result is consistent with the curvature effect on wave propagation established by V. S. Zykov,[2,3] J. J. Tyson and J. P. Keener,[4] and J. P. Keener.[5] Applying the theory of symmetry groups of differential equations[7,8] to the Burgers' equation, we obtain some group-invariant solutions in analytical form. They gives a classification of patterns, and contains the special solutions corresponding to expanding circle patterns, rotating spiral waves.

To extend the theory of wave-fronts in one spatial dimension to two spatial dimension, we introduce a change of variables to a moving coordinate system in which the coordinate axes move with the wave front:[6] $(x, y, t) \to (s, \xi, t)$, where s is the arclength coordinate along a level curve of the boundary layer, ξ is the distance coordinate from this curve, and thus they are locally orthogonal coordinates. We assume that the curvature of the level curve is small (the radius of curvature is much larger than width of a boundary layer); in two-dimensional cross-section, approximate solutions of Eq. 2 are close to solutions of the one-dimensional equation; the variation of u along the arclength coordinate is small; and the form of wave fronts varies slowly. We now introduce a stretched coordinate transformation and perturbation expansions for a small parameter ϵ,

$$\eta = \epsilon^{1/2}s, \quad \xi = \xi, \quad \tau = \epsilon t, \tag{4}$$

$$\kappa = \epsilon\tilde{\kappa}, \quad \alpha = \epsilon^{1/2}\tilde{\alpha}, \quad \beta = \beta_0 + \epsilon\tilde{\beta}, \quad \omega = \epsilon^{3/2}\tilde{\omega},$$

$$u = u^{(0)} + \epsilon u^{(1)} + \cdots, \tag{5}$$

where α is the tangential velocity, β the normal velocity, and κ the curvature of the curve, and ω denotes the rotation frequency of the local coordinate system. Substitution of these expression into Eq. 2 on the moving coordinates, yields the system of equations for each order in ϵ. The zero-order equation is corresponding to Eq. 3, and hence an eigenvalue is $\beta_0 = c$. A solvability condition of the first-order equation gives $\tilde{\kappa} = \tilde{\beta}/D$. Applying these results to the geometrical relations, such as Serret-Frenet formulas, we derive the equation for $\tilde{\alpha}$ which describes asymptotic dynamics of wave fronts,

$$\tilde{\alpha}_{\eta\tau} = (\tilde{\alpha}\tilde{\alpha}_{\eta} + D\tilde{\alpha}_{\eta\eta})_{\eta}. \tag{6}$$

If we get an explicit solution of Eq. 6, then it yields $\tilde{\kappa} = \tilde{\alpha}_\eta / c$. The curvature $\tilde{\kappa}$ determines an instantaneous curve for fixed τ, and temporal evolution of it represents formation process of the curve.

To obtain explicit solutions of Eq. 6, we use the method of the symmetry group of differential equations[7–9] and thereby obtain group-invariant solutions which are characterized by their invariance under some symmetry group of the equation. These solutions are found by solving a system of ordinary differential equations which are often known for their theoretical or applicative interest. We list some of natural equations corresponding to group-invariant solutions of Eq. 6.

1. *Arc or Circle*: $\kappa = -1/ct, \quad t > 0.$

   The curvature κ is independent of arclength s and it represents an arc or circle with radius ct.

2. *Spiral*: $\kappa = \dfrac{2D}{c} \dfrac{d^2}{ds^2} \log Ai\left[\left(\dfrac{c\omega}{2D^2}\right)^{1/3} s\right], \quad s > 0, \quad Ai$: Airy's function.

   The asymptotic form of this curvature $\kappa \sim \sqrt{\omega/2cs} \quad (s \to +\infty)$ represents the rotating Archimedean spiral with the pitch $2\pi c/\omega$.

3. *Knee*: $\kappa = \dfrac{c\phi^2}{8D} \operatorname{sech}^2\left(\dfrac{c\phi}{4D} s\right).$

   This natural equation constructs a travelling line bent at $s = 0$ with an angle ϕ.

We discussed the dynamics of wave fronts in two-dimensional excitable media. Using the perturbation method, we find that a space-derivative Burgers' equation asymptotically governs the shape of boundary layers in two-dimensional space and their propagation or evolution. Further, applying the theory of symmetry groups to this equation, we obtained group-invariant solutions and their classification. Some classes of them contain special solutions corresponding to the patterns which are of experimental or theoretical interest, e.g. expanding circle patterns, rotating spiral waves.

References

1. J. J. Tyson and J. P. Keener, Physica **D32** (1988) 327, and references cited therein.
2. V. S. Zykov, Biophys. **25** (1980) 329.
3. V. S. Zykov, Biophys. **25** (1980) 906.
4. J. J. Tyson and J. P. Keener, Physica **D21** (1986) 307.
5. J. P. Keener, SIAM J. Appl. Math. **46** (1986) 1039.
6. J. P. Keener, Physica **D31** (1988) 269.
7. L. V. Ovsiannikov, *Group Analysis of Differential Equations*, (Academic Press, New York, 1982).
8. P. J. Olver, *Applications of Lie Groups to Differential Equations*, (Springer-Verlag, New York, 1986).
9. M. Tajiri, S. Kawamoto and K. Thushima, Math. Japonica **28** (1983) 125.

C: Convective Instability

CONVECTION IN GASSES AT ELEVATED PRESSURES

Eberhard Bodenschatz†, Stephen W. Morris†, John R. de Bruyn††, David S. Cannell†, and Guenter Ahlers†

†Department of Physics and Center for Nonlinear Sciences
University of California, Santa Barbara, CA 93106

††Department of Physics, Memorial University
St. John's, Newfoundland, CANADA A1B 3X7

ABSTRACT

We present recent experimental results for Rayleigh-Bénard convection in a thin horizontal layer of a gas under pressure. In this system it is possible to reach large aspect ratios. By changing the pressure, samples with negligible or with very large departures from the Boussinesq approximation can be prepared. In the non-Boussinesq system, a defect-free hexagonal lattice of convection cells forms just above onset. This lattice becomes unstable to rolls when the temperature difference is increased sufficiently. In the "roll" state, the rolls are curved and the system forms stable rotating spirals. By suitably adjusting the sample thickness and the pressure, the system can be made very susceptible to noise and therefore it is possible to observe fluctuating convection rolls close to but below the first bifurcation to convection.

Most convection experiments involving flow visualization have been carried out using ordinary liquids like water and oils of various viscosities. These classical systems have yielded a great deal of experimental information;[1] but they have limitations imposed by the typical properties of the liquids. One important limitation is the range of the Prandtl number σ which can be investigated. In particular, the interesting values $\sigma \simeq 1$ are not covered. These are accessible in gasses and in liquid helium. Gasses at ambient temperature have the advantage over liquid helium[2-4] that flow visualization is relatively simple. In the cryogenic environment visualization has not been accomplished so far. Thus convection in gasses can yield interesting information which is complementary to that attainable in classical liquids. Indeed early experiments have used air at ambient pressure to explore this important range.[5,6] More recent experiments have employed Argon at elevated pressures.[7,8] Another experimental limitation encountered in classical fluids is imposed by long time scales of the experiments. These are associated in part with the typically rather large layer thicknesses d which have to be used. The practical lower limit of d is set by the necessity for the system to have an achievable critical temperature difference ΔT_c if convection near onset is to be studied. We have[9]

$$d = (\kappa \nu R_c / \beta g \Delta T_c)^{1/3} \quad , \tag{1}$$

where κ, ν, and β are the thermal diffusivity, kinematic viscosity, and isobaric thermal expansion coefficient respectively, g is the acceleration of gravity, and where $R_c = 1708$ is the critical Rayleigh number. Substituting typical properties of liquids and a reasonable value of ΔT_c yields values for d which are of the order of several mm. This in turn leads to vertical thermal diffusion times $t_v \equiv d^2/\kappa$ of the order of 10^2 s. Since many pattern-adjustment processes are diffusive, the lateral diffusion times often are relevant in typical experiments; these are larger than t_v by a factor L^2, where $L \times d$ is the lateral size of the experimental cell. If equilibration times are to be kept below several hours, then the aspect ratio L essentially is limited to $L \lesssim 10$. Larger systems will have extremely long equilibration times, and indeed some experiments with a run time of one month under constant externally imposed conditions have been carried out in an attempt to explore these long time scales.[10] Obviously, one soon encounters practical time limitations.

In compressed gasses the situation is quite different and more flexible. Here we have approximately pressure independent and relatively small transport coefficients. Since $\kappa = \lambda/\rho C_P$ and $\nu = \eta/\rho$ where λ and η are the thermal conductivity and shear viscosity respectively, and where ρ is the density, Eq. (1) can be written as

$$d = (\lambda \eta R_c / \rho^2 C_P \beta g \Delta T_c)^{1/3} \quad . \tag{2}$$

Since ρ is roughly proportional to the pressure P, the sample thickness can be varied over a wide range by changing P while keeping ΔT_c at a reasonable value. In particular, d can be made as much as an order of magnitude smaller than is the case for typical liquids. This yields values of t_v in the range of one to ten seconds, and permits the investigation of large aspect ratios on reasonable time scales. Experiments can be carried out sufficiently slowly to achieve lateral equilibration of diffusive processes in samples with L as large as 100 or more. If it is desirable to minimize d, gasses of high atomic or molecular mass M are particularly advantageous since ρ is approximately proportional to M. In our experiments we have used primarily CO_2 because it has a relatively large value of M, is readily available in high purity, and is inexpensive. Table 1 below shows some typical values of ΔT_c and d for a few fluids and at various pressures. It also provides information on the effective strength F_0 of thermal noise and on the size Q of the departures from the Boussinesq approximation ($Q = 0$) which will be discussed in more detail below.

It is worth keeping in mind that the experimentalist pays a price for making his sample extremely thin. If the Rayleigh number R is to be spatially uniform, then the thickness of the fluid layer also has to be very uniform. Thus, for instance, a sample cell used in our laboratory[11] had a thickness of $d \simeq 0.05$ cm which was uniform to 1 μm. Although 1 μm variation over a sample diameter of about 10 cm is very little indeed, it nonetheless corresponds to a variation of the Rayleigh number by 0.6% and in that sense is not very satisfactory for experiments which have a temperature resolution and stability[11] corresponding to 0.002% of R. For that reason we have recently developed a new experimental cell in which the top and bottom plates are optically flat and in which the spacing between them can be uniform over the sample diameter to 0.1 μm. This cell is only just now becoming operational.

Table 1. Parameters at the onset of Rayleigh-Bénard convection for various fluids at 25°C, and for electroconvection (EC).

fluid	Pressure [bar]	ΔT_c [°C]	d [mm]	$10^9 \times F_0$	Q
Water	1	5	2.6	2	-0.90
N_2	45	5	1.2	540	0.26
CO_2	10	5	2.0	170	0.25
CO_2	30	5	0.76	1400	0.66
CO_2	45	1	0.81	2000	0.15
CO_2	45	5	0.47	3400	0.76
CO_2	45	30	0.26	6200	4.71
Xe	30	5	0.50	2400	0.68
Xe	45	1	0.48	3600	0.22
Xe	45	5	0.28	6100	1.10
Xe	45	30	0.15	11000	6.95
EC	1	-	0.01	40000	—

Rayleigh-Bénard convection is usually discussed in the approximation of temperature-independent fluid properties (except for the density where it enters into the buoyancy term) which was made by Oberbeck[12] and by Boussinesq[13] (OB). For an OB fluid, the pattern which forms just above the critical temperature difference ΔT_c consists of rolls, and in steady state the roll amplitudes grow continuously from zero[14] as ΔT increases beyond ΔT_c. However, in the non-OB case, the bifurcation is expected to become hysteretic, and the first stable pattern which forms is predicted to consist of a hexagonal arrangement of convection cells.[15–19] Early experiments[20–25] have indeed confirmed some of these predictions. It can be seen from Table 1 that the critical temperature difference for a gas can be varied over a wide range. When ΔT_c is large, the fluid properties vary significantly over the sample thickness and deviations from the OB approximation become noticeable. The parameter Q listed in the last column of the table is defined elsewhere[18,26] and gives a measure of the departure from the OB approximation when $\Delta T = \Delta T_c$.

Here we report on an experimental investigation of non-OB effects[11] which utilized the advantages of convection in gasses. It had much higher stability and resolution than previous work, and could be carried out on a much larger sample than had been used in the past. Our sample of CO_2 had a circular horizontal cross section and a thickness $d = 0.52\ mm$. Its diameter was 8.94 cm and the pressure was 22 bar. As discussed above, the variation in the sample thickness was less than 1 μm. The temperature at the top was 12.8 ^oC. The critical temperature difference was about 29 ^oC. The temperature stability and resolution was approximately 0.5 m^oC or 0.002% of ΔT_c. The pressure was regulated and kept constant to about 10^{-4} bar or 5 ppm. For this sample, t_v was 0.8 sec, and the Prandtl number σ was about 0.9. The large variation of the temperature in the vertical direction led to significant variations of all the fluid properties and yielded the hexagonal state at onset. A digitally enhanced shadowgraph method[27] was used to visualize the flow pattern.

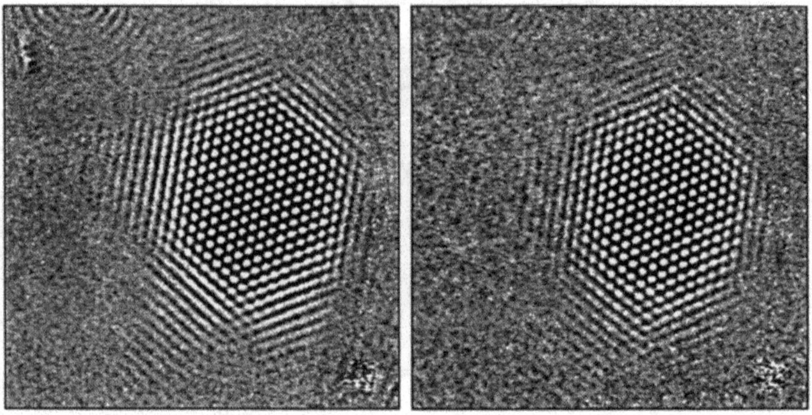

Fig. 1. Patches of hexagonal cells. The left figure is a transient which appears essentially at $\epsilon = 0$ when changing ϵ quasi-statically in steps $\Delta\epsilon = 3 \times 10^{-5}$.The right figure is a steady state for $\epsilon = -1.92 \times 10^{-3}$. Its localized nature is attributable to slow spatial variations in the sample thickness. After one more step in ϵ of -3×10^{-5}, convection ceases altogether.

The left portion of Fig.1 is an image which illustrates the evolution of a hexagonal lattice observed in an experiment where $\epsilon \equiv \Delta T/\Delta T_c - 1$ was raised just barely (by about 10^{-5}) above the critical value for the convective onset. It shows the early transient after a local region of hexagonal convection had just formed. Six fronts of rolls are travelling away from the patch of hexagons. As they propagate further into the region without a pattern, they spread the hexagonal lattice by virtue of their superposition.

If ϵ is reduced sufficiently, the hexagonal pattern ceases to exist and only the conduction state is a solution of the equations of motion. The (stable) pattern found in the experiment for a negative ϵ-value just above this point is shown in the right portion of Fig. 1. In this case there is a limited spatial region of hexagonal cells because of the very slight variations in the sample thickness (less than 1 μm) discussed above which cause local variations of ϵ.[11] Under these conditions the fronts are stationary. They are also somewhat less pronounced than in the transient case (left figure). The existence of convection at negative ϵ confirms the hysteresis[11] predicted by the theory.[18]

When ϵ is increased beyond zero, the entire sample fills with convection cells which are arranged in a hexagonal pattern. A "lattice" of about 5000 such cells is shown in Fig. 2 for $\epsilon = 0.06$. It is noteworthy that the hexagonal lattice in Fig. 2 contains no defects. It apparently represents, at least for length scales of order the sample diameter, a state of broken translational symmetry. There has been some discussion in the past about whether genuine broken symmetries, such as those in equilibrium crystals, exist for dissipative non-equilibrium systems.[28]

Very large dissipative structures without defects have been produced in electro-convection (EC) in nematic liquid crystals.[29] Those systems close to onset are essentially one-dimensional, however, in the sense that the convection rolls have a strongly preferred orientation which is imposed by the boundary conditions at the top and bottom of the sample. In Rayleigh-Bénard convection (RBC), the roll or hexagon orientation is completely degenerate, and the pattern could have any angular orientation unless the lateral walls prevent this. One might think that this isotropic two-dimensional system is more likely to form defects; and thus it is interesting that very large defect-free patterns can be produced for RBC as well as for EC. A quantitative study of the extent to which long-range order prevails in patterns such as that in Fig. 2 should be interesting. One would like to know, for instance, whether the widths of the Fourier peaks corresponding to the hexagonal pattern are determined entirely by the sample size, or whether there is any noticeable intrinsic broadening. Some broadening associated with fluctuations induced by thermal noise would of course be expected; but whether this effect will be measurable remains an open question at this time.

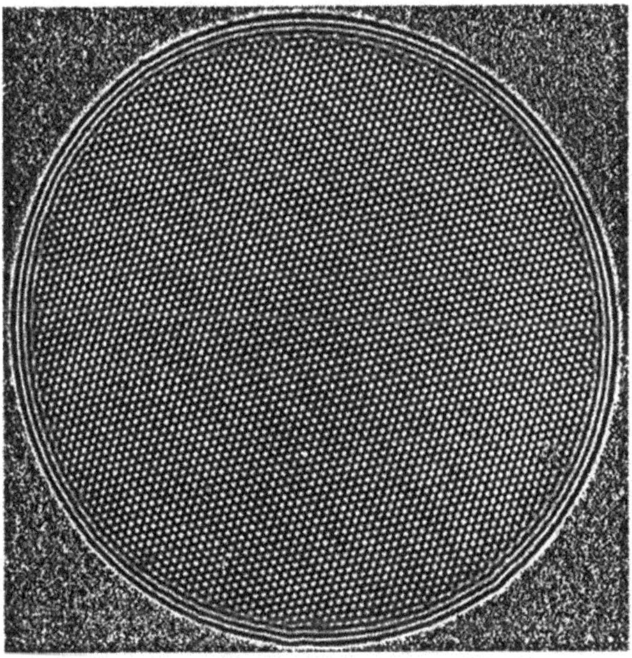

Fig. 2. A defect-free hexagonal lattice of convection cells obtained with a non-Boussinesq sample. For this example, $\epsilon = 0.06$. The small blemish near the lower right corner is a stain on the bottom plate which has no noticeable influence on the fluid flow. After Ref. 11.

For even larger temperature differences, the theory predicts that convection rolls become stable even in the non-OB case. There then follows a range of ϵ

over which both rolls and hexagons are expected to be stable. At even higher ΔT, hexagons lose their stability and only rolls can exist. In the experiment a transition from hexagons to rolls was observed when ϵ was greater than about 0.12. The final state which was reached after transients had died out consisted of an n-armed spiral. So far we have found spirals with $n = 1, 2, ..., 13$. An example for $n = 1$ is shown in Fig. 3. The n arms of the spirals terminate in n dislocations which can be located in the sample interior or near the sidewall. When they are located in the interior, the spiral is surrounded by concentric rolls. The spirals can have either chirality. They are stable states; a two-armed one was kept in the sample cell for two weeks ($1.5 \times 10^6 t_v$) under constant external conditions. An interesting property of the spirals is that they rotate so as to emit a radial outward-travelling wave of rolls. During the rotation, the tips and dislocations rotate at the same angular speed, thus maintaining the spiral structure. Although a systematic study has not yet been performed, a typical rotation frequency is $\omega \cong 10^{-3}$ when time is scaled by the vertical thermal diffusion time. We regard the spiral flow as particularly interesting since it is ubiquitous in pattern-forming non-equilibrium systems. Spirals are encountered in such diverse areas as chemical reactions with spatial variation and biological systems.[30]

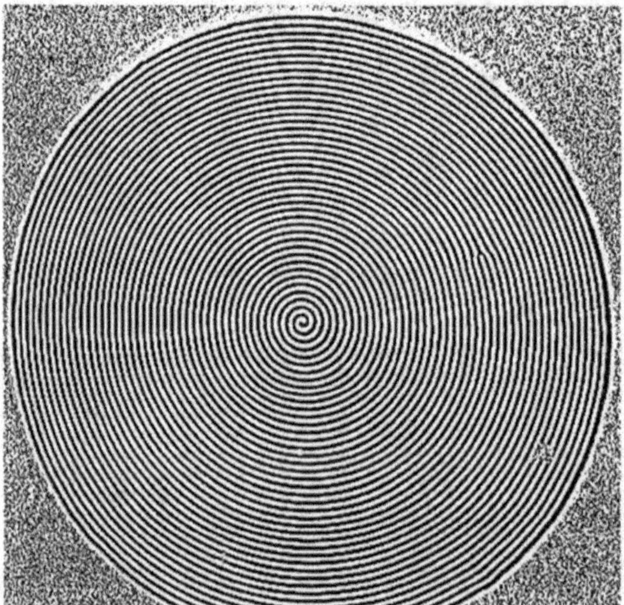

Fig. 3. A one-armed spiral at $\epsilon = 0.12$. Notice that the spiral arm terminates in one defect near the right edge of the image. This structure is a stable state; the spiral tip and the defect rotate synchronously thus preserving the structure.

Another great opportunity offered by convection in gasses is the study of external noise on the system just below the bifurcation to convection. Bifurcations, and the nature of the patterns which evolve beyond them in nonlinear dissipative systems, are usually discussed in terms of deterministic equations for the macroscopic variables[9,31] which neglect the "microscopic" degrees of freedom of the system. In the ideal deterministic system there is a sharp bifurcation point $R = R_c$ at which an exchange of stability occurs between the spatially uniform state and the state with spatial variation. However, if the system is subjected to a stochastic force, then even below the bifurcation there should be fluctuations away from the uniform state.

Although it was recognized some time ago that stochastic forces associated with the microscopic degrees of freedom due to the thermal motion of the atoms or molecules of a system should in principle be taken into consideration,[32–38] it seemed that their effect would be unobservably small for macroscopic systems where typical dissipative energies are many orders of magnitude larger than the microscopic energy k_BT. Nonetheless, it has been possible recently to observe *directly* the fluctuations, which are the response of the system to the stochastic field, below the first bifurcation in hydrodynamic systems.[39,40] In the absence of nonlinear saturation effects, the fluctuation amplitudes diverge at the bifurcation point. In practice these amplitudes become sufficiently large very close to the bifurcation to permit quantitative measurements[39] when a favorable system is chosen. The effective strength of thermal noise for a typical hydrodynamic instability can be estimated by adding Langevin terms to the hydrodynamic equations.[32] It is related to the ratio of k_BT to a characteristic dissipative energy E_d of the system.[33,34,36,41] In the case of RBC, this energy may be estimated from the mass ρd^3 of a volume of fluid equal to the cube of the layer thickness d, and from the square of the velocity ν/d corresponding to the kinematic viscosity ν. Thus one has $E_d \cong \rho d\nu^2$. The exact strength of the noise which is relevant depends on the physical property of the system which is examined; but for the comparison of different experimental systems it is useful to look at the ratio $F_0 = k_BT/\rho d\nu^2$.

We see that the effective noise strength becomes large particularly when $\nu = \eta/\rho$ is small, *i.e.* in systems of small shear viscosity η and high density. This can be realized in gasses of high atomic or molecular mass at elevated pressure, and in liquid helium.[3,4,42] In Table 1 values of F_0 are given for a few fluid systems which are readily usable in experiments.

A system particularly favorable for the observation of thermally induced fluctuations is electro-convection[29] (EC) in a thin layer of a nematic liquid crystal to which a voltage is applied. In this case, the dissipative energy $E_d \cong dK$ is determined by an average K of the Franck constants[43,44] and by the layer thickness d. It turns out that K is two orders of magnitude smaller than $\rho\nu^2$ for water, and nearly as small as $\rho\nu^2$ for the heavy gasses under pressure (see Table 1). The advantage offered by the small Franck constants for the case of *Rayleigh-Bénard* convection in nematics was recognized already very early by Graham.[35] In addition, for EC the critical voltage for the instability is, to lowest order, independent of the layer thickness. Thus d can be made quite small, thereby further enhancing the effective noise strength. In practice it is possible to carry out quantitative experiments with d as small as 0.01 mm. Then, the effective thermal noise-intensity in EC can be several times larger than it is for the most favorable known RBC cases and four orders of magnitude larger than for RBC in water. This is illustrated by the last row in Table 1.

In the top row of Fig. 4 we show shadowgraph visualizations of fluctuations in a layer of CO_2 with $d = 0.05\ mm$ and $P = 30\ bar$. The four images were taken at time intervals of $75 \times t_v$ and for $\epsilon \simeq -4 \times 10^{-4}$. A careful comparison shows that each image differs from the others. Indeed, they should differ if they are the result of fluctuations; for the typical lifetime of a fluctuation should be $\tau_0/\epsilon \simeq 125 t_v$.

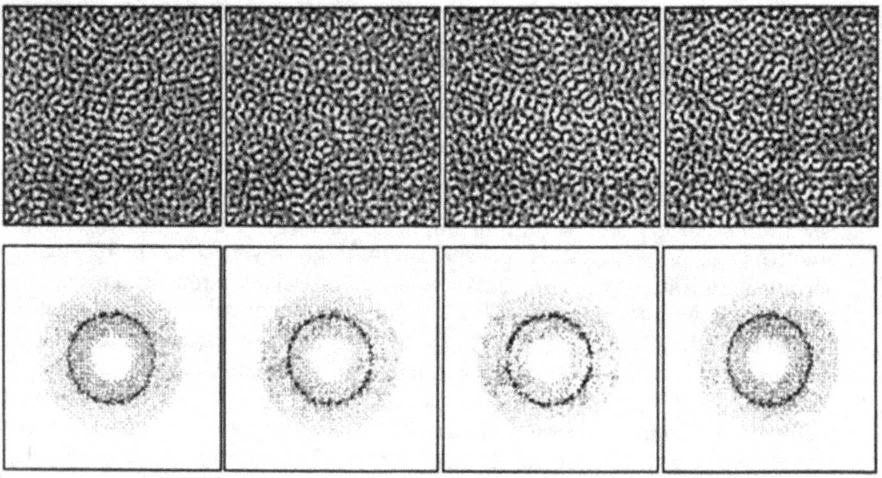

Fig. 4. Top Row: Digitally enhanced shadowgraph images of fluctuations in the Rayleigh-Bénard system. Bottom Row: Magnitudes of the Fourier transforms of the images above as grey levels, as a function of the horizontal wave vector.

The bottom row of Fig. 4 shows the Fourier transforms of the images in the top row. The transform magnitude is given by grey levels as a function of the horizontal wavevector. Each image yields a transform with the largest values located on a circle of radius close to the critical wavevector at the onset of convection. The distribution around the circle in the Fourier transform is, when averaged over many images, quite uniform. This shows that there is no preferred direction in the horizontal plane for the experimental system, consistent with expectations from theory. In this respect the RB system differs from EC, where there is a preferred direction imposed by the alignment of the director and where fluctuations would lead to a transform consisting of only two spots with a finite width. A quantitative study of the fluctuations in the RB system is now under way in our laboratory.

ACKNOWLEDGEMENTS

This work was supported by the U.S. Department of Energy through Grant DE-FG03-87ER13738. One of us (EB) acknowledges support from the Deutsche

Forschungsgemeinschaft. Two of us (S.W.M. and J.R.de B.) acknowledge support from the Natural Sciences and Engineering Research Council of Canada.

REFERENCES

1. A large literature pertaining to this field has evolved. Particularly useful as introductions to early work are the reviews by E.L. Koschmieder, *Adv. Chem. Phys.* **26** (1974) 177; and in *Order and Fluctuations in Equilibrium and Nonequilibrium Statistical Mechanics,* XVIIth International Solvay Conference, ed. by G. Nicolis, G. Dewel, and J.W. Turner (Wiley, NY, 1981), p. 168; and by F. Busse, in *Hydrodynamic Instabilities and the Transition to Turbulence,* edited by H.L. Swinney and J.P. Gollub (Springer, Berlin, 1984), p. 97; and in *Rep. Prog. Phys.* **41** (1978) 1929.
2. The use of liquid helium for the study of Rayleigh-Bénard convection has numerous other advantages, which have been discussed in Refs. 3 and 4.
3. G. Ahlers, *Phys. Rev. Lett.* **33** (1974) 1185.
4. G. Ahlers, in *Fluctuations, Instabilities, and Phase Transitions*, edited by T. Riste (Plenum, NY, 1975), p. 181.
5. See, for instance, Ref. 6, and references therein.
6. G.E. Willis, J.W. Deardorff, and R.C.J. Sommerville, *J. Fluid Mech.* **54** (1972) 351.
7. V. Croquette, *Contemp. Phys.* **30** (1989) 113.
8. V. Croquette, *Contemp. Phys.* **30** (1989) 153.
9. S. Chandrasekhar, *Hydrodynamic and Hydromagnetic Stability*, (Oxford University Press, London, 1961).
10. G. Ahlers, D. S. Cannell, and V. Steinberg, *Phys. Rev. Lett.* **54** (1985) 1373.
11. E. Bodenschatz, J.R. de Bruyn, G. Ahlers, and D.S. Cannell, *Phys. Rev. Lett.* **67** (1991) 3078.
12. A. Oberbeck, *Ann. Phys. Chem.* **7** (1879) 271.
13. J. Boussincsq, *Théorie Analytique de la Chaleur*, vol.2. (Gauthier-Villars, Paris, 1903).
14. A. Schlüter, D. Lortz, and F. Busse, *J. Fluid Mech.* **23** (1965) 129.
15. E. Palm, *J. Fluid Mech.* **8** (1960) 183.
16. L. A. Segel and J. T. Stuart, *J. Fluid Mech.* **13** (1962) 289.
17. E. Palm, T. Ellingsen, and B. Gjerik, *J. Fluid Mech.* **30** (1967) 651.

18. F. Busse, *J. Fluid Mech.* **30** (1967) 625.

19. S. H. Davis and L. A. Segel, *Phys. Fluids* **11** (1968) 478.

20. C. Q. Hoard, C. R. Robertson, and A. Acrivos, *Int. J. Heat Mass Transfer* **13** (1970) 849.

21. M. Dubois, P. Bergé, and J. Wesfried, *J. Phys.* **39** (1978) 1253.

22. F.M. Richter, *J. Fluid Mech.* **89** (1978) 553.

23. R.W. Walden and G. Ahlers, *J. Fluid Mech.* **109** (1981) 89.

24. S. Ciliberto, E. Pampaloni, and C. Perez-Garcia, *Phys. Rev. Lett.* **61** (1988) 1198.

25. S. Ciliberto, P. Coullet, J. Lega, E. Pampaloni, and C. Perez-Garcia, *Phys. Rev. Lett.* **65** (1990) 2370.

26. G. Ahlers, *J. Fluid Mech.* **98** (1980) 137.

27. See, for instance, V. Steinberg, G. Ahlers, and D.S. Cannell, *Phys. Script.* **32** (1985) 534.

28. P.W. Anderson, in *Order and Fluctuations in Equilibrium and Nonequilibrium Statistical Mechanics*, XVIIth International Solvay Conference, edited by G. Nicolis, G. Dewel, and J. W. Turner (Wiley, NY, 1981)

29. For a recent review of pattern formation in electro-convection of nematic liquid crystals, see I. Rehberg, B.L. Winkler, M. de la Torre-Juarez, S. Rasenat, and W. Schöpf, *Festkörperprobleme-Advances in Solid State Physics* **29** (1989) 35.

30. See, for instance, the numerous papers in *Nonlinear Wave Processes in Excitable Media*, edited by A.V. Holden, M. Markus, and H.G. Othmer (NATO ASI Series Vol. 244, Plenum, New York, 1991); and in *Waves and Patterns in Chemical and Biological Media*, edited by H.L. Swinney and V.I. Krinsky, *Physica D* **49**, (1991) 1.

31. A.C. Newell, T. Passot, and M. Souli, *Phys. Rev. Lett.* **64** (1990) 2378.

32. L.D. Landau and E.M. Lifshitz, *Fluid Mechanics* (Addison-Wesley, Reading MA, 1959).

33. V.M. Zaitsev and M.I. Shliomis, *Zh. Eksp. Teor. Fiz.* **59** (1970) 1583 [English translation: *Sov. Phys. JETP* **32** (1971) 866].

34. R. Graham, *Phys. Rev. A* **10** (1974) 1762.

35. R. Graham, in *Fluctuations, Instabilities, and Phase Transitions*, edited by T. Riste (Plenum, NY, 1975), p. 215.

36. J.B. Swift and P.C. Hohenberg, *Phys. Rev. A* **15** (1977) 319.

37. G. Dewel, P. Borkmans, and D. Walgraef, *J. Phys. C: Solid State Phys.* **12** (1979) L491.

38. D. Walgraef, G. Dewel, and P. Borkmans, *Phys. Rev. A* **21** (1980) 397.

39. I. Rehberg, S. Rasenat, M. de la Torre-Juarez, W. Schöpf, F. Hörner, G. Ahlers, and H.R. Brand, *Phys. Rev. Lett.* **67**, 596 (1991); and unpublished.

40. E. Bodenschatz, S.W. Morris, G. Ahlers, and D.S. Cannell, to be published.

41. P.C. Hohenberg and J.B. Swift, preprint.

42. R.P. Behringer and G. Ahlers, *J. Fluid Mech.* **125** (1982) 219.

43. P.G. de Gennes, *The Physics of Liquid Crystals* (Clarendon Press, Oxford, 1973).

44. S. Chandrasekhar, *Liquid Crystals*, (Cambridge University Press, 1977).

Electrohydrodynamic Convection in Nematics: The Homeotropic Case

L. Kramer, A. Hertrich and W. Pesch
Physikalisches Institut der Universität Bayreuth
D-Bayreuth, Germany

Abstract

Electrohydrodynamic convection in homeotropically oriented layers is discussed within the general framework of convective instabilities in nematics. We present calculations of the threshold for the direct transition to convection, which occurs at large wave numbers for materials with slightly negative and positive dielectric anisotropy. Interesting scenarios are predicted especially when a (weak or strong) destabilizing magnetic field is applied in addition.

1. Introduction

In recent years considerable progress has been made in understanding various features of electrohydrodynamic convection (EHC) in **planarly** aligned nematic liquid crystal layers [1,2,3,4]. Thus one can state that at least for sufficiently thick layers (thickness about $d > 50\mu$, depending also on the conductivity) and not too high frequencies ω of the applied ac voltage the threshold and the pattern at threshold as obtained from a linear stability analysis of the full nematodynamic equations are essentially understood for the standard material MBBA. One of the most interesting features is the possibility of a symmetry-breaking transition between "normal rolls" that are perpendicular to the pre-aligned orientation of the director and "oblique rolls" with decreasing frequency along the threshold curve. Besides the wave vector such detailed features as the coherence length and relaxation time turn out to be in good agreement with experiments [1]. In addition effort has been put into the investigation of the weakly nonlinear analysis which is applicable slightly above threshold [1,5,6]. To lowest order one has to determine only one nonlinear coefficient (in addition to the linear parameters) to establish the universal description in terms of the Ginzburg-Landau (-Amplitude) equation [1]. The most spectacular success in this framework is presumably the quantitative description of the dynamics of defects (dislocations) in normal-roll patterns [7,4,8]. The stability analysis of roll solutions has been carried to higher order where nonpotential effects, especially mean flow, are included [6]. One generally finds with increasing voltage a zig-zag destabilization of normal rolls, which is influenced substantially by mean flow, as in most experiments [2,9]. It thus seems likely that mean-flow effects are also responsible for the observed transition to spatio-temporal chaos, which usually comes later [9]. However at this time such a scenario has only been presented in phenomenological models [10]. Moreover for (unrealistic) free boundary conditions one finds spatio-temporal chaos already at the first threshold [5].

It should be mentioned that some features observed especially in thin layers remain mysterious, most prominently the appearance of traveling rolls [11] and a weakly inverted (or backward) bifurcation at threshold [12] (the standard theory always gives a stationary pattern and a forward bifurcation [1]). Presumably electrochemical effects not included in the theory are responsible for these discrepancies.

Very recently a similarly exhaustive analysis has been presented for Rayleigh-Bènard convection (RBC) in planarly aligned nematic layers [13]. Here a stabilizing magnetic field can play the role of a secondary control parameter, similar to the frequency in EHC. The scenario exhibits some additional features like a transition from oblique to parallel rolls at high fields along the threshold curve and occurrence of tricritical points where the bifurcation changes between forward and backward. Moreover at low fields the secondary destabilization of rolls takes place via the skewed-varicose instability leading presumably to spatio-temporal chaos. Experiments are presently going on that exhibit some of the above features [14].

A very different situation is encountered in **homeotropically** aligned layers because here one has isotropy in the plane of the layer similar to RBC in simple fluids [15]. By applying a magnetic field in the plane of the layer an anisotropy can, however, be turned on smoothly. RBC in the homeotropic geometry was investigated in the past under adverse (heating from above [16]) and normal (heating from below [17]) conditions, and various features like an oscillatory instability and backward bifurcations could be understood in some cases [18]. Very recently a new general analysis has been performed with emphasis on the influence of a magnetic field [19]. In adverse convection the experimentally observed rectangular pattern at zero planar field could be explained. Applying a weak magnetic field one first has rolls and then a secondary transition to rectangles (slightly distorted squares) above threshold. In normal convection the codimension-2 point, where oscillatory convection transforms into steady one, was analyzed in detail. One has some analogy with convection in binary fluids [20] with the separation ratio replaced by the more easily tunable magnetic field.

EHC with homeotropic alignment has attracted up to now only little activity, but the destabilization mechanisms have been discussed before qualitatively [21,22,23]. One aim of this communication is to direct attention to this interesting system. Two very different situations are encountered, that of negative dielectric anisotropy (and not too near to zero, if $\alpha_3 < 0$) as in MBBA (case F), and that of positive (and very slightly negative) dielectric anisotropy ϵ_a and $\alpha_3 < 0$ (case C). In case F the homeotropic alignment becomes unstable at a critical voltage $V_f = \pi\sqrt{\frac{k_{33}}{\epsilon_0|\epsilon_a|}}$, where the director starts to reorient and acquire a planar component (bend-Fréedericksz-transition). Ideally the director bends homogeneously in the plane of the layer, i. e. the planar component is pointing uniformly in an arbitrary direction. The isotropy in the plane is therefore spontaneously broken. Clearly large-scale variations and defects can spoil the ideal Fréedericksz deformation and the temporary application of a planar (or slightly tilted) magnetic field might help to achieve the ideal situation. With increasing voltage V a fully planar situation is gradually approached except in small boundary layers.

It is not surprising then that with further increase of V a secondary instability to EHC takes place as a consequence of the conventional Carr-Helfrich-mechanism [24] in analogy to the case of planar anchoring. Recently the threshold voltage has been calculated in analogy to the case of planar anchoring as a function of the material parameters [25]. As an interesting result it is found that oblique rolls, i.e. rolls oriented obliquely with respect to the bending axis of the director, prevail in most situations. Since there are two degenerate oblique directions a superposition leading to a rectangular pattern is also possible.

Going on to the nonlinear problem it was found that weakly nonlinear solutions can be constructed for normal rolls in the usual way but not for oblique rolls without planar magnetic field. The reason is that oblique rolls excite a rotation of the planar component of the Fréedericksz distorted director, which is a soft mode of the system. Presumably in both cases the system goes directly into a complex spatio-temporal chaotic state similar to the case of planarly aligned nematics with free boundary conditions [5]. In the oblique-roll case one may in addition have a permanent local rotation of the director. This state of weak turbulence should be suppressible by a planar magnetic field.

2. Direct Transition to EHC (Case C)

We now discuss case C, where a direct convective destabilization of the homeotropic alignment occurs [21]. The layer is taken parallel to the x-y-plane. One has a bend-twist mode and a bend-splay mode, which remain uncoupled in the presence of a planar magnetic field that is either parallel or perpendicular to the wave vector of the mode. We have determined the stability boundary of the undisturbed state using the full nematodynamic equations [1,21,26,27]. The z-dependence was treated rigorously in the framework of a Galerkin expansion whereas the time dependence of the ac driving was dealt with in lowest-order Fourier approximation [1]. This scheme can be handled analytically when the Galerkin expansion is also truncated after the first mode ("test function approximation"). The results for the neutral curve $V_0(q)$ (the wave vector q is chosen in the x-direction) which separates stable ($V < V_0$) from unstable ($V > V_0$) modes can be written for the bend-splay mode as follows

$$V_0^2(q) = \frac{\pi^2 k_{33}}{\epsilon_0\epsilon_\perp} \frac{\left(1 - h_x^2 + q^2\frac{k_{11}}{k_{33}}\right)\left(\sigma_b^2 + \omega^2\tau_0^2\epsilon_b^2\right)}{(1+q^2)\left[q^2 A(q) - \epsilon_a' B(q)\right]} , \tag{1}$$

where

$$\sigma_b = q^2 + 1 + \sigma_a' , \qquad \sigma_a' = \frac{\left(\sigma_\parallel - \sigma_\perp\right)}{\sigma_\perp} , \tag{2}$$

$$\epsilon_b = q^2 + 1 + \epsilon_a' , \qquad \epsilon_a' = \frac{\left(\epsilon_\parallel - \epsilon_\perp\right)}{\epsilon_\perp} , \tag{3}$$

$$A(q) = \frac{\left(\alpha_2 - \alpha_3 q^2\right)\sigma_b\left(\sigma_a' - \epsilon_a'\right)}{\lambda_1^4\eta_1 + (\alpha_1+\alpha_3+\alpha_4+\alpha_5)q^2 + \eta_2 q^4} I , \tag{4}$$

$$B(q) = \sigma_b\left(1 + \sigma_a'\right) + \omega^2\tau_0^2\epsilon_b\left(1+\epsilon_a'\right) . \tag{5}$$

Here $h_x = \frac{H_x}{H_f}$ is the reduced magnetic field in x-direction with the Fréedericksz field $H_f = \frac{\pi}{d}\sqrt{\frac{k_{33}}{\mu_0\chi_a}}$, $\tau_0 = \frac{\epsilon_0\epsilon_\perp}{\sigma_\perp}$ is the charge relaxation time and $\eta_1 = \frac{1}{2}(\alpha_4+\alpha_5-\alpha_2)$, $\eta_2 = \frac{1}{2}(\alpha_3+\alpha_4+\alpha_6)$ are shear viscosities (α_i = Leslie viscosity coefficients). $\epsilon_\parallel$, $\epsilon_\perp$ and $\sigma_\parallel$, $\sigma_\perp$ denote the components of the uniaxial dielectric and conductivity tensors (see e. g. [1]). The wavenumber q is measured in units $\frac{\pi}{d}$. As test functions one may use Chandrasekhar functions for the velocity potential [28,29] and trigonometric functions with the appropriate boundary conditions. Then one has $\lambda_1 = 1.5056$ and $I = 0.97267$. For stress free boundary conditions the problem is solved exactly with trigonometric functions for all variables. Then one would have $\lambda_1 = I = 1$ [25].

From Eq. (1) one can obtain almost quantitatively (by comparison with our rigorous numerical calculations) all features of the primary destabilization of the homeotropic alignment. It simplifies further in some relevant cases For $\epsilon_a' < 0$ and $h_x^2 < 1$ the Fréedericksz threshold V_f^2 is reproduced at $q = 0$. Then the curvature of $V_0^2(q)$ at $q = 0$ is positive. Furthermore for the case of small $|\epsilon_a'|$ the high-q limit of Eq. (1) becomes relevant. To see this we note that a spatially periodic destabilization of the homeotropic alignment involves the torque $\alpha_3\frac{\partial v_z}{\partial x}$ around the y-axis that must overcome the stabilizing contribution $\alpha_2\frac{\partial v_x}{\partial z}$. From the incompressibility condition div $\mathbf{v}$ = 0 one then easily estimates the corresponding wave number $q^2 \sim \frac{\alpha_2}{\alpha_3}$ (in reduced units) [21,22,23], which is of the order of 100 for MBBA. This result is strictly valid when there are no additional electric torques, i. e. for $\epsilon_a' = 0$. The influence of a nonzero ϵ_a' is discussed below.

For $q^2 \gg 1$, Eq. (1) reduces to

$$V_0^2(q) = \frac{\pi^2 k_{33}}{\epsilon_0\epsilon_\perp} \frac{\left(1 - h_x^2 + q^2\frac{k_{11}}{k_{33}}\right)\left(1+\omega^2\tau_0^2\right)q^2}{q^2 C - \frac{|\alpha_2|}{\eta_2}\left(\sigma_a' - \epsilon_a'\right)} , \tag{6}$$

where

$$C := \frac{|\alpha_3|}{\eta_2}\left(\sigma_a' - \epsilon_a'\right) - \epsilon_a'\left[1 + \sigma_a' + \omega^2\tau_0^2\left(1+\epsilon_a'\right)\right] . \tag{7}$$

We now check when the high-q limit is really justified. Clearly $V_0^2 \to +\infty$ for $q^2 \to \infty$, if the denominator in Eq. (6) is positive, i. e. if $C > 0$ holds. For MBBA $\frac{\alpha_2}{\eta_2} = -4.6$ $(\eta_2 > 0)$, $\frac{\alpha_3}{\eta_2} = -0.046$ and $\sigma'_a = 0.5$ [30] and these ratios are rather typical for low molecular-weight nematics. ϵ'_a on the other hand can vary from negative to positive. Since $\frac{\alpha_3}{\alpha_2}$ is very small the condition $C > 0$ is satisfied only for values of ϵ'_a below the slightly positive value $\frac{|\alpha_3|\sigma'_a}{\eta_2(1+\sigma'_a)}$ (for $\omega = 0$). When $C > 0$ the denominator in (6) decreases with decreasing q^2 and passes through zero at some value q_0^2, which is still large, if ϵ'_a is not too negative. Then $V_0^2(q)$ must have a minimum at some $q_c^2 > q_0^2$. For not too large magnetic fields ($h_x^2 \ll q^2$) one easily finds from Eq. (6).

$$q_c^2 = 2\frac{|\alpha_2|}{\eta_2}\frac{(\sigma'_a - \epsilon'_a)}{C} \quad , \tag{8}$$

leading to

$$V_c^2 = 4\,\frac{\pi^2 k_{11}}{\epsilon_0\epsilon_\perp}\frac{\left(1+\omega^2\tau_0^2\right)}{C^2}\frac{|\alpha_2|}{\eta_2}\left(\sigma'_a - \epsilon'_a\right) \; . \tag{9}$$

This is the threshold for electroconvection. V_c becomes lower than V_f if ϵ'_a is above a very slightly negative value (−0.00037 for MBBA) so that EHC occurs only in a rather small range of values of ϵ'_a. For $\epsilon'_a = 0$ Eqs. (8) and (9) reduce to

$$q_c^2 = 2\frac{\alpha_2}{\alpha_3} \; , \qquad V_c^2 = 4\,\frac{\pi^2 k_{11}}{\epsilon_0\epsilon_\perp}\frac{\left(1+\omega^2\tau_0^2\right)|\alpha_2|\,\eta_2}{\sigma'_a\alpha_3^2} \; . \tag{10}$$

In Fig. 1 the relevant parts of the neutral curve are plotted for MBBA for various values of ϵ'_a. Although the curves were produced by the full numerical procedure, the analytical formulas give nearly the same results.

The influence of frequency and various material parameters on the threshold is easily seen from Eq. (9). V_c^2 decreases with increasing σ'_a and with decreasing ω. From (1) one sees that the influence of a magnetic field on the EHC threshold is small as long as $h_x^2 \ll q_c^2$. One here assumes, however, that the wave vector remains parallel to the field (roll structure perpendicular to **H**: normal rolls). We have checked numerically that oblique rolls have indeed a higher threshold.

3. Discussion

The neutral curve Eq. (1) is obtained from the condition that the total growth rate of modes, i. e. real and imaginary part, is zero. It is thus assumed that one has a stationary bifurcation leading (normally) to a stationary pattern. We have checked numerically that this assumption is correct, so that Hopf bifurcations play no role at threshold.

Going on to the nonlinear behavior one could have a forward bifurcation, leading to a continuous increase of the pattern amplitude, or a backward bifurcation, leading to

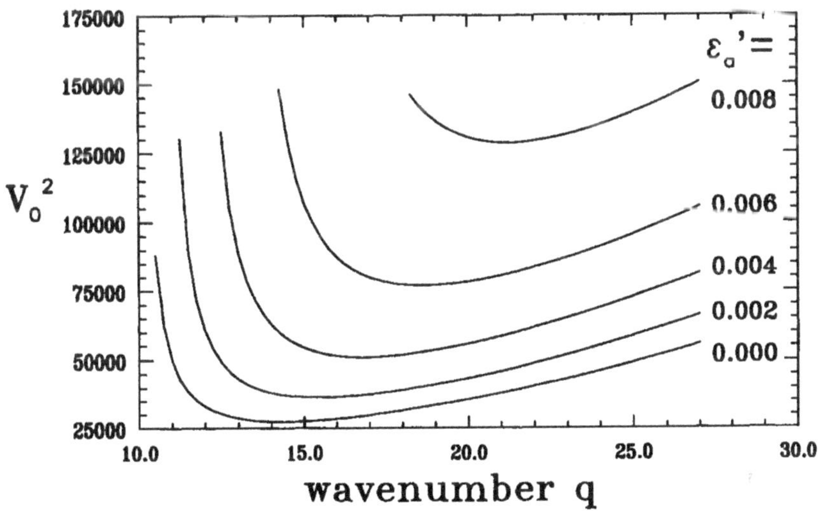

Figure 1: The neutral curve is plotted for $\omega\tau_0 = 0.15$ and various values of ϵ_a' (otherwise MBBA parameter). The minimum corresponds to the threshold V_c (measured in Volt) for electroconvection. V_c diverges with increasing values of ϵ_a' at $\epsilon_a' \approx 0.03$.

an abrupt change. For a forward bifurcation a small-amplitude expansion is useful. In the isotropic case, i. e. without a planar magnetic field, one could have rolls, squares or hexagons slightly above threshold. With a planar field one will definitely have normal rolls (oblique rolls, with a possibility of rectangles, were excluded, see last section). Work on the determination of the nature of the bifurcation and the pattern slightly above threshold is in progress. Actually it appears quite likely that the bifurcation will turn out to be backward because the Carr-Helfrich mechanism should become more efficient when the director is distorted away from the homeotropic alignment.

Let us finally discuss the case $\epsilon_a' > 0$ in the presence of a strong magnetic field h_x $(= \frac{H_x}{H_f})$. For $h_x > 1$ and low voltages the system is magnetically driven into the Fréedericksz deformed state. In the range $\frac{\pi^2 k_{11}}{\epsilon_0\epsilon_\perp}\left(h_x^2 - 1\right) < V^2 < V_c^2$ the homeotropic configuration is restabilized and for $V^2 > V_c^2$ one has EHC. When $\frac{\pi^2 k_{11}}{\epsilon_0\epsilon_\perp}\left(h_x^2 - 1\right) > V_c^2$ holds, restabilization is not possible and instead with increasing voltage a secondary destabilization of the Fréedericksz distorted state has to occur similar to the case of negative ϵ_a (case F, see Sec. 1). The case where restabilization occurs at voltages only slightly larger then destabilization at V_c may be accessible to some simple analysis. Actually at large magnetic fields the minimum of $V_0^2(q)$ at V_c^2 vanishes altogether and the large q-branch decreases monotonically to zero with decreasing q. In this case one clearly also has a secondary destabilization of the Fréedericksz distorted state.

There appear to exist only few experiments for the system investigated in our work

[22,31]. Short wave-length structures with a rather disordered square pattern have been reported there. The threshold was found to be somewhat smaller and the wavelength larger than predicted by theory. The situation should be clarified by further experiments that hopefully will be stimulated by our work.

Acknowledgements: We wish to thank W. Decker and M. Kaiser for help in the programming and useful discussions. Financial support by the Deutsche Forschungsgemeinschaft (Graduiertenkolleg 'Nichtlineare Spektroskopie und Dynamik' and Sonderforschungsbereich 213, Bayreuth) is gratefully acknowledged.

References

[1] W. Zimmermann and L. Kramer, Phys. Rev. Lett. **56** , 2655 (1986). E. Bodenschatz, W. Zimmermann, and L. Kramer, J. Phys. France **49** , 1875 (1988). L. Kramer, E. Bodenschatz, W. Pesch, W. Thom, and W. Zimmermann, Liquid Crystals **5** , 699 (1989).

[2] R. Ribotta, A. Joets, and L. Lin, Phys. Rev. Lett. **56** , 1595 (1986). A. Joets and R. Ribotta, J. Phys. France **47** , 595 (1986).

[3] S. Kai and W. Zimmermann, Prog. Theor. Phys. Suppl. **99** , 458 (1989). S. Kai, M. Kohno, and N. Chizumi, Phys. Rev. A **40** , 6554 (1989).

[4] S. Rasenat, V. Steinberg, and I. Rehberg, Phys. Rev. A **42** , 5998 (1990). I. Rehberg, B. Winkler, M. de la Torre Juarez, S. Rasenat, and W. Schöpf, Adv. Solid State Physics **29** , 35 (1989). S. Rasenat, E. Braun, and V. Steinberg, Phys. Rev. A **43** , 5728 (1991).

[5] E. Bodenschatz, M. Kaiser, L. Kramer, W. Pesch, A. Weber, and W. Zimmermann, in P. Coullet and P. Huerre, editors, *New Trends in Nonlinear Dynamics and Pattern Forming Phenomena: The Geometry of Nonequilibrium*, Plenum Press, (1989). NATO ASI Series. M. Kaiser, W. Pesch, and E. Bodenschatz. submitted to Physica D.

[6] W. Pesch, W. Decker, Q. Feng, M. Kaiser, L. Kramer, and A. Weber, in F. H. J.M. Coron and J. Ghidaglia, editors, *Defects, Singularities and Patterns in Nematic Liquid Crystals: Mathematical and Physical Aspects*, Kluwer Academic Publishers, (1991). NATO ASI Series.

[7] E. Bodenschatz, W. Pesch, and L. Kramer, Physica D **32** , 135 (1988); Phys. Rev. Lett. **64** , 2588 (1990). E. Bodenschatz, A. Weber, and L. Kramer, J. Stat. Phys. **64**, 1007 (1991).

[8] E. Braun and V. Steinberg, Europhys. Lett. **15** , 167 (1991).

[9] E. Braun, S. Rasenat, and V. Steinberg, Europhys. Lett. **15** (6), 597 (1991). S. Nasuno and S. Kai, Europhys. Lett. **14** , 779 (1991). S. Nasuno, O. Sazaki, S. Kai, and W. Zimmermann, preprint (1991).

[10] S. Sasa, Prog. Theor. Phys. **83** , 824 (1990); preprint (1991).

[11] S. Kai and K. Hirakawa, Prog. Theor. Phys. Suppl. **64** , 212 (1978). A. Joets and R. Ribotta, Phys. Rev. Lett. **60** , 2164 (1988). I. Rehberg, S. Rasenat, and V. Steinberg, Phys. Rev. Lett. **62** , 756 (1989).

[12] I. Rehberg, S. Rasenat, M. de la Torre Juarez, W. Schöpf, F. Hörner, G. Ahlers, and H. Brand, Phys. Rev. Lett. **67** , 596 (1991).

[13] Q. Feng, W. Pesch, and L. Kramer, preprint (1991).

[14] L. Berge, G. Ahlers, and D. Cannell. private communication.

[15] F. Busse, in H. Swinney and J. Gollub, editors, *Hydrodynamic Instabilities and the Transition to Turbulence*, Springer, (1986).

[16] E. Dubois-Violette, P. Pieranski, and E. Guyon, Phys. Rev. Lett. **30** , 736 (1973).

[17] E. Dubois-Violette, G. Durand, E. Guyon, P. Manneville, and P. Pieranski, in L. Liebert, editor, *Solid State Physics, Suppl.14*, Academic Press, (1978).

[18] H. Lekkerkerker, J. Phys. France, Lett., **38** , 277 (1977). E. Guyon, P. Pieranski, and J. Salan, J. Fluid Mech. **93** , 65 (1979). J. Salan and E. Guyon, J. Fluid Mech. **126** , 13 (1983).

[19] Q. Feng, W. Decker, W. Pesch, and L. Kramer, preprint (1992).

[20] W. Schöpf and W. Zimmermann, Europhys. Lett. **8** , 41 (1989).

[21] P. de Gennes, *The Physics of Liquid Crystals*, (Clarendon, Oxford 1974).

[22] L. M. Blinov, *Electrooptical and Magnetooptical Properties of Liquid Crystals*, (Wiley, New York 1983).

[23] S. Pikin, *Structural Transformations in Liquid Crystals*, (Gordon and Breach, New York 1991).

[24] W. Helfrich, J. Chem. Phys. **51** , 4092 (1969).

[25] A. Hertrich, W. Decker, W. Pesch, and L. Kramer, preprint (1992).

[26] S. Chandrasekhar, *Liquid Crystals*, (Cambridge University Press, London 1977).

[27] M. Stephen and J. Straley, Rev. Mod. Phys. **46** , 617 (1974).

[28] S. Chandrasekhar, *Hydrodynamic and Hydromagnetic Stability*, (Clarendon, Oxford 1961).

[29] R. Clever and F. Busse, J. Fluid Mech. **65** , 625 (1974).

[30] H. de Jeu, W. Claassen, and A. Spruyit, Mol. Cryst. Liq. Cryst. **37** , 269 (1976).

[31] M. I. Barnik, L. M. Blinov, M.F. Grebenkin, S. A. Pikin, and V. G. Chigrinov, Zh. Eksp. Teor. Fiz. **69** , 1080 (1975).

ORDER PARAMETER EQUATIONS FOR PATTERNS

by Alan C. Newell*

Patterns of an almost periodic nature appear all over the place. One sees them in cloud streets, in sand ripples on flat beaches, in the morphology of plants and animals, in chemically reacting media, on weather maps, in geological formations, in interacting laser beams, on the surface of thin buckling shells, in the grid scale instabilities of numerical algorithms. This review deals for the most part with the class of problems into which these examples fall, namely with pattern formation in spatially extended, continuous, dissipative systems which are driven far from equilibrium by an external stress. Under the influence of this stress, the system can undergo a series of symmetry breaking bifurcations or phase transitions and the resulting patterns become more and more complicated, both temporally and spatially, as the stress is increased. The goal of theory is to provide a means of understanding and explaining these patterns from a macroscopic viewpoint that both simplifies and unifies classes of problems which are seemingly unrelated at the microscopic level.

Convection in a large aspect ratio horizontal layer of fluid heated from below is the grandaddy of canonical examples used to study pattern formation and behavior in spatially extended systems. For low values of the vertical temperature difference, which is the external stress parameter in this case and whose nondimensional measure is called the Rayleigh number R, heat is transported across the layer by conduction, the fluid remains at rest and possesses symmetries of time and horizontal space translation (in the limit of an infinite horizontal geometry), rotation, reflection (at least about vertical axes). If the conditions on the top and bottom boundaries are that the shear stresses are zero, then the system is also Galilean invariant with respect to horizontal velocities. If the boundaries are rigid, this symmetry is no longer present. At a certain value R_c of the external stress, the stability of the conduction state is lost and various new shapes and structures that involve nonzero velocity fields which advect the heat are preferentially amplified. In the case of an infinite horizontal geometry, these structures are often cellular rolls whose horizontal projections are stationary or travelling wave patterns which break one or more of the symmetries enjoyed by the unstable conductive state. For example, a stationary pattern of peri-

*Arizona Center for the Mathematical Sciences, The University of Arizona, Tucson, AZ 85721 odic convecting rolls breaks both the continuous translation symmetry in the direction perpendicular to the roll axes and the rotational symmetry.

What is important to understand is that the pattern which finally emerges is not uniquely identified by a linear stability investigation of the simple state because of degeneracies. The set of states which are equally (or almost equally) linearly amplified (e.g. the convection rolls) or which remain neutrally stable (e.g. the zero horizontal velocity state when the boundary conditions are stress free rather than rigid) as the stress parameter exceeds critical can contain many modes. In the convection example, while there is a single wavenumber k_c which is most amplified, any linear superposition of rolls of wavelengths $\lambda = 2\pi k_c^{-1}$ with arbitrary orientations is equally admissable from linear stability considerations. Therefore linear theory cannot discriminate between competing configurations. Rather, it is a combination of *external biases*, the effects of horizontal boundary conditions in convecting fluids, geometric imperfections in elastic shells and the *nonlinear coupling* between various competing configurations, each of which is almost equally likely from the point of view of its linear growth rate, that determines the final outcome.

As a first step we should like to reduce the dimension of the system by choosing a coordinate system (a basis) which clearly separates the modes which are dynamically active from those which play a passive role. Specifically, we want to divide the space S of all possible fields into two subsets, which we will call $\mathcal{A}$ and $\mathcal{P}$. $\mathcal{A}$ connoting *active*, will contain all these modes that tap directly into the external stress source and are bona fide competitors in the battle for survival. $\mathcal{P}$, connoting passive, will contain all the nonstarters. They will be present in the final state, but only to the extent that they are regenerated by nonlinear interactions of the active modes. They are slaved in the sense that their amplitudes are determined algebraically in terms of the amplitudes of the members of $\mathcal{A}$. They have no direct access to the external stress source. Sometimes, the division of S into $\mathcal{A}$ and $\mathcal{P}$ is unambiguous and the identification of who belongs where is straightforward. Other times, the dividing line is less sharp and it is these situations which are most challenging.

Near onset, that is when the value R of the stress parameter is close to the critical value R_c at which the transition takes place, the choice of $\mathcal{A}$ and $\mathcal{P}$ can be made by examining the linear stability of the state about to destabilize. If the spectrum of the stability operator (which corresponds to the growth rates of various configurations) is discrete, then the division into $\mathcal{A}$ and $\mathcal{P}$ is clear. $\mathcal{A}$ contains those modes whose growth rates are positive, zero or weakly negative for R close to R_c and $\mathcal{P}$ contains all modes which are strongly damped. The coordinates of $\mathcal{A}$ are simply the amplitudes of the discrete modes or configurations belonging to $\mathcal{A}$. They are called *amplitude order parameters* because their relative values tell us about the degree of order and structure in the system. For R close to its critical value R_c, the amplitudes of the passive modes in the set $\mathcal{P}$ will very quickly relax to a manifold determined algebraically as a balance between each linearly decaying passive mode and its regeneration by nonlinear interactions involving members of $\mathcal{A}$. This relation is given by a graph

$$\mathcal{P} = \mathcal{P}(\mathcal{A}) \,, \tag{1}$$

called the Center Manifold. The idea is that no matter where the system begins in its phase space, it very quickly relaxes to the Center Manifold $\mathcal{P}(\mathcal{A})$. On this manifold, the amplitudes will evolve on a time scale proportional to $|R - R_c|^{-1}$ according to a set (usually finite) of coupled ordinary differential equations, called ***amplitude equations***

$$\frac{d\mathcal{A}}{dt} = G(\mathcal{A}) \tag{2}$$

often associated with the names of Landau [1], Stuart [2] and Watson [3] and Schlüter, Lortz and Busse [4]. Their form is universal in the sense that $G(\mathcal{A})$ consists of terms which are simply products of active amplitudes. All relevant information about the underlying microscopic system is contained in coefficients. The original miroscopic fields can be reconstructed to a good approximation by combining the solution of (2) with the configurations which are the basis vectors for $\mathcal{A}$, together with the graph (1). Since solving (2) is much easier than solving the original microscopic equations, considerable simplification in both qualitative and quantitative understanding has been achieved.

If the spectrum of growth rates is continuous (on almost so, in the sense that nearest neighbors are closer than $\sqrt{R - R_c}$), then things are more subtle. In some cases (in others, the problem is still open), we overcome the ambiguity of continuous bands of competing modes by introducing as order parameters the ***envelopes*** of the various short scale mode shapes. The envelopes depend slowly on space and time, in contrast to the amplitudes used in the discrete spectrum case which depended only on time. For the continuous spectrum situation, $\mathcal{A}$ is infinite dimensional and the envelopes obey a class of coupled nonlinear ***partial*** differential equations often called the Newell-Whitehead-Segel [5, 6] or Ginzburg-Landau equations. 1 The advantage of this simplification is that the envelope order parameters depend only on slow scales and satisfy equations which, like the amplitude equations, have a universal structure. Their form depends principally on symmetry considerations and not on microscopic details. Again, in principle, the asymptotic state of the system can be reconstructed from the solutions of these equations and the corresponding analog of the graph (1.1), but, as we shall see, the notion of a Center Manifold is on a much shakier footing. There are many open questions.

Far from onset, it is not at all clear how to decompose the space of all configurations into active and passive modes although, in recent years, there have

1Historical note: while Landau wrote down amplitude equations (o.d.e.'s) to describe the post-bifurcation behavior of unstable modes and, together with Ginzburg in their theory of superconductivity introduced the notion of a space dependent order parameter (which again satisfies and o.d.e. in x) to describe behavior near the phase transition, they never actually wrote down any of these envelope equations nor stressed their universality.

been major efforts to generalize the notion of a center manifold to what is called a global center manifold or inertial manifold. An inertial manifold is a finite dimensional Euclidean space in the phase space of the system which contains the attractor (over a reasonably large parameter range!) and to which orbits are rapidly attracted. Unfortunately, in most cases to which this notion has been applied, the lack of appropriate coordinates has precluded its utility as an analytical tool since the use of inefficient coordinates means that the embedding space is far too big an object to deal with easily. To date, therefore, most pattern dynamicists have resorted to simpler and more manageable means of description which go some distance towards identifying what the correct coordinatization of patterns should be.

The first approach is linked permanently with the name of Busse [7]. The idea is to follow a branch of solutions which begins at R_c to finite values of $R - R_c$ and examine the linear stability of the finite amplitude solutions. In particular, Busse and his coworkers have exhaustively investigated the stability of a family of straight parallel convective rolls parametrized by the Rayleigh number R, wavenumber k and Prandtl number $\frac{\nu}{\kappa}$, the ratio of molecular and thermal diffusivities. For a fixed Prandtl number, the domain of stable solutions in R, k space is called the *Busse balloon*. The nature of the instabilities at its borders has been documented and the predictions of the theory have proven to be very valuable in understanding much observed behavior.

But natural convective patterns, that is those which arise from noise, are rarely a small variation of a straight roll pattern on a global scale. Instead, a combination of the property of rotational invariance, which makes rolls of all orientations equally likely, and the boundary conditions, which choose the roll direction locally, conspire to give fairly complicated pattern textures, consisting of patches of curved rolls almost circular in shape, and defects such as foci, grain boundaries, disclinations and dislocations. The challenge is to develop a macroscopic field-particle theory that takes advantage of some underlying periodic structure of the microscopic field and that, at the same time, can handle the pattern defects which are singularities of (and the price paid for) the macroscopic description. The key observation which leads to a theoretical description and tractable analysis is that locally and almost everywhere (away from defects), the pattern consists of almost straight rolls whose radius of curvature (of the order of the box size L) is much larger than their local wavelength $\lambda = 2\pi k^{-1}$. Therefore one can locally approximate the field $w(x, y, z, t)$ (velocity temperature, pressure, electric) as a 2π-periodic function

$$w = f(\theta; z, A) \tag{3}$$

of a phase variable $\theta(x, y, t)$ whose gradient

$$\vec{k} = \nabla\theta\ , \tag{4}$$

the wavenumber, is slowly varying, i.e. $\nabla\vec{k} = 0(\Gamma^{-1})$, $\Gamma = \frac{d}{L}$, the inverse aspect ratio. If the underlying planform is square or hexagonal, then one must use a multiphase description $w = f(\theta_1, \theta_2, \theta_3, \cdots; z)$ of the field. If the planform involves travelling waves, then the appropriate phase θ must include a fast time dependence, $\nabla\theta = \vec{k}, -\theta_t = \omega$ where both $\vec{k}$ and ω are slowly varying.

In (3), z represents the directions of finite extent and A the amplitude which is slaved to the wavenumber k. When $\vec{k}$ is a constant wavevector, (3) is exactly the finite amplitude roll solution which is the starting point of Busse's stability analysis. When $\vec{k}$ varies slowly, (3) is not an exact solution and corrections to f at orders $\Gamma^{-n}, n \geq 1$, must be calculated. These corrections $w_n, n \geq 1$, satisfy inhomogeneous linear equations, $Lw_n = g_n$ where L is a linear operator in θ with coefficients depending on $f(\theta)$ which is singular because of various symmetries enjoyed by $f(\theta)$ (e.g. the system's underlying translational invariance means that if $f(\theta)$ is a solution so is $f(\theta + \theta_0)$ for arbitrary θ_0). The solvability conditions give rise to an equation for $\vec{k}$, called the phase diffusion equation, $\vec{U}$ a mean drift field, and A, the amplitude, which are the finite amplitude analogues of the equations for the complex envelope order parameters valid near onset. The idea to treat slowly varying patterns in this way comes from the seminal work of Whitham [8] on the modulation of finite amplitude waves. It has been applied in the context of reaction diffusion systems by Howard and Kopell [9] and to convection patterns by Cross and Newell [10] and Newell, Passot and Souli [11]. Far from onset, therefore the relevant dependent variables are the phase (whose gradient is the local wavevector), the amplitude and mean drift fields associated with almost periodic patterns. We say relevant variables rather than order parameters because we want to reserve that particular term to connote the coordinates associated with marginally stable states which belong to the kernel of an appropriate stability operator. For the finite amplitude behavior seen well above threshold, the amplitude will only be an order parameter locally near defects and the mean drift field is a manifestation of the nonlocal nature of the phase diffusion equation rather than an independent order parameter. We also emphasize that, because of defects, the wavevector $\vec{k}$ cannot be considered as a globally defined vector field. It is more correct to think of it as a director field. More about this later.

We summarize the four categories, (i) near onset, discrete spectrum (ii) near onset, continuous spectrum (iii) far from onset, finite amplitude rolls and their stability and (iv) far from onset, modulated roll patterns, with Table 1 and in each box list the names of people who have made important contributions to each of these approaches.

	$R - R_c$ small	$R - R_c$ of order one
$\Gamma = 0(1)$ or periodic, quasiperiodic	Amplitude equations Landau ('44) Malkus-Veronis ('58) Stuart-Watson ('60) Schlüter, Lortz and Busse ('62) Busse ('67)	Finite Amplitude rolls and their stability Busse('65-90).
$\Gamma \gg 1$	Envelope equations Newell-Whitehead ('69, '71, '74) Segel ('69) Stewartson-Stuart ('71) Pomeau-Manneville ('79)	Modulated rolls Whitham ('67) Howard-Kopell ('79) Cross-Newell ('84) Newell-Passot-Souli ('90)

Table 1

The idea to use an envelope order parameter for systems with continuous spectra came directly from the use of envelopes to describe the general propagation properties of nonlinear wave envelopes [Benney and Newell(15)] Indeed the conservative part of the complex Ginzburg-Landau equation is nothing other than the nonlinear Schrödinger equation and one of the most interesting aspects of pattern dynamics is how the dispersive wave and nonlinear focusing character of the system gives rise to a much richer set of behaviors than one would expect from the gradient flow component associated with diffusion and saturable nonlinearities.

It should also be stressed that experiments, involving greatly improved techniques of parameter control and data acquisition and analysis, have played a vital role in the increase of understanding. In particular, we mention the work of Busse and Whitehead [16] on convection patterns and defects, Swinney and Gollub [17] on the Taylor-Couette problem, Ahlers and Behringer [18] and Ahlers, Cannell and Steinberg [19] on large aspect ratio convection, Libchaber [20] on convection in small aspect ratio boxes, Croquette and Pocheau [21] and Heutmaker and Gollub [22] on natural convection patterns, and Kolodner [23], Ribotta and Joets [24], Rehberg [25] and Steinberg [26] on travelling wave convection and localized structures in convecting binary fluid mixtures and nematic liquid crystals. Indeed one of the gratifying features of the field of pattern dynamics is that progress has been made by the collective and friendly-competitive efforts of the computational, experimental and theoretical scientist.

ACKNOWLEDGEMENT. Research support from AFOSR Contract FQ8671-900589 and NSF grants DMS8922179 and DMS9021253 are gratefully acknowledged.

REFERENCES

[1] Landau, L., *C. R. Acad. Sci. USSR* **44**, 311 (1944).

[2] Stuart, J. T., *J. Fluid Mech.* **9**, 353 (1960).

[3] Watson, J., *J. Fluid Mech.* **9**, 371 (1960).

[4] Schlüter, A., Lortz, D., Busse, F., *J. Fluid Mech.* **23**, 129 (1965).

[5] Newell, A. C., Whitehead, J. A., *J. Fluid Mech.* **38**, 179 (1969); Proc. IUTAM Conference on Instabilities in Continuous Systems, Herrenalb, 1969, Ed. H. Leipholz (Berlin: Springer-Verlag), 284-289 (1971).

[6] Segel, L. A., *J. Fluid Mech.* **38**, 203 (1969).

[7] Busse, F. H., Topics in Applied Physics **45**, 96 (1985).

[8] Whitham, G. B., Linear and Nonlinear Waves (Wiley Interscience) (1975).

[9] Howard, L. N., Kopell, N., *Stud. Appl. Math.* **56**, 95 (1977).

[10] Cross, M. C., Newell, A. C., *Physica* **10D**, 299 (1984).

[11] Newell, A. C., Passot, T., Souli, M., *J. Fluid. Mech.* **220**, 187-252 (1990).

[12] Malkus, W., Veronis, G., *J. Fluid Mech.* **4**, 225 (1958).

[13] Stewartson, K., Stuart, J. T., *J. Fluid Mech.* **48**, 529 (1971).

[14] Pomeau, Y., Manneville, P., *J. Physique Lett.* **40**, 609 (1979).

[15] Benney, D. J., Newell, A. C., *Journal of Mathematics and Physics* **46**, 113-139 (1967). (Presently *Stud. Appl. Math.*)

[16] Busse, F. H., Whitehead, J. A., *J. Fluid Mech.* **47**, 305 (1971); *J. Fluid Mech.* **66, 67** (1974).

[17] Fenstermacher, P., Swinney, H., Gollub, J., *J. Fluid Mech.* **94**, 103-129 (1979).

[18] Ahlers, G., Behringer, R., *Phys. Rev. Lett.* **40**, 712 (1978).

[19] Ahlers, G., Steinberg, V., Cannell, D. S., *Phys. Rev. Lett.* **54**, 1373 (1985).

[20] Libchaber, A., Fauve, S., Laroche, C., *Physica* **7D**, 73-84 (1983).

[21] Croquette, V., Pocheau, A., *J. Phys.* (Paris) **45**, 35 (1984).

[22] Heutmaker, M. S., Gollub, J. P., *Phys. Rev. A* **35**, 242 (1987).

[23] Kolodner, P., Surko, C. M., *Phys. Rev. Lett.* **61**, 842 (1988); Kolodner, P. Bensimon, D., Surko, C. M., *Phys. Rev. Lett.* **60**, 1723 (1988).

[24] Ribotta, R., Joets, A., *J. Physique* **47**, 595-606 (1986).

[25] Rehberg, I., Busse, F. H., Proceedings of Les Houches Workshop, March 1987, Eds. J. E. Wesfried, H. R. Brand, P. Mannesville, G. Albinet and N. Boccara (Berlin, Heidelberg: Springer-Verlag) (1988).

[26] Moses, E., Steinberg, V., *Phys. Rev. A* **43**, 707 (1991).

EHD INSTABILITY OF A NEMATIC UNDER A TRANSVERSE ELECTRIC FIELD

V.A.RAGHUNATHAN, P.R.MAHESWARA MURTHY AND N.V.MADHUSUDANA
Raman Research Institute,
Bangalore 560080, India

ABSTRACT

The electrohydrodynamic instability of a homogeneously aligned nematic subject to a transverse AC electric field parallel to the bounding plates is analysed theoretically. The threshold field and wavevector are calculated using the Galerkin approximation. The resulting ***phase diagram*** has many similarities with that of the standard geometry. However, in the present geometry two distinct solutions are possible in both the conduction and dielectric regimes. We have also conducted an experimental investigation in this geometry. There is broad agreement between the observed and theoretical trends.

1. Introduction

Nematic liquid crystals with negative or weakly positive dielectric anisotropy ($\Delta\epsilon$) and non-zero conductivity anisotropy ($\Delta\sigma$) exhibit electroconvective instabilities under the action of an external electric field (see e.g., Blinov[1]). In the widely studied geometry, a thin planar aligned nematic layer is subjected to a field applied across its thickness. A bend fluctuation of the director field leads to space charge formation due to the anisotropy of the conductivity and electroconvective instabilites set in above a critical value of the electric field.[2] Under AC excitation two regimes of instability are observed, depending on the frequency of the applied field. The one occurring at lower frequencies is called the conduction regime and the other the dielectic regime. In the conduction regime the space charges oscillate with the applied field, with the director distortion being essentially stationary. In the dielectric regime, on the other hand, the director distortion oscillates with the field and the space charges remain stationary. These instabilities are fairly well understood in terms of the Orsay model.[3] More recently, calculations[4–6] including the flexoelectric effect[7] have shown that at low frequencies in the conduction regime the director field develops additional twist distortion resulting in the formation of oblique rolls.

In this paper we describe electroconvective instabilities in a different geometry, with the field applied in the plane of the nematic layer and normal to the undistorted director. Though there have been some experimental studies[8,9] on EHD instabilities in this geometry, no theoretical analysis of this problem has hitherto been reported. In this case the space charges are created by a periodic twist-bend distortion of the director field. Many features of the instabilities in this geometry are similar to those described earlier. However, there are some important differences. In the earlier case, with the field applied across the layer thickness, the flexoelectric effect destroys the symmetry of the variables in the problem about the midplane of the layer[6,10] while in the present case these symmetries are conserved. As a result two types of solutions are possible in the present case, depending on the symmetry of the variables about the midplane of the layers.

2. Theory

Consider a nematic layer of thickness d and of infinite lateral extent in the XY plane. The undistorted director is along X and the electric field $\mathbf{E}$ is applied along Y. Above a critical value of the field EHD instabilities develop in the medium giving rise to a periodic twist distortion of the director field described by the angle φ. The wavevector of this distortion is assumed to be along X, i.e., the convective rolls are along Y. The flexoelectric effect would lead to an additional distortion of the diretor field out of the XY plane, described by the angle θ. Thus in the distorted state the director is given by $\mathbf{n} = (\cos\theta\cos\varphi, \cos\theta\sin\varphi, \sin\theta)$. Near the onset of the instability the system is described by the following set of linearized equations.

The Poisson eqn., $\nabla.\mathbf{D} = 4\pi Q$, gives

$$\epsilon_{\|}\frac{\partial E_x}{\partial x} + \epsilon_{\perp}\frac{\partial E_z}{\partial z} + \Delta\epsilon E_y\frac{\partial\varphi}{\partial x} + 4\pi(e_1 + e_3)\frac{\partial^2\theta}{\partial x\partial z} - 4\pi Q = 0 \tag{1}$$

where $\epsilon_{\|}$ and $\epsilon_{\perp}$ are the principal dielectric constants of the medium, $\Delta\epsilon = \epsilon_{\|} - \epsilon_{\perp}$, e_1 and e_3 are the flexoelectric coefficients and Q, the space charge density.

The charge conservation eqn. $\dfrac{\partial Q}{\partial t} + \nabla.\mathbf{J} = 0$, leads to

$$\frac{\partial Q}{\partial t} + \sigma_{\|}\frac{\partial E_x}{\partial x} + \sigma_{\perp}\frac{\partial E_z}{\partial z} + \Delta\sigma E_y\frac{\partial\varphi}{\partial x} = 0 \tag{2}$$

where $\sigma_{\|}$ and $\sigma_{\perp}$ are the principal electrical conductivities and $\Delta\sigma = \sigma_{\|} - \sigma_{\perp}$. Torque balance eqns. along Y and Z give,

$$\gamma_1\frac{\partial\theta}{\partial t} + \alpha_3\frac{\partial v_x}{\partial z} + \alpha_2\frac{\partial v_z}{\partial x} - K_3\frac{\partial^2\theta}{\partial x^2} - K_1\frac{\partial^2\theta}{\partial z^2} + (e_1 - e_3)E_y\frac{\partial\varphi}{\partial z} + e_1\frac{\partial E_x}{\partial z} + e_3\frac{\partial E_z}{\partial x} = 0 \tag{3}$$

$$\gamma_1\frac{\partial\varphi}{\partial t} + \alpha_2\frac{\partial v_y}{\partial x} - K_3\frac{\partial^2\varphi}{\partial x^2} - K_2\frac{\partial^2\varphi}{\partial z^2} - \frac{\Delta\epsilon}{4\pi}(E_y^2\varphi + E_yE_x) - (e_1 - e_3)E_y\frac{\partial\theta}{\partial z} = 0 \tag{4}$$

where the α's are the Leslie viscosity coefficients, $\gamma_1 = \alpha_3 - \alpha_2$, $\mathbf{v}$ is the velocity and K_1, K_2 and K_3, the splay, twist and bend elastic constants, respectively.
The three components of the eqn. of motion,

$$\frac{\partial p}{\partial x} - (\alpha_1 + \alpha_4 + \alpha_5 + \alpha_6)\frac{\partial^2 v_x}{\partial x^2} - \frac{1}{2}(\alpha_4 + \alpha_6 - \alpha_3)\frac{\partial^2 v_z}{\partial x\partial z} - \frac{1}{2}(\alpha_4 + \alpha_6 + \alpha_3)\frac{\partial^2 v_x}{\partial z^2} - \alpha_3\frac{\partial^2\theta}{\partial z\partial t} = 0 \tag{5}$$

$$\frac{1}{2}(\alpha_4 + \alpha_5 - \alpha_2)\frac{\partial^2 v_y}{\partial x^2} + \frac{1}{2}\alpha_4\frac{\partial^2 v_y}{\partial z^2} + \alpha_2\frac{\partial^2\varphi}{\partial x\partial t} + E_yQ = 0 \tag{6}$$

$$\frac{\partial p}{\partial z} - \frac{1}{2}(\alpha_4 + \alpha_5 + \alpha_2)\frac{\partial^2 v_x}{\partial x \partial z} - \frac{1}{2}(\alpha_4 + \alpha_5 - \alpha_2)\frac{\partial^2 v_z}{\partial x^2} - \alpha_4 \frac{\partial^2 v_z}{\partial z^2} - \alpha_2 \frac{\partial^2 \theta}{\partial x \partial t} = 0 \quad (7)$$

where p is the pressure. Note that in the above three eqns. we have dropped the inertial terms as these are negligible at the frequencies of interest.[3]

The Maxwell relation $\nabla \times \mathbf{E} = 0$, leads to

$$\frac{\partial E_x}{\partial z} - \frac{\partial E_z}{\partial x} = 0 \quad (8)$$

The incompressibility condition $\nabla . \mathbf{v} = 0$ gives

$$\frac{\partial v_x}{\partial x} + \frac{\partial v_z}{\partial z} = 0 \quad (9)$$

These eqns. have to be solved with the boundary conditions, viz.,

$$(\theta, \varphi, Q, E_x, E_z, v_x, v_y, v_z, p) = 0 \quad at \quad Z = 0, d.$$

The X-dependence of these quantities in the above eqns. can be rigorously satisfied by sinusoidal functions. The Z-dependences are taken into account in a Galerkin approximation by choosing trial function solutions. We eliminate the pressure p from eqns. (5) and (7) and assume the following solutions for the remaining variables.

$$\begin{aligned} &\varphi = \varphi(t) sin\ qx\ sin\ pz, && \theta = \theta(t) sin\ qx\ sin\ 2pz, \\ &Q = Q(t) cos\ qx\ \ sin\ pz, && E_z = E_z(t) cos\ qx\ sin\ 2pz, \\ &E_x = E_x(t) sin\ qx\ sin\ pz, && v_z = v_z(t) cos\ qx\ sin\ 2pz, \end{aligned}$$

$$v_x = v_x(t) sin\ qx\ sin\ pz, \qquad v_y = v_y(t) cos\ qx\ sin\ pz \qquad with\ \ p = \pi/d \quad (10)$$

Substituting these solutions in eqns. (1-9) and on eliminating E_x, E_z, v_x, v_y and v_z we get the following three eqns.

$$\frac{\partial Q}{\partial t} + \frac{1}{\tau}Q + \sigma_H E_y q \varphi - A\theta = 0 \quad (11)$$

$$\frac{\partial \varphi}{\partial t} + \frac{1}{T_\varphi}\varphi - BE_y Q + 2C_1 E_y I_1 \theta = 0 \quad (12)$$

$$\frac{\partial \theta}{\partial t} + \frac{1}{T_\theta}\theta - C_2 E_y I_2 \varphi + DQ = 0 \quad (13)$$

where

$$\frac{1}{\tau}=\frac{4\pi\sigma_c}{\epsilon_c},\qquad \sigma_H=\sigma_c\left(\frac{\Delta\sigma}{\sigma_c}-\frac{\Delta\epsilon}{\epsilon_c}\right),\qquad A=\frac{8\pi\sigma_c}{\epsilon_c}(e_1+e_3)pqI_1\ ,$$

$$\frac{1}{T_\varphi}=\left(K_3+K_2S^2-\frac{\Delta\epsilon}{4\pi}\frac{E_y^2}{q^2}\frac{\epsilon_\perp(1-2S^2I_1I_2)}{\epsilon_c}\right)\frac{q^2}{\eta_1},\qquad B=\left(\frac{2\alpha_2}{\eta+\alpha_4S^2}+\frac{\Delta\epsilon}{\epsilon_c}\right)\frac{1}{q\eta_1}$$

$$C_1=\left(\frac{\Delta\epsilon(e_1+e_3)}{\epsilon_c}-(e_1-e_3)\right)\frac{p}{\eta_1},\qquad \frac{1}{T_\theta}=\left(K_3+4K_1S^2-\frac{8\pi(e_1+e_3)^2S^2I_1I_2}{\epsilon_c}\right)\frac{q^2}{\eta_2}$$

$$C_2=\frac{C_1}{\eta_2}\eta_1,\qquad D=\frac{4\pi(e_1+e_3)SI_2}{\epsilon_c.\eta_2},\qquad \sigma_c=\sigma_{\|}-2\sigma_\perp S^2I_1I_2,\qquad \epsilon_c=\epsilon_{\|}-2\epsilon_\perp S^2I_1I_2$$

$$\eta_1=\gamma_1-\frac{2\alpha_2^2}{\eta+\alpha_4S^2},\qquad \eta=(\alpha_4+\alpha_5-\alpha_2)$$

$$\eta_2=\left(\gamma_1-\frac{(\alpha_2-4\alpha_3S^2)(\alpha_2+2\alpha_3S^2I_1I_2)}{\beta}\right)$$

$$\beta=(\alpha_4+\alpha_6+\alpha_3)S^4I_1I_2+2(\alpha_4-\alpha_6+\alpha_3)S^2-2(\alpha_1+\alpha_6+\frac{1}{2}\eta)S^2I_1I_2+\frac{1}{2}\eta$$

$$S=p/q,\qquad I_1=-\frac{4}{3\pi}\qquad \textit{and}\qquad I_2=\frac{8}{3\pi}$$

Another possible set of solutions is obtained if we replace sin pz in (10) by sin $2pz$ and vice versa. The equations equivalent to eqns. (11-13) are obtained in this case by redefining the following quantities.

$$A'=\frac{4\pi\sigma_c}{\epsilon_c}(e_1+e_3)pqI_2$$

$$\frac{1}{T'_\varphi}=\left(K_3+4K_2S^2-\frac{\Delta\epsilon}{4\pi}\frac{E_y^2}{q^2}\frac{\epsilon_\perp(1-2S^2I_1I_2)}{\epsilon_c}\right)\frac{q^2}{\eta'_1}$$

$$B'=\left(\frac{2\alpha_2}{\eta+4\alpha_4S^2}+\frac{\Delta\epsilon}{\epsilon_c}\right)\frac{1}{q\eta'_1},\qquad C'_1=\frac{C_1\eta_1I_2}{2\eta'_1I_1}$$

$$\frac{1}{T'_\theta}=\left(K_3+K_1S^2-\frac{8\pi(e_1+e_3)^2S^2I_1I_2}{\epsilon_c}\right)\frac{q^2}{\eta'_2}$$

$$D'=\frac{8\pi(e_1+e_3)SI_1}{\epsilon_c.\eta'_2},\qquad C'_2=\frac{2C'_1\eta'_1I_1}{\eta'_2I_2},\qquad \eta'_1=\gamma_1-\frac{2\alpha_2^2}{\eta+4\alpha_4S^2}$$

$$\eta'_2=\gamma_1-\frac{(\alpha_2-\alpha_3S^2)(\alpha_2+2\alpha_3S^2I_1I_2)}{\beta'}$$

$$\beta'=4(\alpha_4+\alpha_6+\alpha_3)S^4I_1I_2+\frac{1}{2}(\alpha_4-\alpha_6+\alpha_3)S^2-2(\alpha_1+\alpha_6+\frac{1}{2}\eta)S^2I_1I_2+\frac{1}{2}\eta$$

and all other parameters as defined after eqn. (13).

It is clear from eqns. (11-13) that under AC excitation either Q and θ oscillate with the field or φ oscillates with the field. Thus we have two regimes of instability as in the case with E along Z.[3,9]

Eqns. (11-13) were solved for square-wave excitation,[11] using the following values

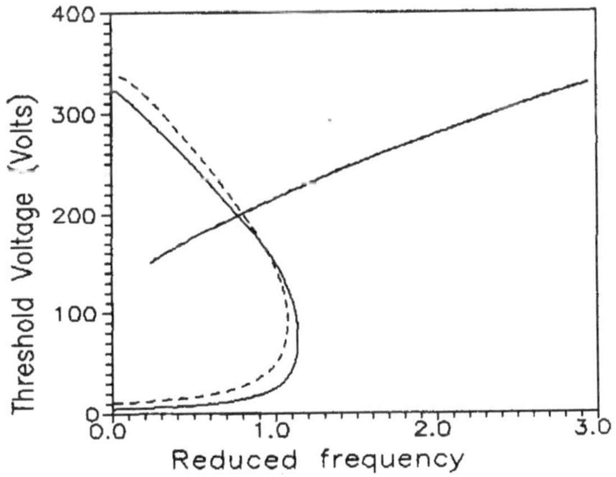

Figure 1: Variation of the threshold voltage with reduced frequency obtained from the calculations using MBBA parameters. d = 50 μm, lateral separation between the electrodes is 60 μm.

of the material parameters of MBBA: $\epsilon_{\parallel} = 4.75$, $\epsilon_{\perp} = 5.25$, $\sigma_{\parallel} = 1 \times 10^{-9}$ ohm $^{-1}$ cm^{-1}, $\Delta\sigma/\sigma_{\perp} = 0.5$, K$_1$ = 6.1 x 10^{-7} dyne, K$_2$ = 4x10^{-7} dyne, K$_3$ = 7.3 x 10^{-7} dyne, α_1 = 0.065 P, α_2 = -0.775 P, α_3 = -0.012 P, α_4 = 0.832 P, α_5 = 0.463 P, α_6 = -0.342 P, e$_1$ + e$_3$ = -7x10^{-4} e.s.u. and e$_1$ - e$_3$ = 1x10^{-4} e.s.u. Fig. 1 shows the critical voltage for the onset of the instability as a function of the reduced frequency $f/f_c(f_c - 2\sigma_{\parallel}/\epsilon_{\parallel})$, obtained from the calculations. The solid lines correspond to the solution set 1 and the dashed lines to set 2. In the dielectric regime the critical voltage for set 2 is only marginally greater than that for set 1. It should be noted here that due to its flexoelectric origin, θ is negligible in the conduction regime at all but very low frequencies of the applied field. The larger φ distortion and larger viscous dissipation present in solution set 2, therefore, leads to a larger threshold voltage. In the dielectric regime, on the other hand, a larger φ distortion can lead to a larger relaxation frequency 1/T$_\varphi$. The threshold of the dielectric regime is determined by the condition that the frequency of the applied field $\approx 1/T_\varphi$. This may be the reason for the thresholds for solution sets 1 & 2 to have nearly equal values in this regime.

Fig. 2 gives the critical wavevector as a function of the reduced frequency. As is clear from these two figures, the general features of this instability are very similar to those in the case with E along Z.

3. Experimental Studies

The material used in the experiments was a room temperature nematic mixture containing 49% (2-methyl phenyl)-bis 4-n-butyl benzoate and 51% 1-n-propyl-4[4-n-ethoxy phenyl] cyclohexane, which were obtained from Hoffmann-La Roche. Homo-

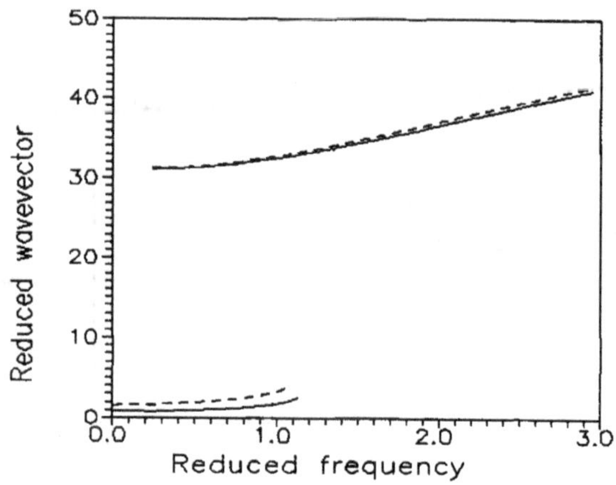

Figure 2: Variation of the reduced wavevector q/p with reduced frequency obtained from the calculations.

geneous alignment was obtained by unidirectional rubbing of polyimide coated glass plates. Stainless steel wires of 50 μm diameter were used to apply the transverse electric field. They also served as spacers. The lateral separation between the wires was about 400 μm. The onset of the instability was determined by the appearance of the EHD pattern in the cell under a polarizing microscope. As mentioned earlier, the distortion of the director field near the onset is predominantly twist. In view of the Mauguin criterion a twist distortion cannot be detected near the threshold. However, at slightly higher voltages, observations clearly indicate that twist distortion is quite predominant, neighbouring convective cells having twist of opposite sense. On the other hand, the splay-bend distortion produced by the flexoelectric coupling can be detected optically even very close to the onset of the instability. However, the threshold voltage measured by us may be an overestimate, especially at higher frequencies where the splay-bend distortion is small. The high frequency dielectric regime could not be observed as an irregular flow developed in the cell at high voltages needed at these frequencies. Figs. 3 & 4 show the threshold voltage and wavevector in the conduction regime as a function of the frequency of the applied field. The initial decrease in the threshold voltage with increasing frequency may be attributed to the screening of the applied field by the impurity ions present in the sample. The material parameters of the mixture used in the experiments are not known. Consequently, we cannot directly compare the experiment with our calculations. However, comparison of Figs. 3 & 4 with Figs. 1 & 2 clearly indicates the broad agreement between theory and experiment.

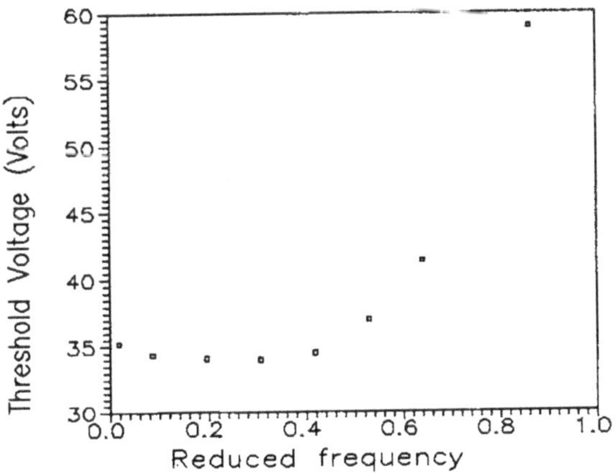

Figure 3: Experimentally observed variation of the threshold voltage with reduced frequency.

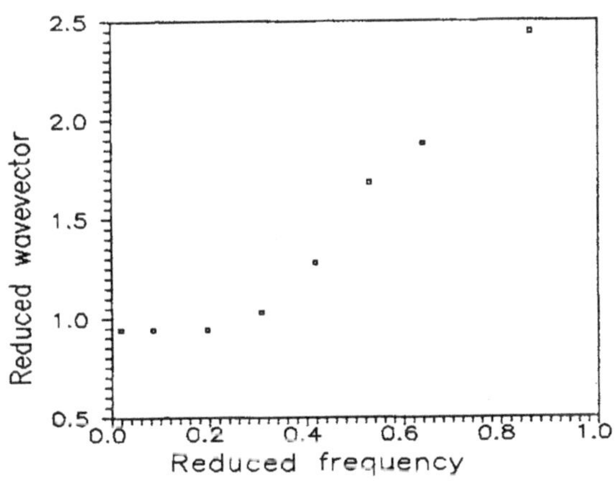

Figure 4: Experimentally observed variation of the reduced wavevector with reduced frequency.

References

[1] L.M.Blinov, *Electro-Optical and Magneto-Optical Properties of Liquid Crystals (Wiley, 1983).*

[2] W.Helfrich, *J. Chem. Phys.* **51** *(1969) 4092.*

[3] E.Dubois-Violette, P.G. de Gennes and O. Parodi, *J. Phys. Paris)* **32** *(1971) 305.*

[4] N.V.Madhusudana, V.A.Raghunathan and K.R.Sumathy, *Pramana-J.Phys.* **28** *(1987) L311.*

[5] N.V.Madhusudana and V.A.Raghunathan, *Mol. Cryst. Liquid Cryst. Lett.* **5** *(1988) 201.*

[6] L.Kramer, E.Bodenschatz, W.Pesch, W.Thom and W.Zimmermann, *Liq. Crystals* **5** *(1989) 699.*

[7] R.B.Meyer, *Phy. Rev. Lett.* **22** *(1969) 918.*

[8] R.Williams, *J.Chem.Phys.* **56** *(1971) 147.*

[9] I.G.Chistyakov and L.K.Vistin, *Kristallografiya* **19** *(1974) 195.*

[10] V.A.Raghunathan and N.V.Madhusudana, *Pramana-J.Phys.* **31** *(1988) L163.*

[11] I.W. Smith, Y.Galerne, S.T.Lagerwall, E.Dubois-Violette and G.Durand, *J. Phys. (Paris)* **36** *(1975) C1-237.*

NONLOCAL AMPLITUDE EQUATIONS

E. Knobloch

Department of Physics, University of California
Berkeley, CA 94720, U.S.A.

Abstract

The coupled complex Ginzburg-Landau (CCGL) equations describing the competition between counterpropagating wavetrains in 1+1 dimensions require a modification when the group velocity is $O(1)$. The correct equations are coupled by mean field coupling and are called MFGL equations. The connection between the two sets of equations is established, and the origin of the different predictions for the stability of standing wavetrains against large scale spatial modulation is explained.

1. INTRODUCTION

Amplitude equations are asymptotically exact equations that describe the evolution of weakly nonlinear patterns in extended systems. The equations describe the large scale modulation of patterns that are periodic in both space and time, and are derived by multiple scale analysis of the governing field equations. They typically arise in one of two ways: as evolution equations near a symmetry-breaking instability in extended systems, in which case the strength of the nonlinearity is related to the supercriticality of the system, or as evolution equations for weakly nonlinear dispersive waves in conservative systems, in which case the strength of the nonlinearity is related to the amplitude of the initial disturbance that excites the waves in the first place.

The resulting equations are usually partial differential equations, or a set of partial differential equations, if the amplitudes are coupled to mean flows generated by inhomogeneities in the wave pattern. Explicitly nonlocal equations also arise, however. Such equations have been derived for the Taylor vortex flow in the narrow gap approximation[1], and for the ferromagnetic resonance above the threshold for the Suhl instability[2]. Both equations are derived near the onset of a steady state instability, and share a common structure because of reflection and translation invariance in one direction in both systems. More recently Knobloch and De Luca[3,4] noted that the usual derivation of the coupled complex Ginzburg-Landau equations (hereafter CCGL equations) describing the evolution of wave patterns in spatially extended systems undergoing a symmetry-breaking Hopf bifurcation fails when the group velocity c_g of the waves is of order one. In this case the correct equations in one dimension are a pair of Ginzburg-Landau equations with mean-field (i.e., nonlocal) coupling. In this sense the equations are a natural extension of the equation derived by Hall and Elmer. In the following we refer to these equations as MFGL equations. For a very recent generalization of the MFGL equations to the oscillatory instability of propagating fronts in gasless combustion and reaction-diffusion systems, see refs. 5, 6.

The difference between the CCGL equations and the MFGL equations is not merely academic. For example, as noted already by Knobloch and De Luca[3], the two sets of equations make *different* predictions for the stability of a train of standing waves with respect to spatial modulation. In this article we explore in detail the origin of this difference, and establish the connection between the two sets of equations. In addition we emphasize that the differences in their predictions make the applicability of each set to a particular system amenable to experimental or numerical investigation. Our analysis, however, points strongly to the MFGL equations as the appropriate equations for systems with a finite group velocity. It should be mentioned that the two sets of equations do agree on the stability properties of uniform travelling wavetrains, since in this case the group velocity term can be eliminated by going into an appropriate moving frame.

In section 2 we derive, in some generality, the MFGL equations for both dissipative and conservative systems. Their connection to the CCGL equations is explored in section 3. In section 4 we discuss the stability of a constant amplitude standing wave with respect to spatial modulation within the framework of both sets of equations, filling in some of the details not given in ref. 3.

2. DERIVATION OF THE EQUATIONS

We consider a scalar dispersive system on a line of the generic form

$$L\left(\frac{\partial}{\partial t}, \frac{\partial}{\partial x}; \epsilon^2\right)\psi = N[\psi], \quad 0 < \epsilon \ll 1, \tag{2.1}$$

where L and N are linear and nonlinear operators, both of which can depend on space and time derivatives, and suppose that it is equivariant under translations $x \to x + \ell$ and reflections $x \to -x$. Under these conditions equation (2.1) has solutions in the form of spatially uniform wavetrains of wavelength $2\pi/k$ travelling in either direction along the x-axis. In the following we choose k to be either the critical wavenumber selected by the instability (dissipative systems) or the excitation mechanism (dispersive conservative systems), and derive equations describing the effect of nearby wavenumbers on the competition in the nonlinear regime between two such wavetrains travelling in opposite directions. The nearby wavenumbers are excited along with the basic wavenumber k and lead to large-scale spatial modulation of the wavetrains.

Equations describing the interaction of two small amplitude counter-propagating wavetrains can be derived by the method of multiple scales. We introduce the hierarchy of slow scales by

$$X_1 = \epsilon x, \quad T_1 = \epsilon t, \tag{2.2a}$$

$$X_2 = \epsilon^2 x, \quad T_2 = \epsilon^2 t, \tag{2.2b}$$

and let

$$\psi(x,t;\epsilon) = \epsilon\psi_1 + \epsilon^2\psi_2 + ..., \tag{2.3}$$

where

$$\psi_j \equiv \psi_j(t, T_1, T_2; x, X_1, X_2). \tag{2.4}$$

Then

$$L = L_0\left(\frac{\partial}{\partial t}, \frac{\partial}{\partial x}; 0\right) + \epsilon\left[L_1\frac{\partial}{\partial T_1} + L_2\frac{\partial}{\partial X_1}\right] + \frac{1}{2}\epsilon^2 \times$$
$$\left\{L_{11}\frac{\partial^2}{\partial T_1^2} + 2L_{12}\frac{\partial^2}{\partial T_1 \partial X_1} + L_{22}\frac{\partial^2}{\partial X_1^2} + 2L_1\frac{\partial}{\partial T_2} + 2L_2\frac{\partial}{\partial X_2} + 2L_3\right\} + O(\epsilon^3), \tag{2.5}$$

where L_{jk} are partial derivatives of L with respect to its three arguments, evaluated at $(\partial/\partial t, \partial/\partial x; 0)$. We assume that all such derivatives are formally $O(1)$. Then at $O(\epsilon)$ one recovers the linear problem

$$L_0\left(\frac{\partial}{\partial t}, \frac{\partial}{\partial x}; 0\right)\psi_1 = 0 \tag{2.6}$$

with the general solution

$$\psi_1 = A^+(T_1, T_2, X_1, X_2)\exp i(kx - \omega t) + A^-(T_1, T_2, X_1, X_2)\exp i(kx + \omega t) + c.c. \tag{2.7}$$

Here $\omega \equiv \omega(k)$ is given by the *dispersion relation*

$$\ell_0(i\omega, ik; 0) = 0. \tag{2.8}$$

At $O(\epsilon^2)$ one finds

$$L_0\psi_2 + [L_1\frac{\partial}{\partial T_1} + L_2\frac{\partial}{\partial X_1}]\psi_1 = N_2[\psi_1, \psi_1]. \tag{2.9}$$

Since the nonlinear terms $N[\psi_1, \psi_1]$ cannot lie in the kernel of L_0, equation (2.9) has a periodic solution ψ_2 if and only if the following solvability conditions hold:

$$[\ell_1(i\omega, ik; 0)\frac{\partial}{\partial T_1} + \ell_2(i\omega, ik; 0)\frac{\partial}{\partial X_1}]A^- = 0, \tag{2.10a}$$

$$[\ell_1(-i\omega, ik; 0)\frac{\partial}{\partial T_1} + \ell_2(-i\omega, ik; 0)\frac{\partial}{\partial X_1}]A^+ = 0. \tag{2.10b}$$

Here, as in (2.8), the scalars $\ell_j^{\pm} \equiv \ell_j(\mp i\omega, ik; 0)$ are defined by $L_j \exp i(kx \mp \omega t) = \ell_j(\mp i\omega, ik; 0)\exp i(kx \mp \omega t)$. It follows that

$$A^{\pm} = A^{\pm}(\xi^{\pm}, T_2, X_2), \tag{2.11}$$

where

$$\xi^{\pm} = c_g T_1 \mp X_1, \tag{2.12}$$

and c_g is the group velocity:

$$c_g = -\ell_2(i\omega, ik; 0)/\ell_1(i\omega, ik; 0). \tag{2.13}$$

Note that, by assumption, the group velocity is $O(1)$.

With the solvability conditions satisfied we can now solve equation (2.9) for ψ_2. We obtain

$$\psi_2 = B^+ e^{i(kx-\omega t)} + B^- e^{i(kx+\omega t)} + c.c. + \Psi_2(x, t; A^+, A^-, \bar{A}^+, \bar{A}^-), \tag{2.14}$$

where Ψ_2 denotes $L_0^{-1} N_2[\psi_1, \psi_1]$ and contains 10 terms.

At $O(\epsilon^3)$ one obtains the equation

$$L_0\psi_3 + \left[L_1\frac{\partial}{\partial T_1} + L_2\frac{\partial}{\partial X_1}\right]\psi_2 + \frac{1}{2}\left\{L_{11}\frac{\partial^2}{\partial T_1^2} + 2L_{12}\frac{\partial^2}{\partial T_1 \partial X_1} + L_{22}\frac{\partial^2}{\partial X_1^2} + 2L_1\frac{\partial}{\partial T_2} + 2L_2\frac{\partial}{\partial X_2} + 2L_3\right\}\psi_1 = N_2[\psi_1, \psi_2] + N_2[\psi_2, \psi_1] + N_3[\psi_1, \psi_1, \psi_1]. \tag{2.15}$$

If the nonlinear operator N depends on spatial derivatives additional terms are present on the right side of this equation. In either case the solvability condition for ψ_3 now yields

$$\left[\ell_1^-\frac{\partial}{\partial T_1} + \ell_2^-\frac{\partial}{\partial X_1}\right]B^- + \frac{1}{2}\left\{\ell_{11}^-\frac{\partial^2}{\partial T_1^2} + 2\ell_{12}^-\frac{\partial^2}{\partial T_1 \partial X_1} + \ell_{22}^-\frac{\partial^2}{\partial X_1^2} + 2\ell_1^-\frac{\partial}{\partial T_2} + 2\ell_2^-\frac{\partial}{\partial X_2} + 2\ell_3^-\right\}A^- = \alpha A^-|A^-|^2 + \beta A^-|A^+|^2 \tag{2.16}$$

with a similar equation obtained by the substitution

$$A^- \to \bar{A}^+, \quad B^- \to \bar{B}^+. \tag{2.17}$$

In terms of the comoving variables $\xi^{\pm}$ equation (2.16) becomes

$$2\ell_1^- c_g \frac{\partial B^-}{\partial \xi^+} + \gamma^- \frac{\partial^2 A^-}{\partial \xi^{-2}} + \left\{\ell_1^-\frac{\partial}{\partial T_2} + \ell_2^-\frac{\partial}{\partial X_2} + \ell_3^-\right\}A^- = \alpha A^-|A^-|^2 + \beta A^-|A^+|^2 \tag{2.18}$$

with a similar equation for B^+. Here

$$\gamma^{\pm} = \frac{1}{2}(\ell_{11}^{\pm} c_g^2 \mp 2\ell_{12}^{\pm} c_g + \ell_{22}^{\pm}) \tag{2.19}$$

and α, β are two complex coefficients. Note that $\gamma^{\pm} = \bar{\gamma}^{\mp}$.

The final step in the derivation is the observation that the expansion for $\psi(x,t)$ remains *asymptotic* for $(x,t) = O(\epsilon^{-2})$ if and only if $B^{\pm}$ do not contain any secular terms in $\xi^{\pm}$. If this requirement is not met the ordering assumed in equation (2.3) would be violated at $(x,t) = O(\epsilon^{-2})$. The resulting solvability conditions on $B^{\pm}$ take the form

$$\gamma^{-}\frac{\partial^2 A^-}{\partial \xi^{-2}} + \left\{\ell_1^- \frac{\partial}{\partial T_2} + \ell_3^-\right\}A^- = \alpha A^-|A^-|^2 + \beta A^- < |A^+|^2 >, \tag{2.20a}$$

$$\gamma^{+}\frac{\partial^2 A^+}{\partial \xi^{+2}} + \left\{\ell_1^+ \frac{\partial}{\partial T_2} + \ell_3^+\right\}A^+ = \bar{\alpha} A^+|A^+|^2 + \bar{\beta} A^+ < |A^-|^2 >, \tag{2.20b}$$

where

$$< |A^+|^2 > \equiv \frac{1}{P^+}\int_0^{P^+} |A^+|^2 d\xi^+, \tag{2.21a}$$

$$< |A^-|^2 > \equiv \frac{1}{P^-}\int_0^{P^-} |A^-|^2 d\xi^-, \tag{2.21b}$$

and $P^{\pm}$ ($\leq \infty$) are the assumed periods of $A^{\pm}$ in $\xi^{\pm}$. In writing equations (2.20) we have, without loss of generality, set all derivatives with respect to X_2 equal to zero.

Equations (2.20) are the desired evolution equations for the amplitudes $A^{\pm}$. Observe that in contrast to, say, the coupled complex Ginzburg-Landau equations, these equations are coupled only through mean-field coupling, i.e., because $c_g = O(1)$ one wavetrain "sees" the counter-propagating wavetrain only through its mean. There are two particularly interesting cases to consider.

(i) *A dissipative system near the onset of an oscillatory instability.* In this case $\ell^{\pm}$ can be absorbed in the remaining coefficients and equations (2.20) become

$$\frac{\partial A^-}{\partial T_2} = \{\mu + \beta < |A^+|^2 >\}A^- + \gamma\frac{\partial^2 A^-}{\partial \xi^{-2}} + \alpha A^-|A^-|^2, \tag{2.22a}$$

$$\frac{\partial A^+}{\partial T_2} = \{\bar{\mu} + \bar{\beta} < |A^-|^2 >\}A^+ + \bar{\gamma}\frac{\partial^2 A^+}{\partial \xi^{+2}} + \bar{\alpha} A^+|A^+|^2, \tag{2.22b}$$

where μ, α, β and γ are complex and $\mu = O(\lambda - \lambda_c)$, where λ is the bifurcation parameter. In addition $\mu_i \equiv \Im\mu$ can be absorbed in the evolution of $A^{\pm}$. We may therefore without loss of generality assume that μ is real, and treat it as the bifurcation parameter. The resulting equations will be referred to as the MFGL equations.

(ii) *A stable conservative system.* Such a system can support small but finite amplitude waves, whose interaction is again described by equations (2.20). In this case, however, the coefficients α, β and γ are purely imaginary, and the bifurcation parameter does not enter. After a suitable rescaling one obtains a pair of nonlinear Schrödinger equations with mean field coupling:

$$-2i\frac{\partial A^-}{\partial T_2} + \left(\frac{\partial^2\omega}{\partial k^2}\right)\frac{\partial^2 A^-}{\partial\xi^{-2}} = \alpha A^-|A^-|^2 + \beta A^- < |A^+|^2 >, \tag{2.23a}$$

$$+2i\frac{\partial A^+}{\partial T_2} + \left(\frac{\partial^2\omega}{\partial k^2}\right)\frac{\partial^2 A^+}{\partial\xi^{+2}} = \alpha A^+|A^+|^2 + \beta A^+ < |A^-|^2 >, \tag{2.23b}$$

where α, β are now real (see ref. 4).

The case of a steady state instability is a special case of equations (2.20) in which all the coefficients are real:

$$\frac{\partial A}{\partial T_2} = \{\mu + \beta < |A|^2 >\}A + \gamma\frac{\partial^2 A}{\partial x^2} + \alpha A|A|^2, \tag{2.24}$$

as derived in refs. 1 and 2. It should be emphasized, however, that in this case the origin of the nonlocal term is not propagation of the wavetrain, but is instead associated with induced mean flow, or the nonlocality of the demagnetization field, respectively.

3. THE CONNECTION BETWEEN CCGL AND MFGL EQUATIONS

When $c_g = O(\epsilon)$, there is no need to introduce the time scale $T_1 \equiv \epsilon t$. In this case the procedure employed in the preceding section leads to the CCGL equations:

$$A^-_{T_2} - vA^-_{X_1} = \mu A^- + \gamma A^-_{X_1X_1} + \alpha|A^-|^2A^- + \beta|A^+|^2A^-, \tag{3.1a}$$

$$A^+_{T_2} + vA^+_{X_1} = \bar{\mu}A^+ + \bar{\gamma}A^+_{X_1X_1} + \bar{\alpha}|A^+|^2A^+ + \bar{\beta}|A^-|^2A^+, \tag{3.1b}$$

where $v \equiv c_g/\epsilon = O(1)$. These equations are a pair of local equations.

To explore the connection between these equations and the MFGL equations consider the limit of large v (i.e., of $c_g = O(1)$). Introducing the faster time $T_1 \equiv T/\epsilon$, the CCGL equations take the form

$$A^-_{T_1} - c_gA^-_{X_1} = \epsilon\{\mu A^- + \gamma A^-_{X_1X_1} + \alpha|A^-|^2A^- + \beta|A^+|^2A^-\}, \tag{3.2a}$$

$$A^+_{T_1} + c_gA^+_{X_1} = \epsilon\{\bar{\mu}A^+ + \bar{\gamma}A^+_{X_1X_1} + \bar{\alpha}|A^+|^2A^+ + \bar{\beta}|A^-|^2A^+\}. \tag{3.2b}$$

Introducing the variables $\xi^\pm \equiv c_gT_1 \mp X_1$, these equations become

$$2c_gA^-_{\xi^+} = \epsilon\{\mu A^- + \gamma\left(\frac{\partial}{\partial\xi^+} - \frac{\partial}{\partial\xi^-}\right)^2 A^- + \alpha|A^-|^2A^- + \beta|A^+|^2A^-\}, \tag{3.3a}$$

$$-2c_gA^+_{\xi^-} = \epsilon\{\bar{\mu}A^+ + \bar{\gamma}\left(\frac{\partial}{\partial\xi^+} - \frac{\partial}{\partial\xi^-}\right)^2 A^+ + \bar{\alpha}|A^+|^2A^+ + \bar{\beta}|A^-|^2A^+\}. \tag{3.3b}$$

The equations are now in standard form for the application of the averaging theorem (see, e.g., ref. 7). The theorem considers nonautonomous systems of the form

$$\dot{x} = \epsilon f(x,t) + \epsilon^2 g(x,t;\epsilon), \quad x \in R^n, \quad 0 \leq \epsilon \leq \epsilon_c, \tag{3.4}$$

with both f, $g \in C^r$ $(r \geq 1)$ and periodic in time with period P. Then there exists a near-identity change of coordinates that puts the system (3.4) into the form

$$\dot{y} = \epsilon \bar{f}(y) + \epsilon^2 h(x,t;\epsilon), \tag{3.5}$$

where

$$\bar{f}(y) \equiv \frac{1}{P}\int_0^P f(y,t)dt \tag{3.6}$$

and h is a new P-periodic function.

Since, at leading order, $A^- = A^-(\xi^-)$, $A^+ = A^+(\xi^+)$, both of which are assumed to be periodic functions of their arguments, the averaging theorem yields at once the equations

$$2c_g\tilde{A}^-_{\xi^+} = \epsilon\{\mu\tilde{A}^- + \gamma\frac{\partial^2\tilde{A}^-}{\partial\xi^{-2}} + \alpha|\tilde{A}^-|^2\tilde{A}^- + \beta < |\tilde{A}^+|^2 > \tilde{A}^-\}, \tag{3.7a}$$

$$-2c_g\tilde{A}^+_{\xi^-} = \epsilon\{\bar{\mu}\tilde{A}^+ + \bar{\gamma}\frac{\partial^2\tilde{A}^+}{\partial\xi^{+2}} + \bar{\alpha}|\tilde{A}^+|^2\tilde{A}^+ + \bar{\beta} < |\tilde{A}^-|^2 > \tilde{A}^+\}. \tag{3.7b}$$

Here $\tilde{A}^\pm$ and $A^\pm$ are related by a near-identity transformation. The MFGL equations (2.22) are now recovered on noting that

$$\frac{1}{\epsilon}\frac{\partial}{\partial\xi^\pm} = \frac{1}{2}\left(\frac{1}{c_g}\frac{\partial}{\partial T_2} \mp \frac{\partial}{\partial X_2}\right), \tag{3.8}$$

and dropping tildes together with the inessential derivatives with respect to X_2.

4. STABILITY OF STANDING WAVES

Since the CCGL and MFGL equations take quite different form it is important to see whether or not they also make qualitatively different predictions. In this section we study the stability of uniform trains of standing waves with respect to spatial modulation within both sets of equations, and uncover important differences between the two systems.

(i) *The MFGL equations*

Within both the MFGL and CCGL equations a uniform train of standing waves takes the form

$$A^+ = Re^{-i\Omega T_2}, \quad A^- = Re^{i\Omega T_2}, \tag{4.1}$$

where

$$\Omega = \mu_i + (\alpha_i + \beta_i)R^2, \tag{4.2a}$$
$$0 = \mu_r + (\alpha_r + \beta_r)R^2. \tag{4.2b}$$

To study the stability of such a solution we let

$$A^+ = Re^{-i\Omega T_2}(1+a), \quad A^- = Re^{i\Omega T_2}(1+b) \tag{4.3}$$

and linearize the MFGL equations in a, b. We find

$$a_{T_2} = \gamma a_{\xi^+\xi^+} + R^2\alpha(a+\bar{a}) + R^2\beta \langle b+\bar{b} \rangle, \tag{4.4a}$$

$$b_{T_2} = \bar{\gamma} b_{\xi^-\xi^-} + R^2\bar{\alpha}(b+\bar{b}) + R^2\bar{\beta} \langle a+\bar{a} \rangle\,. \tag{4.4b}$$

At this stage one must distinguish two cases, when a and b are independent of the comoving coordinates $\xi^\pm$, and when they are not. In the first case the perturbations are spatially uniform and the averaging implied by $\langle \dots \rangle$ becomes trivial. The resulting equations constitute four coupled linear equations for a, $\bar{a}$, b and $\bar{b}$. These equations have solutions proportional to $\exp sT_2$, where

$$s = 2R^2(\alpha_r \pm \beta_r),\ 0,\ 0. \tag{4.5}$$

The eigenvalue proportional to $\alpha_r + \beta_r$ describes stability with respect to amplitude perturbations (i.e., changes in R); the second eigenvalue, proportional to $\alpha_r - \beta_r$, describes stability with respect to spatially uniform travelling wave perturbations. In the following we assume that

$$\alpha_r + \beta_r < 0, \quad \alpha_r - \beta_r < 0 \tag{4.6}$$

so that the wave train is *stable* with respect to spatially uniform perturbations. The zero eigenvalues correspond, respectively, to translation invariance and phase shift invariance. It is these neutral modes that can cause instability when spatially nonuniform perturbations are considered. These take the form

$$a = a_+(T_2)e^{iq^+\xi^+} + \bar{a}_-(T_2)e^{-iq^+\xi^+}, \quad b = b_+(T_2)e^{iq^-\xi^-} + \bar{b}_-(T_2)e^{-iq^-\xi^-}. \tag{4.7}$$

The averages $\langle \dots \rangle$ now vanish, provided we restrict attention to $q^\pm = 2\pi/N^\pm p^\pm$, where the $N^\pm$ are integers. In this case equations (4.4) *decouple*, and it is sufficient to consider equation (4.4a). The resulting equations for $a_\pm(T_2)$ are given by

$$\dot{a}_+ = -\gamma q^{+2} a_+ + R^2\alpha(a_+ + a_-), \tag{4.8a}$$

$$\dot{a}_- = -\bar{\gamma} q^{+2} a_- + R^2\bar{\alpha}(a_+ + a_-), \tag{4.8b}$$

where the dot denotes differentiation with respect to T_2. Because of this decoupling the analysis that follows is essentially identical to that for the single Ginzburg-Landau equation (2.24) with real coefficients[2].

The two eigenvalues s are given by the *dispersion relation*

$$s^2 + 2s(\gamma_r q^{+2} - R^2\alpha_r) + |\gamma|^2 q^{+4} - 2R^2 q^{+2}(\alpha_r\gamma_r + \alpha_i\gamma_i) = 0. \tag{4.9}$$

Thus

$$s = 2R^2\alpha_r + O(q^{+2}), \quad s = -\left(\frac{\alpha_r\gamma_r + \alpha_i\gamma_i}{\alpha_r}\right)q^{+2} + O(q^{+4}). \tag{4.10}$$

Similar results follow from equation (4.4b). Observe that as $q^+ \to 0$ the first eigenvalue does *not* reduce to the amplitude eigenvalue in (4.5). However, since the conditions (4.6) imply that $\alpha_r < 0$, the corresponding eigenvalue remains stable. This discontinuity in the amplitude eigenvalue as $q^+ \to 0$ was first noted by Elmer[2]. The second eigenvalue takes the place of one of the zero eigenvalues in (4.5) and is responsible for the possibility of a long wavelength instability. Since $\alpha_r < 0$, we see that the wavetrain will be unstable to spatial modulation whenever

$$\alpha_r\gamma_r + \alpha_i\gamma_i > 0, \tag{4.11}$$

as obtained by Knobloch and De Luca[3].

The above computation is readily extended to two other situations of interest, when the wavenumbers $q^\pm$ of the perturbation are not commensurate with the periods $P^\pm$, or when the basic wavetrain is spatially nonuniform. In the latter case one starts with solutions of the form

$$A^\pm = Re^{ik^\pm\xi^\pm \mp i\Omega T_2} \tag{4.12}$$

with $k^\pm \neq 0$. As a consequence the condition (4.11) for modulational instability is modified, and one now has

$$\alpha_r\gamma_r + \alpha_i\gamma_i + \frac{4k^{\pm 2}}{\alpha_r^2 R^2}|\alpha|^2\gamma_r^2 > 0. \tag{4.13}$$

Such wavetrains are therefore *more* susceptible to modulational instability than uniform wavetrains.

(ii) *The CCGL equations*

The corresponding calculation of the stability of the uniform wavetrain (4.1) within the CCGL equations is a little more involved because no decoupling takes place. In this case the perturbation equations are

$$a_{T_2} - va_{X_1} = \gamma a_{X_1X_1} + R^2\alpha(a + \bar{a}) + R^2\beta(b + \bar{b}), \tag{4.14a}$$

$$b_{T_2} + vb_{X_1} = \bar{\gamma} b_{X_1X_1} + R^2\bar{\alpha}(b + \bar{b}) + R^2\bar{\beta}(a + \bar{a}). \tag{4.14b}$$

To solve these equations, we let

$$a = a_+(T_2)e^{iqX_1} + \bar{a}_-(T_2)e^{-iqX_1}, \quad b = b_+(T_2)e^{iqX_1} + \bar{b}_-(T_2)e^{-iqX_1}. \tag{4.15}$$

Then

$$\dot{a}_+ - iqva_+ = -\gamma q^2 a_+ + R^2\alpha(a_+ + a_-) + R^2\beta(b_+ + b_-), \tag{4.16a}$$
$$\dot{a}_- - iqva_- = -\bar{\gamma} q^2 a_- + R^2\bar{\alpha}(a_+ + a_-) + R^2\bar{\beta}(b_+ + b_-), \tag{4.16b}$$
$$\dot{b}_+ + iqvb_+ = -\bar{\gamma} q^2 b_+ + R^2\bar{\alpha}(b_+ + b_-) + R^2\bar{\beta}(a_+ + a_-), \tag{4.16c}$$
$$\dot{b}_- + iqvb_- = -\gamma q^2 b_- + R^2\alpha(b_+ + b_-) + R^2\beta(a_+ + a_-), \tag{4.16d}$$

where the dot denotes differentiation with respect to T. These four coupled constant coefficient equations have solutions proportional to $\exp sT_2$, where s satisfies the *quartic* dispersion relation

$$\begin{aligned}
&(s - iqv + \gamma q^2 - R^2\alpha)(s - iqv + \bar{\gamma} q^2 - R^2\bar{\alpha})(s + iqv + \bar{\gamma} q^2 - R^2\bar{\alpha}) \\
&\times (s + iqv + \gamma q^2 - R^2\alpha) - R^4|\alpha|^2\{(s - iqv + \gamma q^2 - R^2\alpha)(s - iqv + \bar{\gamma} q^2 - R^2\bar{\alpha}) \\
&+ (s + iqv + \bar{\gamma} q^2 - R^2\bar{\alpha})(s + iqv + \gamma q^2 - R^2\alpha)\} + R^8|\alpha|^4 \\
&- R^4[\beta(s + iqv + \bar{\gamma} q^2) + \bar{\beta}(s + iqv + \gamma q^2)][\beta(s - iqv + \bar{\gamma} q^2) + \bar{\beta}(s - iqv + \gamma q^2)] \\
&= 0
\end{aligned} \tag{4.17}$$

For fixed v and small q this relation takes the form

$$\begin{aligned}
&s^4 - 4R^2(\alpha_r + O(q^2))s^3 + 4R^4(\alpha_r^2 - \beta_r^2 + O(q^2))s^2 + 4R^2q^2s[2R^2\alpha_r(\alpha_r\gamma_r + \alpha_i\gamma_i) \\
&- 2R^2\beta_r(\beta_r\gamma_r + \beta_i\gamma_i) - \alpha_r v^2 + O(q^2)] + 4R^4(\alpha_r^2 - \beta_r^2 + O(q^2))v^2q^2 = 0
\end{aligned} \tag{4.18}$$

and the eigenvalues are

$$s = 2R^2(\alpha_r \pm \beta_r) + O(q^2), \quad \pm ivq + \sigma q^2 + O(q^3), \tag{4.19}$$

where

$$\sigma = -\gamma_r - \left(\frac{\alpha_r\alpha_i - \beta_r\beta_i}{\alpha_r^2 - \beta_r^2}\right)\gamma_i. \tag{4.20}$$

In view of (4.6) we see that the wavetrain is modulationally *unstable* whenever

$$\alpha_r\gamma_r + \alpha_i\gamma_i > \frac{\beta_r}{\alpha_r}(\beta_r\gamma_r + \beta_i\gamma_i) \tag{4.21}$$

(cf. equation 4.11). These results were first obtained by Coullet *et al* [8] using a different method.

Observe that if we now take the limit $v \to \infty$ as suggested by the discussion in section 3, we find that the growth rate (4.20) of the instability remains unchanged, while its frequency tends to infinity. The latter is a reflection of the fact that the equations are written with respect to the wrong time scale. Observe also that the condition (4.21) *differs* from the corresponding condition for the modulational instability of standing wave solutions of the MFGL equations. To understand the origin of the this difference, consider *first* the limit $v \to \infty$ in the dispersion relation (4.17), *followed* by the approximation that q is small. If we set $s = iqv + \sigma$, where $\sigma = O(1)$ in v, we obtain

$$\sigma^2 + 2\sigma(\gamma_r q^2 - R^2\alpha_r) + |\gamma|^2 q^4 - 2R^2 q^2(\alpha_r\gamma_r + \alpha_i\gamma_i) = O(\frac{1}{v}). \qquad (4.22)$$

(cf. equation 4.9). As before the presence of instability requires a real positive eigenvalue, and for small q this is again $O(q^2)$. We have then

$$s = iqv - \left(\frac{\alpha_r\gamma_r + \alpha_i\gamma_i}{\alpha_r}\right)q^2 + O(\frac{1}{v}). \qquad (4.23)$$

Again the frequency approaches infinity as $v \to \infty$, but now the condition for modulational instability is precisely the condition (4.11) obtained from the MFGL equations. It is thus clear that the two limits $v \to \infty$, $q \to 0$ do *not* commute, and that this is the origin of the different stability conditions for standing waves solutions of the CCGL and MFGL equations.

5. DISCUSSION

In this article we have derived a pair of amplitude equations governing the evolution of counter-propagating wavetrains in translation- and reflection-invariant systems near a symmetry-breaking Hopf bifurcation. These equations are coupled via mean-field terms and are therefore nonlocal. This type of coupling arises whenever the group velocity of the waves is of order one relative to their amplitudes. This is the typical situation close to the onset of the instability. The resulting MFGL equations differ from the better known local equations called the coupled complex Ginzburg-Landau (CCGL) equations. The latter are local because the group velocity is assumed to be of order of the wave amplitude[3]; in this case the advection effects which give rise to the mean field coupling in the MFGL equations are sufficiently weak that a wavetrain responds to the local amplitude of the counter-propagating wavetrain, rather than to its mean. The presence of the mean field coupling alters the stability properties of standing waves with respect to long wavelength spatial modulation. This effect is quantitative, and should be detectable in numerical simulation of standing waves in continuous systems undergoing a Hopf bifurcation from a trivial (i.e., translation- and reflection-invariant) state. In particular, the stability of such waves should change as parameters are varied to increase the group velocity from zero. Thus the MFGL equations predict new phenomena, and are not merely a different description of the dynamics already described by the CCGL equations. The profound differences that are possible between the dynamics in local and nonlocal evolution equations are beautifully illustrated in ref. 9.

In this paper we discussed in some detail the connection between the MFGL and CCGL equations and traced their difference in their predictions to the mathematical observation that the limits $c_g/\epsilon \to \infty$, and $q \to 0$ do not commute. Here ϵ is the wave amplitude and hence is related to the supercriticality $(\lambda - \lambda_c)^{1/2}$ of the system, and q is the wavenumber of the spatial perturbation to which the standing waves are subjected. The stability criterion (4.11) should then apply whenever $c_g/(\lambda - \lambda_c)^{1/2} \gg 1$, while the conditions (4.20) should be valid for $c_g/(\lambda - \lambda_c)^{1/2} \lesssim 1$.

As already mentioned the stability properties of uniform travelling wavetrains are the same within both sets of equations[3]. Stability properties of spatially nonuniform (i.e., modulated) wavetrains can also be readily obtained. Here there is one qualitative difference between the CCGL and MFGL equations. The latter admit spatially *quasiperiodic* solutions, while the former do not. The calculations of the stability properties of such states are pursued in refs. 5, 6.

Acknowledgement: I acknowledge helpful discussions with F.J. Elmer, A.C. Newell, H. Riecke and M. Silber prior to the writing of this article.

References

1. P. Hall 1984 Phys. Rev. A **29**, 2921-2923.

2. F.J. Elmer 1988 Physica D **30**, 321-342.

3. E. Knobloch and J. De Luca 1990 Nonlinearity **3**, 975-980.

4. E. Knobloch and J.D. Gibbon 1991 Phys. Lett. A **154**, 353-356.

5. B.J. Matkowsky and V. Volpert 1991 Physica D, to appear.

6. B.J. Matkowsky and V. Volpert 1992 Random and Computational Dyn., to appear.

7. S. Wiggins 1990 *Introduction to Applied Nonlinear Dynamical Systems and Chaos*, New York: Springer-Verlag.

8. P. Coullet, S. Fauve and E. Tirapegui 1985 J. Phys. Lettres **46**, L787-791.

9. J.M. Ball, P.J. Holmes, R.D. James, R.L. Pego and P.J. Swart 1991 J. Nonlin. Sci. **1**, 17-70.

STABILITY DIAGRAM IN ELECTROHYDRODYNAMIC CONVECTION OF NEMATICS

Satoru Nasuno,[1] Osamu Sasaki,[1] Shoichi Kai[1], and Walter Zimmermann[1,2]

1) *Department of Electrical Engineering, Kyushu Institute of Technology, Tobata, Kitakyushu 804, Japan*
2) *Institut für Festkörperforschung, Theorie III, Forschungszentrum Jülich 5170 Jülich, Fed. Rep. Germany*

Abstract

We report experimental results on the stability diagram for periodic roll structures in electrohydrodynamic convection of nematics. The stability is investigated as a function of the wavenumber of rolls and the amplitude of applied electric field for several values of the applied frequency. We find that the secondary instability which one meets naturally increasing the voltage is the zigzag type for low frequency and the skewed varicose type for high frequency.

1. Introduction

Electrohydrodynamic convection (EHC) in a thin layer of nematic liquid crystals has been investigated intensively in recent years [1, 2], because it shows a rich variety of patterns and bifurcations due to the presence of two independent parameters, i.e. the amplitude and the frequency of external electrical field. However the mechanisms of the bifurcations in this system are still poorly understood, except for the first one from homogeneous rest state to stationary periodic rolls [3]. In this paper we report the experimental study on the stability of periodic normal-roll structure. The stability is investigated as a function of the wavenumber of rolls and the amplitude of applied electric field for several values of the applied frequency. We find that the dominant destabilizing mechanism at the second bifurcation is the zigzag instability for low frequency and the skewed varicose instability for high frequency.

2. Experimental Setup

The geometry of liquid crystal cell used in the present experiments is the usual one. The cell is consisting of two parallel transparent electrodes separated by a Mylar sheet of thickness $d = 25\mu\text{m}$. It is filled with *N*-*p*-methoxybenzylidene-*p*-butylaniline (MBBA) doped with 0.01wt.% of tetra-*n*-butylammonium bromide. The horizontal size of the cell is $800d \times 800d$. The inner glass surfaces of the cell were coated with

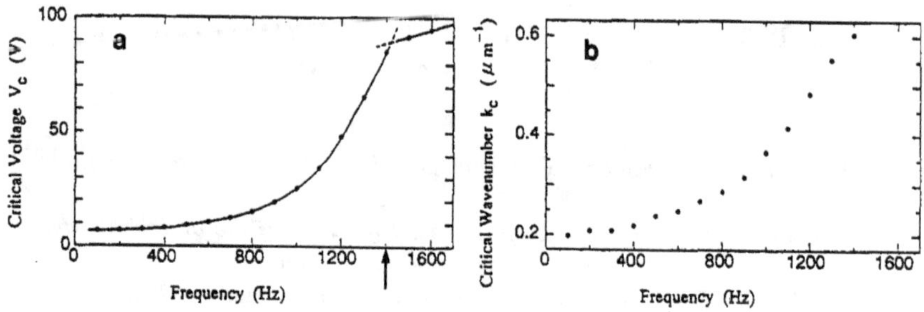

Figure 1: (a) The critical voltage V_c for the onset of convection and (b) the critical wave number k_c, as a function of the external frequency f. At the onset of convection we observe normal oriented rolls in the whole conduction regime (below the cut-off frequency f_c=1400Hz which is indicated by the arrow in (a)).

polyvinyl alcohol and rubbed in one direction $\hat{x}$ to achieve planar alignment. The temperature of the sample is held constant at 30.00 ± 0.01°C. The ac electric field applied across the liquid crystal layer are provided by the digital wave synthesizer system. In Fig.1 we present the critical voltage V_c for the onset of convection and the critical wave number k_c of the roll pattern at the onset of convection as function of external frequency f.

The convection patterns were visualized under a microscope with polarized light. Visualized images are taken with a CCD camera and stored on video tape. Images are typically taken from the central 0.1% of the cell area. The recorded images were analyzed by use of a digital image processor system with 512 × 512-pixels spatial resolution and 8-bit gray scale resolution.

3. Experimental Procedure

A detailed quantitative measurement on the stability of normal rolls requires the control on the wavenumber of rolls as well as the voltage and the frequency of the external field. This can be done efficiently by utilizing the frequency dependence of the critical wavenumber k_c [4]. Briefly, the experimental procedure to investigate the stability of a normal rolls with a given wavenumber k^* at a chosen frequency f^* and a chosen voltage V^* is as following : First, we prepare a normal rolls with a given wavenumber k^* by setting the frequency f so as to $k_c(f) = k^*$ and increasing the voltage slightly above the onset of convection. Then both the frequency and the voltage are jumped simultaneously to the aimed values, f^* and V^*. In this way we can put the normal roll with a desired wavenumber k^* at a chosen frequency f^* and a chosen voltage V^* and test its stability. Repeating this procedure for many sets of parameters we can study the stability of normal rolls as a function of k, V and f. In the present study, the voltage and the wavenumber is measured in normalized units $\epsilon = (V^2 - V_c^2(f))/V_c^2(f)$ and $q = k/k_c(f)$, respectively.

4. Experimental Results

The behavior of the system after the voltage-frequency jump depends on parameters and is linked with the stability of a given initial state. Clearly, one can say that a given state is stable if the pattern remains unchanged, and unstable if some pattern evolution leading the system away from a given initial state takes place. In the present experiment, the stable case is observed when ϵ is adequately small and when the wavenumber is adequately close to k_c. In this case, indeed, the pattern remains unchanged, except that its amplitude is adjusted to more preferred value. On the other hand, for unstable case, we observed two distinctive kinds of pattern evolutions :

1) evolutions with modulation of initial pattern,

2) evolutions without any modulations.

In this chapter, We first give the qualitative overview of these instability processes, and then proceed to the stability diagrams obtained in the present experiments.

4.1 Pattern evolutions through modulational instabilities

In the case of 1), the evolution always begin with the long-wavelength modulation of the initial rolls. The observed modulations are classified into three characteristic types. Typical examples are shown in Fig.2, 3 and 4. The 2D-spatial Fourier spectra calculated from these patterns reveal that such characteristic deformations are manifestations of the growth of unstable modes with the wavevector $\vec{k}_o \pm \vec{k}_u$, where $\vec{k}_o$ corresponds to the wavevector of the initial pattern, and can be distinguished each other by the relative direction of the wavevector $\vec{k}_u$ with respect to $\vec{k}_o$, as is described below. (We note that although the measured spectra show also other growing peaks, all these peaks can be analyzed by using the $\vec{k}_o$-mode and the $\vec{k}_u$-mode.) In most case, the peaks at $\vec{k}_o \pm \vec{k}_u$ are broader than the main one at $\vec{k}_o$, which indicates that the nearby modes also grow. Therefore in the present experiment we determine the type of secondary instability from the relative direction of wavevector $\vec{k}_u$ of the *fastest-growing* mode together with the direct observation of pattern evolution. In the following, attention has been focused on the initial stage of the pattern evolutions, because our main aim here is to investigate the distabilizing mechanism of normal rolls. The final states at the end of pattern evolutions will be described elsewhere [5].

a. the Eckhaus instability :

The first type of observed modulational instability appears as a compression and dilatation of rolls along the longitudinal direction [see Fig.2]. If the initial wavenumber is adequately far from k_c, this instability develops the over-compressed or over-dilated regions in which local effective wavelengths are farther away from k_c than the original one. The amplitude of the flow approaches zero at these over-compressed (over-dilated) regions, and eventually the elimination (nucleation) of roll pairs take place. Corresponding 2D Fourier spectra taken during the distabilization shows the

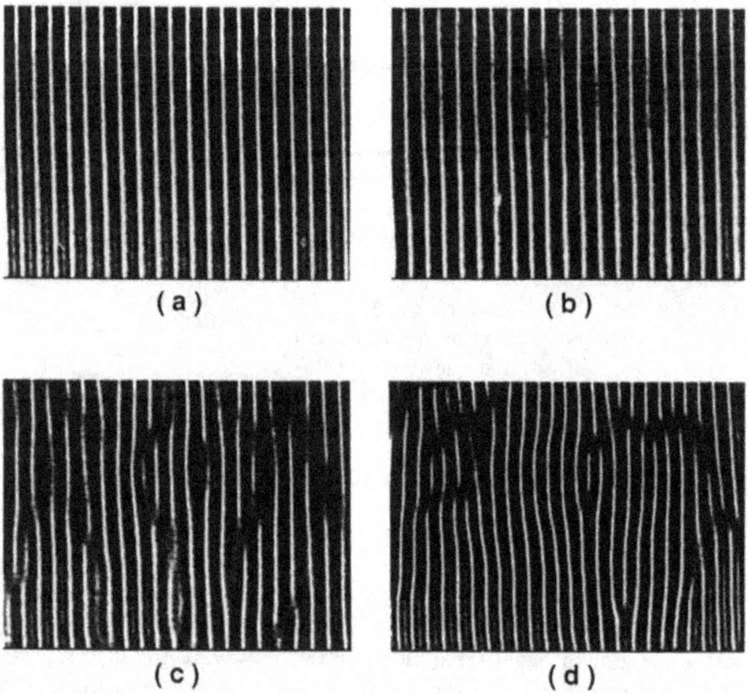

Figure 2: Typical pattern evolution initiated by the Eckhaus instability [$k/k_c = 0.83$, $\epsilon = 0.12$, f = 580Hz]. Only the early stage of the evolution after the frequency-voltage jump is shown : (a)0sec, (b)4.8sec, (c)6sec, and (d)7.2sec after the frequency-voltage jump. Note that the original uniform rolls receive the compression/dilatation modulations.

development of new peaks at $\vec{k}_o \pm \vec{k}_u$, where $\vec{k}_u$ is parallel to $\vec{k}_o$. Apparently these features are manifestation of Eckhaus instability [6]. However, in contrast to pure one-dimensional Eckhaus instability, the modulations are not homogenous along the y-direction, and thus the nucleation and elimination of roll pairs are usually accompanied with an appearance of defects, as is shown in Fig.2.

b. the zigzag instability :

The second type leads a long wavelength zigzag-like modulations of the rolls along their axis as is shown in Fig.3. The rolls are tilted at two symmetrical oblique angles $\pm\theta$ with respect to original axis and the domains with different sign of tilt angle are separated sharply by the narrow transition regions, i.e. grain boundaries. In 2D Fourier space, this instability can be characterized by the appearance of new peaks at $\vec{k}_o \pm \vec{k}_u$, where $\vec{k}_u$ is perpendicular to the $\vec{k}_o$. From these features we refer this modulational instability as the zigzag instability by the analogy with that occurring

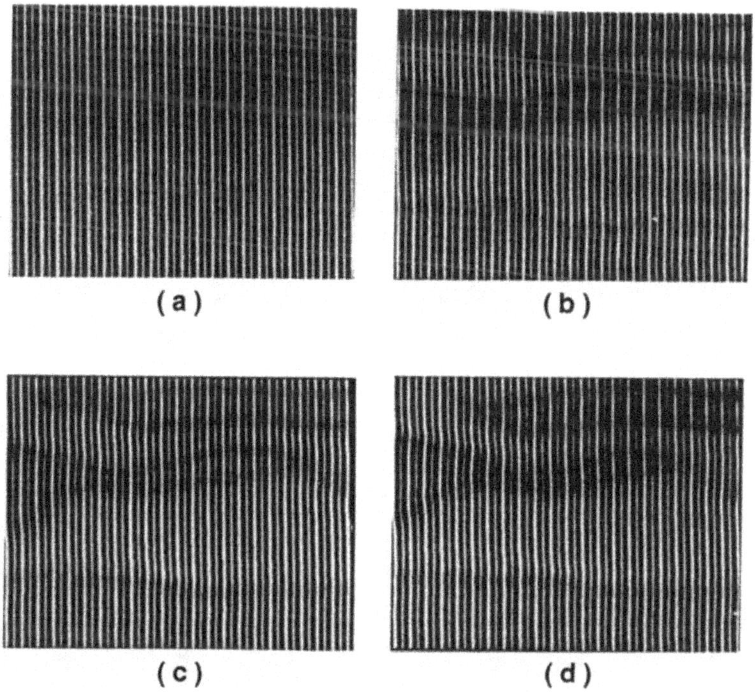

Figure 3: Typical pattern evolution initiated by the zigzag instability [k/k_c = 1.0, ϵ = 0.27, f = 400Hz]. Only the early stage of the evolution is shown: (a)0sec, (b)10sec, (c)13.3sec, and (d)16.7sec after the frequency-voltage jump. Note that the primary rolls are tilted with zigzag shape. In this case the growth of zigzag perturbation saturates at around (b) and in the subsequent stage only the quite slow dynamics of grain boundary can be seen.

in the isotropic convection system [7]. In contrast to the Eckhaus instability, this instability sometimes saturates leaving a stable zigzag roll structure, whereas in most case it leads the nucleation of defects at the grain boundaries.

c. the skewed varicose instability :

The third type appeared as the skew wavy deformations of original rolls, as shown in Fig.4. In the whole sample we observed generally two symmetrical skew deformations, however in different areas. Corresponding Fourier spectra taken during the distabilization show the appearance of new peaks at $\vec{k}_o \pm \vec{k}_u$, where $\vec{k}_u$ is oblique to $\vec{k}_o$. As the modulations grow, it tends to pinch the rolls and to create defects pairs. According to these features we refer this instability as the skewed varicose instability by the analogy with that occurring in the isotropic convection system [7].

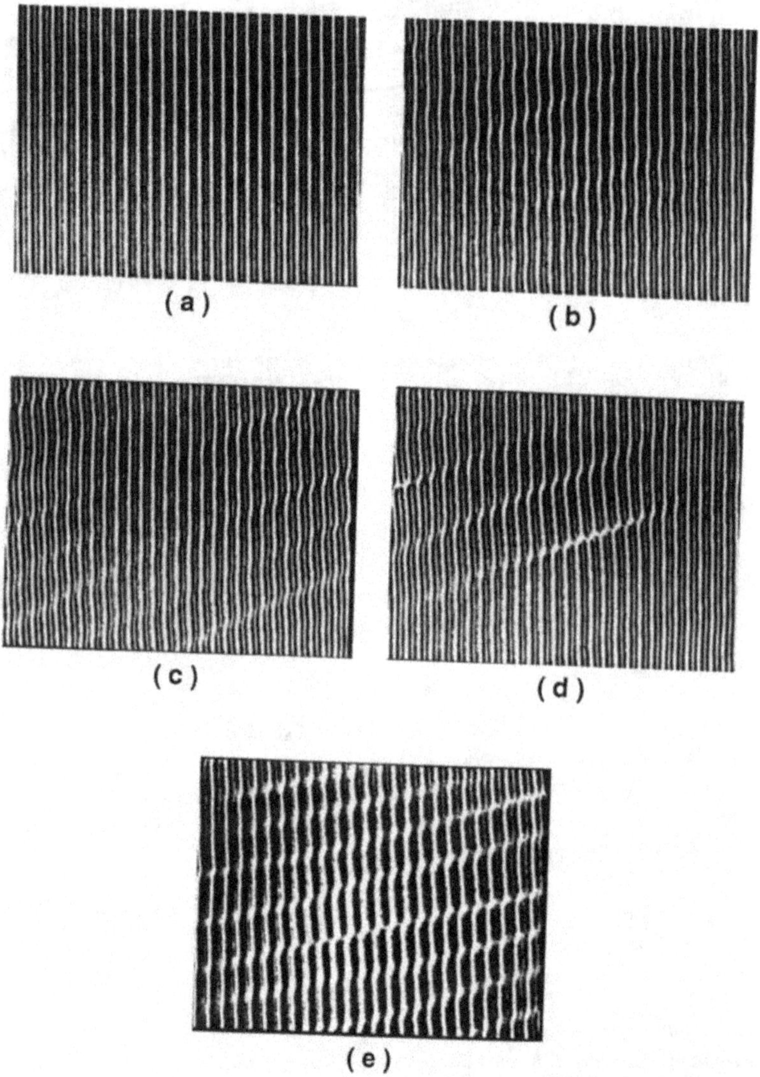

Figure 4: Typical pattern evolution initiated by the skewed varicose instability on the normal rolls [k/k_c = 1.04, ϵ = 0.22, f = 700Hz]. Only the early stage of the evolution is shown: (a)0sec, (b)18sec, (c)24sec, and (d)42sec after the frequency-voltage jump. (e) shows the skewed varicose modulation observed in another sample under similar experimental configuration. Note that the original uniform rolls receive the periodic skew modulation.

4.2 Pattern evolutions without modulational instabilities

In the case of 2), the pattern evolution is more simpler than those described above. After the voltage-frequency jump, the initial pattern begins to decay exponentially without any deformations, in contrast to the modulational instabilities mentioned above. This kind of evolution is observed when the initial wavenumber is sufficiently far from the critical one or when $\epsilon < 0$. For $\epsilon < 0$, the resulting state is always the homogeneous no flow state. On the other hand, for $\epsilon > 0$, a new roll pattern with a mean wavenumber closer to k_c appears after the decay of the initial pattern. Such pattern evolutions are predicted to occur under the neutral stability boundary for the onset of convection. By measuring the decay rate one can determine exactly the neutral curve [8], although it is not done in the present experiment.

4.3 Stability diagrams

The quantitative results for the stability of normal rolls are summarized in Fig.5, in which the stability of normal rolls is shown in (q, ϵ)-plane for three different values of f. The symbols represent the experimentally determined stability of normal rolls: closed circles indicate the stable points, and open circles, open squares and open diamonds indicate the points at which the Eckhaus, the zigzag and the skewed varicose instability take place as the *fastest-growing* unstable mode, respectively. Crosses are the absolutely unstable points at which the initial modes decay exponentially without any deformation. Based on these experimental points given in Fig.5 we draw the smooth curves to guide the eye. In Fig.5(a) we have only a few data points for $k < k_c$, because of lower limit of initial wavenumber achieved by the voltage-frequency jump method.

As is seen from Fig.5, for all frequencies measured in the present experiments the Eckhaus instability dominates in the lower part of the the stability diagrams. However the instability which dominates in the upper part of the stability diagrams strongly depends on the frequency.

At low frequency f=400Hz, the dominant instability which gives the the upper boundary of the stable region is the zigzag instability, as is seen from Fig.5(a). We always observed the zigzag type instability in the region in-between the line ZZ and the line SV. On the other hand above the line SV both the zigzag type and the skewed varicose type instability are observed as the most rapidly growing mode. It is not fully reproducible above the SV-curve whether the zigzag or the skewed varicose mode grows fastest. Moreover above the SV-curve we sometimes observed complicated competition between these two modes during the distabilization process. These facts indicate that above the line SV both the zigzag and the skewed varicose mode have positive growth rate and the growth of these unstable modes is strongly influenced by initial infinitesimal fluctuations. The y component of the wavevector of the fastest-growing zigzag mode tends to decrease as an initial state approaches to the line ZZ from the above. At higher frequency f=580Hz [fig.5(b)], the region where

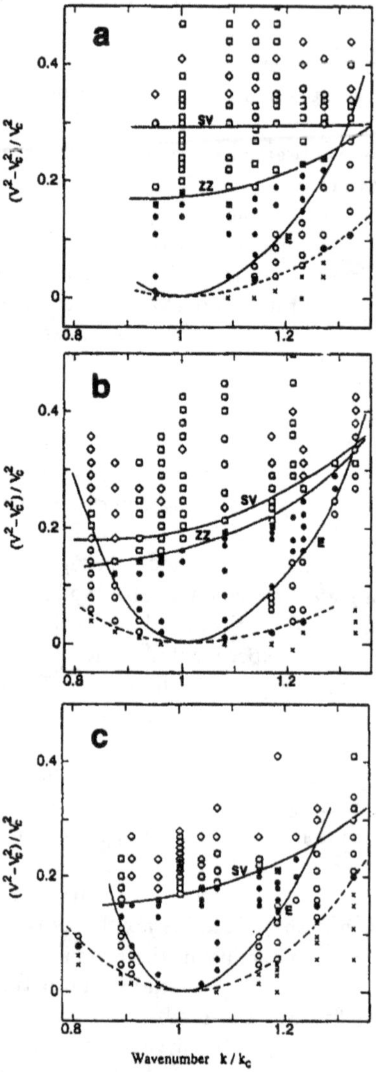

Figure 5: The stability of normal rolls as a function of the reduced voltage $\epsilon = (V^2 - V_c^2)/V_c^2$ and the reduced wave number k/k_c is plotted for three frequencies 400Hz(a), 580Hz(b) and 700Hz(c). Symbols represent experimental data points and the lines are guides to the eye. Closed circles refer to linearly stable and crosses refer to absolutely decaying normal rolls. Open circles, open squares and open diamonds refer to voltage wave numbers pairs where the normal rolls become unstable against Eckhaus, zigzag and skewed varicose instabilities, respectively.

the skewed varicose type instability takes place is shifted to lower ϵ in comparison with at f=400Hz, although the zigzag type instability is still the dominant one.

At still higher frequency f=700Hz [fig.5(c)], the zigzag instability seems to be no longer the dominant one. We observed the skewed varicose type instability even immediately above the upper boundary of the stable region. Nearing to the line SV, both x and y component of the modulational wavevector of the skewed varicose mode tend to decrease. However quantitative measurements of the modulational wavevector are difficult to make at this frequency, because the corresponding peaks in the measured Fourier spectra in the early distabilizing stage were broader than at 400Hz and 580Hz. In addition, close to the line SV the maximum amplitude of new peaks are usually small and this also make it sometimes technically difficult to decide whether the observed instability is of the zigzag type or of the skewed varicose type. Therefore we believe that the open squares close to the line SV are also instabilities of the skewed varicose type and the dominant instability at this frequency is the skewed varicose type.

5. Discussion

In a previous experiment with a fixed frequency f/f_c=0.5 the skewed varicose boundary is found to locate slightly above the zigzag one for the wavenumbers close to k_c [9]. On the basis of the present observations this type of results is expected to be obtained in the crossover frequency regime [see Fig.5(b)]. It is also interesting to compare our results with the recent experimental results by E. Braun et al. [10]. They performed the similar experiment but only for low applied frequencies, f/f_c=0.3 and 0.4, and show that the dominant instability for large values of ϵ is the zigzag type. This observation is qualitatively consistent with the present results obtained at low frequency [see Fig.5(a)].

Recent theoretical results on the stability of normal rolls show qualitative agreement with our measurements [11, 12, 13, 14]. The calculations were done at the one hand on two-dimensional anisotropic model systems [12] and directly in the framework of the electrohydrodynamic equations [13, 14]. In both special attention was paid to the mean flow effects. In these calculations also the zig-zag instability and the skewed varicose instability are dominant in the upper part of the stability diagram. However there seems up to now no quantitative agreement between our experimental results and the theoretical calculations on the fundamental equations [14]. The lack of the quantitative agreement is perhaps related to the incompleteness of the fundamental equations [15].

In the present experiment we observed three types of secondary instabilities, i.e. the Eckhaus, the zigzag and the skewed varicose instability. All three instabilities are also relevant for the stability of straight convection rolls in Rayleigh-Bénard convection in simple isotropic fluids [7]. Rayleigh-Bénard convection in simple fluids has isotropy in the fluid layer in contrast to the intrinsic anisotropic behavior in EHC. In isotropic systems the zigzag instability always takes place for $k < k_c$ and is understood as a wavelength adjusting mechanism for straight convection rolls, because the length of the effective wavevector always gets closer to k_c through the zigzag deformation. In EHC the zigzag instability occurs also for $k > k_c$ as is shown in Fig.5 and triggers the transition from normal rolls to zigzag-like pattern consisting of oblique rolls [5] Therefore in EHC the zigzag instability acts more like wavevector adjusting mechanism. In isotropic systems it is believed that if the aspect-ratio is large the skewed varicose instability triggers the onset of weak turbulence [16]. This is not the case in EHC, moreover, the skewed varicose instability leads in most of the cases from the normal rolls to a new ordered pattern [5, 17]. These differences are obviously related to the different symmetries between both systems.

6. Conclusion

We have measured the stability of normal oriented rolls and found that three kinds of secondary instabilities take place, i.e. the Eckhaus, the zigzag and the skewed varicose instability. We also found that the secondary instability which one meets naturally by increasing the voltage strongly depends on the applied frequency : it is the zigzag instability for low frequency, whereas the skewed varicose instability for high frequency.

Acknowledgment : We would like to thank H.R. Brand, W. Pesch, S. Sasa and M. Sano for discussions. W.Z. is grateful to the Kyushu Institute of Technology (KIT) for a Fellowship in 1989 where this cooperation has been started and to the Kajima foundation which made a second stay at the KIT and Kyushu University in 1991 possible.

References

[1] L.M. Blinov, *Electro-Optical and Magneto-Optical Properties of Liquid Crystals* (Wiley, New York, 1983).

[2] W. Zimmermann in MRS Bulletin **XVI** (1991) 46, and references therein.

[3] E. Bodenschatz, W. Zimmermann and L. Kramer, J. Phys. (Paris) **49** (1988) 1875.

[4] G. Goren, I. Procaccia, S. Rasenat, and V. Steinberg, Phys. Rev. Lett. **63** (1989) 1237.

[5] S. Nasuno, O. Sasaki, S. Kai, and W. Zimmermann, submitted to Phys. Rev. A.

[6] W. Eckhaus, *Studies in Nonlinear Stability Theory* (Springer, Berlin, 1965).

[7] F.H. Busse, Rep. Prog. **41** (1978) 1929; and in *Hydrodynamic Instabilities and Turbulence* , edited by H.L. Swinney and J.P. Gollub, (Springer, Berlin, 1984).

[8] M. Lowe and J. P.Gollub, Phys. Rev. Lett. **55** (1985) 2575.

[9] S. Nasuno and S. Kai, Europhys. Lett. **14** (1991) 779.

[10] E. Braun, S. Rasenat and V. Steinberg, Europhys. Lett. **15** (1991) 597.

[11] E. Bodenschatz, M. Kaiser, L. Kramer, A. Weber and W. Zimmermann, in *New Trends in Nonlinear Dynamics*, edited by P. Coullet and P. Huerre, NATO ASI Series (Plenum, New York, 1991).
[12] S. Sasa, Prog. Theor. Phys. **83** (1990) 824.
[13] M. Kaiser, W. Pesch and E. Bodenschatz, *Mean Flow Effects in the Electro-Hydrodynamic Convection in Nematic Liquid Crystals* , preprint (1990).
[14] W. Pesch (private communication).
[15] W. Zimmermann in *Defects, singularities and patterns in nematic liquid crystals* , edited by J.M. Coron, F. Helen, J.M. Ghidaglia , NATO ASI Series, Kluwer, Dordrecht 1991; H.R. Brand und H. Pleiner, Phys. Rev. **A35** (1987) 3122 ; W. Zimmermann, H.R. Brand and H. Pleiner (in preparation).
[16] A. Pocheau, V. Croquette, and P. Le Gal, Phys. Rev. Lett. **55** (1985) 1094.
[17] S. Nasuno, et al. (in preparation) ; W.Zimmermann et al. (in preparation).

Oscillatory Instabilities of 2-D Periodic Pattern in EHD Convection

M. Sano[1], K. Sato[1] and B. Janiaud[1,2]

(1) *Research Institute of Electrical Communication, Tohoku University, Sendai 980, Japan*
(2) *Permanent address: Laboratoire de Physique Statistique, École Normale Supérieure, 24 rue Lhomond, 75031 Paris cedex 05, France*

Abstract

We studied experimentally the two-dimensional periodic pattern, created by electro-hydrodynamic convection of a nematic liquid crystal. Two different kind of modes of oscillation can arise. We also observed a localized pattern of traveling wave fronts emitted from a self-organized wave center. This pattern has a life cycle and can be explained by the interaction between the waves and the lattice.

1 Introduction

Pattern formation by electro-hydrodynamic (EHD) convection [1] in nematic liquid crystals has been studied as a prototype for spatio-temporal instabilities and chaos in large aspect ratio systems [2, 3, 4, 5]. The crucial advantage of EHD convection, compared to other systems such as thermal convection, is that we can work with a short time scale and large aspect ratio very easily. These properties are essential for studying the long time behavior of spatio-temporal chaos.

Here we present an experimental study of the dynamics of a two-dimensional periodic pattern created by EHD convection where we have observed new types of dynamical behavior related to the Hopf bifurcation of dissipative structures. We found two kinds of oscillating modes with different symmetries. Furthermore one of these oscillating modes can couple to deformations of the underlying two-dimensional convection lattice. This coupling gives rise to a localized wave pattern with a self-organized oscillating center, which finally collapses by spontaneously nucleating topological defects. This behavior leads to spatiotemporally erratic behavior and is a new type of defect turbulence.

After describing the experimental setup and general features of EHD convection, we will focus on the grid pattern and its oscillatory instabilities. The third part will be devoted to the target pattern and its dynamics. In the last part we will discuss the relevant points of amplitude and phase equations for the description of these instabilities.

2 Experimental Setup

We use a nematic liquid crystal, N-(p-methoxybenzylidine)-p-butylaniline (MBBA) doped with 0.01 wt% of tetra-n-butylammonium bromide to increase and control the electrical conductivity. It is sandwiched between two parallel glass plates with transparent electrodes (ITO). The surface of the transparent electrodes is coated with polyvinyl alcohol and rubbed to achieve an uniform and unidirectional planar alignment of the nematics along the $\hat{x}$ direction. The size of the cell is 2cm×2cm and its depth d_0 is 25 μm or 50 μm (aspect ratio 400 × 400 or 800 × 800). The temperature of the cell is regulated to 25± 0.01 °C by a circulating water flow and electrical heating with feedback control. The samples are stable for more than one year with very good reproducibility. An alternating voltage is applied between the two electrodes. The cutoff frequency is 1400 Hz. We observe the convection pattern under a microscope using a shadowgraph technique with light polarized parallel to the nematic director. The intensity $V(x,y)$ of the visualized image is measured with a CCD camera mounted on the microscope and digitized with a resolution of 512×480 pixels and 256 grey levels.

In figure 1, we show a phase diagram of the pattern which results from the variation of the applied voltage and frequency. The open circles represent the critical voltages for the transition from the conduction state to the stationary convection state of parallel rolls with their axis perpendicular to the alignment of the nematic director (Williams domains). Increasing the voltage leads to defect turbulence (fluctuating Williams domains) via secondary instabilities. From this state the grid pattern appears via a percolation phenomenon [6]. It can be seen in the region encircled by bullets in figure 1.

3 The Grid Pattern and Oscillatory Modes

3.1 The Stationary Grid Pattern

The grid pattern is a convection pattern with a two-dimensional periodic cellular structure, which appears at a relatively high voltage. If we take only the fundamental Fourier modes, the basic stationary pattern $V_0(x,y)$ corresponding to the angle of the director of the nematic or the vertical component of the velocity at middle height

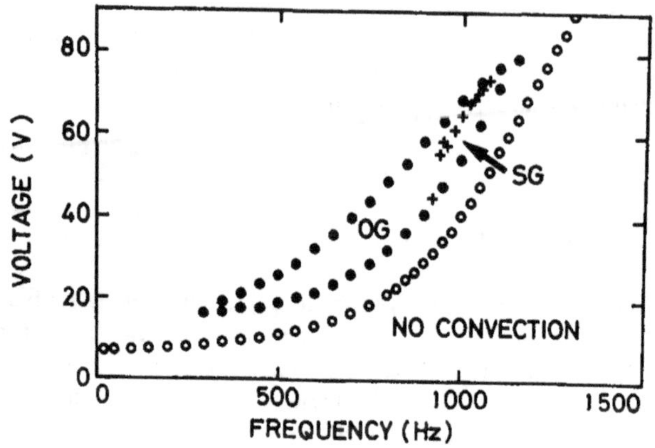

Figure 1: Phase diagram of EHD convection as a function of voltage and frequency. Open circles denote the onset of convection (Williams domains). The percolated grid pattern is observed inside the region encircled by bullets. SG denotes the stationary grid pattern and OG the oscillating grid pattern.

of the sample, can be written as a superposition of two oblique rolls: $V_0^{\pm}(x,y) = \cos(k_x x \pm k_y y)$,

$$\begin{aligned} V_0(x,y) &= V_0^+(x,y) + V_0^-(x,y) \\ &= U_0\cos(k_x x + k_y y) + U_0\cos(k_x x - k_y y) = 2U_0\cos(k_x x)\cos(k_y y). \quad (1) \end{aligned}$$

When we change the control parameter crossing the boundary from SG to OG in figure 1, the oscillatory instabilities occur. We found two different types of oscillating modes. The first type is related to amplitude modulation (breathing) of the basic pattern and the second to its phase modulation. These two modes can coexist in the OG region.

3.2 Breathing Mode

This mode appears and dominates in the lower part of OG region in fig.1. As shown in fig.2, the original rectangular lattice oscillates between left inclined parallelograms and right inclined parallelograms. We suggest that this mode can be expressed by amplitude modulations of the basic two oblique rolls. Their amplitude now oscillate in time:

$$V(x,y,t) = (1 + B^+\sin(\omega_b t))V_0^+(x,y) + (1 + B^-\sin(\omega_b t + \theta))V_0^-(x,y), \quad (2)$$

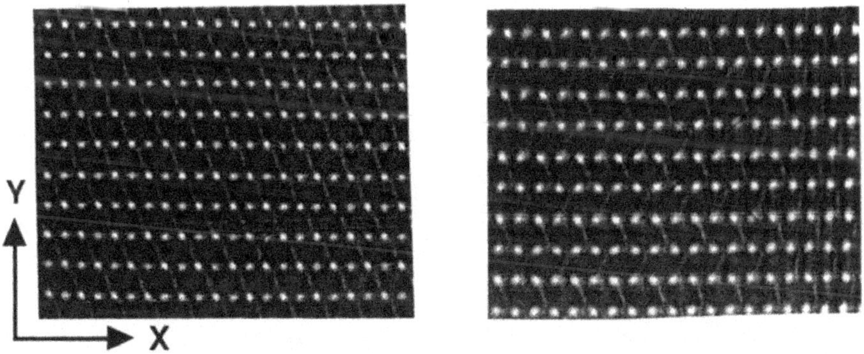

Figure 2: Magnification of the grid pattern showing the oscillation of the lattice. The anchoring direction of the nematic is $\hat{x}$. Left: breathing mode. Right: phase wave mode (zig-zag oscillation)

where θ is the phase difference between the two oscillating modes, and $B^{\pm}$ can have a slow modulation in space and time. Near the treshold the oscillation is almost symmetric in left and right directions, thus implying; $B^{+} = B^{-}$ and $\theta = \pi$. Each oscillating term has the same symmetry as the corresponding basic oblique roll under parity change $S : x \to -x$.

A Hopf bifurcation of a mode with the same parity as the basic pattern is called breathing mode [7].

3.3 Phase Wave Mode

In the upper part of the OG region in fig.1 a second type of oscillating mode exists and dominates. In this mode, minima and maxima of the pattern oscillate back and forth in phase in $\hat{x}$-direction, and out of phase in $\hat{y}$-direction. It is an optical mode (zig-zag oscillation) in the $\hat{y}$-direction as shown in fig.2. The characteristic period of the oscillations is close to the relaxation time T_0 of the nematic director [8]. We have confirmed that the mean frequency ω_p is proportional to $1/d_0^2$ by varying the thickness d_0 from $25\mu m$ to $200\mu m$. And it is almost half of the frequency of the breathing mode, ω_b. This mode can be expressed as a phase modulation of the basic pattern by putting $\Psi(y,t) = A\sin(\omega_p t)\sin(k_y y)$, where A can have slow modulations in space and time:

$$\begin{aligned} V(x,y,t) &= V_0(x + \Psi(y,t), y) \\ &\simeq V_0(x,y) + \Psi(y,t)\sin(k_x x + k_y y) + \Psi(y,t)\sin(k_x x - k_y y) \\ &= V_0(x,y) + A\sin(\omega_p t)\sin(k_x x) sin(2k_y y). \end{aligned} \tag{3}$$

In this mode the oscillating term and the basic pattern have opposite symmetry under parity change, and the perturbation is the second harmonics of the y-component. We call this mode the phase wave mode because of its modulation type. An oscillation mode with a parity opposite to the one of basic pattern is called vascillation mode [7, 9].

4 The Target Pattern

4.1 General Features

In the oscillating region of the grid pattern we observed waves, creating spiral and target patterns (see. fig.3) [3]. Target patterns with pacemakers have also been seen in oscillatory and excitable media such as the Belousov-Zhabotinsky (B-Z) reaction [10]. But in the present system, waves can interact with the underlying lattice so we expect richer behavior. In our experiment the oscillation mode in the target and spiral pattern is always a phase wave mode. The waves visualized in fig.3 are long wavelength modulations of the phase wave mode: between a dark and a light region there is a phase variation of $\pi/2$.

The localized target pattern spontaneously generates with a wave emitting center where the lattice is always deformed. The shape of the target pattern depends on the voltage: it is localized in space for low voltages in the OG region (see fig.1) but extended and interacting with other targets for higher voltage. The frequency of the waves as well as the deformation of the lattice in the target pattern grows in time. Finally we see the nucleation of defects of the grid pattern in the center of the target driving it to collapse. The cycle of creation, deformation and collapse leads to an erratic behavior.

In order to clarify the mechanism of the instability, we measured the short time scale behavior of the waves, the lattice deformation, and the long time behavior of the target pattern.

4.2 Dynamics of the Target Pattern

To understand the short time behavior of the waves, we plotted their spatiotemporal evolution along a line parallel to the $\hat{x}$ axis passing through the center of the target pattern as shown in fig.4. The waves are emitted from the center of the target pattern and their amplitude decays towards the edge creating a localized pattern.

We measured the local lattice wavevectors, k_x and k_y, around the target pattern by complex demodulation. The value of k_y is spatially constant within measurement error. But the value of k_x is decreased in the target and reaches a minimum value at the core [6]. This lattice deformation pattern coincides with the elliptical wave

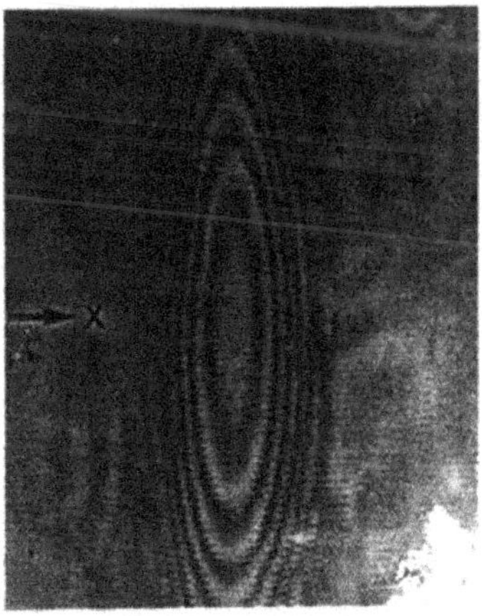

Figure 3: Snapshot of the shadowgraphic image of a target pattern created by the oscillation of the underlying two-dimensional cellular structure (lattice).

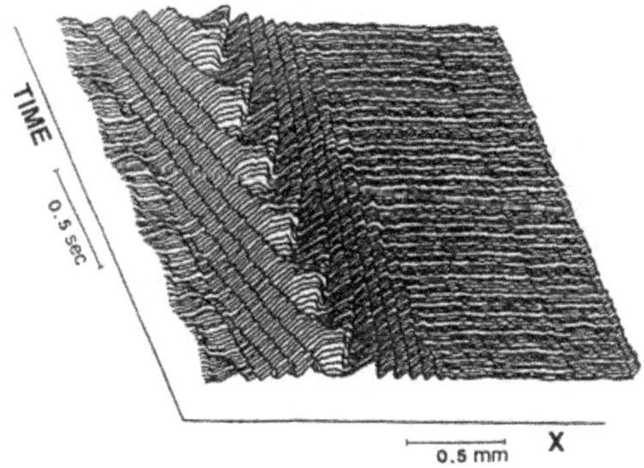

Figure 4: Spatio-temporal evolution of the wave intensity along the $\hat{x}$ axis through the center of the target pattern. The wave intensity is measured as the temporal variation of the light intensity caused by the oscillation of lattice.

pattern. In other words, the underlying lattice is locally expanded in the $\hat{x}$-direction due to the presence of the target pattern but is uniform in the $\hat{y}$-direction.

In figure 5, we define $\delta k_x(x)$ as the difference between the local wavenumber $k_x(x)$ and the spatially averaged wavenumber $< k_x >$. The envelope of the wave amplitude $A(x)$ is measured by averaging 256 successive images of wave intensities and is plotted in arbitrary units. Both $\delta k_x(x)$ and $A(x)$ curves show localized structure and their shapes are strongly correlated, indicating that the wave amplitude and the lattice deformation are coupled slow variables.

For the long time behavior of this localized structure, we measured the evolution of the local lattice wavenumber in the center of the target pattern, δk_c and the frequency of the oscillation as a function of time. The data are shown in fig.6 where time is normalized by the nematic director relaxation time T_0. The local wavenumber decreases, i.e. the lattice expands and the frequency of phase wave increases. During the same time the oscillation amplitude increases.

4.3 Collapse and Creation of Target Pattern

Later, δk_c jumps upward and gradually relaxes to a value near zero ($k_c \rightarrow < k_x >$). This sudden jump in the δk_c curve is caused by the nucleation of topological defects (extra rolls or edge dislocations) and the collapse of the target pattern (see fig.7).

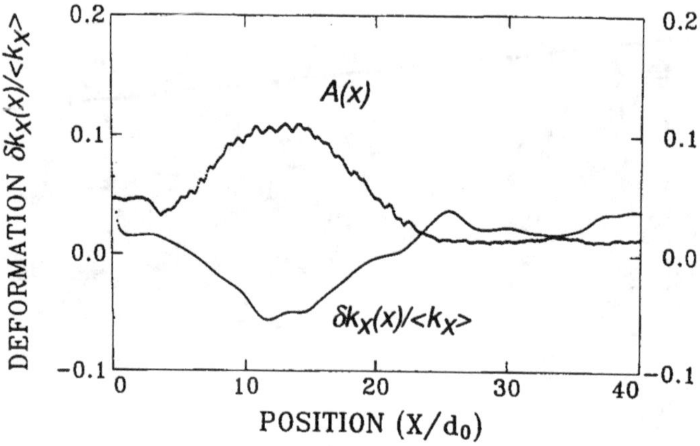

Figure 5: Profile of the lattice deformation $\delta k_x(x)/<k_x>$ and the wave amplitude $A(x)$ along a line parallel to the $\hat{x}$ axis passing through the center of the target pattern. x is normalized by the thickness of the cell, d_0.

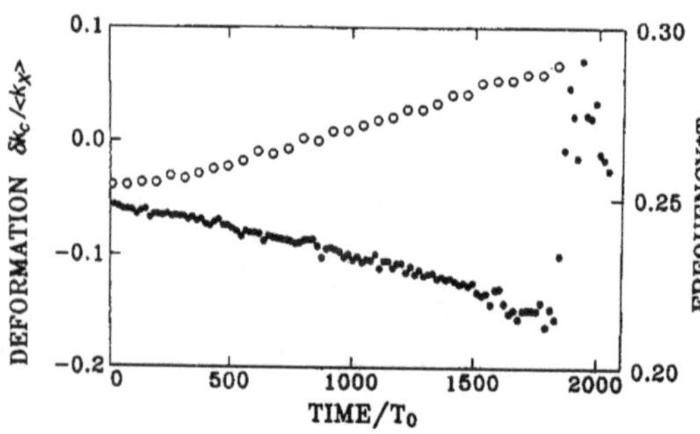

Figure 6: Temporal evolution of the lattice deformation δk_c (solid circles) and the wave frequency (open circles) at the core of the target pattern.

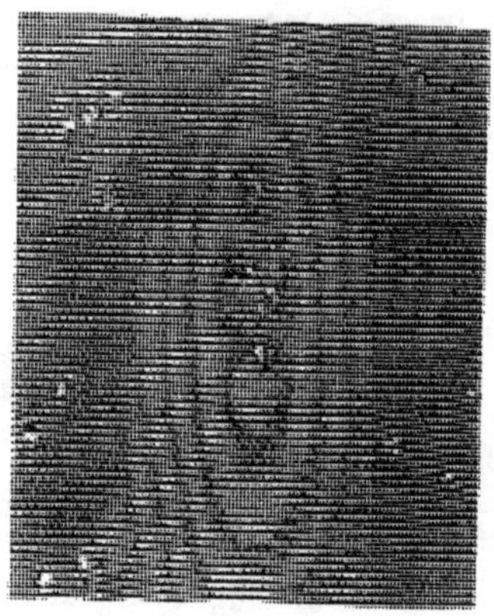

Figure 7: Snapshot of target just after collapse and nucleation of dislocations (extra rolls).

Measurements of several collapsing processes show the same value for the lowest δk_c just before the collapse. We conclude that there exists a threshold in the local wavenumber for the nucleation of dislocations [6]. We measured also the threshold for the Eckhaus instability of the lattice. For a given frequency and voltage the collapsing wavevector is beyond the Eckhaus critical wavevector, presumably because the deformation is localized, pushing further the phase stability band [11]. Just after the collapse many dislocations are nucleated. They migrate towards the outside of the core and gradually annihilate pairwise to recreate a rather uniform lattice. In this nearly uniform state, new target patterns can appear spontaneously from fluctuations of the lattice wavevector.

This cycle of creation, growth, and collapse repeats persistently. As a result the system exhibits a new type of defect turbulence or spatiotemporal chaos.

5 Discussion

In this section we discuss the symmetries of the two different oscillatory modes, a possible mechanism for creation of target pattern and nucleation of topological defects. The essential mechanism of the oscillatory instabilities as well as that leading to the target pattern can be explained by a one-dimensional model [7]. However some phenomena such as the elliptical shape, the presence of spiral waves and the different modes of oscillation are two-dimensional.

5.1 Oscillating Modes

Let us consider a one-dimensional model, for simplicity, to see the difference of symmetry between the two oscillating modes:

$$\begin{aligned} V(x,t) &= (1+B(x,t)\sin(\omega_b t))V_0(x+\Psi(x,t)+\phi) \\ &\simeq V_0(x)+B(x,t)\sin(\omega_b t)V_0(x)+\Psi\partial V_0/\partial x+\phi\partial V_0/\partial x \qquad (4) \\ &= V_0(x)+V_b(x,t)+V_p(x,t)+V_t(x). \qquad (5) \end{aligned}$$

The first term denotes the basic pattern, the second the amplitude modulation (breathing) mode, the third the phase modulation mode, and the fourth comes from the x-translational invariance of the system. Without loss of generality we can assume that the basic pattern is an even function, implying $SV_0 = V_0$ under parity change $S : x \to -x$. The breathing mode has the same symmetry as the basic pattern but not the phase wave and the translation mode, because $\partial V_0/\partial x$ is an odd function. For the linear operator associated to the stability of the basic pattern [7], the third term has eigenvalues $\pm i\omega_p$, the translational mode an eigenvalue 0. And since two modes with different eigenvalues can not have the same eigenfunction, a spatially uniform phase modulation Ψ is prohibited in one dimension. A simple way to satisfy this requirement is that Ψ is a periodic function of x, e.g. $\Psi(x,t) = \sin(k_x x/2)\sin(\omega_p t)$. In two dimensions, an uniform phase modulation in $\hat{x}$ direction is possible because of a y dependence of Ψ. In fact the phase wave mode, $V_p(x,y,t) = A\sin(\omega_p t)\sin(k_x x)\sin(2k_y y)$, is orthogonal to the translational mode. In this experiment, phase wave mode creates the target pattern thus implying the existing coupling between oscillation mode and lattice deformation, while the breathing mode does not. Why one mode has a coupling to the lattice deformation and not the other is still an open question.

5.2 Instability of the Target Pattern

The self-organized pacemaker may be due to an amplifying feedback interaction between the lattice deformation and the waves. Coullet and Iooss [7] proposed coupled equations to describe the Hopf bifurcation of a one-dimensional periodic pattern from general symmetry considerations;

$$\partial_t A = \mu A - (1+i\alpha)|A|^2 A + \xi_1 \partial_x^2 A + \xi_2 \partial_x \phi A \quad (6)$$
$$\partial_t \phi = \partial_x (A^2) + i\beta(\overline{A}\partial_x A - A\partial_x \overline{A}) + \partial_x^2 \phi, \quad (7)$$

where α and β are real, ξ_1 and ξ_2 are complex constants. Sakaguchi and Daviaud et al. [9, 12] derived coupled phase equations from the Eqs.(6) and (7);

$$\partial_t \psi = \partial_x^2 \psi + (\partial_x \psi)^2 + \xi k \quad (8)$$
$$\partial_t k = \partial_x^2 k - \beta' \partial_x^2 \psi, \quad (9)$$

where ψ is the phase of oscillatory mode A, and k is the local wavenumber shift from the optimal one of the basic periodic pattern defined by $k = \partial\phi/\partial x$. If the value of k is locally changed, because of coupling term ξk in Eq.(8), the oscillation phase ψ changes locally. This local variation of ψ reacts to k through the coupling term $\beta'\partial_x^2\psi$ in Eq.(9). If these two coupling terms amplify the fluctuations, an instability takes place. The condition for the instability, given by the linear stability analysis is $\beta'\xi > 0$.

To check the validity of the equations, we artificially changed the lattice constant, and we have experimentally confirmed that the lattice deformation modifies the frequency shift of the waves, $\partial\psi/\partial t$. Expansion of lattice leads to an increased frequency as shown in fig.8. The slope of the linear fit gives the coupling coefficient ξ. It turns out that ξ is negative. This results is consistent with the observations shown in fig.6.

If the other coupling coefficient β' is negative, uniform state becomes unstable, and localized pattern can appear [12]. Hence the following will happen: if the lattice expands at the core, the frequency of oscillation is higher inside the target pattern than outside. When frequency synchronization occurs, the waves look as if they are emitted outwards from the core. This is the coupling from the lattice to the oscillation. Conversely, a local phase advance of the wave can modify the lattice constant. We are now conducting experiments to check the existence of the coupling from the

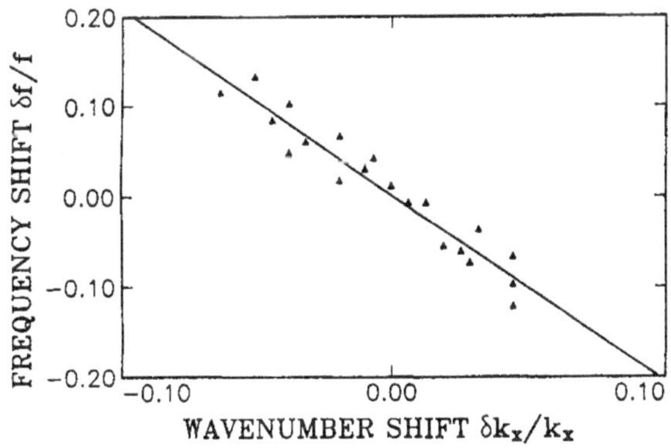

Figure 8: The frequency of the spontaneous phase wave modes as a function of wavenumber k_x of uniform grid pattern. The initial state of the basic grid pattern was controlled and initial value of uniform k_x was varied by a frequency jump method. Here the phase ψ of the wave is almost uniform in space.

oscillation to the lattice, and to find β'. This coupling shows that the growth of the deformation of the lattice is not simply an instability of the spatial structure like the Eckhaus or skewed varicose instabilities: the waves play an important role. However, the Eckhaus instability may be responsible for the nucleation of dislocations during the final collapse. When the local wavenumber is pushed further beyond the Eckhaus boundary, the creation of new rolls (dislocations) is expected. For the description of nucleation, it is necessary to consider the complex amplitude U_0 instead of phase ϕ [12].

6 Conclusion

We have found two kinds of oscillatory instabilities of a 2-D periodic structure in EHD convection. We have analyzed them as amplitude or phase modulation of the basic pattern. For one of these oscillating modes, we observed a self-organized structure which consisted of a lattice deformation and waves emited from its center. It originates from an almost uniform state through the growth of small fluctuations. As the structures grow, the lattice expands at the core and the frequency of oscillation increases proportionally. Coupled Ginzburg-Landau equations can explain the interaction between the lattice deformation and the waves, and hence the instability.

7 Acknowledgements

We are grateful to T. Kawagishi, H. Kokubo, S. Nasuno and Y. Sawada for a fruitful collaboration, and to P. Coullet, Y. Kuramoto, J. Lega, T. Mizuguchi, H. Sakaguchi and S. Sasa for important insights. We are also grateful to J. A. Glazier and F. Graner for valuable comments and careful reading of the manuscript. One of us (B. Janiaud) would like to thank the Japanese Ministry of Education (Monbusho) for financial support.

References

[1] E. Dubois-Violette, P. G. de Gennes and O. Parodi, J. Physique, **32** (1971) 305; P.G. de Gennes, *The Physics of Liquid Crystals* (Clarendon, Oxford, 1973).

[2] A. Joets and R. Ribbota, J. Physique, **47** (1986) 595.

[3] S. Nasuno, M. Sano and Y. Sawada, J. of Phys. Soc. Japan, **58** (1989) 1875.

[4] S. Kai and K. Yamaguchi, Japanese J. of Appl. Phys., **14** (1975) 1653. S. Kai and K. Hirakawa, Prog. Theor. Phys. Suppl., **64** (1978) 212.

[5] L. M. Blinov, J. de Physique, **C3** (1979) 247, and A. N. Chuvyrov and V. G. Chigrinov, Sov. Phys. JETP, **60** (1984) 101.

[6] M. Sano, K. Sato, S. Nasuno and H. Kokubo, preprint; and further quantitative results will be published elsewhere.

[7] P. Coullet and G. Iooss, Phys. Rev. Lett., **64** (1990) 866.

[8] We approximate the relaxation time T_0 of the nematic director by its value in zero electric field [1], $T_0 = Cd_0^2$ where C is determined by physical constants of MBBA. We thus estimate $T_0 = 41.1\ ms$ ($165\ ms$) for the 25 μm cell (50 μm cell).

[9] F. Daviaud, J. Lega, P. Berge, P. Coullet and M. Dubois, to appear in Physica D.

[10] Y. Kuramoto, *Chemical Oscillations, Waves, and Turbulence* (Springer, Berlin, 1984).

[11] L. Krammer and W. Zimmermann, Physica **16D** (1985) 221; R. Heinrichs, G. Ahlers, and D. S. Cannel, Phys. Rev. Lett. **56** (1986) 1794.

[12] H. Sakaguchi, to appear in Prog. Theor. Phys.

Erratum to Oscillatory Instabilities of 2-D Periodic Pattern in EHD Convection

M. Sano, K. Sato and B. Janiaud

In the discussion part, subsection "Instability of the Target Pattern", we discuss the formation of a localized pattern from coupled phase equations labelled (8) and (9). After correction they should read:

$$\partial_t \psi = D_\psi \partial_x^2 \psi + \xi k + N.L \qquad (8)$$

$$\partial_t k = D_\phi \partial_x^2 k - \beta' \partial_x^2 \psi + N.L, \qquad (9)$$

where $N.L$ denotes non linear terms. The condition for the instability, given by the linear stability analysis, becomes $\beta'\xi / D_\psi D_\phi > 0$.

But since we know that the waves and the lattice before collapse are phase stable so that D_ψ and D_ϕ are positive, the discussion about the sign of coefficients ξ and β' is not changed.

PATTERN FORMATION OF INSTABILITIES IN SPHERICAL COUETTE FLOW

Karl Bühler
Institut für Strömungslehre und Strömungsmaschinen, Universität(TH)
W 7500 Karlsruhe 1, Kaiserstraße 12, Germany

ABSTRACT

The instabilities of the flow between two concentric rotating spheres are studied during the laminar-turbulent transition along different routes. The structure development of the secondary instabilities is observed by flow visualization.

1. Introduction

The spherical Couette flow is an interesting subject to study bifurcation phenomena and structure development in supercritical flow regions. An overview of theoretical and experimental results for steady supercritical flows is given by Bühler[1] . The spherical gap flow is realized by rotating the inner sphere about the vertical axis while the outer transparent one is at rest. The geometry is determind by the dimensionless gap width $\sigma = s/R_1$ with $s = R_2 - R_1$ as distance between the two concentric spherical surfaces. The Reynolds number is defined as $Re = R_1^2 \omega_1 / \nu$, where ν is the kinematic viscosity of the fluid, in this case silicone oil, and ω_1 is the angular velocity of the inner sphere. Small aluminium flakes were used to visualize the flows.

2. Results

The supercritical flow is non-unique and depends for a fixed geometry σ on the acceleration from rest to the final Reynolds number Re. In Fig.1 two different modes with two Taylor vortices (left) and four vortices (right) are shown. The large red arrow marks the equatorial plane and the distance between

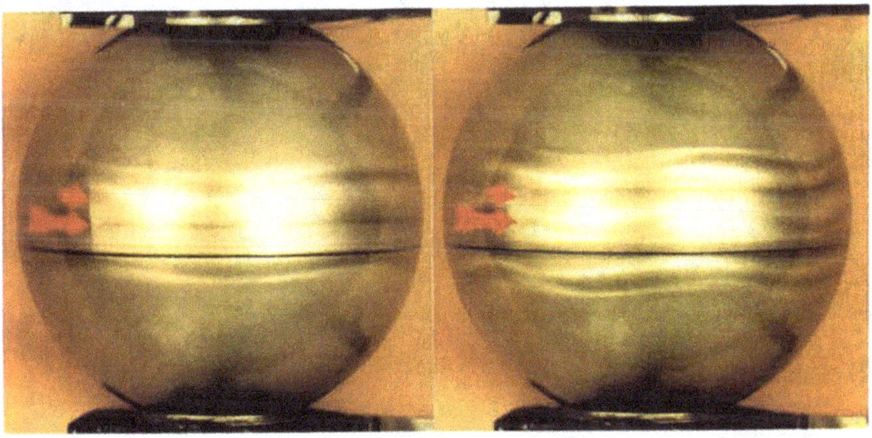

Figure 1: Spherical Taylor vortex flow with superimposed wavy-mode, gap width σ= 0,154, one vortex in each hemisphere at Re = 2150 (left) and the flow with two vortices in each hemisphere at Re = 1900 (right).

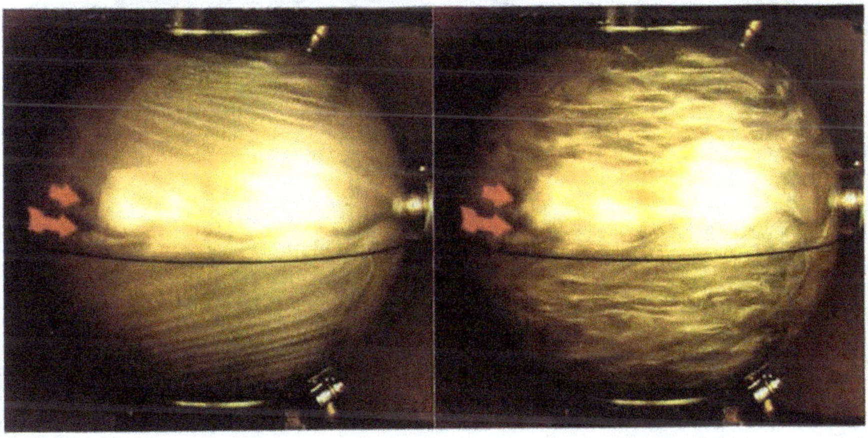

Figure 2: Supercritical flow without Taylor vortices, gap width σ=0,178, shear instabilities at Re=5400 (left) and turbulent motion with a wavy interface at the equator at Re=7800 (right).

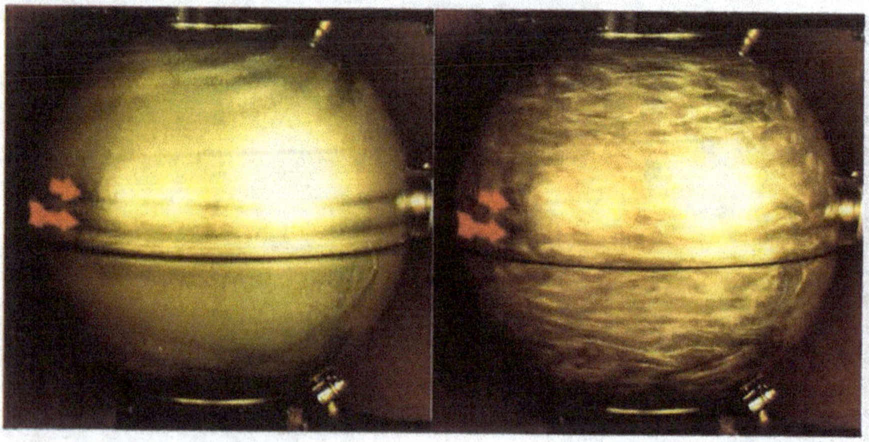

Figure 3: Supercritical flow with one Taylor vortex in each hemisphere, gap width σ=0.178, spiral vortices near the poles at Re=7800 (left) and turbulent flow with Taylor vortices at Re=10700 (right).

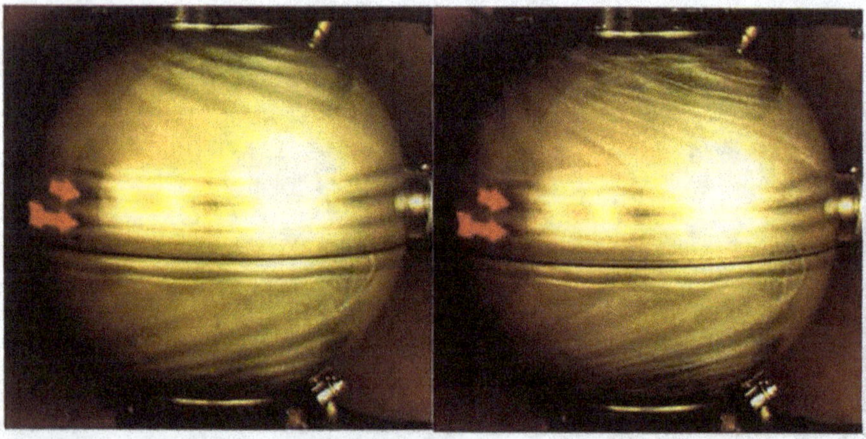

Figure 4: Supercritical flow with two Taylor vortices in each hemisphere, gap width σ=0.178, time-dependent shear waves near the poles at Re=5600 (left) and stochastic motion starting near the poles at Re=6400 (right).

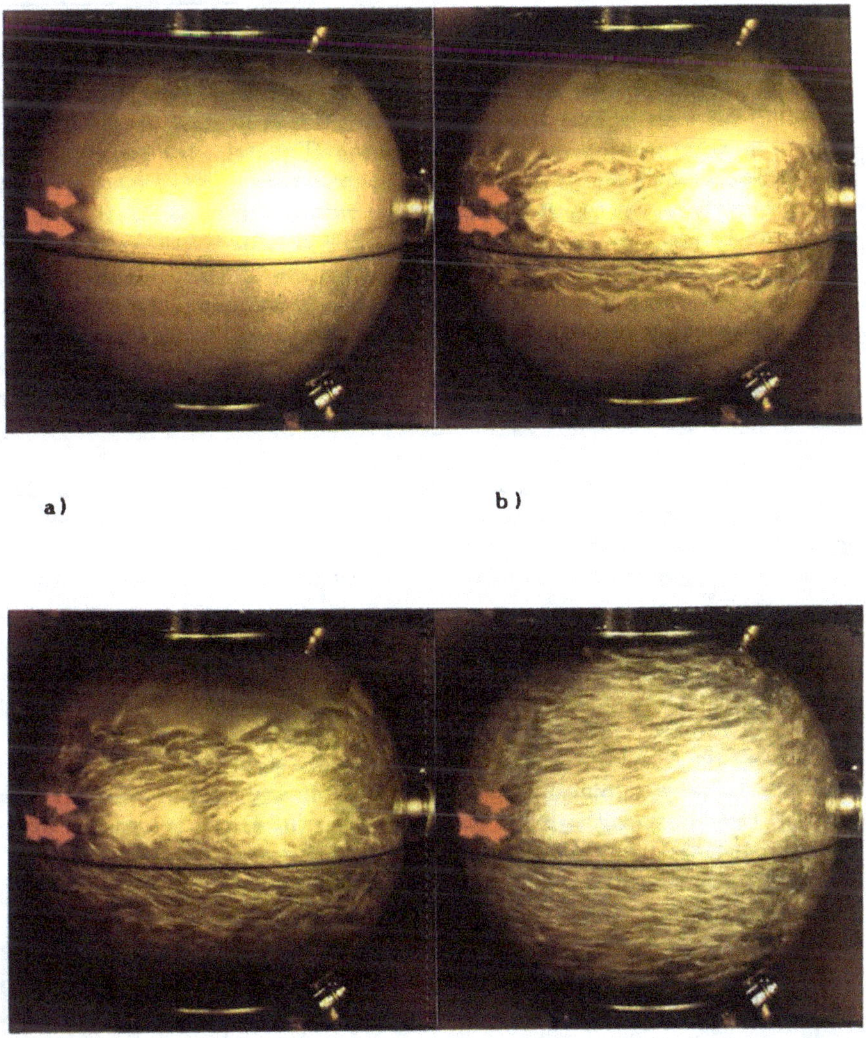

a) b)

c) d)

Figure 5: Spreading turbulence front after a sudden start of the inner sphere to Re= $2{,}5 \cdot 10^4$, gap width σ=0,178, the time after the sudden start increases from a) to d).

the two arrows is proportional to the gap width s. The vortex boundary are modulated periodically in circumferential direction. By increasing the Reynolds number the flow restabilizes and these wavy modes disappear.

A supercritical state without vortices is realized in Fig.2 by rapidly accelerating the inner sphere from rest. Threedimensional shear waves appear as spirals in both hemispheres. Their interaction at the equator gives rise to a wavy interface which moves circumferentially. The state in Fig.2 on the right is achieved by increasing the Reynolds number further. This stochastic motion begins near the poles by the interaction of different wave systems. The flow is now turbulent but the wavy interface at the equator is still preserved. The supercritical state with two Taylor vortices near the equator in Fig.3 (left) is realized by slowly accelerating the inner sphere from rest. In contrast to the time-dependent shear waves seen in Fig.2 we now observe steady spiral vortices near the poles similar to Stuart vortices on a rotating disk. Also clearly visible is a secondary instability structure within the vortices. The Taylor vortices exist also in turbulent motion at higher Reynolds number as can be seen in Fig.3 (right).

The state with four Taylor vortices at higher Reynolds number is shown in Fig.4. Time-dependent shear waves exist between the vortices and the poles, giving rise to wavy vortex boundaries. The stochastic motion begins near the poles by the interaction of different wave systems as can be seen in Fig.4 on the right.

The picture series in Fig.5 shows a turbulence front spreading from the equator to the poles following a sudden-start of the inner sphere. These pictures demonstrates, that for this flow it is possible for regions of order and chaos to exist simultaneously. After the fronts have reached the poles, the motion is fully turbulent without evidence of any Taylor vortex type structure. This spreading process is described analytically by Bühler and Zierep[2]. In this paper the dynamical behaviour of the instabilities is analyzed by spectral measurements of the local wall shear stress. An extension to the case of spherical gap flows with a superimposed mass flux in meridional direction is investigated by Bühler[3].

3. References

1. K. Bühler, Symmetric and asymmetric Taylor vortex flow in spherical gaps. Acta Mech. 81, (1990) 3-38.
2. K. Bühler and J. Zierep, Dynamical instabilities and transition to turbulence in spherical gap flows, in Advances in Turbulence, ed. G.Comte-Bellot and J.Mathieu (Springer,Berlin,1987) 16-26.
3. K. Bühler, Dynamical behaviour of instabilities in spherical gap flows: Theory and experiments, Eur. J. Mech., B/Fluids 10, n° 2-Suppl.,(1991) 187-192.

THEORY OF TRANSIENT PATTERNS IN THE SPLAY FREEDERICKSZ TRANSITION OF NEMATICS

Agnes Buka
Central Research Institute for Physics
of the Hungarian Academy of Sciences
P.O.B. 49, H-1525 Budapest, Hungary
and
Lorenz Kramer
University of Bayreuth, Institute of Physics
P.O.B. 101251, W-8580 Bayreuth, Germany

ABSTRACT

A nonlinear theoretical approach of transient patterns arising during the splay Freedericksz transition and having a wave vector perpendicular to the initial director alignment is presented. The threshold of the patterning process (ε_c) depends on the time evolution of the homogeneous realignment. The theory delivers a range of ε_c between the early and late time values.

1. INTRODUCTION

In ordinary low molecular-weight nematic liquid crytals the magnetically or electrically driven splay Freedericksz transition leads to a distortion of the director that is homogeneous in the plane of the sample and varies only perpendicularly to it (z-axis). It now appears that when the transition is induced by a sufficiently high field-jump there always occur transient patterns during the realignment process [1,2,3]. The patterns are fairly irregular, but they can be described by a superposition of wavevectors that are usually predominantly parallel to the undistorted director (x-axis) or perpendicular to it (y-axis).

The x-patterns (or oblique variants) were the first to be observed with a magnetic field [4,5,6,7] and explained on the basis of the nematodynamic equations by the fact that due to backflow linear destabilising modes with nonzero wavenumber (perpendicular to x or oblique depending on the conditions) are the most rapidly growing ones [4,6]. Only recently has it become clear that in the electrically driven case often y-patterns occur [1,2,3] which can apparently not be understood on the basis of the linear theory. However, a theory where the patterning process was treated superimposed on the homogeneous Freedericksz realignment so that nonlinear coupling terms arise could provide a qualitative explanation of the phenomenon

[8]. The decisive mechanism for this pattern formation is connected with the field distortion due to strong dielectric (high frequency) or conductive (low frequency) anisotropies. The large anisotropies also lead to a drastic increase of the threshold for the formation of x-patterns [3]. The theory explains why in the magnetically driven case, where the field distortion is always small, only x-patterns (or oblique ones) are observed.

In this work we present a new version of the nonlinear theory representing a different limiting case, compare it with the previous one [8] and discuss the results in the context of what is known about these transient patterns experimentally.

2. NONLINEAR THEORY

We start from the nonlinear nematodynamic equations [9,10] and first consider the homogeneous Freedericksz realignment process. We neglect flexoelectric effects and the time derivatives acting on the velocities i.e. fast modes. Then we apply small and space-dependent variations around the homogeneous deformation in linear approximation. The modal solutions are harmonic in the x and y directions and characterized by a wavevector (q, p). The time dependence of the variables (electric potential, components of $\mathbf{n}$ and $\mathbf{v}$) after switching on the external field is not simply exponential, like in the linear theory. The full problem is extremely complicated, so we only treat the case $q = 0$ and make an expansion in terms of N_z, the z-component of the homogeneous director realignment at the center of the slab up to N_z^2. It is an adequate approximation because in the experiments the pattern starts to build up while N_z is still rather small.

We treat the z dependence together with the realistic rigid boundary conditions by applying the test function approximation as used in [10,11]. For simplicity we set here the Leslie coefficient $\alpha_3 = 0$, which is a good approximation for our type of problem except for $(q,p) \to 0$, see [4]. We measure time in units of the director relaxation time $\tau_d = \gamma_1 d^2/(K_{11}\pi^2)$, wave numbers in units of π/d and introduce $\varepsilon = V^2/V_{th}^2 - 1$ where $V_{th} = \pi\sqrt{K_{11}/(\varepsilon_o \varepsilon_a)}$ is the Freedericksz threshold. We use dimensionless quantities: $\sigma'_a = \sigma_a/\sigma_\perp$, $\varepsilon'_a = \varepsilon_a/\varepsilon_\perp$ and $K'_{ii} = K_{ii}/K_{11}$.

At early times of the realignment process our approximation is valid for arbitrary ε. In ref. [8] the problem was solved in this context for early times. The result for n_z, the modulation of the director has the form

$$n_z = N_p e^{\sigma t}[1 + N_o^2 c_3(e^{2\varepsilon t} - 1)]cos(z) \tag{1}$$

where N_o and N_p are the (extremely small) initial amplitudes of the $p = 0$ (homogeneous) and $p \neq 0$ modes respectively. σ and c_3 are functions of p^2. The term proportional to N_o^2 represents the nonlinear correction.

The results showed that the coupling with the homogeneous process can lead to a faster growth of modes with $p \neq 0$ when the dielectric or conductive anisotropies are sufficiently large. Then the maximum of the "growth speed" vs. p at $p = 0$ would transform with increasing ε into a minimum and give rize to two new maxima at $\pm p_{max}$. The onset ε_c of this scenario can be obtained by an expansion in p^2:

$$\sigma = c + \sigma_1 p^2 \quad and \quad c_3 = -c_3^0 + c_3^1 \, p^2 \tag{2}$$

with σ_1 negative (for $K'_{22} > 0.3$, see below) and c_3^o, c_3^1 positive. The threshold (ε_c) defined in this way depends on the time t_1 chosen during the evolution process (or equivalently the amplitude of the homogeneous realignment). It is easy to see from eqs.(1),(2) that ε_c tends to infinity for $t_1 \to 0$ and ε_c tends to zero (Freedericksz threshold) for $t_1 \to \infty$ (of course the approximation then breaks down).

For parameters $N_o = 0.025$, $K'_{22} = 0.615$, $K'_{33} = 1.615$ and $\alpha_4/\gamma_1 = 1$ the onset of pattern formation ε_c was calculated in ref.[8] for $t_1 = 1/(2\varepsilon) ln(\mu/N_o^2 c_3^o)$ chosen in such a way that the nonlinear correction at $p = 0$ has become equal to μ . The result for $\mu = 0.5$ is shown in Fig.1 as a function of ε'_a in the high-frequency limit, curve (a).

We now present an alternate calculation where the small-p expansion is made from the beginning on. Then we can discard the restriction to early times keeping only the one of small N_z^2. The theory is thus valid at small t for arbitrary ε and at arbitrary t for small ε (as long as the Landau expansion is valid: $\varepsilon << g$, see below). The results for n_z can be represented in the form

$$n_z = [V_3(t) + p^2 W_3(t) + ...] \cos(z) \tag{3}$$

where

$$V_3 \propto \partial_t N_z(t), \qquad N_z = \frac{N_o e^{\varepsilon t}}{[1 + N_o^2 (g/\varepsilon)(e^{2\varepsilon t} - 1)]^{1/2}} \tag{4}$$

$$W_3 \propto [w_3^o t + \frac{w_3^1}{g} \ln(1 + N_o^2 \frac{g}{\varepsilon} e^{2\varepsilon t})] \partial_t N_z(t) \tag{5}$$

$$w_3^o = \sigma_1 = \frac{64}{(3\pi)^2} \frac{(1 - K'_{22})^2}{4K'_{22} + \varepsilon} - K'_{22} \tag{6}$$

$$\begin{aligned} w_3^1 = \varepsilon c_3^1 = \; & -\beta + S(\varepsilon + 1)(1/8 + \beta/\varepsilon'_a) \\ & + (64/45\pi^2) A \, [S(\varepsilon + 1) + 3(1 - K'_{22}) + K'_{22} - K'_{33} + 8\beta\varepsilon] \\ & + (8/45\pi^2) B \, [S(\varepsilon + 1)(8 - \alpha/\varepsilon'_a) + 8(K'_{22} - K'_{33}) + \alpha] \\ & + (4/3\pi)^2 AB \, [6g - \varepsilon - 4(K'_{33} - K'_{22}) - (16/15\pi)^2 \alpha\varepsilon] \end{aligned} \tag{7}$$

$$A = \frac{1 - K'_{22}}{4K'_{22} + \varepsilon} \qquad B = \frac{1 - K'_{22}}{4K'_{22} + 3\varepsilon} \tag{8}$$

Here $N_z(t)$ gives the time evolution of the amplitude of the homogeneous realignment for small ε (Landau expansion) with a (small) initial value N_o. $S = (\sigma'_a +$

$\varepsilon_a'\tau_o^2\omega^2)/(1+\tau_o^2\omega^2)$ with the charge relaxation time τ_o characterizes the anisotropy, $g = \frac{2}{3}\varepsilon c_3^o = [K_{33}'+S+\varepsilon(S+\frac{3}{2})]/2$ is the nonlinear coefficient in the Landau expansion (with ε corrections) and $\alpha = 16(5/\pi^2-1)\gamma_1/\alpha_4$, $\beta = 3(1-6/\pi^2)\gamma_1/(16\alpha_4)$. The quantities V_3 and W_3 are only known up to a prefactor related to initial conditions.

In principle eqs.(3)-(7) describe the full time-dependent process. At early times the logarithm in (5) can be expanded leading to the result of eq.(1) with (2). For late times the logarithm tends to $2\varepsilon t$, and the term in the square bracket becomes $[w_3^0+2w_3^1(\varepsilon/g)]t$. The first (=linear) term is negative (for $K_{22}' > 0.3$), and the second (=nonlinear) term is always positive.

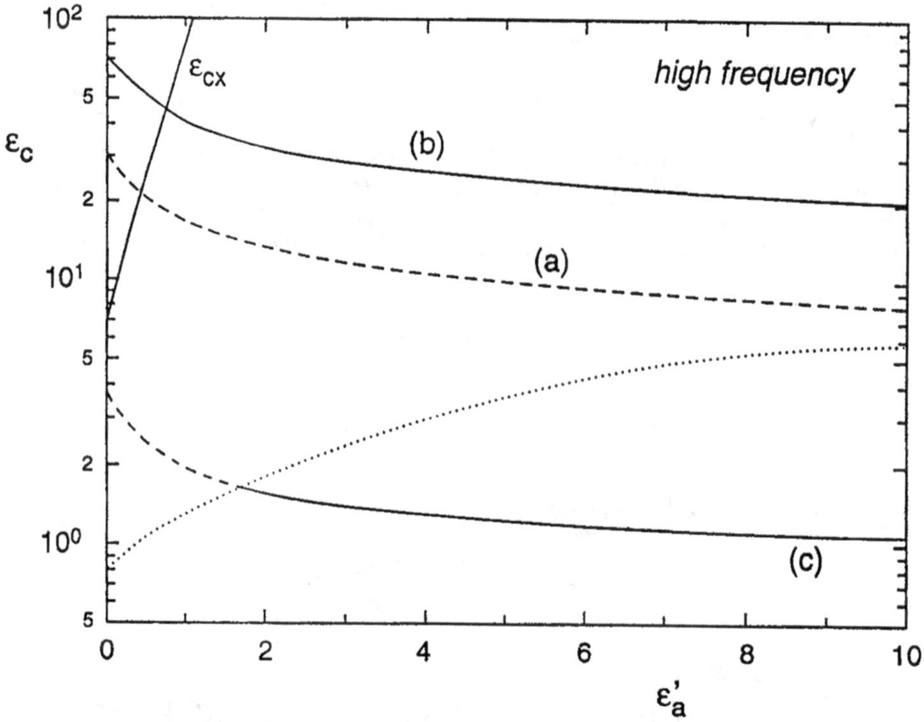

Figure 1: *The onset of the pattern formation is shown versus the relative dielectric anisotropy for x-structures: ε_{cx} and for y-structures: (a) and (b) - at early times, (c) - at late times. The dotted line represents the condition $\varepsilon = g$.*

In Fig.1 we have plotted the threshold values ε_c evaluated from the condition $W_3 = 0$, eq.(5), for the same early time t_1 used before (curve (b)) and for a late time (curve (c)) in the high-frequency limit ($S = \varepsilon'_a$). The fact that (a) and (b) do not coincide shows that t_1 is really already outside the early-time approximation. The dotted curve represents the condition $\varepsilon = g(\varepsilon = 0)$ which should give an estimate of the upper bound for the range of applicability of the Landau expansion except for early times.

3. DISCUSSION

One of the benefits of the new analysis is that it predicts a positive lower bound for ε_c that is reached for $t_1 \to \infty$. The experimental estimates for ε_c [1,2] appear to be in this region or somewhat higher, but certainly far below the early-time estimate (b) in Fig.1. Since the patterns do start to be observable quite early there remains a quantitative discrepancy. This could be connected with the visualization method by the shadowgraph technique, which puts a strong bias on larger wavenumbers because it is sensitive to gradients of the refractive index (in fact, the homogeneous mode is not observable at all). It thus seems important to repeat the experiments using another visualization technique before proceeding with the theory. It appears also at this point not very useful to compare the theoretical time behavior of the transient patterns with the experimentally measured contrast because the sensitivity depends strongly on the momentary homogeneous distortion.

In spite of these quantitative problems the theory can explain many aspects of the competition between linear ($x-$) and nonlinear ($y-$)patterns. In Fig.1 we have included the threshold for the linear patterns [3] which shows that $x-$patterns cannot appear for large anisotropy, in contrast to the case of small anisotropy where x-patterns are observed either in the magnetically driven transition or in a material with a small (positive) value of ε'_a. Since at low frequencies the conductive anisotropy becomes dominant one can even generate a crossover between the two structures with the same material by merely changing the frequency [3].

References

[1] A.Buka, M.de la Torre Juarez, L.Kramer, and I.Rehberg, Phys.Rev.**A40**, 7427 (1989)

[2] B.L.Winkler, H.Richter, I.Rehberg, W.Zimmermann, L.Kramer, and A.Buka, Phys.Rev.**A43**, 1940 (1991)

[3] A.Buka, and L.Kramer, J. Phys. (Paris), in press. (1992)

[4] E.Guyon, R.Meyer, and J.Salan, Mol.Cryst.Liq.Cryst. **54**, 261 (1979)

[5] E.F.Carr, Mol.Cryst.Liq.Cryst. **34**, 159 (1977)

[6] A.J.Hurd, S.Fraden, F.Lonberg, and R.B.Meyer, J.Phys. (Paris) **46**, 905 (1985)

[7] S.Fraden, A.J.Hurd, R.B.Meyer, M.Cahoon, and D.L.D.Caspar, J.Phys.(Paris) Colloq. **46**, C3-85 (1985)

[8] A.Buka, and L.Kramer, Phys.Rev.A. in press, (1992)

[9] L.M. Blinov, Electrooptical and Magnetooptical Properties of Liquid Crystals, Wiley, New York (1983)

[10] E.Bodenschatz, W.Zimmermann, and L.Kramer, J.Phys. (Paris) **49**, 1875 (1988)

[11] L.Kramer, E.Bodenschatz, W.Pesch, W.Thom, and W.Zimmermann, Liquid Crystals **5**, 699 (1989)

[12] D.W.Allender, R.M.Hornreich, and D.L.Johnson, Phys.Rev.Lett. **59**, 2654 (1987)

[13] F.Lonberg, and R.B.Meyer Phys.Rev.Lett. **55**, 718 (1985)

PARALLELS BETWEEN PATTERN FORMATION IN HIGH-POWER FERROMAGNETIC RESONANCE AND FLUID CONVECTION

F.J. ELMER*
Institute for Pure and Applied Physical Sciences
University of California, San Diego

Abstract

In high-power ferromagnetic resonance (FMR) a ferromagnetic insulator is driven by a microwave field. If the strength of the microwave exceeds a certain threshold the uniform state will lose its stability through parametrically excited spin waves. The pattern formation above the threshold will be compared with the pattern formation in convective systems. The comparision will be done in the framework of amplitude equations. The main parallels are that both systems have a nonlocal degree of freedom (magnetostatic potential in FMR, pressure in fluid convection) and in both systems a uniform state becomes linearly unstable through perturbations with finite wavelength.

1. Introduction

The study of patterns in fluid convection (FC) is one of the main fields of pattern formation in dissipative systems. Simple fluids, binary mixtures and complex fluids like liquid crystals driven by buoyancy (Rayleigh-Bénard setting), centrifugal forces (Taylor-Couette setting) or electric fields (EHC setting for liquid crystals) have been intensively investigated experimentally and theoretically.

It is typical for all FC systems with large aspect ratios that a non-convective state becomes unstable if the driving force exceeds a certain threshold. Above the threshold a convective pattern occurs which is built up by a lattice of elementary convective cells. Amplitude equations (Ginzburg-Landau equations) describing a convective cell only by a complex amplitude are fruitful tools to study the pattern dynamics near the threshold.

*Permanent address after Jan. 1992: Institut für Physik, Universität Basel, Klingelbergstr. 82, CH-4056 Basel, Switzerland

The aim of this paper is to introduce a totally different kind of dissipative system, high-power ferromagnetic resonance (FMR), which shows much parallels in its pattern formation compared with FC.

In a high-power FMR experiment a uniform microwave (with magnetic field $\mathbf{h}$) drives a ferromagnetic insulator in a strong, uniform, static, magnetic field $\mathbf{H}_0$. The strong static field ensures that the whole sample is in one ferromagnetic domain. Typical experimental settings are transverse pumping (i.e. $\mathbf{h} \parallel \mathbf{H}_0$) and parallel pumping (i.e. $\mathbf{h} \perp \mathbf{H}_0$). Samples are usually spheres, disks or films. For the comparision with Rayleigh-Bénard systems we choose similiar geometry, namely disks or films.

Forty years ago, Bloembergen and co-workers[1] found experimentally instabilities in FMR. Later on Suhl[2] showed that these instabilities are caused by parametric resonance of spin waves. In the last decade FMR instabilities were investigated experimentally and theoretically as an example of a system showing deterministic chaos[3]. But nearly nothing is known about the patterns which emerge above the instability threshold.

2. Equation of Motion

The dynamics of the magnetization $\mathbf{M}$ is governed the Landau-Lifshitz equation[4]

$$\frac{1}{\gamma}\partial_t\mathbf{M} = -\mathbf{M}\times\mathbf{H}_{eff} - \frac{g}{M_0}\mathbf{M}\times(\mathbf{M}\times\mathbf{H}_{eff}) \quad \text{with} \quad |\mathbf{M}(\mathbf{r},t)| = M_0, \tag{1}$$

where γ is the gyromagnetic ratio and g the dimensionless damping parameter. In the magnetostatic limit (i.e. electromagnetic wave propagation can be neglected) $\mathbf{H}_{eff}$ of an isotropic ferromagnet is

$$\mathbf{H}_{eff} = \mathbf{H}_0 + \mathbf{h}(\omega\, t) + D\,\nabla^2\mathbf{M} - \nabla\Phi_M, \tag{2}$$

where the third term is the exchange field and the fourth term is the dipolar field produced by the magnetostatic potential Φ_M which obeys the Poisson equation

$$\Delta\Phi_M = 4\pi\,\mathbf{M}. \tag{3}$$

Clearly the Landau-Lifshitz equation is totally different from the Navier-Stokes equation which is at the heart of any set of equations of motion for FC. But there is the striking similarity that for incompressible fluids the pressure is likewise a nonlocal degree of freedom fulfilling a Poisson equation. Contrary to fluid systems the dissipation in FMR is very low. For example, $g \approx 10^{-5} - 10^{-3}$ for the standard substance YIG (Yttrium iron garnet)[5].

3. Main Instabilities

In FC as well as in FMR a *uniform* state will be destabilized by a perturbation with a *finite* wavenumber if the driving force exceeds a certain threshold. But in

contrast with FC, in FMR the critical wavenumber is not determined by a film thickness but by a *parametric resonance* condition of the bulk spin waves:
Transverse pumping[2]:

first order instability: $\omega_{\mathbf{k}_c} = \omega/2,\ \theta_c \approx \pm\pi/4$

second order instability: $\omega_{\mathbf{k}_c} \approx \omega,\ \theta_c = 0$

Parallel pumping[6]:

first order instability: $\omega_{\mathbf{k}_c} = \omega/2,\ \theta_c = \pi/2$

Here $\omega_{\mathbf{k}}$ is the ferromagnetic spin wave dispersion[5] and θ is the angle between $\mathbf{k}$ and $\mathbf{H}_0$. The curvature of neutral curve in the minimum is of order g $(1/g)$ if $\mathbf{k}$ is variied parallel (perpendicular) to the manifold given by $\omega_{\mathbf{k}} = n\,\omega/2$.

4. Amplitude Equations

The amplitude equation for second order tranverse pumping with $\mathbf{H}_0$ parallel to the film plane (film coordinates: $x \perp \mathbf{H}_0$ and $y \parallel \mathbf{H}_0$) is

$$\begin{aligned} \frac{1}{g}\,\partial_t A &= \epsilon A + \frac{1}{g^2}\,(\omega_{ky}\,\partial_y - \frac{i}{2}\,\omega_{kxx}\,\partial_x^2)^2 A + \alpha\,\partial_x^2 A \\ &\quad - b\,|A|^2 A + b_x\,\partial_x\Psi\cdot A + b_y\,\partial_y\Psi\cdot A \\ (1+4\pi M_0/H_0)\,\partial_x^2\Psi + \partial_y^2\Psi &= c_x\,\partial_x|A|^2 + c_y\,\partial_y|A|^2, \end{aligned} \tag{4}$$

where A is the amplitude of a standing spin wave in y direction, Ψ is the magnetostatic potential, and $\epsilon \propto h - h_c$. All parameters are real. It should be noted that a similiar amplitude equation holds for small gap Taylor-Couette systems[7].

For first order instability we get a similar set of equations but for two amplitudes. In the case of parallel pump or $\mathbf{H}_0$ perpendicular to the film plane the equations for A and Ψ are decoupled. Coupling terms occurs in higher order perturbation theory. This is similar to the case of Rayleigh-Bénard convection[8].

5. Secondary Instabilities

Stationary patterns with constant amplitudes of Eqs. 4 are given by

$$\Psi = 0,\ \ A = \sqrt{(\epsilon - [\omega_{ky}\Delta k_y + \omega_{kxx}(\Delta k_x)^2/2]^2/g^2 - \alpha(\Delta k_x)^2)\,/b}\; e^{i(\Delta k_x x + \Delta k_y y)}. \tag{5}$$

They bifurcate supercritically if $b > 0$. For $\Delta k_x = 0$ they become *Eckhaus* unstable if $R^2 < 2(\omega_{ky}\Delta k_y/g)^2/(b - b_y c_y)$ or *zigzag* unstable if $\alpha g^2 + \omega_{ky}\omega_{kxx}\Delta k_y < 0$. This is similar to FC. But the *skewed-varicose* instability which is caused by the *nonlocal* degree of freedom (i.e. mean flow)[8] does not exist in FMR. Instead we get a *long wavelength amplitude* instability[9] if

$$b_x c_x > \sigma b \quad \text{or} \quad b_y c_y > 0 \quad \text{or} \quad (b_x c_y + b_y c_x)^2 > 4(\sigma b - b_x c_x)(b - b_y c_y), \tag{6}$$

where $\sigma = 1 + 4\pi M_0/H_0$.

In order to study the pattern formation caused by this kind of instability we discuss a simplified form of Eqs. 4 and add a quintic term:

$$\partial_t A = \epsilon A + \partial_x^2 A - |A|^2 A + \partial_x \Psi \cdot A - r|A|^4 A \tag{7}$$

$$\partial_x^2 \Psi = c\partial_x |A|^2 \tag{8}$$

For periodic boundary conditions $\partial_x \Psi$ in Eq. 7 can be eliminated by: $\partial_x \Psi = c\left(|A|^2 - \frac{1}{L}\int_0^L |A|^2 dx\right)$. The stationary state with constant amplitude is given by $A = R\exp(ikx)$ with $\epsilon - k^2 - R^2 - rR^4 = 0$. For $r > 0$ it is stable if

$$\epsilon > \frac{c^2 - 1 + 16rk^2 + (c+1)\sqrt{(c-1)^2 + 16rk^2}}{8r}. \tag{9}$$

The long wavelength instability occurs for $c > 1$. But the constant amplitude patterns become stable again if $\epsilon > (c^2 - 1)/(4r)$ (see Fig. 1). For large L there

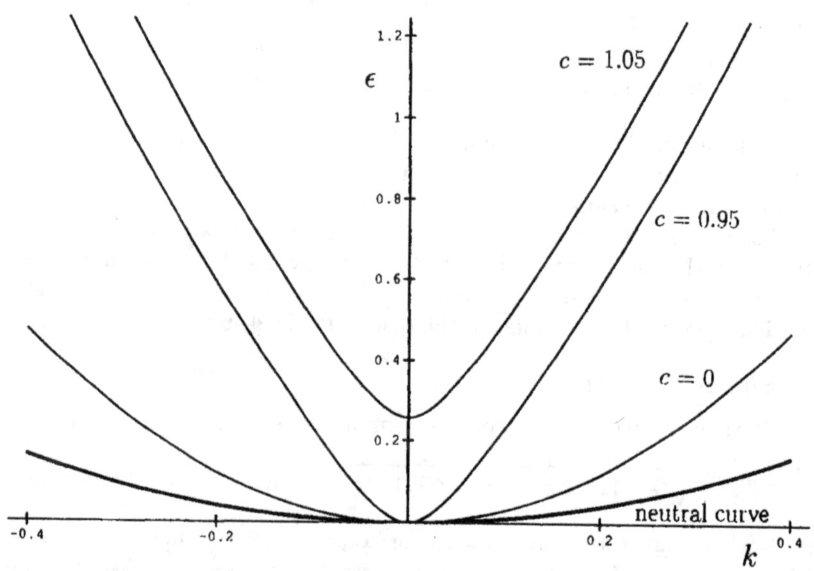

Figure 1: Stability regime of the constant-amplitude solutions for $r = 0.1$.

exists stable two-domain patterns for $-3(c-1)^2 < 16r\epsilon < 3(c-1)(3c+1)$. In the first domain $A = \sqrt{3(c-1)/(4r)} \neq 0$ and the second domain $A = 0$. The length of the first domain is $L[3(c-1)^2 + 16r\epsilon]/[12c(c-1)]$ and increases linearly with ϵ. Note the two bistabilities (see Fig. 2). For details of the bifurcation scenario near $\epsilon = 0$ see Ref. 9.

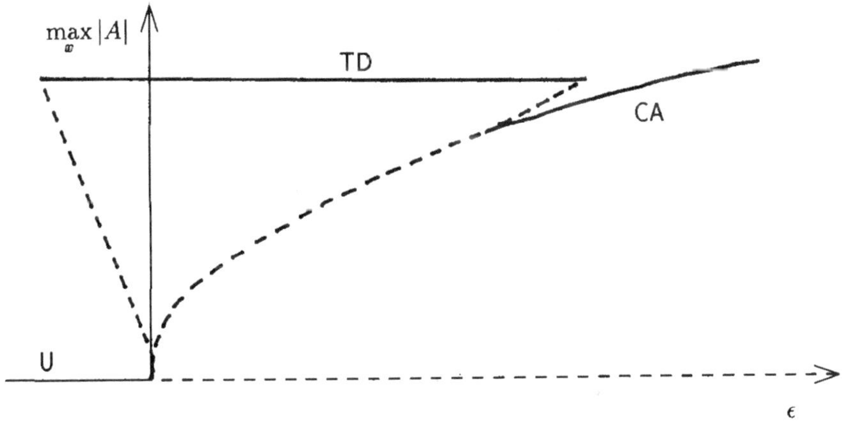

Figure 2: Bifurcation scenario for $c > 1$ and $r > 0$ (schematically). Solid (dotted) lines indicate stable (unstable) solutions. U uniform state (A=0), CA constant-amplitude state, TD two-domain state.

6. Acknowledgements

I would like to thank H. Suhl for fruitful discussions about pattern formation in magnetic systems. I also thank L. Kramer, A.C. Newell and Y. Pomeau for useful informations about fluid convection. This research was supported by the Volkswagen Stiftung.

7. References

1. N. Bloembergen and R.W. Damon, *Phys. Rev.* **85** (1952) 699; N Bloembergen and S. Wang, *Phys. Rev.* **93** (1954) 72.
2. H. Suhl, *J. Phys. Chem. Solids* **1** (1957) 209.
3. early papers are K. Nakamura, S. Ohta, and K. Kawasaki, *J. Phys.* **C15** (1982) L143; G. Gibson and C. Jeffries, *Phys. Rev.* **A29** (1984) 811.
4. see e.g. A.I. Akhiezer, V.G. Bar'yakhtar, and S.V. Peletminskii, *Spin Waves*, (North-Holland, Amsterdam, 1968).
5. see e.g. M. Sparks, *Ferromagnetic relaxation theory*, (McGraw-Hill, New York, 1964).
6. F.R. Morgenthaler, *J. Appl. Phys.* **31** (1960) 95S; E. Schlömann, J.J. Green, U. Milano, *J. Appl. Phys.* **31** (1960) 386S.
7. P. Hall, *Phys. Rev.* **A29** (1984) 2921.
8. see e.g. A.C. Newell, in *Propagation in Nonequilibrium Systems*, eds. J.E. Wesfreid, H. Brand, P. Manneville, G. Albinet, and N. Boccara, (Springer, Berlin, 1988).
9. F.J. Elmer, *Z. Phys.* **B68** (1987) 105; F.J. Elmer, *Physica* **D30** (1988) 321.

TRAVELING-WAVE CONVECTION IN BINARY FLUID MIXTURES AND SPATIO-TEMPORAL STRUCTURE

Yoshifumi Harada, Yasuou Masuno, Katsumi Sugihara, Kouya Nomura
Department of Applied Physics, Faculty of Engineering, Fukui University, Bunkyo 3-9-1, Fukui 910, Japan

and

Hideo Yahata
Department of Material Science, Faculty of Science Hiroshima University, Kagamiyama 1-3-1, Higashi Hiroshima 724, Japan

ABSTRACT

We present an experimental evidence of the existence of one-dimensional nonlinear localized traveling wave convection near the saddle-node point in which fast and slow traveling wave overturning convective rolls and a motionless state coexist in a region of arbitrary length in a large aspect slot channel, separated from the rest by a front that can be motionless. We have found, for the first time, the new two confined state in which different three stable state (two confined state and convection free state) coexist each other without any rapid quenching of reduced Rayleigh number r. Computer simulation is also studied using the two-dimensional MAC scheme for integrating the equation of motion for the velocity, the temperature, the concentration and pressure of a fluid contained in a rectangular cell.

1. Introduction

One dimensional traveling-wave (TW) convection in binary fluid mixtures provides a model system which produce temporally evolving patterns in a single spatial dimension. One of the most fascinating phenomena revealed by these experiments is the existence of a variety of confined state, in which TW's are observed in one spatial region of the experimental cell, while the rest of the fluid is motionless [1-3]. This self-organized confined structure exists in convecting binary fluid mixtures as a steady state and the various analogies with second order phase transitions and symmetry-breaking and mode selection mechanisms have been carefully investigated just above such a convective onset. The well known distinct type of stable confined state have been observed thus far in one-dimensional geometry. These are categorized following four type: (1)"weakly nonlinear" confined state[1,2] (2) "fast confined state of arbitrary length" [2] and was observed below the onset of convection in a strongly hysteric regime in a annular cell, (3) "stationary pulse" [3] observed in a band of Rayleigh numbers straddling the onset of convection, (4) "neutrally stable fronts" [4] confined state.

It has proved to be a particularly useful model for understanding nonlinear dynamics of a variety of weakly nonlinear TW patterns in one dimensional geometry convection cell[5]. During the last several years significant developments towards theoretical understanding of the problem have been made. However, it has not been possible to avoid complicated three dimensional behavior in experiments using large-aspect-ratio wider rectangular container. Therefore, experiments on convection in wider rectangular container have not been useful for testing the predictions of one-dimensional models which produce complex behavior. Recent quasi-one dimensional Rayleigh Benard experiments was successfully studied in the long cell with narrow gaps[6] . We present here observation of strongly nonlinear TW's overturning convective rolls using one dimensional slot channel(cell) with large aspect ratio $\Gamma_x = 46.06$ in binary fluid mixtures.

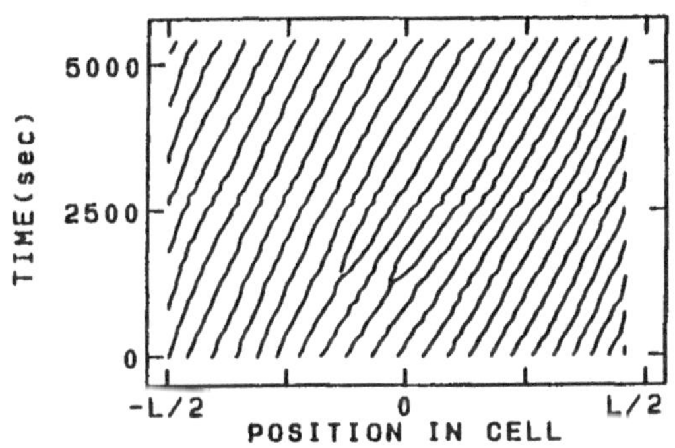

Fig. 1
Space-time representation of convective flow pattern. The line show the position of the downflow roll boundaries.

2. Transition from Mixing State to Localized State

2.1 Slot Convection in Ethanol-Water Mixtures

We have described our apparatus elsewhere. The cell is a long and narrow slot channel formed by plexiglas sidewall that are clamped between a copper plated mirror polished bottom and top plates. It has a depth d=0.521 cm, a width L_y =0.30 cm, and a length L_x =24.0 cm so that the aspect ratio is Γ_x = 46.06, Γ_y = 0.57 respectively. The working fluid was 8 wt.% solution of ethanol in water at a mean temperature of 25.0 °C, and had vertical diffusion time $\tau_v \equiv d^2/\kappa$ = 219 ~ 221 sec, where κ is the thermal diffusivity. The convective flow patterns along the long sidewall are visualized with the shadowgraph technique due to Bérge and Dubois[6].

In this paper, the experiments are begun by rasing r sufficiently large amount of the value to r = 5.93 in order to obtain the fully mixing state for the stationary over turning convective(SOC) rolls. From here, we decrease r to r = 5.42, we observe undulation mode of TW until r = 5.14. We further decrease Rayleigh number r to r = 4.49, we observe uniform state of the stable linear TW. In this state, the cell is filled with uniformly spaced rolls which propagate in unidirection of cell. At near the saddle-node point between r = 4.27 ~ 3.72 we have observed two spatiotemporal defects near the center of the cell as shown in Fig. 1. In this case, there appear long-wavelength-rolls near the left end wall and short-wavelength-roll near the right end wall.

We decrease further Rayleigh number very slowly, these spatiotemporal defect states is continued stably until r = 3.72. When the Rayleigh number is decreased further to r = 3.61, these roll pairs patches on the left and right hand-side separate and expand each other as shown in Fig. 2.

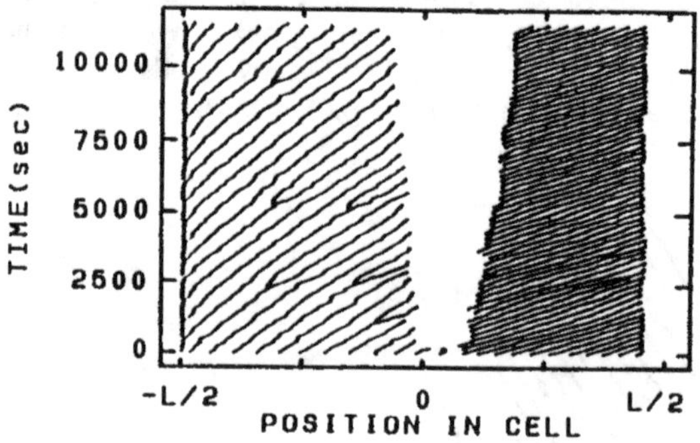

Fig. 2
Transient produced by creating two patches of propagating confined rolls. The patches expand and separately each other.

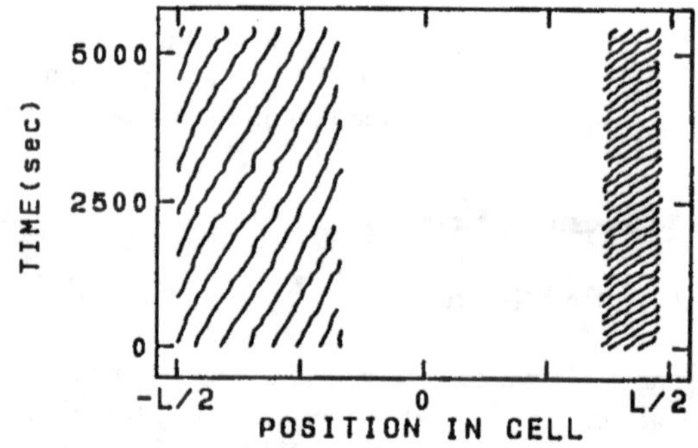

Fig. 3
Coexistence of the confined states (FCS & SCS) and space-time evolution of one dimensional nonlinear TW's.

If we prepare a uniform state and then r to the region near the saddle-node point, we find after the $272\tau_v$ a localized conducting region begins to slowly invade the slot channel occupied by the convecting flow. This process is accompanied by the generation of defects and adjustments of wavenumber and frequency. Finally, we have isolated right traveling fast and slow confined state (FCS and SCS) with both endwall of the cell near the saddle-node point as shown in Fig. 3.

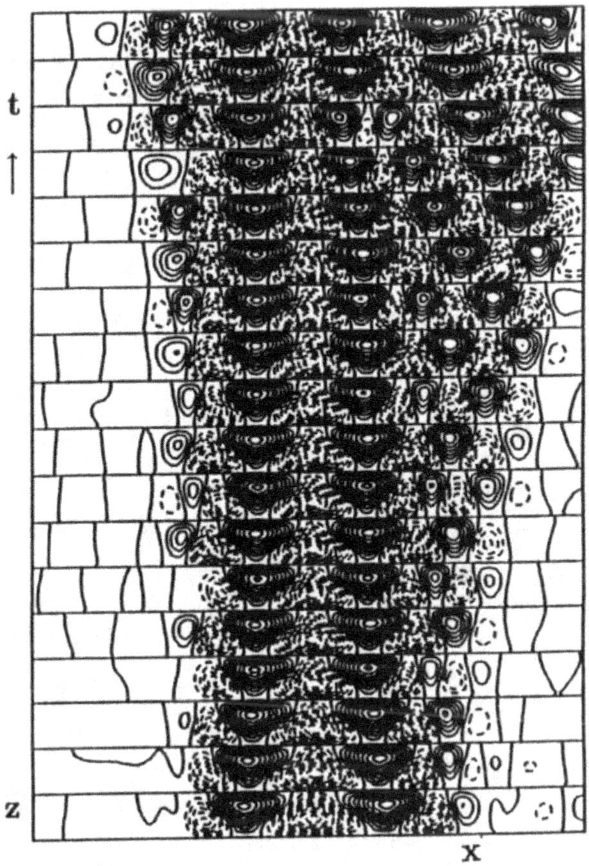

Fig. 4 Evolution of the temperature field of convection θ (not including the linearly distributed field due to thermal conduction) at the third stage of with $r = 2.10$. Localized convection state is gradually dominated by the extended state. Each block represents the isopleths of the temperature in the space(x,z) where the full and the dashed lines correspond to the positive and the negative values of θ respectively. Time proceeds in the upward direction where time difference between the successive two blocks is $16d^2/\nu$.

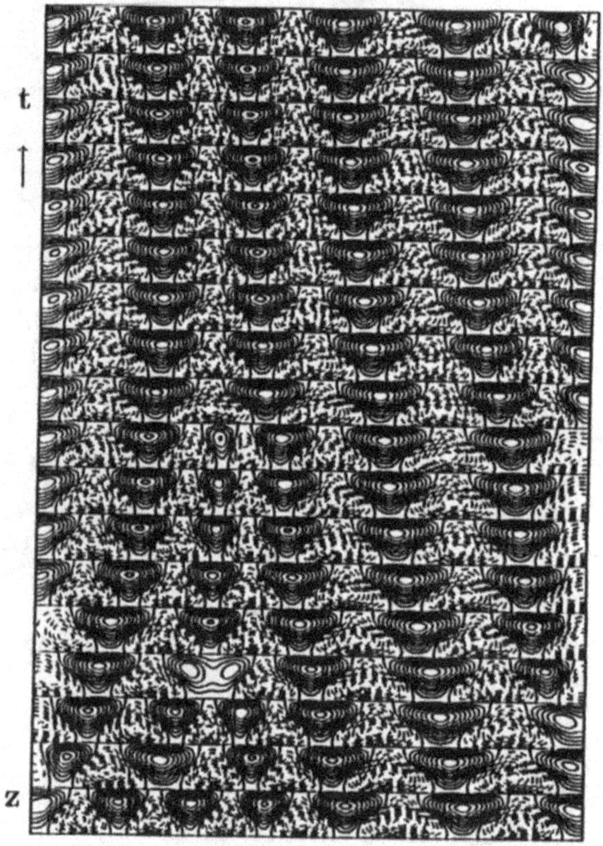

Fig. 5 Evolution of the temperature field of convection θ at r = 2.10 appearing soon after the final stage given in Fig. 4. Annihilation of the roll pair occurs as a result of their collision. Each isopleth is depicted in the same manner as in Fig. 4. Time proceeds in the upward direction where time difference between the successive two blocks in $32d^2/\nu$.

2.2 Computer Simulations of Convection Motion in Ethanol–Water Mixtures

Evolution of convection in a binary fluid mixtures is also studied with the aid of computer simulation. Let the aspect ratios of the rectangular container of the fluid be Γ_x and Γ_y with $\Gamma_x \gg \Gamma_y$. Under this condition we assume that the spatial pattern of convection is uniform in the y direction, and thereby reduce the problem to the two–dimensional one defined on the domain $\{ (x,y) \mid 0 < x < \Gamma_x , 0 < z < 1 \}$. We employed the MAC scheme to integrate the resulting two–dimensional equations of motion for the velocity, the temperature, the concentration and the pressure. In conforming with the experiment, we here give an example of the computational results for the case representing the 8 wt.–% ethanol in water at 14°C. Hence, the fluid parameters are chosen as the Prandtl number $\sigma = 13.8$, the Lewis number $L = 0.01$ and the separation ratio $\psi = -0.47$. The aspect ratio is chosen as $\Gamma_x = 12$ (due to computational convenience) and the domain is uniformly discretized into 192 and 16 meshes in the x and z directions respectively. We numerically integrated the finite difference equations for the Rayleigh number $R = 2.1R_c$ ($R_c = 1708$) starting from an initial condition consisting of an array of nearly periodic roll with small amplitudes.

In a similar way to other cases, evolution develops three successive stages of spatio–temporal structure different from one another. In the first stage, the amplitude exhibits a linear oscillatory growth resulting from the Hopf bifurcation. The second stage is characterized by successive emergence of the counter–propagating waves of rolls. The localized convection roll pattern accompanied by the quiescent state on both sides does not stably persist and gradually turns into the extended convection state Fig.4. However, the ensuing extended state is also unstable in the sense that collision of the two rolls travelling in the opposite directions gives rise to their annihilation in pairs Fig.5. At present, we cannot exclude the possibility that the system shown by Fig. 4 and Fig. 5 still remains in a long transient state before some other steady state. In addition, computation for the system with a large aspect ratio is needed to make more definitive comparison with the available experimental data.

3. Acknowledgements

We would like to thank P. Bérge and M. Dubois for helpful suggestions, to H. Brand and V. Steiberg for many fruitful stimulating conversations and suggestions and valuable discussions.

4. References

1. E. Moses, J. Fineberg, and V. Steinberg, Phys. Rev. A35, (1987)2757.
2. P. Koldner, D. Bensimon, and C.M. Surko, Phys. Rev. Lett. 60, (1988)1723.
3. J.J. Niemela, G. Ahlers, and D.S. Cannell, Phys. Rev. Lett. 64, (1990)1365.
4. P. Koldner, Phys. Rev. A42, (1990)2475.
5. V. Steinberg and E. Kaplan, NATO ASI Series(1991), Norway, April, 1991.
6. M. Dubois, R. Da Silva, F. Daviaud, P. Berge and A. Petrov, Europhys. Lett. 8, (1989) 135.

ON THE EXPERIMENTAL REALIZABILITY OF CONVECTION IN GAS MIXTURES AT MODERATE POSITIVE SEPARATION RATIOS

Stefan J Linz
Department of Engineering Sciences and Applied Mathematics
Northwestern University, Evanston, Illinois 60208-3125

Instabilities to convection in binary gas mixtures are quite different than in liquid mixtures provided that the modulus of the separation ratio and the Dufour number are large. In order to support which binary gas mixtures might be appropriate candidates for ones with positive separation ratios we consider a dilute mixture of H_2 in CO_2. We provide estimates of the ranges of accessible separation ratios and Dufour numbers. We also study the relevance of barodiffusion. Finally, we suggest a simple demonstration experiment to prove the relevance of the Dufour effect in gas mixtures.

1. Introduction

It was pointed out recently[1−5] that a strong Dufour effect can dramatically change instability behavior to convection in binary fluid mixtures. The strength of the Dufour effect is measured by the product of Lewis number L, Dufour number Q and separation ratio ψ. In gas mixtures[1,5] L and Q are quite large and so the basic question to be answered is whether it is possible to reach in gas mixtures combinations whose separation ratios are such that the contrast to liquid mixtures can be observed experimentally. However, knowledge about the accessible ranges of the separation ratio in gas mixtures is poor. The purpose of this short communication is to suggest a possible candidate of a binary gas mixture where large *positive* separation ratios and large Dufour numbers can be realized. Beyond that we discuss the role of barodiffusion[6] and suggest a simple experiment which shows the influence of the Dufour effect on the stability thresholds to convection.

2. Governing equations

We consider a horizontal layer of a binary fluid mixture of height d enclosed between plane parallel boundaries and subject to the gravitational field $-g\mathbf{e}_z$. In general, this system is described by field equations[7] for the baricentric velocity $\mathbf{u} = (u, v, w)$, the temperature T, and the mass concentration C of the lighter component. We assume that the fluid mixture is still Boussinesq in the sense that density variations are only significant in the buoyancy term and that its thermodynamic and hydrodynamic parameters[7] like thermal diffusivity κ, thermal (solutal) expansion coefficient α (β), mass diffusivity D, specific heat c_p, thermodiffusion ratio k_T, barodiffusion ratio k_p and kinematic viscosity ν, can be considered to be constant. However, we do not neglect barodiffusion[6,7] in the concentration flux. Then, we end up with the following system of governing equations

$$(\partial_t + \mathbf{u}\cdot\nabla)\mathbf{u} = -\nabla p - \hat{g}(\rho/\rho_0)\mathbf{e}_z + \nu\nabla^2\mathbf{u}, \tag{1}$$

$$(\partial_t + \mathbf{u}\cdot\nabla)T = -\nabla\cdot\mathbf{J_e} + Q\nabla\cdot(\psi\mathbf{J_c}), \tag{2}$$

$$(\partial_t + \mathbf{u}\cdot\nabla)C = -\nabla\cdot\mathbf{J_c}, \tag{3}$$

and $\nabla\cdot\mathbf{u} = 0$. The diffusive heat and concentration fluxes, $\mathbf{J_e}$ and $\mathbf{J_c}$, are given by

$$\mathbf{J_e} = -\nabla T \quad \text{and} \quad \mathbf{J_c} = -L\left[\nabla C - \psi\nabla T + (\Gamma/\hat{g})\nabla p\right]. \tag{4a, b}$$

In writing Eqs.(1)—(4) we have non-dimensionalized length by layer thickness d; time by d^2/κ; temperature by $\kappa\nu/\alpha g d^3$; concentration by $\kappa\nu/\beta g d^3$; and the quotient of pressure and mass density p by κ^2/d^2. We have introduced a non-dimensional gravitational acceleration $\hat{g} = gd^3/\kappa^2$, the Prandtl number $\sigma = \nu/\kappa$, the Lewis number $L = D/\kappa$, the separation ratio $\psi = -\beta k_T/\alpha T_0$. The Dufour number[1] Q and the barodiffusion number[6] Γ are defined as

$$Q = (T_0\alpha^2/c_p\beta^2)\,\Pi, \quad \text{and} \quad \Gamma = (g^2 d^4 \beta^2/\nu\kappa)\,\Pi^{-1}, \tag{5a, b}$$

where $\Pi = (\partial\mu/\partial\tilde{C})^0_{T_0,p_0}$ denotes the variation of the chemical potential[7] of the mixture with the (dimensionalized) concentration $\tilde{C}$ $(0 < \tilde{C} < 1)$, taken at the reference point. To close this system of equations we use the linearized equation of state for the mass density $(\rho/\rho_0) = 1 - (\sigma/\hat{g})\left[T - T_0 + C - C_0\right]$.

3. Relation between Dufour number and Barodiffusion number

The Dufour and barodiffusion numbers are solely determined by classical thermodynamics and can be calculated if the equation of state is known, e.g. as a virial expansion. However, there is a simple, general relation between barodiffusion number and Dufour number. Inspecting Eq.(5a) and Eq.(5b) shows immediately that Q and Γ are not independent, but that their product is given by

$$Q\Gamma = g^2\alpha^2\kappa^{-2}c_p^{-1}\nu^{-1}T_0 d^4. \tag{6}$$

Thus, for a given setup, i.e. fixed r.h.s. of Eq.(6), the larger the Dufour numbers, the smaller the barodiffusion numbers and vice versa. For a typical gas mixture at room temperature and a layer height of $d = 1cm$ the r.h.s. of (6) is of the order 10^{-3}. Since the Dufour number in an ideal gas mixture approximation[5] diverges if $\tilde{C} \to 0$ or $\tilde{C} \to 1$, we find that the barodiffusion contribution vanishes if $\tilde{C} \to 0$ or $\tilde{C} \to 1$. Thus, the more dilute the mixture, the smaller the barodiffusion.

4. Estimates for accessible ranges of separation ratio and Dufour number

Our focus is to realize large moduli of the separation ratio $\psi = -\beta k_T/\alpha T_0$. For a gas mixtures at room temperature and not too high pressure the thermal expansion coefficent α can be approximated by its ideal gas value $1/T_0$, so that $\psi \simeq -\beta k_T$. Let us first discuss β. As a guide for its magnitude an ideal gas mixture approximation[5] is appropriate (at least for not too high pressures), leading to

$$\beta = \frac{m_2 - m_1}{m_1 + \tilde{C}(m_2 - m_1)}, \tag{7}$$

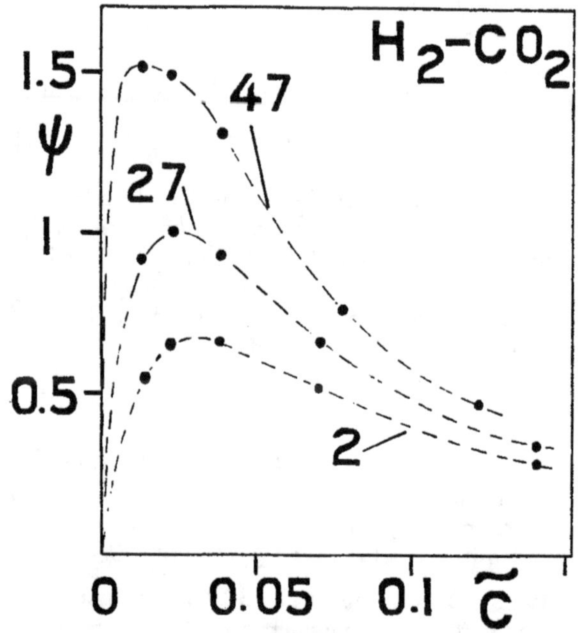

Figure 1. Dependence of the separation ratio ψ on the mean concentration $\tilde{C}$ for H_2/CO_2 mixtures. The points are extrapolated values from experiments[8] (see text for details). The dashed lines are guides for the eyes. The numbers at the curves denote the pressure (in atm) at which the experiment was carried out.

where m_1 and m_2 denote the particle masses of component 1 and 2 The mass concentration $\tilde{C}$ is by definition the ratio of the mass density of the *lighter* component 1 and the total mass density. The main point is that for small $\tilde{C}$, i. e. dilute mixtures, β can be approximated by $m_2/m_1 - 1$. This shows that to get large β the mass ratio of the heavier and the lighter component has to be large. So combinations with H_2 might be most promising. Let us now turn to k_T, which is a combination of phenomenological Onsager coefficients, and so is not immediately determined by classical thermodynamics. We suggest here that a dilute mixture of H_2 in CO_2 might be a good candidate to reach large positive separation ratios. In the literature one can find measurements[8,9] of the thermal diffusion factor α_T for this mixture, measured with respect to the heavier component. α_T is related to k_T by $k_T = -x_1(1 - x_1)\alpha_T$ where x_1 is the mole fraction of the lighter component which is related to the concentration by $\tilde{C} = x_1/[x_1 + (1 - x_1)M_2/M_1]$. Here M_1 and M_2 are the molar masses of component 1 and 2.

In Figure 1 we show the dependence of the separation ratio ψ of H_2/CO_2 on the mean concentration $\tilde{C}$ for a fixed temperature of $T_0 = 360K$ and three different pressures 2, 26, and 47 atm. The dots are extrapolations of Becker's[8] measurements

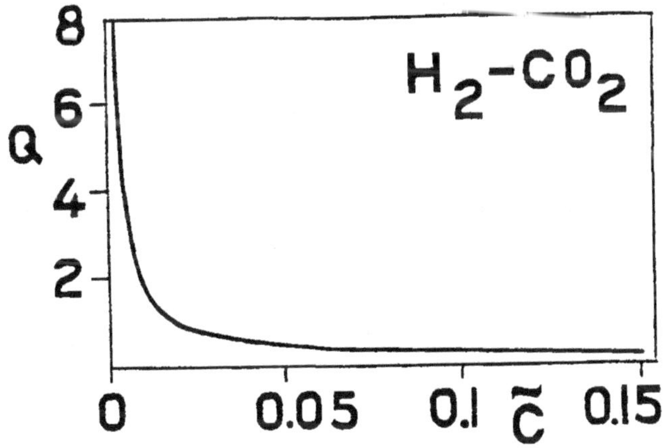

Figure 2. Dependence of the Dufour number Q on the mean concentration $\tilde{C}$ in H_2/CO_2 mixtures, based on an ideal gas mixture approximation for the solutal expansion coefficient β and the variations of the chemical potential of the mixture with mean concentration.

of α_T using eq.(7) to evaluate the solutal expansion coefficient. The dashed lines are guides for the eyes and represent the expected behavior of $\psi(\tilde{C})$. Note that we assume that ψ vanishes as $\tilde{C}$ approaches zero, a fact which cannot be read off from the measured values. However, Becker's[8,9] results of α_T do not show any indication for a divergence as x_1 approaches zero, and so ψ should vanish as x_1 or $\tilde{C}$ goes to zero. Another argument is that the strength of the Dufour effect, $QL\psi$, remains finite only if $\psi \to 0$ as $\tilde{C} \to 0$ or 1. Already for a low pressure of 2 atm the separation ratio can be varied over a considerably wide range of positive values with a maximum of about $\psi = 0.65$ at about $\tilde{C} = 0.035$. Moreover, ψ can be enhanced by enhancing the mean pressure. The pressure dependence of ψ is somewhat stronger than linear. For higher pressures the maximum of ψ is shifted to lower mean concentrations. Note that the $\tilde{C}$ —dependence of ψ in H_2/CO_2 resembles qualitatively the one of ethanol/water[10], in particular there is also an extremum at small concentrations. However, the sign of ψ is different.

In Figure 2 we show the dependence of the Dufour number Q on the mean concentration $\tilde{C}$ for H_2/CO_2 mixtures using the ideal gas mixture approximation reported in Ref.5. In this approximation Q is independent of the mean pressure and diverges as $\tilde{C} \to 0$ and decays rapidly to value of about $Q = 0.25$ for $\tilde{C} = 0.15$. Significant effects caused by barodiffusion have to be expected if ψ is close to zero and if ψ crosses zero at finite mean concentrations (as e.g. for ethanol/water[6]). However, in H_2/CO_2 small values of ψ can only be realized if $\tilde{C} \to 0$ and in this limit also the barodiffusion number becomes very small. So we conclude that in H_2/CO_2 barodiffusion can be neglegted.

We conclude that in H_2/CO_2 mixtures the realization of large separation ratios and Dufour numbers should be possible simultaneously. We hope that our estimates trigger detailed experimental investigations on H_2/CO_2 similar as it was done for ethanol/water in a fine paper by Kolodner et al.[10].

5. A simple way to show the influence of the Dufour effect experimentally

Until now there has been no simple experimental evidence for the relevance of the Dufour effect in convection of binary gas mixtures. A simple demonstration might be as follows: from Figure 1 and Figure 2 we read off that one can realize the *same* separation ratio with two mean different concentrations, say $\tilde{C}_a$ and $\tilde{C}_b$. However, the Dufour numbers Q_a and Q_b associated with $\tilde{C}_a$ and $\tilde{C}_b$ are different, as can be seen in Figure 2. If the Dufour effect is not negligible, then one has to expect two different stability thresholds belonging to the two different concentrations, but the same ψ. Using thermally-driven convection[5] between impermeable plates as demonstration system and assuming the most simple case of free slip, permeable boundary conditions[5], we find that the difference in the onset of stationary convection is given by

$$[r^c_{stat}(\psi, Q_a)]^{-1} - [r^c_{stat}(\psi, Q_b)]^{-1} = (1+\psi)\psi^2(Q_a - Q_b), \tag{8}$$

where $r^c_{stat}(\psi, Q)$ is the reduced critical stationary stability threshold given in Ref. 5. Thus, it is directly proportional to the differences in the Dufour numbers, $Q_a - Q_b$ and independent of the Lewis number. As an example, one finds for a pressure of 2atm and $\psi = 0.5$ for the right hand side of eq.(8) a value of about 0.56 which should be experimentally observable. The actual size varies only slightly if realistic impermeable boundary conditions are used provided that the *experimentally* realized critical wave numbers are almost equal. To incorporate also large differences in the wave numbers the FSI approximation of Ref.5 can be used.

Acknowledgments

This work was supported by Deutsche Forschungsgemeinschaft (Li 498/1-1&2) and a NASA grant with the Microgravity Sciences and Applications Program. The author thanks S.H. Davis, M. Lücke, and H. Riecke for helpful discussions.

References

1. S.J. Linz, Phys. Rev.**A40** (1989), 7175.
2. W. Hort, S.J. Linz, and M. Lücke, in *Nonlinear evolution of spatio-temporal structures in dissipative continuous systems* (F.H. Busse and L. Kramer, eds.) NATO ASI Series **B225** (1990),149.
3. N.D. Stein, Phys. Rev.**A43** (1991), 768.
4. S.J. Linz, Phys. Rev.**A45** (1992), 1262.
5. W. Hort, S.J. Linz, and M. Lücke, Phys. Rev.**A45** (1992), in press.
6. S.J. Linz and M. Lücke, Phys. Rev.**A36** (1987), 3997.
7. For a derivation of the basic equations see L.D. Landau and I.M. Lifshitz, *Fluid Mechanics*, (Pergamon Press 1959).
8. E.W. Becker, Z. Naturforsch. **5a** (1950), 457.
9. see also: R. Haase, Z. phys. Chem.(Leipzig) **196** (1950), 219.
10. P. Kolodner, H. Williams, and C. Moe, J. Chem. Phys. **88** (1988), 6512.

COMPETING WAVE LENGTHS IN PATTERN FORMING SYSTEMS

Atsushi Ogawa[1], Walter Zimmermann[2], Kyozi Kawasaki[1] and Toshihiro Kawakatsu[1]

[1]*Department of Physics, Kyushu University 33, Fukuoka 812, Japan*
[2]*IFF Theorie III, Forschungszentrum Jülich, 5170 Jülich, Germany*

ABSTRACT

We investigate wave-length competition in 2-dimensional pattern forming systems with uniaxial anisotropy. Spatially periodic modulation of the external driving force leads to an interplay between this and the wavelength of the convection rolls. In the vicinity of the transition point from normal to oblique oriented convection rolls the external modulation leads to new two-dimensional extended patterns, namely a rectangular, a skewed varicose, an undulated and an incommensurate pattern. All of them are stable against long wavelength perturbations in a finite parameter range.

1. Introduction

Wave-length competition leads to the well known commensurate - incommensurate transition phenomena for adsorption at solid surfaces.[1] One may ask the question what are the typical effects of wave-length competition in driven systems in contrast to such phenomena at thermal equilibrium. Non-equilibrium systems which have a natural wave-length similar to the lattice constant at the solid surface are for instance Rayleigh-Bénard convection (RB)[2] and electroconvection (EHC),[3] where one has at the onset of convection a periodic arrangement of parallel convection rolls. In the adsorption problem the second wavelength is the equilibrium lattice constant of the adsorped material and in the convection systems the second wavelength is introduced by periodic modulation of the external driving force namely the applied electric field in electroconvection or the temperature gradient in Rayleigh-Bénard convection. Lowe *et al.* observed in EHC similar commensurate-incommensurate transition phenomena as in adsorption problems.[4] Moreover Coullet was able to formulate this wave length competition in EHC for resonant cases in a one-dimensional phenomenological model and he found also that this model reduces in some limit (phase approximation) to the continuum limit of the well known Frenkel Kontorova model.[5]

However the convective systems are - especially EHC - quasi two-dimensional systems. EHC has in addition an uniaxial anisotropy in the plane of the fluid layer (which we call the x-direction) and belongs therefore to a different symmetry class than Rayleigh-Bénard convection in simple fluids.[2] The convection rolls in EHC or also in Rayleigh-Bénard convection in nematics can orient perpendicularly or obliquely to the prefered direction with a wave vector $\vec{k}_c = (k_{xc}, 0)$ (normal rolls) or $\vec{k}_c = (k_{xc}, \pm k_{yc})$ (oblique rolls).[6,7] In the previous investigation[5] a one-dimensional description was used in modeling the interplay of normal rolls with its wave vector $\vec{k}_c = (k_{xc}, 0)$ parallel to wave vector describing the externally imposed modulation $\vec{k}_m = (k_m, 0)$. Here we concentrate especially on the situation where we have obliquely oriented convection rolls with the wave vector $\vec{k}_c = (k_{xc}, \pm k_{yc})$ nonparallel to the modulation wave vector $\vec{k}_m$. Existence of two degenerate oblique rolls and the nonparallel arrangement of the two wave vectors is one of the essential differences to the previous investigations[4,5] and is also the important prerequisite for our predicted new two-dimensional extended patterns. In addition we concentrate here on the transition region for oblique to normal oriented convection rolls. The transition point is called the Lifshitz-point. The reader should refer our forthcoming publications for more details.[8]

2. Amplitude Equation

Here we write the relation between the modulation wave number k_m and the natural wave number k_{xc} of normal rolls as $mk_m = n(k_{xc} + q)$ where m and n are integers and the misfit wave number q is assumed to be $k_{xc} \gg q$ (almost resonant condition). To investigate possible patterns and their stabilities, we start from an amplitude equation which can be derived in a systematic way by separating fast and slowly varying variables. Thus we introduce a complex amplitude $A(x, y, t)$ which is defined as $\vec{u}(x, y, z, t) = A(x, y, t)e^{ik_{c}x}\vec{U}_0(z) + c.c.$ where $\vec{u}$ is the physical quantities such as the flow fields near the instability threshold and $\vec{U}_0$ represents variation of the physical quantities across the fluid layer. The explicit form of the amplitude equation near the Lifshitz-point under the modulation field is[9]

$$\partial_t A = (\partial_x{}^2 - iZ\partial_x\partial_y{}^2 + W\partial_y{}^2 - \partial_y{}^4 + \mu - |A|^2)A + \alpha e^{inqx} A^{*n-1}, \tag{2.1}$$

where μ is a control parameter which measures the distance from the instability threshold, α is proportional to the amplitude of the external modulation and Z is the material parameter($0 < Z < 2$). W is the second control parameter proportional to the distance from the normal-oblique roll transition point. A slightly oblique rolls appear for $W < 0$ when $\alpha = 0$. It is important to note that Eq.(2.1) can be rewritten in a variational form using a Lyapunov functional H which is also called a potential.

$$\partial_t A = -\frac{\delta H}{\delta A^*}, \tag{2.2}$$

$$H \equiv \int dxdy[|\partial_x A|^2 + Z\Im\{(\partial_x A^*)(\partial_y{}^2 A)\} + W|\partial_y A|^2 + |\partial_y{}^2 A|^2$$
$$-\mu|A|^2 + \frac{1}{2}|A|^4 - \frac{\alpha}{n}(e^{-inqx}A^n + e^{inqx}A^{*n})], \tag{2.3}$$

where $\Im\{\cdots\}$ means the imaginary part of $\{\cdots\}$. Due to the existence of the potential, the dynamics is purely relaxational and we can expect that emerging patterns are stationary.

3. Solutions and Stability Analysis

We make an ansatz for Eq.(2.1) which allows besides original wave vectors $\vec{k}_m$ and $\vec{k}_c$ a new one (Q,P):

$$A = e^{iqx} \sum_{l,j=-N,-M}^{N,M} e^{i(lQx+jPy)}\left(F_{l,j} + v_{l,j}(x,y,t)\right) \tag{3.1}$$

where $F_{l,j}$ are real variables, and $v_{l,j}$ are small perturbations which are taken to be zero for calculating the stationary solutions and they are assumed to be small for investigating the linear stability behavior of the stationary solutions.

Qualitative features of the solution properties of Eq.(2.1) can be already found within a 3-mode approximation

$$A = e^{iqx}(F_0 + F_1 e^{i(Qx+Py)} + F_2 e^{i(Qx-Py)}) \tag{3.2}$$

Using such an approximation we can treat various types of patterns which look like those in Fig.1 in the real space. For example we have in the case $Q = 0$ a normal roll pattern ($F_0 \neq 0$, $F_1 = F_2 = 0$), a rectangular pattern ($F_0 = 0$, $|F_1| = |F_2|$), a skewed varicose type pattern ($F_0 = 0$, $|F_1| \neq |F_2|$), an undulated pattern ($F_0 \neq 0$, $F_1 = -F_2$)[10], and for $Q \neq 0$ an oblique roll pattern and a new incommensurate pattern ($F_0 \neq 0$, $F_1 = F_2$).

We can investigate the linear stability of these solutions by substituting the fiollowing form for $A(x,y,t)$ into Eq.(2.1):

$$A = F_0 + v_0 + (F_1 + v_1)e^{i(Qx+Py)} + (F_2 + v_2)e^{i(Qx-Py)}, \tag{3.3}$$

where $v_i(x,y,t)$ $(i = 0,1,2)$ are complex variables and are assumed to be

$$v_i = u_{ia}e^{\sigma t+i(Kx+Ly)} + u_{ib}e^{\sigma^* t-i(Kx+Ly)}. \tag{3.4}$$

where u_{ia} and u_{ib} are constants. Linearized equations for u_{ia} and u_{ib} determine the growth rate σ of long wave length perturbations. This procedure allows us to calculate the local stability properties of the patterns. Furthermore we can compare values of the potential in Eq.(2.3) for the various patterns which give the global stability properties.

In the case of $n = 2$, for example, the first pattern that appears immediately above threshold is the rectangles (Fig.1(a)) and at large values of the control parameter μ the skewed varicose type pattern (Fig.1(b)) is the globally stable one.

The skewed varicose pattern is a superposition of oblique rollsand rectangles. For $q \neq 0$ in the transition region between these two patterns none of both is stable and we expect higher commensurate or even incommensurate patterns. These two new patterns are unstable in the case without modulation ($\alpha = 0$). In the case of $n = 3$, besides normal rolls and oblique rolls, the undulation pattern(Fig.1(c)) can be locally and globally stable in the range of not too large μ and relatively large α.[8] The undulations are only locally stable in small parameter ranges but they never have the lowest potential in the unforced case $\alpha = 0$.[10]

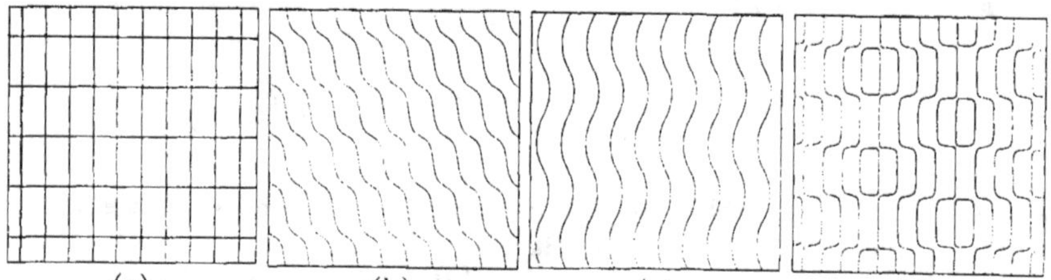

(a) (b) (c) (d)

Fig.1 Solutions by 3-mode approximation; (a)rectangular, (b)skewed varicose, (c)undulated and (d)new incommensurate pattern. Zero contours of $\Re\{A(x,y)\exp(ik_{xc}x)\}$ are drawn.

4. Computer Simulations

The full numerical solution of Eq.(2.1) confirms the qualitative scenarios we obtained by the 3-mode approximation (Eq.(3.2)), with possibly different parameter regions characterizing various stable solutions. Starting with random initial conditions we show two snapshots of our simulation showing transient patterns on the route to rectangles (Fig.2(a)) and to undulations (Fig.2(b)). Fig.2(a) shows for n=2 a typical defect pair in a rectangular pattern and the final state is a regular rectangular pattern. In Fig.2(b) we show for n=3 a transient pattern finally developing into undulations like that in Fig.1(c). We chose parameter sets in such a way that the rectangular pattern (Fig.2(a)) exists in the 3-mode approximation. The transient time brfore reaching the regular patterns can be quite long depending on the actual parameters.

5. Conclusion

We have predicted, for anisotropic convective systems like electroconvection and Rayleigh-Bénard in nematics, four stable solutions extended in two dimensions. These are the rectangular, skewed varicose, undulated and incommensurate patterns as shown in Fig.1. Qualitatively these solutions can be covered by simple 3-mode approximations to Eq.(2.1). The independent numerical solution of Eq.(2.1) are qualitatively in agreement with those of 3-mode approximation and they allow in addition to resolve the interesting dynamics from random initial conditions to the final states as shown in Fig.2.

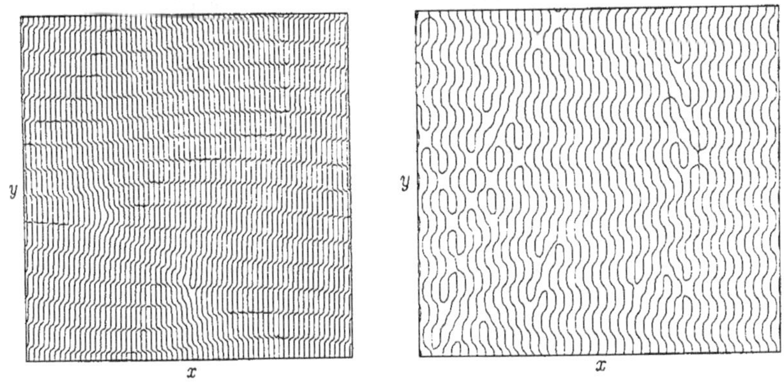

(a) (b)

Fig.2 Snap shots in the transient regime; (a)$n = 2$, $W = -0.8$, $q = -0.2$, $Z = 1.0$, $\mu = -0.25$, $\alpha = 0.1$, (b)$n = 3$, $W = -1.5$, $q = 0.1$, $Z = 0.8$, $\mu = 1.0$ and $\alpha = 0.9$. Zero contours of $\Re\{A(x,y)\exp(ik_{xc}x)\}$ are drawn.

6. Acknowledgement

This work was supported by the Scientific Research Fund of the Ministry of Education, Science and Culture of Japan. W.Z. is grateful to the Kyushu Institute of Technology for a Fellowship in 1989 and to the Kajima foundation which made a second stay at the KIT and Kyushu University in 1991 possible.

7. References

1. P. Bak, *Rep. Prog. Phys.* **45** (1982) 587; V. L. Pokrovsky and A. L. Talapov, *Theory of Incommensurate Crystals*, (Harwood, Amsterdam, 1984); and for a more recent review see W. Selke, *Spatially modulated structures in systems with competing interactions* in *Phase Transitions and Critical Phenomena* Vol.15, ed. C. Domb and M. Lebowitz, (preprint 1992).
2. F. H. Busse, in *Hydrodynamic Instabilities and the Transition to Turbulence*, 2nd ed., ed. H. L. Swinney and J. P. Gollub, (Springer 1986).
3. For a recent overview see W. Zimmermann, MRS Bulletin **16** (1991) 46.
4. M. Lowe, J. P. Gollub and T. Lubensky, *Phys. Rev. Lett.* **51** (1983) 786; M. Lowe and J. P. Gollub, *Phys. Rev.* **31** (1985) 3893.

5. P. Coullet, *Phys. Rev. Lett.* **56** (1986) 724; P. Coullet and D. Repaux, *Europhys. Lett.* **3** (1987) 573.
6. R. Ribotta, A. Joets and L. Lei, *Phys. Rev. Lett.* **56** (1986) 1595.
7. W. Zimmermann and L. Kramer, *Phys. Rev. Lett.* **55** (1985) 402; E. Bodenschatz, W. Zimmermann and L. Kramer, *J. Physique.* **49** (1988) 1875.
8. W. Zimmermann, A. Ogawa, S. Kai, K. Kawasaki and T. Kawakatsu (submitted for publication); A. Ogawa, W. Zimmermann K. Kawasaki and T. Kawakatsu (in preparation).
9. A. Ogawa, K. Kawasaki, W.Zimmermann and T. Kawakatsu, in *Dynamics and Patterns in Complex Fluids*, ed. A. Onuki and K. Kawasaki, (Springer 1990), p. 78.
10. W. Pesch and L. Kramer, *Z. Phys.* **63** (1986) 121; E. Bodenschatz, M. Kaiser, L. Kramer, W. Pesch, A. Weber and W. Zimmermann, in *New Trends in Nonlinear Dynamics*, ed. P. Coullet and P. Huerre, NATO ASI Series (Plenum, New York, 1991).

Time Evolution After Phase Instabilities in the Ginzburg-Landau Equation

Hidetsugu Sakaguchi
Department of Physics, College of General Education, Kyushu University, Fukuoka 810, Japan

abstract

We study time evolution after the Eckhaus instability and the zigzag instability in the Newell- Whitehead equation and the Benjamin-Feir instability in the complex Ginzburg-Landau equation.

1. Introduction

A periodic structure sometimes emerges as a dissipative structure when a nonequilibrium parameter is increased. Phase dynamics is an effective method to investigate slow dynamics and stability of the periodic structure. In some cases the phase dynamics becomes unavailable since the phase gradient becomes too large and diverges. We study time evolution after the phase instability for the Newell-Whitehead equation and the complex Ginzburg-Landau equation.

2. Newell-Whitehead equation

The Newell-Whitehead equation is an amplitude equation for the spatially periodic pattern[1]. It is written as

$$\frac{\partial W}{\partial t}=\mu W-|W|^2W+\left(\frac{\partial}{\partial x}-\frac{i}{\sqrt{2\pi}}\frac{\partial^2}{\partial y^2}\right)^2 W \ . \qquad (1)$$

This equation has a family of stationary solutions:

$$W_0=\sqrt{\mu-Q^2}\exp(iQx), \qquad \text{for } Q^2<\mu \qquad (2)$$

where Q is wavenumber of the periodic solution. If we express W as $W=\{\sqrt{\mu-Q^2}+u(\phi)\}\exp\{iQx+i\phi(x,y,t)\}$ and substitute it to the Newell-Whitehead equation, we obtain a linear phase equation:

$$\frac{\partial\phi}{\partial t}=\frac{\mu-3Q^2}{\mu-Q^2}\frac{\partial^2\phi}{\partial x^2}+\sqrt{\frac{2}{\pi}}\,Q\,\frac{\partial^2\phi}{\partial y^2}\,. \tag{3}$$

When μ is smaller than $3Q^2$, the diffusion constant in the x-direction is negative. We call the instability the Eckhaus instability. When Q is negative, the diffusion constant in the y-direction is negative. We call the instability the zigzag instability.

The time evolution immediately after the Eckhaus instability is described by a nonlinear phase equation[2,3]:

$$\frac{\partial\phi}{\partial t}=D_2\frac{\partial^2\phi}{\partial x^2}+g\frac{\partial\phi}{\partial x}\frac{\partial^2\phi}{\partial x^2}-D_4\frac{\partial^4\phi}{\partial x^4}\,, \tag{4}$$

where $D_2=(\mu-3Q^2)/(\mu-Q^2)$, $g=-4\mu Q/(\mu-Q^2)^2$ and $D_4=2Q^2/(\mu-Q^2)^3$. If a small sinusoidal wave is given to Eq.(4) as an initial condition, the sinusoidal wave grows up to a saw-toothed wave of large amplitude. Phase gradient becomes large and eventually it leads to divergence, and the phase description becomes unavailable. In the Newell-Whitehead equation the amplitude is depressed where the phase gradient is large and then a phase-slip occurs. At the instant of the phase-slip the amplitude goes to zero and the phase cannot be defined. As a result of this process the wavenumber is changed to a smaller value. Figure 1 shows a snap-shot of ReW for $\mu=2.94Q^2$, in which the wavenumber is changing from $Q=\pi/10$ to $Q'=3\pi/40$.

Behaviors immediately after the zigzag instability is described by a nonlinear phase equation[4]:

$$\frac{\partial\phi}{\partial t}=D_2\frac{\partial^2\phi}{\partial y^2}+g\left(\frac{\partial\phi}{\partial y}\right)^2\frac{\partial^2\phi}{\partial y^2}-D_4\frac{\partial^4\phi}{\partial y^4}\,, \tag{5}$$

where $D_2=\sqrt{2/\pi}\cdot Q$, $g=(3\mu-7Q^2)/\{\pi(\mu-Q^2)\}$ and $D_4=1/(2\pi)$, if we consider only the y-direction. When g is positive, a stable zigzag pattern is obtained. But when g is negative, the phase gradient becomes large and it leads to divergence. Figure 2 shows a time evolution for $\mu=1.5Q^2$ in the Newell-

Whitehead equation. The zigzag pattern grows up to a grid-like pattern and reconnection of the roll pattern occurs and finally an oblique roll pattern obtained. The phase-slip occurs at the instant of the reconnection.

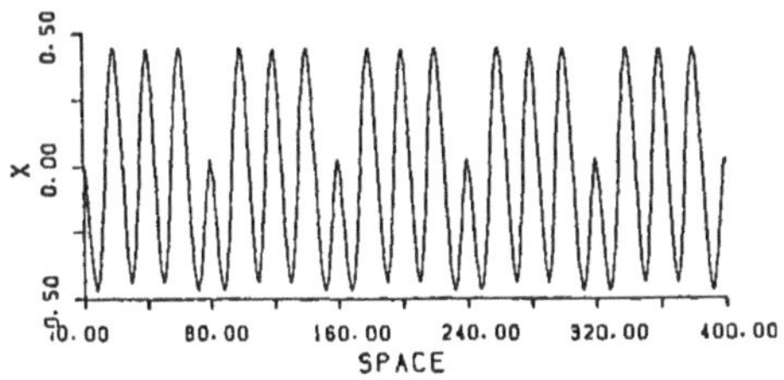

Fig.1 Wavenumber changing process.

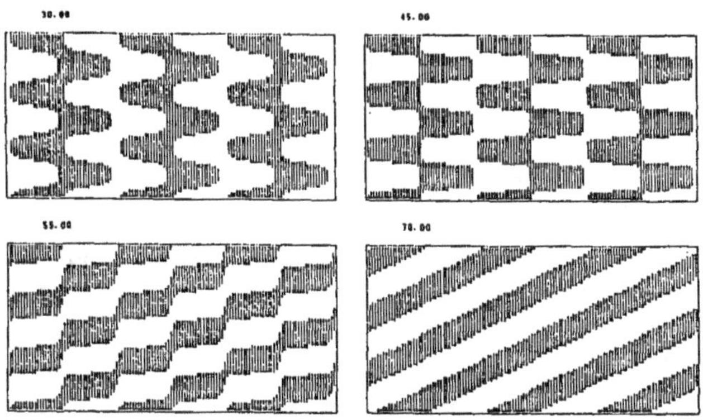

Fig.2 Time evolution after the zigzag instability.

3. Complex Ginzburg-Landau equation

The complex Ginzburg-Landau equation is an amplitude equation for the Hopf bifurcation.[5]

$$\frac{\partial W}{\partial t} = (1+ic_0)-(1+ic_2)|W|^2W+(1+ic_1)\frac{\partial^2 W}{\partial x^2} \quad . \tag{6}$$

This equation has a uniformly oscillating solution.

$$W_0=e^{i\omega t}, \quad \omega=c_0-c_2. \tag{7}$$

If we assume W as $W=\{1+u(\phi)\}\exp\{i\omega t+i\phi(x,t)\}$ and substitute it to Eq.(7), we obtain a nonlinear phase equation called the Kuramoto-Sivashinsky equation[5]:

$$\frac{\partial\phi}{\partial t}=D_2\frac{\partial^2\phi}{\partial x^2}+g_1\left(\frac{\partial\phi}{\partial x}\right)^2-D_4\frac{\partial^4\phi}{\partial x^4}, \tag{8}$$

where $D_2=1+c_1c_2, g_1=c_2-c_1$ and $D_4=c_1{}^2(1+c_2{}^2)/2$. When $1+c_1c_2$ is negative, a phase instability occurs. It is called the Benjamin-Feir instability. The Kuramoto-Sivashinsky equation exhibits a phase turbulence. In the phase turbulence regime the phase distribution is random but the amplitude is nearly 1. When $1+c_1c_2$ is decreased even more, the phase fluctuation becomes large and a next higher order term becomes necessary. A nonlinear phase equation of the next order is

$$\frac{\partial\phi}{\partial t}=D_2\frac{\partial^2\phi}{\partial x^2}+g_1\left(\frac{\partial\phi}{\partial x}\right)^2-D_4\frac{\partial^4\phi}{\partial x^4}+g_2\left(\frac{\partial\phi}{\partial x}\right)^2\frac{\partial^2\phi}{\partial x^2}, \tag{9}$$

where $D_2=1+c_1c_2, g_1=c_2-c_1$, $D_4=c_1{}^2(1+c_2{}^2)/2$ and $g_2=-2(1+c_2{}^2)$. Eq.(9) exhibits a phase turbulence similar to Eq.(8) above a critical value of D_2. When D_2 is decreased below the critical value, the phase gradient becomes very large at some points and it leads to divergence. In the complex Ginzburg-Landau equation the large phase gradient induces depression of the amplitude and a phase-slip occurs. Below the critical value the phase-slip occurs randomly and the fluctuation of amplitude is large. We call it the amplitude turbulence. Figure 3 is a phase diagram of the complex Ginzburg-Landau equation. The line BF indicates the Benjamin-Feir instability, i.e., $1+c_1c_2=0$, and the line PA expresses the transition from the phase turbulence to the amplitude turbulence. The line BD is a boundary line of the amplitude turbulence, i.e., the amplitude turbulence disappears in the left of the line BD. Between the line BD and the line PA the phase turbulence and the amplitude turbulence coexist, i.e., the complex Ginzburg-Landau equation is bistable in the parameter region. When c_1 is larger than -1.4, the transition from the phase turbulence to the amplitude turbulence is discontinuous. On the other hand when c_1 is smaller than -1.4, the transition is continuous. Namely the frequency of the phase-slip increases continuously from 0. Figure 4 shows two snap-shot patterns of $|W|$ at t=2230 (Fig.4a) and t=2235 (Fig.4b) for c_1=-2.0 and c_2=0.97, which parameter is near the transition line PA. Near the transition line PA below $c_1 \sim -1.4$ a spatio-temporal intermittency appears. For the most part of time the system exhibits a phase turbulence like Fig.4b. Intermittently the phase gradient becomes very large at some points and the amplitude is depressed at the points like Fig.4a and then the phase slip occurs. After the processes the system returns to a phase turbulence state.

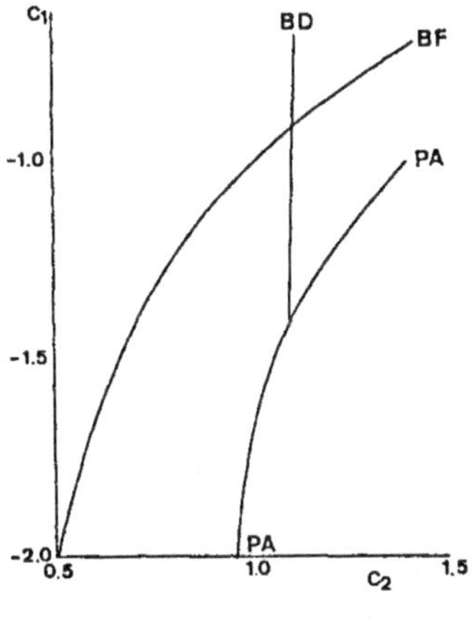

Fig.3

Phase diagram of the complex Ginzburg-Landau equation.

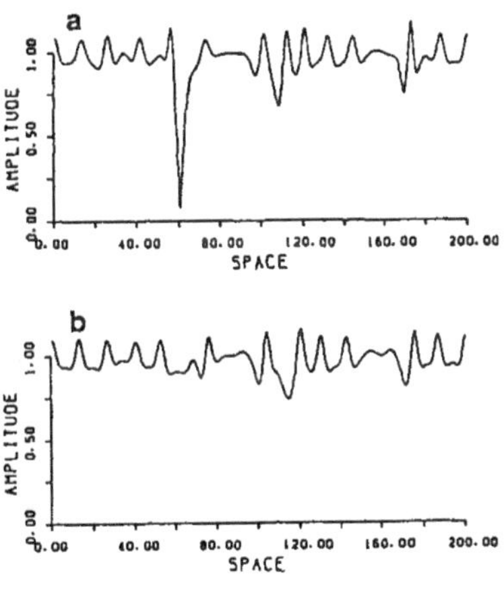

Fig.4 intermittency

References

1. A.C.Newell and J.A.Whitehead, J.Fluid Mech.38(1969)279
2. H.Sakaguchi, Prog.Theor.Phys.84(1990)792
3. H.Sakaguchi, Prog.Theor.Phys.85(1991)927
4. H.Sakaguchi, Prog.Theor.Phys.86(1991)759
5. Y.Kuramoto, Chemical Oscillation, Waves and Turbulence (Springer-Verlag,1984)

DEFECT CHAOS IN 2D ANISOTROPIC SYSTEMS

SHIN-ICHI SASA

Department of Physics, Kyoto University, Kyoto, 606 Japan

ABSTRACT

We present a simple mathematical model which exhibits a transition from stationary periodic patterns to defect chaos in two-dimensional anisotropic systems. Pattern evolution and statistical properties in the regime of defect chaos are also discussed with the aid of numerical simulations of our model equation.

Let us consider a two-dimensional system in which spatial periodic patterns are formed. These systems often exhibit weakly disordered structures such as topological defects i.e. the point singularities in the phase fields associated with the periodic patterns. In the first stage of the ordering process, the periodic patterns are locally formed and a defect appears at the point connecting two periodic patterns with identical orientation but slightly different wavelengths. Since defects are topologically stable in two-dimensional systems, the motion of a defect is well-defined until it vanishes on collision with an anti-defect i.e. the defect with an opposite topological number. Therefore, defects should play important roles in the late stage of ordering process. On the other hand, defects can also be created spontaneously through the instability of periodic patterns. In that case, if unstable periodic patterns are reorganized after the defect creation, the system still retains its unstable nature, and such a state is called *defect chaos* or *defect turbulence.*

Electro-hydrodynamic convection (EHC) of nematic liquid crystals provides an example of defect chaos,[1] which has recently been studied experimentally and and theoretically to a considerable extent. Theoretical approaches are twofold. One is called the method of amplitude equation,[2] and the other approach is a phenomenological one in which some important effects are heuristically taken into the equation. In the latter approach, we regard EHC as two-dimensional pattern forming systems with anisotropy. Pesch and Kramer proposed a simple model equation,[3] an anisotropic version of Swift-Hohenberg equation, in order to describe periodic pattern formations in such systems. Although the Pesch-Kramer model can explain some behaviors such as the normal roll-oblique roll transition, their model can not exhibit the transition to defect chaos. We extended the model on an idea that a large scale horizontal flow should play an important role in instabilities of stationary periodic patterns,[4] and succeeded in describing defect chaos.[5]

Our model equation is written as

$$\dot{W} + (\vec{U} \cdot \vec{\nabla})W = RW - |W|^2 W + \hat{D}W, \tag{1a}$$

$$\hat{D} = -(1 + \triangle)^2 - \eta_1 \partial_y{}^4 - 2\eta_2 \partial_x{}^2 \partial_y{}^2.$$

$$U_i = \sum_j h_{ij}(W^* \partial_i \partial_j^2 W + c.c) - \partial_i p, \tag{1b}$$

$$\vec{\nabla} \cdot \vec{U} = 0. \tag{1c}$$

Here $\triangle = \partial_x{}^2 + \partial_y{}^2$, $\vec{\nabla} = (\partial_x, \partial_y)$, η_1 and η_2 are anisotropic parameters, and R is a control parameter. W is supposed to be a complex field so that the analysis of the stability of stationary roll solutions may be easy. One may interpret Re(W) to be the vertical velocity. Considering the situation in which the formation of rolls with $x-$direction is inhibited, we see that the condition $\eta_1 > 0$ must be satisfied. Pesch and Kramer showed that $\eta_2 = 0$ corresponds to the Lifshitz point and that $\eta_2 > 0$ holds for the normal roll regime. Since we are concerned with the normal roll regime, $\eta_2 > 0$ will be assumed below.

When $R > 0$, Eq.(1) has a family of stationary periodic solutions. Corresponding to values of parameters h_{ij} and η_2, we can obtain several types of stability diagrams for these solutions. It is then shown that the relevant instability of roll solutions is skewed-varicose instability or zigzag instability. We choose parameter values by the requirement that roll solutions should be unstable beyond a finite distance apart from the onset of convection. In this article, we fixed the parameters such as $\eta_1 = 0.6$, $\eta_2 = 0.3$, and $h_{ij} = -1.0\delta_{ij}$. The roll solution with wavenumber $k = 1$ becomes unstable through the zigzag instability at $R_c = 0.3$.

Numerical simulations of Eq.(1) with a system size $L = 63$ were carried out under the realistic boundary conditions. For $R = 1.0$, a roll pattern starts to deform spontaneously and slowly. We can see the formation of defects after the appearance of the zigzag pattern at an early stage. A dynamic steady state is established after a number of creations and annihilations of defects. A time evolution of patterns in the steady state is seen in Fig. 1. This shows that the motion and spontaneous creation of defects are the most important ingredients in our pattern dynamics. The temporal change of the number of defects shows aperiodicity. We thus identify such a state as defect chaos.

(1) (5)

(2) (6)

(3) (7)

(4) (8)

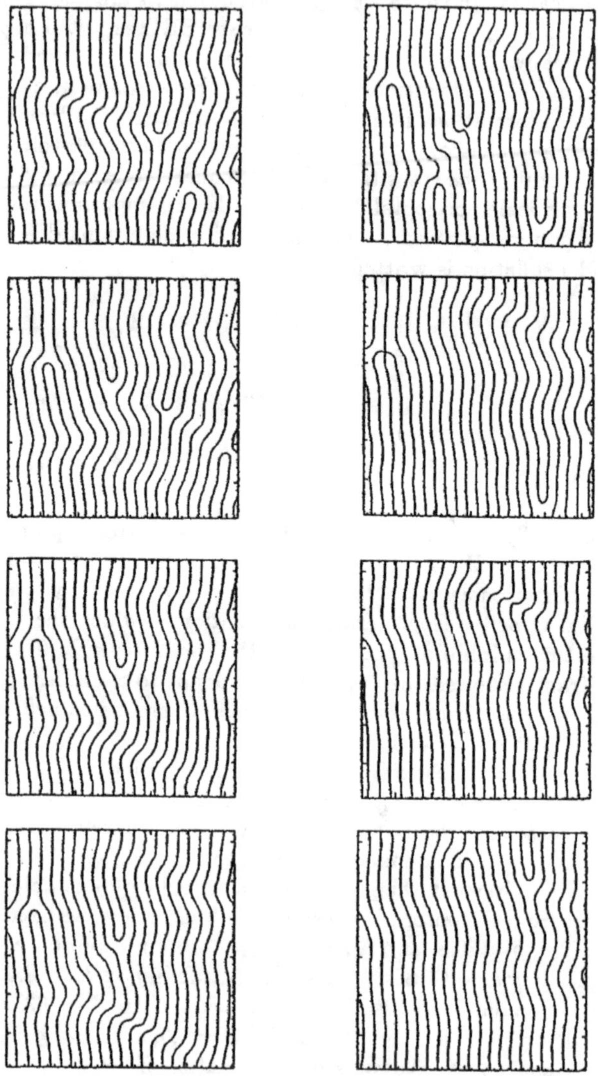

Fig. 1 Pattern evolutions at every 10 time unit; (1) → (8). The contours $Re(W) = 0$.

We will now argue some statistical properties of the temporal variation of the average amplitude over the space which is denoted by $A(t)$. We expect that the low frequency property of the quantity $A(t)$ will be identical to that of the number of the defects. In Fig. 2, the spectra $S(w)$ of the quantity $A(t)$ are shown for $R = 1.0$ and $R = 1.5$. These spectra seem to have two characteristic regimes:

$$\begin{cases} S(w) \sim const. & \text{for } w \ll w_1; \\ S(w) \sim w^{-\alpha} & \text{for } w_1 \ll w \ll w_2. \end{cases} \tag{2}$$

Here $2 < \alpha < 3$ is satisfied.

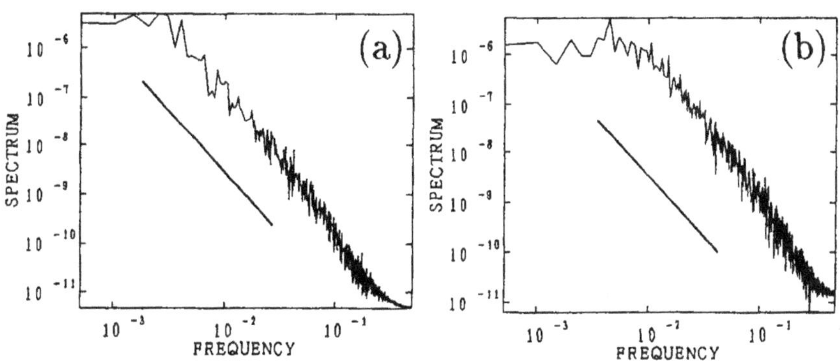

Fig. 2 Log-log plots of spectra $S(w)$ for (a)$R = 1.0$ and (b)$R = 1.5$. These spectra were calculated from the average over five samples. The straight lines show $S(w) \sim w^{5/2}$.

We find further that the cross-over frequency w_1 between the two regimes, which corresponds to the inverse of the correlation time, becomes large with R. This suggests that the correlation time τ diverges at the onset of the zigzag instability such as $\tau \sim \epsilon^{-\alpha}$, where ϵ is a distance from the onset of the zigzag instability.

Let us discuss the physical interpretation for these spectral behavior. The time scale for a pair-annihilation of defects τ_a is estimated as $\tau_a \sim \xi/\bar{v}$, where ξ corresponds to average distance between two adjacent defects, and $\bar{v}$ denotes average magnitude of velocity of a moving defect. Since a roll pattern sustains zigzag-unstable nature, $\bar{v}$ may be proportional to magnitude of phase deformation $\partial_y\phi$ in the form $\bar{v} \sim D_\perp \partial_y \phi$, where $D_\perp$ is a phase diffusion constant in the longitudinal direction of rolls. We thus obtain $\tau_a \sim \xi\epsilon^{-3/2}$ by using the relation $D_\perp \sim \epsilon$ and $\partial_y\phi \sim \epsilon^{1/2}$. On the other hand, the time scale for defect creations τ_c will be given as $\tau_c \sim \xi^2/D_\perp \sim \xi^2\epsilon^{-1}$ from the dimensional analysis. We assume the relation $\tau \sim \tau_c \sim \tau_a$ which means that the correlation time τ is determined by the balance between creation and annihilation processes of defects. This leads

$$\xi \sim \epsilon^{-1/2} \qquad \text{and} \qquad \tau \sim \epsilon^{-2}. \tag{3}$$

These results have not been confirmed by numerical simulations yet.

A number of problems on defect chaos remain to be solved, of which the following two are particularly important: The first question is the value of the control parameter R at which the transition to the chaotic state occurs. This question may be related with the problem of spatio-temporal intermittent transition.[6] The second problem is how to describe the motions of defects in the defect chaos regime. Although many researches on the defect motions were presented,[7] they are concerned with nonchaotic systems.

The author thanks Y. Kuramoto, S. Kai, Y. Sawada, M. Sano and S. Nasuno for stimulating discussions.

References

1. S. Kai and K. Hirakawa, *Prog.Theor.Phys.Suppl.* **64** (1978) 212;
S. Kai, N. Chizumi, and M. Kohno, *Phys.Rev.* **A40** (1989) 6554;
S. Kai and W. Zimmermann, *Prog.Theor.Phys.Suppl.* **99** (1989) 458;
S. Nasuno and S. Kai, *Europhys.Lett.*14 (1991) 779;
E. Braun, S. Rasenat and V. Steinberg, *Europhys.Lett.*15 (1991) 597.
2. E. Bodenschatz, M. Kaiser, L. Kramer, W. Pesch, A. Weber and W. Zimmermann, in *New Trends in Nonlinear Dynamics and Pattern Forming Phenomena*, ed. P. Coullet and P. Huerre, (NATO ASI Series, Plenum Press, 1989);
Q. Feng, W. Pesch and L. Kramer, preprint;
M. kaiser, W. Pesch and E. Bodenschatz, preprint.
3. W. Pesch and L.Kramer, *Z.Phys.*B63 (1986) 121.
4. E.D. Siggia and A. Zippelius, *Phys.Rev.Lett.* **47** (1981) 835;
M.C. Cross and A.C. Newell, *Physica*10D (1984) 299.
5. S. Sasa, *Prog.Theor.Phys.* **83** (1990) 824.
6. K. Kaneko, *Prog.Theor.Phys.* **74** (1985) 1033;
H. Chate and P. Manneville, *Phys.Rev.Lett.* **58** (1987) 112.
7 E.D. Siggia and A. Zippelius, *Phys.Rev.* **A24** (1981) 1036;
Y. Pomeau, S. Zaleski and P. Manneville, *Phys.Rev.* **A27** (1983) 2710;
K. Kawasaki, *Prog.Theor.Phys.Suppl.* **80** (1984) 123;
E. Bodenschatz, W. Pesch and L. Kramer, *Physica*D32 (1988) 135.

NUMERICAL MODELING OF CONVECTIVE MOTION IN GRANULAR MATERIALS

Y-h. TAGUCHI
Department of physics, Tokyo Institute of Technology
Oh-okayama, Meguro-ku, Tokyo 152, Japan

ABSTRACT
A new modeling of convective motion in granular materials is proposed. Pseudo-particles having visco-elastic interaction are introduced in order to model the set of granular materials. The dependence of the convection upon the interaction and the geometry of vessels is discussed.

1 Introduction

Convective motion is one of the important topics in the pattern formation[1], as you see in this conference. In this paper, I would like to report a new origin of convective motion in granular materials which is found by Faraday [2] for the first time. This phenomenon has been researched recently by many physicists [3, 4, 5, 6]. Most of them are experimental works where glass spheres with the diameter $\simeq 1$ mm are used as granular materials. They are packed in a small rectangular vessel having horizontal geometry 10 cm $\times$ a few cm with the depth of a few cm. This vessel is shaken strongly vertically, typically by the loud speaker with a frequency of $10 \sim 10^2$ Hz. The control parameter is an acceleration amplitude Γ, which is equivalent to $b\omega_0^2$ if the vertical motion of bottom obeys $b \cos \omega_0 t$ as a function of time t. If Γ exceeds some critical value Γ_c which is typically a little bit larger than the gravity amplitude, a heap is constructed on the granular surface. Accompanying with it, the convective motion starts inside the granular body. The direction of flow is upwards at the center of bottom and downwards beside the wall. Typically they have a single heap having its top at the center of the vessel.

2 Model

In order to explain the phenomena, we model the granular body as a set of pseudo-particles obeying the equation

$$\ddot{\mathbf{x}}_i = -\sum_{j=1}^{N} \theta(a- \mid \mathbf{x}_i - \mathbf{x}_j \mid) \left(k(\mathbf{x}_i - \mathbf{x}_j - a\frac{\mathbf{x}_i - \mathbf{x}_j}{\mid \mathbf{x}_i - \mathbf{x}_j \mid}) + \eta(\mathbf{v}_i - \mathbf{v}_j) \right) - \mathbf{g}, \qquad (1)$$

where $\theta(x)$ is a step function, N is a total number of particles, $\mathbf{x}_i$ is the position vector of the i-th particle. k and η are the elastic constant and the viscosity coefficient respectively. a is a diameter of a particle, $\mathbf{v}$ is velocity. It introduces the interaction between particles which acts only when they keep contact. If two particles collide with each other in front, they have effective coefficient of restitution $e = \exp(-\eta\pi/\omega)$ and collision time $\Delta t = \pi/\omega$ and $\omega = \sqrt{2k - \eta^2}$. $\mathbf{g}$ is a gravity acceleration. Here the meaning of pseudo-particle is that each particle does not correspond to a granular particle but represents some set of large number of particles. Therefor the 'diameter' a is not that of each granular particle. Elasticity and viscosity are not those of individual particle but something coarse grained. These phenomenological physical property should be measured in the experiment like kinetic viscosity in Navier-Stokes equation.

In order to see the behavior of the modeling, we integrate Eq. 1 with Euler scheme. The time resolution δt is decided at each step such that $\max_i \mid x_i(t+\delta t) - x_i(t) \mid = \Delta d$ where Δd is given properly. First we do two-dimensional calculations. A large number of pseudo-particles are put into the a vessel having horizontal fixed dimension and infinite vertical depth. The interaction between particles and bottom is pure elastic one. When a particle 'collide' with the wall, elastic interaction is the same as that between particles but viscous interaction can take different value as an additional parameter. If the viscous interaction with walls is much smaller than those between particles, the direction of convection is opposite to that of experiment. In order to get the convection pattern which is similar to the experiment, we must increase the viscous interaction with walls large enough [7]. In the following, we choose the large enough viscous interaction with walls to reproduce the direction of convection which is coincident to that in experiments.

In this geometry, the modeling exhibits[8] the phase transition from the rest state to the convective state and surface fluidization which is also observed in the experiments[5]. The scenario to cause the convection is the followings. During $\Gamma \cos\omega_0 t > -g$, powders are packed into the vessel strongly. Therefore they keep contacts with each other and behave like visco-elastic body. This causes not only vertical stress but also horizontal stress. On the other hands, when $\Gamma \cos\omega_0 t < -g$, the vertical stress is released quickly. However, horizontal strain remains because the side wall suppress the horizontal movement. This causes the hump at the center of the surface. Vacancies formed under the hump are filled with the flow from the surrounding particles. The hump is destroyed by the flow along the slope and particles flow down toward sidewalls. The whole process transfers the powder at the bottom upwards, and the powder beside wall fall into the bottom. Total flux is a global transfer of powders which is nothing but convective motion.

3 Dependence upon Interactions

In order to see the dependence upon parameters, we simulate the model with two fixed Δt, 0.5 and 0.9. Another control parameter is e. Viscocity between walls and particles is fixed large enough. The parameters in Eq. 1 are $N = 100, a = 2.0, b = 1.1, \omega_0 = \pi/3, (\Gamma \simeq 1.15), g = 1.0$, and horizontal dimension of the vessel is 26. Here we consider the cell-to-cell flux $\mathcal{J}$ [8] of particles which represents the strength of the convective motion. In order to calculate $\mathcal{J}$, whole space in the vessel is divided into square cells having a dimension of a. And if a particle moves from a cell to another cell, a counter records the number of particle hoppings and directions. $\mathcal{J}$ is calculated by averaging over the record. Fig. 1 shows the dependence of $\mathcal{J}$ upon Δt and e. Although mainly Δd is set to be 0.01, we show some data gotten when

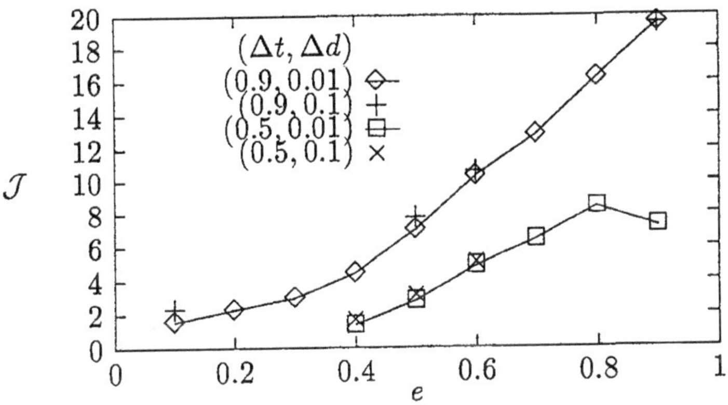

Figure 1: The dependence of $\mathcal{J}$ upon $\Delta t, e$.

$\Delta d = 0.1$ to check convergence. One can see $\mathcal{J}$ is almost independent on Δd. Roughly speaking, $\mathcal{J}$ increases monotonously with Δt and e. This is very natural if one reminds the scenario to the convection. Increasing Δt enable the powders to keep larger elastic energy which causes the convective motion. And if we increase e then the loss of elastic energy is decreased, which results in the increasing of the strength of the convection.

4 Dependence upon the Geometry of the Vessel

Next we report the dependence upon the geometry of the vessel. As reported before[7, 8], the aspect ratio is taken properly, two roll convection is obtained. Then what happens with the different geometry of vessel? Fig. 2 shows the convection pattern which is obtained under too small aspect ratio. One of two rolls vanishes and only one roll remains. Perhaps this corresponds to the one roll convection observed in the experiment[4]. Contrary, for much larger aspect ratio, the convection pattern is shown in Fig. 3. This time clear convection roll vanishes but some coherent

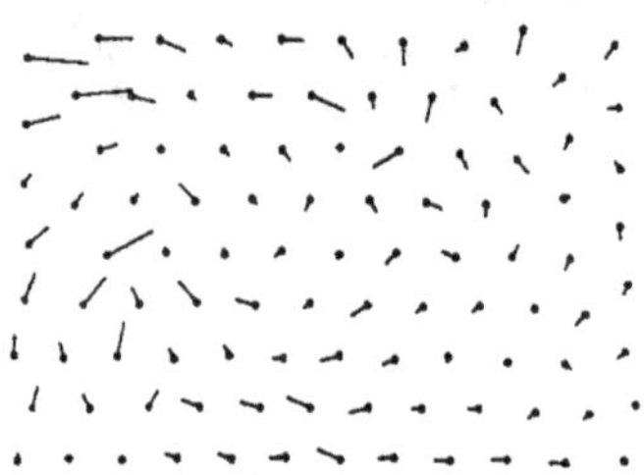

Figure 2: Convection pattern for small aspect ratio. The horizontal dimension of the vessel is 21, $N = 100, a = 2.0, b = 1.1, \omega_0 = \pi/3, (\Gamma \simeq 1.15), \Delta t = 0.5, e = 0.5$. No friction between walls and particles.

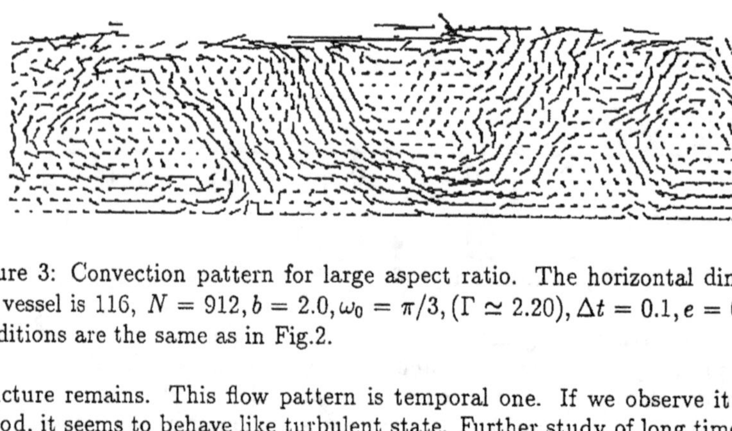

Figure 3: Convection pattern for large aspect ratio. The horizontal dimension of the vessel is 116, $N = 912, b = 2.0, \omega_0 = \pi/3, (\Gamma \simeq 2.20), \Delta t = 0.1, e = 0.8$. Other conditions are the same as in Fig.2.

structure remains. This flow pattern is temporal one. If we observe it over long period, it seems to behave like turbulent state. Further study of long time behavior of convective motion is on progress.

5 Summary

In this paper, we propose a new modeling of convective motion in granular materials. This exhibits the transition from the rest state to convective state. It also reveals surface fluidization phenomena. The dependence of phenomena upon interactions and geometry of vessels are discussed. The convection becomes stronger with increase of Δt and e. And changing aspect-ratio we observe the one-roll configuration and many-roll configurations.

References

[1] P. Manneville, *Dissipative Structure and Weak Turbulence*, (Academic Press,San Diego,1990)

[2] M. Faraday, *Phil. Trans. R. Soc. Lond.* **52**, (1831) 299.

[3] C. Laroche,S. Douady and S. Fauve, *J. Phys. (Paris)* **50**, (1989) 699.

[4] P. Evesque and J. Rajchenbach, *Phys. Rev. Lett.* **61**, (1989) 44.

[5] P. Evesque, E. Szmayula and J. P. Denis, *Europhys. Lett.* **12** (1990) 623.

[6] J. Rajchenbach, *Europhys. Lett.* **16**, (1991) 149.

[7] Y-h. Taguchi, in proceedings of *Slow Dynamics in Condensed Matter*, Fukuoka, Japan, November 1991, to be published.

[8] Y-h. Taguchi, submitted to *Phys. Rev. Lett.*.

Phase-Slip Dynamics in One-Dimensional Convective Systems

Hideki YAMAZAKI† Michael I.TRIBELSKY[1‡]

Satoru NASUNO‡ and Shoichi KAI‡

† Fukuoka Institute of Technology Wajiro Higashi Fukuoka 811-02 Japan
‡ Kyushyu Institute of Technology Tobata Kitakyushu 804 Japan

Abstract

The phase slip process for periodic pattern-forming systems is studeied experimentally and theoretically. The phase slip is a phenomenon which shows instantaneous variation of a phase of a order parameter when and where an amplitude of the order parameter vanishes. The general analytical results of the phase slip process are supplimented by a computer simulation for one-dimensional Ginzburg-Landau equation. Experimental results using one-dimensional electrohydrodynamic convective cells show good agreement with theoretical predictions; the amount of phase shift during one phase slip process, the evolution of a characteristic width of a phase slip core and its value which shows the amplitude is minimum.

1 Introduction

The phsae slip phenomenon is a very common pattern selection process from unstable to stable state, which frequently occurs as instantaneous variation of the phase of the order parameter on a multiple of 2π for one-dimensional (1D) systems. This phenomenon occurs at the point and the moment which the amplitude of an order parameter vanishes. Through the phase slip moment one (or multiple) pair of element is nucleated or annihilated. Such phenomenon has known in various systems, espatially in periodic pattern-forming ones[1−6]. In pattern-forming systems with translational symmetry, the phase slip phenomenon is deeply related to the

[1‡] Present address: *IZMET, P.O. Box 512 SU-125057 Moscow USSR*

Eckhaus instability[7]. The pattern forming system can be observed also in two- (2D) or three-dimensional (3D) systems, however, we restrict our interest within 1D systems. Various experiments have been performed on the Eckhaus instability and the evolution of its dynamics[1-6]. The theory of the phenomenon has also been developed in detail[8-12]. The main feature of the phenonenon in various 1D systems is that a long-wavelength modulation of the periodic pattern leads to the creation or destraction of pair-element. Experimental results showed very good qualitative agreement with theoretical predictions.

Nevertheless, there is no work of the detailed comparison of theoretical results with experimental results from the point of the phase slip phenomenon in 1D systems. Some important facts therefore are yet unknown.[13-16] For instance, while the phase change during one phase slip process in general may be equal to any multiple of 2π, computer simulations[12] show that usually it equals its lowest value 2π. Further it is unknown what kind of physical quantities do characterize the dynamics of the phase slip process. Thus, detailed dynamics of the phase slip is not known in theory and experiment. In this paper we develop the experiment and the general analysis of the phase slip process in order to fill these gaps.

In Sec. 2 we present general analysis of the phase slip process and describe our computer simulation for 1D Ginzburg-Landu equation. In Sec. 3 we describe briefly our experiment. In this study, the electrohydrodynamic instability (EHD) is employed to test the validity of our theoretical consideration in a real system. We used a special convective cell with one dimension geometry. In Sec.4 we present results and discussion of the experiment and the simulation. The conclusion is presented in Sec. 5.

2 Theoretical

Many dynamical phenomena can be described by the following equation,

$$\psi_t = \hat{D}\psi, \tag{1}$$

where $\psi(\vec{r},t)$is a complex function of the space coodinate $\vec{r}$ and $\hat{D}$ is a diffential operator with respect to $\vec{r}$. We have here introduced two real functions, an amplitude R and a phase ϕ with $\psi = R\exp(i\phi)$ and already disscussed elsewhere[15]. We will again however describe briefly on it to understand easier. The main point of our consideration on these is the assumption that on (x,t) plane a phase slip point (t_{ps}, x_{ps}) is a singular one if R and ϕ are taken as independent variables. If we introduce $u = \mathrm{Re}\psi$ and $v = \mathrm{Im}\psi$ instead of them, the phase slip point becomes a regular one. In accordance with this, close to the phase slip point the function $u(x,t)$ and $v(x,t)$ may be expanded into powers of x and t as follows:

$$u = a_u t + b_u x + c_u x^2/2 + \ldots \tag{2}$$

$$v = a_v t + b_v x + c_v x^2/2 + \ldots, \tag{3}$$

Here $a_{u,v}$, $b_{u,v}$ and $c_{u,v}$ are constants.

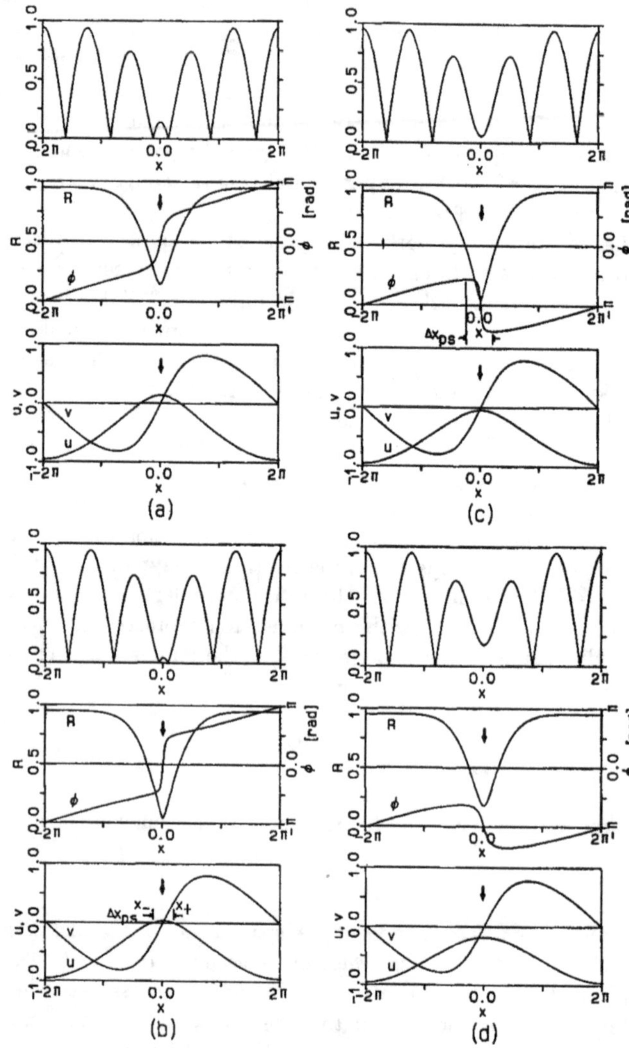

Fig.1. Temporal change of order parameter ψ, spatial variation of amplitude R, phase ϕ of complex order parameter ψ and, real(u) and imaginary(v) parts of ψ obtained as results of numeraical integration (see ref.15). The profile of the order parameter is obtained from an equation $R(x)\cos[q_c x + \phi(x)]$, where $q_c = 1$. (a) $t = -0.8$, (b) $t = 0.2$ (just before the phase slip moment) the value of phase shift equals 2π, (c) $t = 0.2$ (just after the phase slip moment), the value fo phase shift equals zero. (d) $t = 0.7$.

Let us make attention to a region which shows rapid spatio-temporal variation of the phase near the phase slip point $x = x_{ps}$; the phase slip core. A case in which the system has the invariance with respect to arbitarary rotations in the complex plane, the term $b_u x$ may be eliminated from Eq.(2). As a result the curve $u(x, t_{ps})$ becomes tangent to the x-axis. In this case a variation of the value of minimun amplitude, extreme of u(x,t), equals to $|a_u t|$ independently of the sign t; we put t equalls zero at $t = t_{ps}$. Further a variation of x from $x = -\infty$ to $x = \infty$ at any fixed t, depending on the sign of t, results either in (i) the phase shift of 2π, or (ii) gives no phase shift at all (see e.g. Fig.1).

Let us introduce now for the case (i) two quantities $x_{\pm}(t)$ according to the relations $\phi(x_{\pm}, t) = \pm\pi/2$, so that the phase shift between these two points (equal to π) is the half of the total one (2π). A characteristic width of the phase slip core Δx_{ps} may be determined as the distance between x_+ and x_-, see Fig.1a. For the case (ii) the same quantity may be introduced as the distance between the local points of the maximaum and minmum of the curve $\phi(x)$ inside the phsae slip core (Fig.1b). The characteristic width Δx_{ps} for the phase silp in both cases equals to $|a_u t/c_u|^{1/2}$ independently of the sign of t.

Thus we obtained that in the general case phase slip process gives rise to the phase shift equal to 2π. Multiples greater than that of 2π are possible only if at least two coeffients in Eqs.(2)-(3) vanish simultaneously. Then the heigher terms in the expansion must be taken into account.

To illustrate these genaral speculations, a numerical integration was carried out for one-dimensional (1D) Ginzburge-Landau equation, $\psi_t = \psi_{xx} + (1 - |\psi|^2)\psi$, having finite segment $-2\pi \leq x \leq 2\pi$ with periodic boundary conditions. The computer simulation was done by a work station SUN3-260 (SUN Microsystems Inc.), using an explicit integration method with second-order Runge-Kutta method. The integration was done using a grid of 256 points with a grid spacing $\Delta x = 4\pi/256$. The initail conditions were the same as those in the previous publication.[15]

3 Expermental

We employ the electrohydrodynamic convection (EHD) of a liquid crystal as a real experimental system. The details of an experimental setup employed have been already described in Ref.14, 17 and will be also in 18. We used here a 1D convective cell, that is, the thickness of the cell was 50 μm and the area of transparent electrods was 102 μm(y-direction) $\times$ 1 cm(x-direction) where the initial director alignment is in the x direction; the aspect ratios were $\Gamma_y = 2$ and $\Gamma_x = 100$. To obtain uniform aligment of periodic rolls in the x-direction the electrodes were coated with polyvinyl alchohol and then rubbed in the x-direction. We used nematic liquid crystal N-p-methoxybenzylidene-p-butylaniline (MBBA) and doped it 0.01w% of tetra-n-butylammonium bromide. The convection of the fluid was driven an ac field V of low frequency f across the layer. In order to control the wavenumber of the roll pattern at a given value of the control parameter, we used voltage-frequency

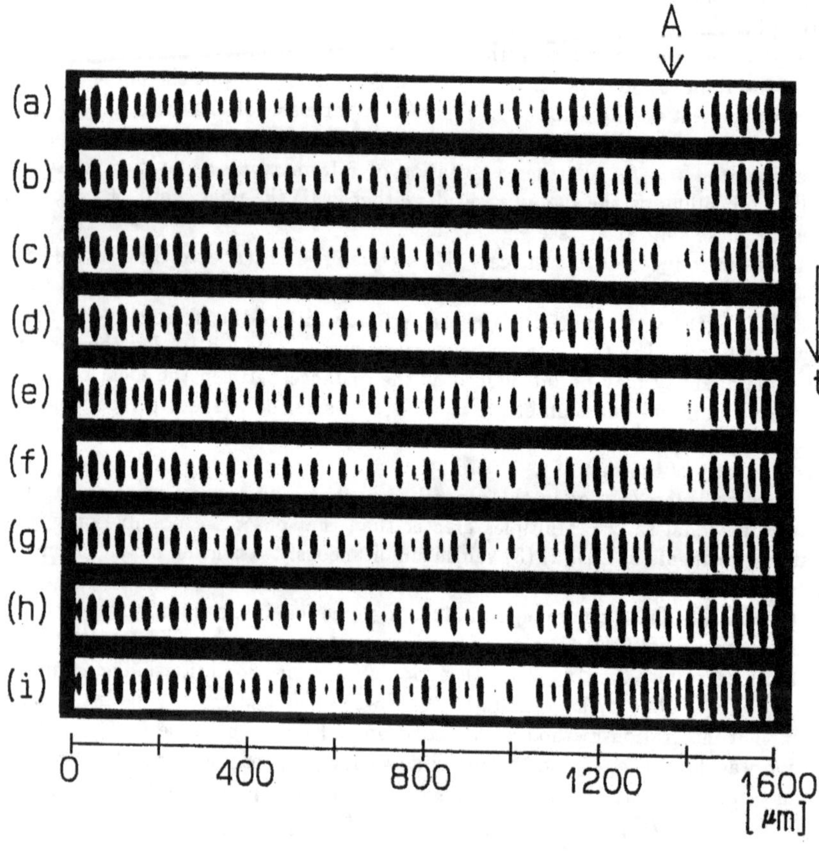

Fig.2. Temporal sequence of a phase slip process in EHD (experiment). Each figure was taken every four seconds from top to bottom. One pair of rolls is nucleated after the phase slip.

$(V-f)$ jump method. The detail of this method and how to realize the pattern selection process such as the Eckhaus instability have been already described in refs.14 and 18. The temperature of the cell was kept constant at 30.00 ± 0.01 °C. The roll patterns were observed under a microscope with polarized light(Nikon OPTIPHOTO). The image of a charge-coupled-device TV camera was recorded by a professonal video editing recorder. The time evolution of the phase slip process was first recorded onto the magnetic tape by the VTR. Then the recorded image was digitized by the high resolution image processer (512 pixels/line and 256 gray levels; ADS Co. PIP-4000). The amplitude R, phase ϕ, real part u and imagimay one v of R were calculated using similar method with that of Goren et. al[13]. The minimum value of R and the characteristic width of a phase slip core Δx_{ps} were obtained after the calculation from experimental data. The analysis of digitized image was done by a personal computer(NEC PC-9801).

4 Results and discussion

Figure 2 shows a time sequence of the phase slip process observed experimentally. As the phase slip process starts, some location of rolls (for example A in Fig.2) begins to vague ($t=-8s$) and becomes the phase slip core. Such a location is not defined and randomely appear. The clearness of the pattern decreases gradually as the phase slip moment t_{ps} ($t=0s$) approaches. After the phase slip a pair of rolls nucleates at the region A ($t=8s$). This process is clearly recognized at two points A and B in a spatio-temporal image in Fig.3, where two phase slips are observed while in Fig.2 the location A is shown. Another case that one pair of rolls is annihilated can be also observed depending on which state is stabler finally.

Time interval between two phase slips is about 30 s and the effective period of the phase slip process is about ±10 s in both cases. The distance between both is about 360 μm and the correlation length here is about 240 μm. We observe no deformation roll already before the second phase slip starts. Thus the influence of the first phase slip on the second one could be very small. In the Eckhaus instability, these happen simultaneously in time and periodically in space with certain periodicity (Eckhaus mode)[14,18]. We pick the first phase slip and describe its details to avoid some influence between a successive phase slip phenomena.

Figure 4 shows an evolution of the roll pattern from initially prepared to finally stable one, obtained from the experiments; ψ, amplitude R, phase ϕ and real(u) and imagimary part(v) of R. The amplitude $R(x)$ did not change even if the curves $u(x)$ and $v(x)$ were rotated with any angle on the u-v plane as described in ref.15. If the curves are rotated, wave-forms of u and v differ from ones without rotation, so that u is not tangent to the x-axis at the minimun amplitude point and v does not intesect the x-axis at that point. The rotation angle in the theoretical case equals zero in Fig.1. Therefore after the rotation angle in the experimental data is calculated by $\tan^{-1}(u/v)$ at the minimun amplitude point, all wave-forms u and v in Fig.4 are drawn by compensation of the rotation to zero.

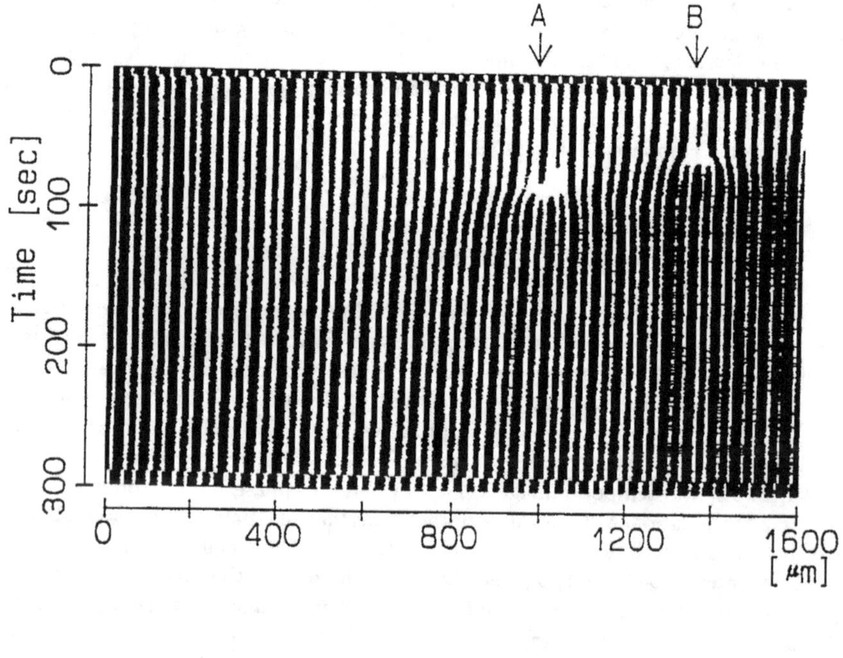

Fig. 3. Spatio-temporal change of image in Fig.2. The origins in time and in space are arbitrary. White lines show bright lines in the image of a roll patten. Interval of two white lines shows the width of one roll. Near $t = 60s$ we can see one phase slip phenomenon. After the phase slip one pair of rolls is nucleated. The second phase slip phenomenon is seen near $t = 90s$. We discuss the first one in text.

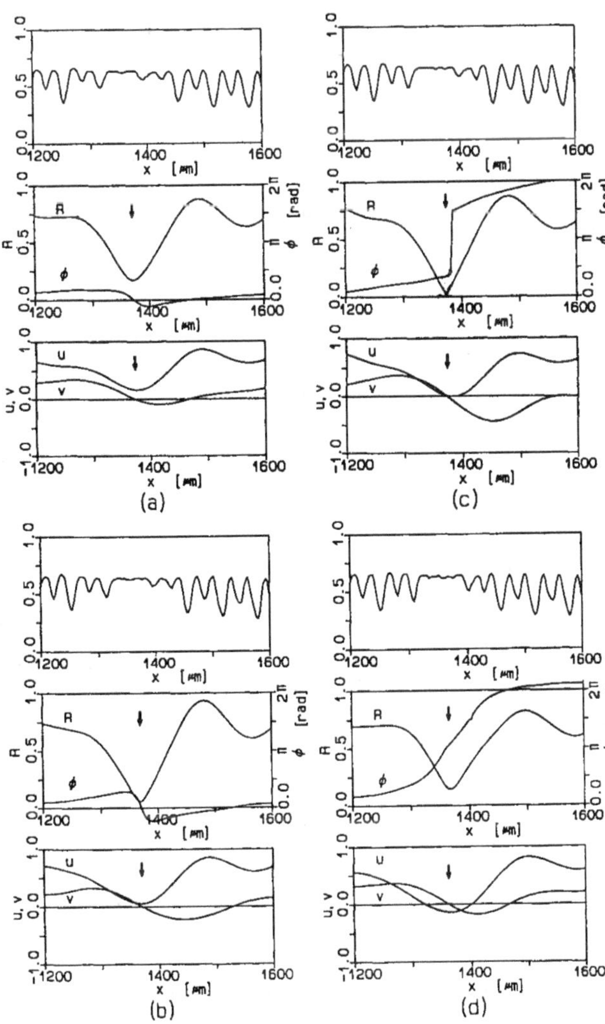

Fig.4. Temporal change of the roll pattern after the jump from an initially prepared pattern to a different pattern stabler, ψ, spatial variation of the amplitude R, phase ϕ of the complex order parameter ψ and, real(u) and imaginary(v) parts of ψ from each set of the above figures, respectively. t=0 means the phase slip time. (a) $t = -2.8s$,(b) $t = -0.8s$ (just before the phase slip moment). Here the phase difference between both edges is zero. (c) $t = 0.2s$ (just after the phase slip moment). The phase difference between both edges is 2π. (d) $t = 2.2s$.

Nearing t_{ps}, the extreme of u approaches to zero at the minimum amplitude point where the lower order term of Eqs.(1) and (2) becomes more important. Just at the phase slip moment u tangents at zero R, then the extreme of u passes over zero. On the other hand the value of v remains zero at the minimum R points (the tangent point of u). The amplitude R is equal to zero at the phase slip moment. When a pair of rolls is created, u moves in opposite direction in a case which a pair of rolls is annihilated (compare Figs.1 and 4). All the aspect of R, u and v in Fig.4 are quite similar to ones in Fig.1 numerically calculated. Figure 1 theoretically shows annihilation process of a pair of rolls by computer simulation and Figure 4 experimentally shows nucleation process of a pair of rolls. Both are essentially the same dynamics. To compare the simulation with the experiment more precisely, the dynamical behavior of its nucleation process is shown in Figs. 5 and 6.

A phase difference between both edges is kept nearly zero at any time before the phase slip moment in the case which one roll-pair is created. This difference changes suddenly from nearly zero to 2π just after the phase slip moment: 2π phase shift occurs as creation of one pair of rolls. The value of the shift 2π is reasonable considering the creation of one pair of rolls. The value 2π is kept at any time after the phase slip moment. On the contrary, decrease of the phase occurs when one pair of rolls is annihilated, i.e. the phase difference is from 2π to zero as shown in Fig.1. During one phase slip process, the absolute value of the phase shift is always equal to 2π as theoretically expected.

As the phase slip moment comes closer, a slope of the phase ϕ increases in the phase slip region and the width of the phase slip core decreases. After the phase slip the distance between zero crossing points of u increases as time passes away from t_{ps}; the width of the phase slip core increases. To see this behavior quantitaively we measured the characteristic width of the phase slip core(see Fig.5) using the method explained in Fig.1. The core size is well proportional to $|t|^{1/2}$ near t_{ps} which is also obtained by the computer simulation[15]. Both slopes before and after t_{ps} are the same.

Figure 6 shows an evolution of the minimum value of R_{min} of the amplitude. R_{min} also relatively fits the thoretical curve ,that is proportional to t. The slope however at $t < t_{ps}$ is smaller than that at $t > t_{ps}$. The slopes are 1.32 and 1.29 for $t < t_{ps}$ and $t > t_{ps}$ respectively. The reason of difference for large R at $t > 0$ between the simulated (dotted line) and the expanded (solid line) results is probably that the expanded one is only available for very small R because higher terms in expansion are not taken into account here. Therefore the simulated results may be more reasonable when R is rather large and well agree with real experimental results, as shown in Fig.6. This aspect of course depends on magnitude of the expansion coefficients. Similarly the difference of the slope is due to the difference of a_u in Eq.(2). However there is no explicit connection of a_u to real physical quantity in EHD. Therefore it is not still clear that the physical reason why the decreasing rate of R_{min} is slower than increasing one. Although the time dependence of R_{min} before the phase slip moment is different from after, one of the characteristic width of the

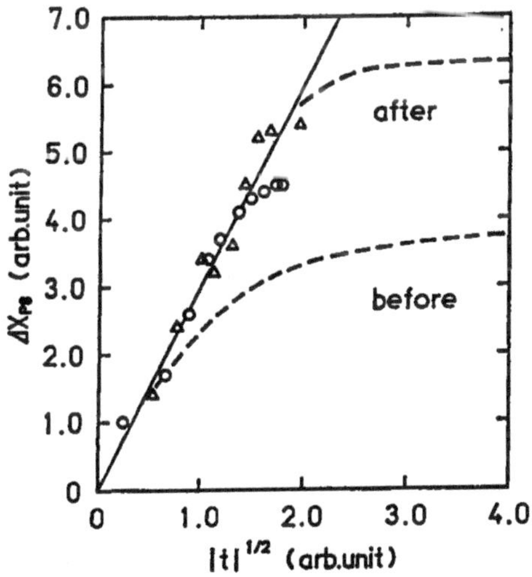

Fig.5. Temporal change of the charateristic width (Δx_{ps}) of a phase slip core. Δ for $t < 0$, $\circ$ for $t > 0$ for experimental data. t_{ps} is shifted into $t = 0$. The solid and the dotted lines respectively show analitical and computational results.

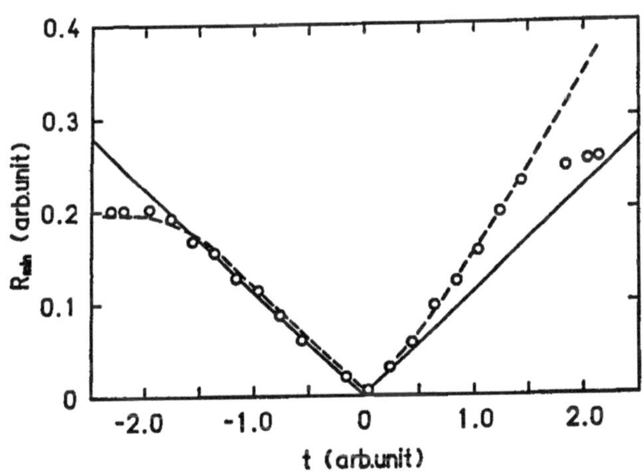

Fig.6. Temporal change of minimun value of amplitude R. The solid line shows theoretical results. Here t in EHD experiment is normalized as shifting t_{ps} into $t = 0$.

phase slip core Δx_{ps} has no difference as shown in Fig.5. This might indicate that our definition of the characteristic width of core is not perfectly appropriate in the experiment and the simulation.

5 Conclusion

We analized experimentally and theoretically the dynamics of the phase slip process in 1D systems. The experiment was made using electrohydrodynamic cell. The temporal changes of the minimum amplitude and the width at the phase slip process in the present experiment quite agree with ones of theoretical analysis. No such a investigation has been done in this viewpoint. We conclude our results as follows. It is proved that the wavenumber selection of periodic pattern-forming systems is made through the phase slip process whose magnitude is 2π, as expected from theoretical prediction. The shift with other magnitude cannot be observed in both theoretical and experimental. The characteristic width of the phase slip core Δx_{ps} changes in proportion to $|t|^{1/2}$. Δx_{ps} after the phase slip has the same tendency of temporal change with one of the before. The minimun amplitude (R_{min}) is propotional to $|t|$ which is also theoretically expected. All these experimantal observations well agree with theoretical predictions. In 2-dimensional system, some non-ideality comes in deviating the experimental results from theoretical one [14]. The dynamical aspect related to the present study in 1D Eckhaus is in progress and will be reported elsewhere[18].

Acknowledgement

We would like to thank Mr. T.Matsuda for experimental assistance. This work is in part supported by the Grant-in-Aid for Scientific Research (No.02452047)from the Ministry of Education, Science and Culture of Japan , and the Kajima Foundation. One of the authors (M.I.T.) wishes to express his appreciation for the visiting scholarship from the Kyushu Institute of Technology.

References

[1] M. Boucif, J. E. Wesfreid and E. Guyon, J. Phys.(Paris) Lett. 45, L413(1984).

[2] M. Lowe and J. P. Gollub, Phys. Rev. Lett. 55, 2575(1985).

[3] W. A. Dominguez-Lerma, D. S. Cannell and G. Ahlers, Phys. Rev. A 34, 4956(1986).

[4] A. Pocheau, V. Croquette, P. Le Gal and C. Poitou, Europhys. Lett. 3, 915(1987).

[5] A. J. Simon, J. Bechoefer and A. Libcharber, Phys. Rev. Lett. 61, 2574(1988).

[6] S. Rasenat, E. Braun and V. Steinberg, Phys. Rev. A 43, 5728(1991).

[7] W. Eckhaus, *Studies in Nonlinear Stability Theory,* Springer, Berlin,(1965).

[8] B. I. Ivlev and N. B. Kopnin, Usp. Fiz. Nauk 142, 435(1984) [Sov. Phys. Usp. 27, 206(1984)]; Adv. Phys. 33, 47(1984).

[9] L. Kramer and W. Zimmermann, Physica D 16, 221(1985).

[10] H. R. Schober, E. Allroth, K. Schoeder and H. Muller-Krumbhaar, Phys. Rev. A 33, 567(1986).

[11] L. Kramer, H. R. Schober and W. Zimmermann, Physica D 31, 212(1988).

[12] S. G. Gol'berg, N. B. Kopnin and M. I. Tribelsky, Zh. Eksp. Theor. Fiz. 94, 289(1988); J. Low. Temp. Phys. 77, 209(1989).

[13] G. Goren, I. Procaccia, S. Rasenat and V. Steinberg, Phys. Rev. Lett. 63, 1237(1989).

[14] S. Nasuno and S. Kai, Europhys. Lett. 14 779(1991).

[15] M.I. Tribelsky, S. Kai and H. Yamazaki, Phys. Rev.A (1992) (in print).

[16] M.I. Tribelsky, S. Kai and H. Yamazaki, Prog.Theor.Phys. 86, 963(1991).

[17] S. Nasuno et al.(in this Proceedings).

[18] S. Nasuno et al.(in preparation).

D: Chaos and Turbulence

ANALYSIS OF THE DIMENSION OF SPACE SERIES AS A METHOD FOR THE INVESTIGATION OF CRITICAL PHENOMENA IN THE GINZBURG–LANDAU MODEL

M.V. Bazhenov, M.I. Rabinovich, A.L. Fabrikant

Institute of Applied Physics, Russian Academy of Sciences
46 Uljanov Str., 603600 Nizhny Novgorod, Russia

Abstract

Spatial disorder in the large box, one–dimensional, complex Ginsburg–Landau problem is investigated quantitatively. The transition from phase to amplitude turbulence is studied in detail. This transition is described by the dimension of the space series, d_s, that estimates the number of normal (independent) medium perturbations forming the chaotic space series. It is found that at a critical point, d_s undergoes a jump whose value is universal, i.e. does not depend on the dimension of the system. Thus the number of perturbations in the medium grows anomalously near the transition point. This behavior is typical for critical phenomena. The space of parameters describing nonlinearity and dispersion is constructed and regions of the existence of different nonstationary regimes are found. In paticular, a new type of pulsations is revealed. There are time–periodic pulsations having a complicated changing spatial structure.

1.The generalized (or complex) Ginzburg–Landau equation (CGLE)

$$\partial\tilde{a}/\partial t = \gamma\tilde{a} + (i\delta - \rho)\tilde{a}|\tilde{a}|^2 + (ik + \nu)\Delta\tilde{a} \tag{1}$$

describes the wave dynamics of many different physical situations (e.g., shear hydrodynamic flows [1], chemical reactions in media with diffusion [2], water [3] and plasma [4] waves, etc.). Its universality has a very simple explanation. This equation can be obtained, independent of the physical origin of the medium or field, as a result of the transition to slow coordinates and time in all cases of a complex amplitude of quasi–monochromatic wave perturbations proportional to $\exp(-iwt + ikx)$. It provides the correct asymptotic behavior when the instability threshold is slightly exceeded in a narrow spectral interval.

The CGLE model is a convenient tool for the investigation of basic phenomena in the nonlinear dynamics of nonequilibrium media. The birth and subsequent evolution of spatio–temporal chaos (turbulence), both in the evolution of a regular initial field distribution and as a result of parameter variation, is one of the principal problems in this field [5].

Numerous computer and analytical investigations have shown that (depending on parameter ratio) both weak turbulence, which is characterized only by irregular phase pulsations, and a regime of strong turbulence may be established in a two–dimensional CGLE system. Localized long–lived structures (spiral vortices) are typical for the regime of strong turbulence. The motion and interaction of these structures determine the properties of the two–dimensional turbulence of the wave field envelope [5–11]. Two-dimensional effects in a CGLE are extremely diverse and attractive. However even in a one–dimensional model the spatio-temporal behavior of the complex amplitude has a great variety of nontrivial effects. Within a one–dimensional model there also exist well-pronounced localized structures whose interaction may lead to different dynamical and turbulent regimes. The investigation of one–dimensional systems is undoubtedly much simpler than two–dimensional system. But even in one dimension much is yet to be understood. In particular, there are no answers to the questions such as: What happens in the phase space of the system when the regime of phase turbulence is replaced by the regime of strong turbulence? What are the governing parameters of this transition and how is it to be described? Our paper is concerned with these problems. Apparently, the information on the dynamics of one–dimensional structures may be useful for the analysis of the properties of stochastic self–modulation in two– and three–dimensional systems.

The architecture of the remainder of this paper is as follows. Section 2 presents an analysis of the peculiarities of localized CGLE solutions in the form of kinks or "holes". These structures are useful in the construction of models for strong turbulence. In Section 3 we propose to use the fractal dimension of the space series as a description of the transformation from various spatio–temporal regimes within CGLE. We also use it to investigate the properties of the "phase turbulence–strong turbulence" transition. Finally, the relation between the fractal dimension of the time series on the one hand, and the space series on the other, is considered in Section 4.

2. The one–dimensional system

$$\partial a/\partial t = Ra - a|a|^2 + \partial^2 a/\partial x^2 + i\alpha\partial^2 a/\partial x^2 + i\beta a|a|^2 \tag{2}$$

with the periodic boundary conditions

$$a(x + 2\pi, t) = a(x, t) \tag{3}$$

is obtained from Eq.(1), for solutions with a spatial period $L/2\pi$, by substituting

$$R = \frac{\gamma L^2}{\nu}, \quad \beta = \frac{\delta}{\rho}, \quad \alpha = \frac{k}{\nu}, \quad a = \tilde{a}L(\rho/\nu)^{1/2} \tag{4}$$

and using new dimensionless time and space coordinates: $t = \nu\, t_{old} \,/\, L^2$ and $x = x_{old} \,/\, L^2$.

Note that the growth rate, R, also determines the space scale of modulation structures, l. Linear stability is compensated by nonlinear dissipation and diffusion, i.e.,

$$R \sim |a|^2 \sim l^{-2} \equiv |a^{-1}\partial^2 a/\partial x^2| \tag{5}$$

With increasing growth rate, the characteristic amplitude of appearing structures grows and their characteristic scale, l, diminishes.

Detailed analytical analysis of the behavior of the field $a(x,t)$ is possible only in limiting cases: in a conservative medium without diffusion $(\alpha, \beta \to \infty)$ [1] or in dissipative media without dispersion $(\alpha, \beta \to 0)$. When $\alpha = \beta = 0$, system (2) transforms into a nonlinear diffusion equation that can be written in a gradient form:

$$\frac{\partial a}{\partial t} = -\frac{\delta F}{\delta a^*}, \tag{6}$$

where $F = -\int \left[R|a|^2 - \frac{1}{2}|a|^4 - (\partial a/\partial x)^2\right] dx$ is a Lyapunov functional. Since the functional, $F(a)$, may only decrease along the trajectory($dF/dt = -\int |\partial a/\partial t|^2 dx \leq 0$), the phase space of the dynamical system (2)-(3) (1-2) may contain only static attractors, i.e. equilibrium states. Consequently, the spatial distribution of the fields corresponding to these attractors may be only regular and is described by the following equation for a nonlinear complex oscillator: $d^2a/dx^2 = a(|a|^2 - R)$.

In particular a stationary kink,

$$a = R\, th(x/\sqrt{2}), \tag{7}$$

that corresponds to the boundaries between domains (i.e. the regions with a homogeneous complex amplitude whose phases differ by p) are of special interest. The amplitude profile in this solution contains a gap – a "hole" [13]. Such "hole" solutions persist for complex coefficients in the Ginzburg–Landau equation [14]. It is significant that one–dimensional structures ("holes") behave like the cores of two–dimensional structures (spirals).

The phenomenology of spatio–temporal chaos within a CGLE has been studied in ample detail. It has been found that if the parameter $p = \sqrt{\alpha\beta} > 1$ is not too high, then the spatially homogeneous distribution of the amplitude becomes unstable, and a regime that is referred to as "phase turbulence" is established [3]. Slow random walks of the phase, φ, of the wave field, $a = A\exp(i\varphi)$, are typical for this regime, as long as the fluctuations of the amplitude, A ,are small and "follow" the phase fluctuations. As the parameter p increases, the behavior of turbulence changes. Regions of abrupt phase variations appear and, respectively, there occur "holes" in the amplitude profile. It was shown in [13] that as the parameter p increases, the "holes" grow in number, move (see [15]), oscillate and interact with one another. The resulting developed "amplitude" turbulence makes a quite complicated picture.

[1] In this case from (2) we have a nonlinear Schrödinger equation (NSE) where the spatial structures are envelope solitons whose amplitude A and velocity V may slowly vary under the action of weak amplification or dispersion [4–12].

It is important to understand the processes that occur in the phase space of the CGLE and to describe them in terms of nonlinear dynamics. It is also essential to find whether the characteristics of the "phase–amplitude" turbulence transition are universal. For the solution of this problem we employ a new approach to the analysis of the phase–to–amplitude turbulent transition. This approach is based on the calculation of the dimension of spatial chaos and on the analysis of the parameter dependence of the variation of this quantity.

3.The correlation dimension of the space series, (d_s), was first introduced in Ref.[16]. When calculating d_s we assume that the space series at a certain moment of time, t, may be considered to be generated by a finite–dimensional translational dynamical system, G_x.

In a one–dimensional medium, d_s is determined similarly to the correlation dimension of the time series (d_t) [17] employing the correlation integral. Assume that the dynamical system , G_x, in an m–dimensional phase space describes the trajectory

$$\mathbf{Y}(x) = \{a(x), a(x+\xi), a(x+2\xi), ..., a(x+(m-1)\xi)\}.$$

We will calculate the dimension using the correlation integral in a standard form:

$$R^m(\varepsilon) = (M-m)^{-2} \sum_i \sum_j H\left[\|\mathbf{Y}_i - \mathbf{Y}_j\| - \varepsilon\right] = < M_\varepsilon^i > /M \tag{8}$$

where $H(\varepsilon)$ is the Heaviside function, $\mathbf{Y}_i$ is a point in the m–dimensional phase space, M denotes the total number of points in the series, and M_ε^i, the number of points in the ε–neighborhood of the i–th point. Because $R^m(\varepsilon) \sim \varepsilon^{d_s}$, the correlation dimension is determined as the ratio $\log[R^m(\varepsilon)]/\log(\varepsilon)$ for small enough ε. Note that correct calculation of the dimension, d_s, needs sufficiently long space series and a proper choice of ξ (see Appendix). The definition (8) is analogous to the definition of the temporal correlation function as a time–averaged number of "fragments" of the series, with the distance between the "fragments" being smaller than ε. In a similar fashion and in analogy with the local dimension that is determined by averaging over time, the value of the correlation dimension of the space series, d_s, in m–dimensional enclosure space, is defined as the $\log[R^m(\varepsilon)]/\log(\varepsilon)$ ratio for sufficiently small ε.

The length of the space series is determined by the actual number of spatial structures. In the case of the boundary value problem (2)–(3), the value $R \gg 1$ corresponds to a long space series. The dimensional characteristics of the turbulent regime that has established as $t \to \infty$, in this case, depend neither on the boundary conditions nor on the length of the system.

4. The boundary value problem (2)–(3) was solved numerically by the Fourier method. The integration algorithm was an apparent generalization of the method of operator exponents described in Ref.[18]. Within this method exact solutions to a linear problem are employed to construct an approximate nonlinear solution, which shortens the computation time significantly as compared to standard grid algorithms. Spatial resolution was determined by the number of Fourier harmonics n. This number lies in the interval $512 < n < 8192$ and depends on the complexity of the regime

and on needed accuracy. Small sinusoidal amplitude perturbations or the spatial amplitude distributions that are established in a definite parameter region were taken as initial conditions.

The dimension of the space series in the solutions of interest can be estimated only if we have extended turbulent fields. However, insufficient computer memory and speed impede the calculations. Even for $n \geq 4096$ harmonics, extremely long time is needed for the computation of solutions. So as to avoid this difficulty, especially in the case of very long space series, the following procedure was used.

The phase trajectory of the system in the established regime lies on an attractor and has stationary statistical properties. Consequently, we can use, for the determination of the dimension, space series taken at different moments of time. We match these series into one. To decrease the inhomogeneities at the matching points, we choose the series where the sequences of k neighboring points coincide and match them in the region of these sequences. The inhomogeneity in the region of matching can be neglected and the joint space series can be considered to be generated by a translational dynamical system, if the number k is larger than the dimension d_s on the stochastic set. In this fashion we can obtain arbitrary long space series and determine the value of dimension that is, actually, averaged over an ensemble of systems having different initial conditions.

The space resolution, Δx, in our numerical solutions was rather high to reveal the characteristic shape of the structures having the size $l : l/\Delta x \geq 10$. For the calculation of the correlation dynamical dimension, the stochastic set of a multidimensional dynamical system was reconstructed in a pseudo–phase space from the space series. A fixed–mass algorithm [19] and Grassberger–Procaccia's procedure [20] were employed to calculate the value d_s.

Proper "delay" of the space series is needed for the construction of embedding space. If the "delay" step, ξ, is too small, the ensemble taken for the calculation of the correlation dimension will represent only the local structure of the solution (a section of a curve). The measured dimension will, consequently, tend to unity. If, on the other hand, the "delay" step is too large, the measured dimension will tend to the dimension of embedding space since the points of the space series are completely uncorrelated.

A "delay" step, ξ, comparable with the correlation length provides correct calculation of dimension. A typical scale (for example, position of the first minimum) of the mutual information function gives the best choice for ξ [21]. The motivation for this is that the information in two successive delay coordinates should be as independent as possible, without making the delay too large. This criterion yields $\delta \cong (3 \div 5)l$. So as to verify this estimate and to determine the "delay" step more accurately we calculated the dependence of the dimension d_s on the "delay" step ξ. The function $d_s(\xi)$, as was to be expected, grows with increasing ξ(see Fig.1). The curve in the figure contains a flat section (a "plateau") which is the interval within which the "delay" step should be chosen. By choosing the value of ξ within this section we obtain a correct value of dimension. Besides, we calculated the dependence of d_s on the length of the space series N. The dimension d_s asymptotically tends to a constant

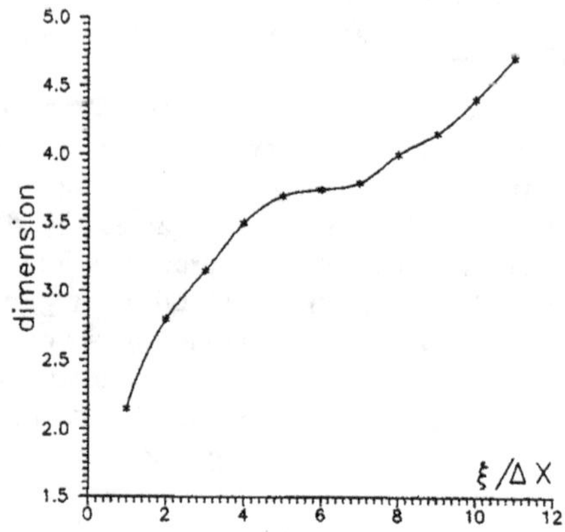

Figure 1: The value of dimension, d_s, calculated for various space delays ξ for $R = 10^4$, $\alpha = 1.0$, $\beta = 1.3$. The number of harmonics used in Fourier method is $n = 2048$.

that corresponds to the true dimension when $N \geq 2000$.

5. We shall now investigate the transformations of the attractor that corresponds to the regime of established spatio–temporal chaos in a long system. To this end we will measure the dimensions d_s (for the space series) and d_t (for the time series). The dependences of fractal dimensions on the product $p = \sqrt{\alpha\beta}$ are given in Fig.2a for fixed $s = \sqrt{\beta/\alpha}$ [2]. Of principal importance here are two phenomena: the jump in dimension when $p = p''$ and hysteresis in the parameter region $p \in (\tilde{p''}, p'')$. To understand their origin we shall consider the variation of d_s and d_t with increasing p.

For small p the dimension is equal to zero, which corresponds to the regime of spatially homogeneous $e^{i\beta t}$–oscillations. When $p > p_1$, the dimension grows up to unity, which corresponds to the onset of a spatially periodic regime, while for $p = p_2$ a double frequency (2–D) regime is established. Further evolution of the system is related to the appearance of beats against the background of quasi–periodic oscillations (see Fig.3a) and to the transition, when $p > p'$, to a chaotic regime. Note that for large values of the parameter R, the region on the plane of the parameters (α, β) that corresponds to periodic and quasi–periodic oscillations is very narrow and is not depicted in Fig.2 [3].

A developed turbulent regime that sets in when $p \geq p'$ is phase turbulence: the amplitude has a value close to $|a| = \sqrt{R}$ (Fig.3b) and its fluctuations adiabatically

[2]The combinations of the parameters α and β are chosen such as to move along the normal to the boundary of the regions of phase and amplitude turbulence (see the text below).

[3]The large–scale structures whose size is comparable with a resonator length (a short system, small R) was studied earlier [22,23].

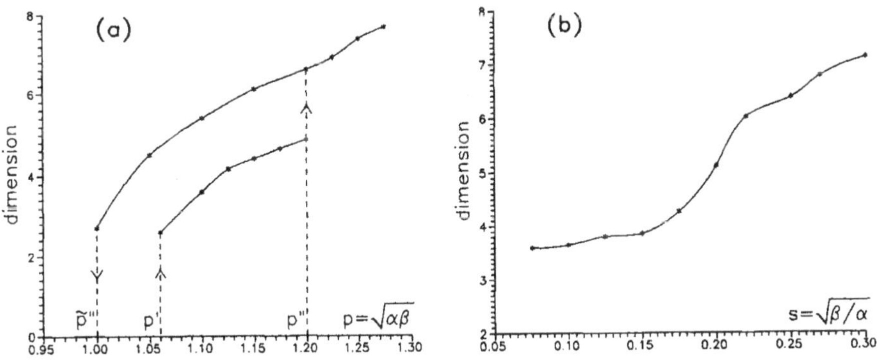

Figure 2: Dimension of space series for $R = 10^4$: a) versus p for $s = 1.15$; b) versus s for $p = 3.0$.

"follow" relatively smooth random walks of the phase. While moving over the parameter plane deep into the region of phase turbulence (with p increasing for $s = const$) up to $p = p''$, the stochastic set corresponding to phase turbulence gets more and more complicated; the dimension of the space series grows monotonically within the interval $3 < d_s \leq 5$.

Past the critical value of the parameter $p = p''$, we observe a qualitatively new behavior: the dimension of the space series increases abruptly up to $d_s \geq 6$. This is accompanied with strong oscillations in the spatial amplitude distribution – amplitude turbulence is established.

Amplitude turbulence in the region $p > p''$ sets in as follows. First, small inital perturbations give rise to a transient regime with random phase oscillations and small amplitude pulsations, like in phase turbulence. Then some rather abrupt phase jumps (which correspond to amplitude peaks) produce domains of amplitude turbulence with well pronounced boundaries (see Fig.4) In the course of time these domains expand, merge and, finally, fill the entire region see (Fig.4 b–d)

The spatial amplitude distribution in the regime of amplitude turbulence may have different characteristic forms. When $\beta \gg 1$, the turbulence typically has narrow impulses ("holes") on the background of weak oscillations near the level $|a| = \sqrt{R}$ (Fig.3c). With increasing p, such "holes" grow in number past the critical point $p = p'$, which, evidently, leads to a further growth in the dimension of the space series. When $\beta \leq 1$, the number of "holes" is sufficiently great immediately past the threshold where amplitude modulation is established, consequently, the spatial picture is visualized as strong oscillations in the interval from zero to a certain maximum (Fig.3d).

In a reverse motion over parameter plane hysteresis is observed: the amplitude turbulence is retained and does not change to the phase turbulence even when $p < p'$.

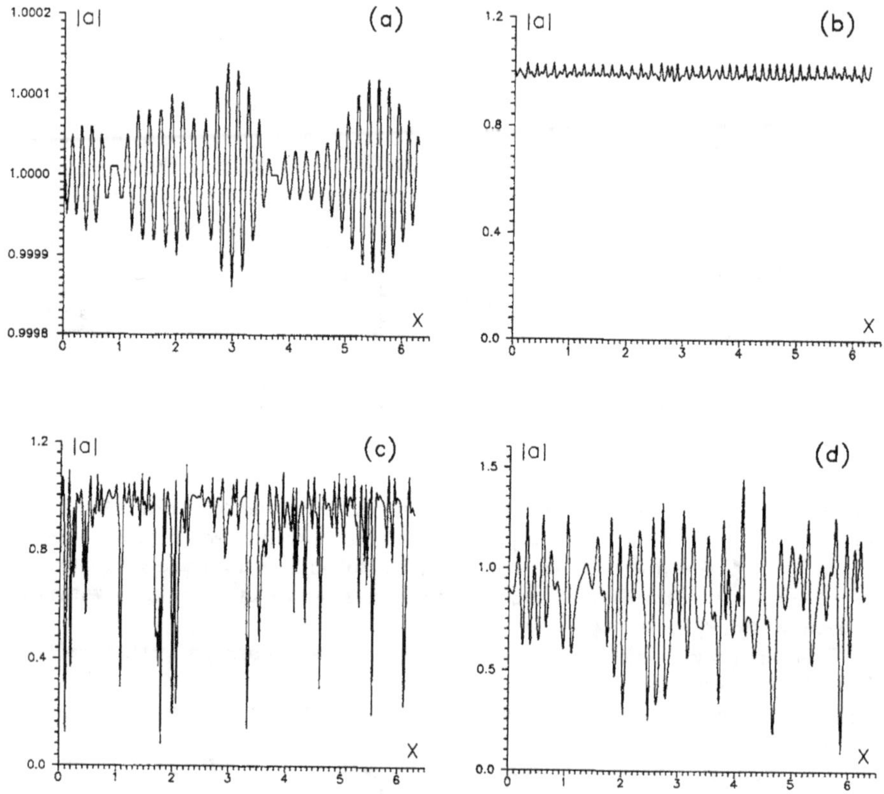

Figure 3: **Typical "snapshots" of amplitude variations for $R = 10^4$: a) $\alpha = 1.03$, $\beta = 1.3$ – quasi-sinusoidal oscillations; b) $\alpha = 1.0$, $\beta = 1.3$ – phase turbulence; c) $\alpha = 1.1$, $\beta = 1.44$ – amplitude turbulence; d) $\alpha = 12.0$, $\beta = 0.75$ – amplitude turbulence.**

Stepwise decrease of dimension occurs when $p = \tilde{p}''$. Thus, the initial conditions determine the regime – developed amplitude turbulence or phase turbulence – to be established in the region $(\tilde{p}'', p'')$. A possible mechanism responsible for the phenomenon described above is as follows. The transition of the motion of the translational dynamical system, G_x, from one stochastic set in the phase space to another corresponds to a sharp change in the value of the space series dimension. The established turbulent regime corresponds to a strange attractor in the phase space . The phase transition occurs at the instant the separatrix manifold of the saddle periodic motion bounding the region of attraction of a low–dimensional ("phase") attractor

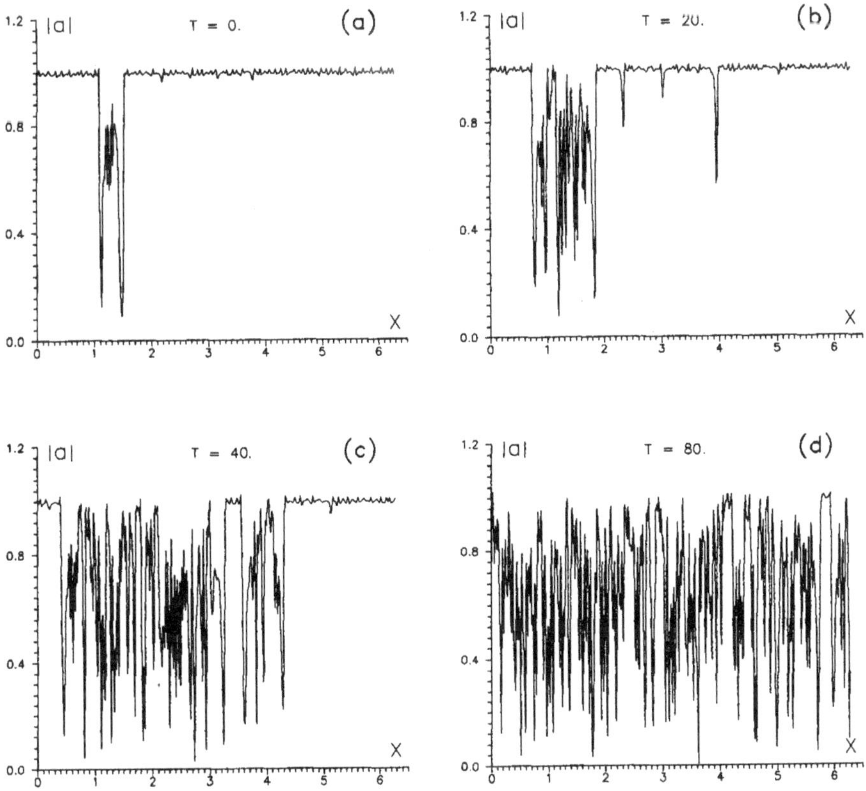

Figure 4: Time evolution of amplitude turbulence domains for $R = 10^4$, $\alpha = 0.4$, and $\beta = 3.6$: a) $t = 0$, b) $t = 20$, c) $t = 40$, and d) $t = 80$.

merges with the latter which, as a result, ceases to be an attractor: now nearly all the trajectories tend to another, "laminar" attractor. Both strange attractors – the phase and the amplitude ones – co-exist in the hysteresis region in the phase space.

Partitioning of the parameter space (α, β) into regions of existence and stability of different spatially disordered regimes is presented in Fig.5 [4]. In region 1 the homogeneous amplitude field is stable. Phase turbulence appears from small initial disturbances in region 2, and amplitude turbulence exists in region 3. The broken curve bounds the region of hysteresis (region 4 in Fig.5), where the solution depends

[4]A similar partitioning of parameter space into regions of different behavior of the correlation function (which is determined by the presence or absence of "holes") was performed in [24].

on initial condition. Note that this region stretches to the negative halfplane of the parameter $\alpha < 0$, where the space series (static or oscillating lattices of randomly located holes) are established.

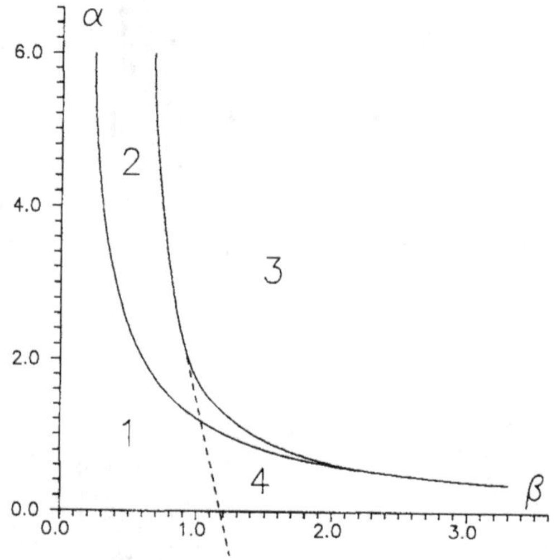

Figure 5: α, β – plane for $R = 10^4$ with regions of different dynamics: 1 – stable monochromatic homogeneous oscillations; 2 – phase turbulence; 3 – amplitude turbulence; 4 – region of hysteresis.

Near the lower boundary of hysteresis region some unexpected phenomena are observed. Here spatial amplitude distribution is a random "comb" that consists of holes with time periodic behavior. It is remarkable, that at first sight different holes oscillate independently. The spatial field distribution appears to be chaotic and acquires various shapes during one time period. However, the amplitude and the shift of each hole always oscillate at equal frequencies and oscillations are phased in at different points. This gives rise to recurrency – the initial (rather complicated) spatial pattern of the field (Fig.6) is reconstructed in one oscillation period.

There is no hysteresis in region 2 (outside region 4). The transition from phase to amplitude turbulence occurs here not by a jump but gradually, in a finite interval: the intensity of phase fluctuations increases near the boundary of region 2. The rapid growth of the dimension is observed in this interval (Fig.2b) due to the appearance of holes.

6. Of principal importance is the problem of the relationship between different quantities describing turbulence: the number of structures, N, the dimension of the space series d_s, and the dimension of the time series d_t. Our analysis indicates that for the one-dimensional CGLE model the average number of pulsed structures, N, grows

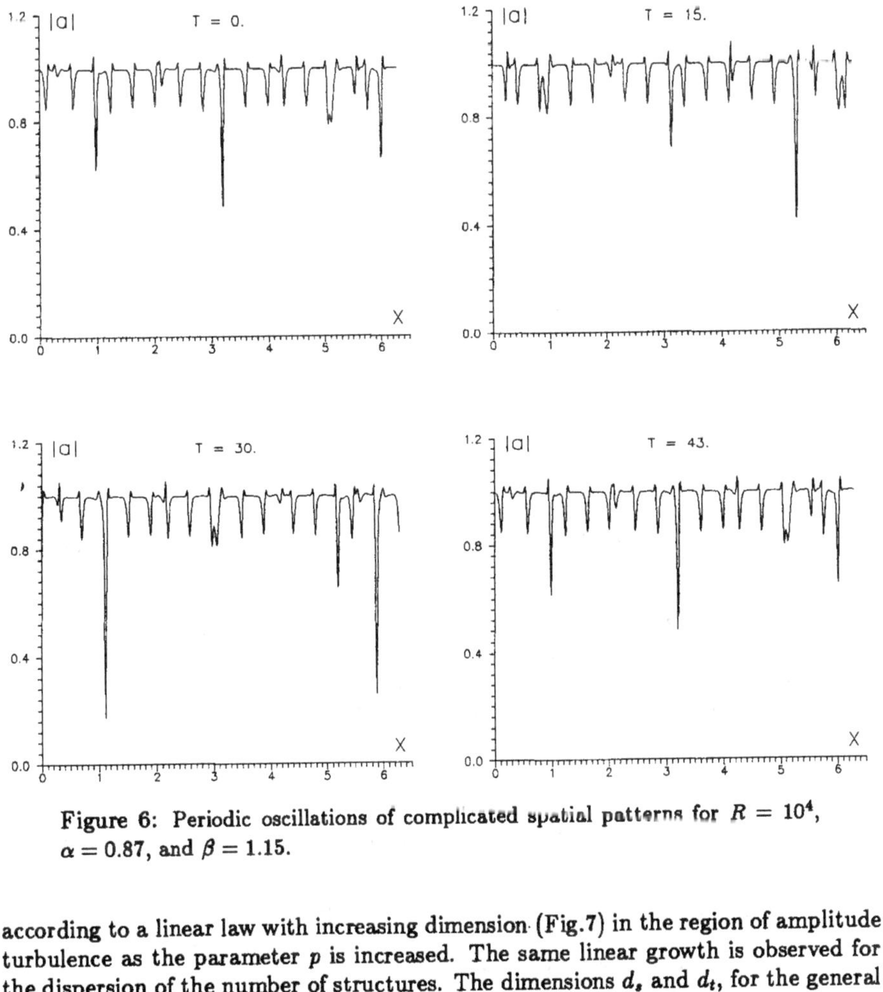

Figure 6: Periodic oscillations of complicated spatial patterns for $R = 10^4$, $\alpha = 0.87$, and $\beta = 1.15$.

according to a linear law with increasing dimension (Fig.7) in the region of amplitude turbulence as the parameter p is increased. The same linear growth is observed for the dispersion of the number of structures. The dimensions d_s and d_t, for the general case, may be related arbitrarily. For instance, for a static disordered field distribution, $d_t = 0$ and $d_s > 2$. On the contrary, for a regular spatial field distribution with chaotic temporal dynamics, d_t may be rather great, while d_s is small. In a large box one-dimensional complex Ginzburg–Landau problem d_t grows with increasing effective length of the system (i.e. with increasing number of structures) and is much higher than the spatial dimension d_s. Within CGLE models that are described by specific dispersion laws there evidently exists a certain relation between d_s and d_t per unit length in the regime of established turbulence.

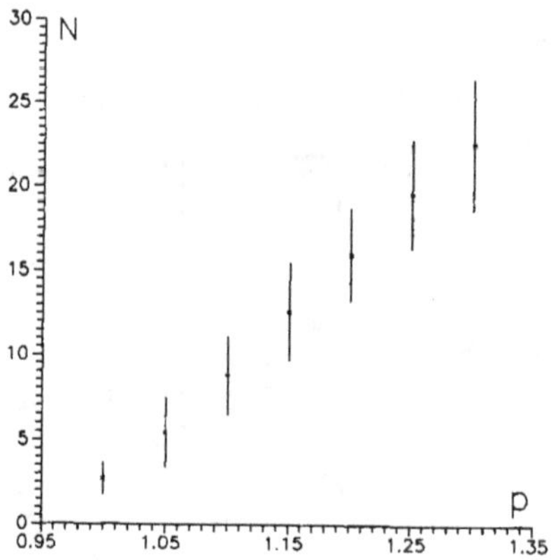

Figure 7: Number of "holes", N, in amplitude turbulence versus p for $R = 10^4$, $s = 1.15$. Vertical lines indicate mean deviation of N.

We would like to note that the jump of the dimension d_s in the motion over the parameter plane across the boundary of region 3 may occur a little later than that of d_t. This effect has a simple explanation. In the transition of the parameter across the critical, value the intermittency of amplitude turbulence occurs: "holes" in the amplitude profile fluctuate, i.e. they disappear and appear at different moments of time. Therefore, if the space series corresponds to the moment, t, when there are no "holes", its dimension may be smaller that at other moments of time.

References

1. A.S. Newell and J.A. Whitehead, *J.Fluid.Mech.* **38** (1969) 279; K. Stewartson and J.T. Stewart, *J.Fluid.Mech.* **48** (1971) 529.

2. Y. Kuramoto, *Progr.Theor.Phys. Suppl.* **64** (1978) 346.

3. A.L. Fabrikant, *Wave Motion* **2** (1980) 355.

4. N.R. Pereira and L. Stenflo, *Phys.Fluids* **20** (1977) 1733.

5. I.S. Aranson, A.V. Gaponov-Grekhov, and M.I. Rabinovich, *Sov.Phys.-JETP* **89** (1985) 92.

6. I.S. Aranson, A.V. Gaponov-Grekhov, M.I. Rabinovich, A.V. Rogal'skii, and R.Z. Sagdeev, *Lattice models in nonlinear dynamics of non-equilibrium media* (Preprint N.163, Inst.Appl.Phys., Acad. Sci.USSR, Gorky 1987).

7. I.S. Aranson and M.I. Rabinovich, *Radiophysics and Quantum Electronics* **29** (1986) 1514.

8. P. Coullet, L. Gil, and J. Lega, *Phys.Rev.Lett.* **62** (1989) 1619; *Physica D* **37** (1989) 91.

9. M. Bartuccelli, P. Constantin, C.R. Doering, J.D. Gibbon, and M. Gisselfalt, *Physica D* **44** (1990) 421.

10. L.M. Pismen and A.A. Nepomnyashchy, *Phys.Rev.* **44** (1991) 2243.

11. G. Goren, I. Procaccia, S. Rasenat,and V. Steinberg, *Phys.Rev.Lett.* **63** (1989) 1237.

12. A.L. Fabrikant, *Sov.Phys - JETPH* **86** (1984) 470.

13. H. Sakaguchi, *Progr.Theor.Phys.* **85** (1991) 417.

14. K. Nozaki and N. Bekki, *J.Phys.Soc.Jap.* **53** (1984) 1581.

15. N. Bekki and K. Nozaki, *Phys.Lett. A* **110** (1985) 133.

16. V.S. Afraimovich, A.B. Ezersky, A.L. Zheleznyak, M.I. Rabinovich, and M.A. Shereshevsky, *Complexity and dimension of disordered states of non-equilibrium media* (Preprint N.291, Inst.Appl.Phys., Acad.Sci.USSR, Nizhny Novgorog, 1991).

17. G. Mayer–Kress, ed., *Dimensions and Entropies in Chaotic Systems* (Springer-Verlag, New York, N.Y.,1986).

18. I.S. Aranson, K.A. Gorshkov, M.I. Rabinovich, and A.D. Junakovsky, *Stochastization and spreading of wave trains in the oscillating potential* (Preprint N.226, Inst.Appl.Phys., Acad.Sci.USSR, Gorky,1991).

19. P. Grassberger, R. Badii,and A. Politi, *J.Stat.Phys.* **51** (1988) 135.

20. P. Grassberger and I. Procaccia, *Phys.Rev.Lett.* **50** (1983) 346.

21. A.M. Fraser and H.L. Swinny, *Phys.Rev.A* **33** (1986) 1134.

22. M.I. Rabinovich and A.L. Fabrikant, *Sov.Phys-JETPH* **77** (1979) 617.

23. L. Sirovich and J.D. Rodrigez, *Phys.Lett. A* **120** (1987) 211; J.D. Rodrigez and L. Sirovich, *Physica D* **43** (1990) 77.

24. B.I. Shraiman, A. Pumir, W. van Saarloos, P.C. Hohenberg, H. Chate, and M. Holen, *Spatio temporal chaos in the one dimensional complex Ginzburg–Landau equation* (Preprint).

SPATIOTEMPORAL INTERMITTENCY: THE CASE OF THE COMPLEX GINZBURG-LANDAU EQUATION

HUGUES CHATÉ

Service de Physique de l'Etat Condensé, Centre d'Etudes de Saclay
91191 Gif-sur-Yvette, France

ABSTRACT

After presenting the problem of the transition to spatiotemporal intermittency in extended systems, a brief overview of recently obtained results is given, which stresses the genericity of this mode of spatiotemporal disorder and shows the interest of an approach based on concepts borrowed from statistical mechanics and critical phenomena. The one-dimensional complex Ginzburg-Landau equation, taken as an ubiquitous model in the field of weak turbulence, is investigated in this context. It is shown that several regions of the parameter space exhibit genuine spatiotemporal intermittency regimes.

1. Introduction to Spatiotemporal Intermittency

Spatiotemporal intermittency is a common phenomenon which has recently been the subject of both theoretical and experimental works in the field of weak turbulence[1]. In its most general —and vaguest— definition, spatiotemporal intermittency refers to all dynamical regimes exhibited by spatially extended systems in which instantaneous snapshots and local time-series both present an intermittent character (as understood for example in fluid mechanics).

Although examples can be found in the behavior of turbulent flows, weakly turbulent situations[2], which possess only a few characteristic scales, offer the best test-bench for understanding spatiotemporal intermittency. This is particularly true with respect to the general conjecture due to Pomeau which suggested an analogy with contamination processes such as directed percolation[3]. One is then led to consider systems with a basic lengthscale and an elementary period (easily decomposable into space-time units), or arrays of locally coupled identical units updated synchronously, in order to facilitate the comparison with cellular automata similar to directed percolation.

Pomeau argued that there exist systems which transit to turbulence in a "subcritical" manner, i.e. with the sudden appearance of localized patches of a disordered state separated by sharp fronts from the otherwise laminar flow (a classic example is given by the Emmons spots in a Blasius boundary layer[4]). These two local states ("laminar" and "turbulent"), which can be seen to lie at a finite distance from each other, do not play symmetric roles in the dynamics: the turbulent state possesses intrinsic fluctuations which allow the local transition "down" to the laminar state, whereas the reverse transition is made impossible by the absence of such fluctuations in the (usually linearly stable) laminar phase. In this case, propagation of disorder can only occur though contamination by the turbulent regions. This constitutes in fact the essential property of directed percolation[5]: there, the absorbing character of the empty state forbids the spontaneous nucleation of the active phase. Although a large variety of models with one absorbing state can be built, it is widely believed that, in the regimes where sustained activity first appears, a universal critical behavior occurs, characterized by a set of critical indices[6]. Pomeau suggested that this universality of directed-percolation-like phase transitions might extend to the transition to turbulence via spatiotemporal intermittency.

The first results obtained on partial differential equations similar to the Kuramoto-Sivashinsky model revealed that spatiotemporal intermittency is frequently observed, and that the transition shares many qualitative aspects with the directed percolation one[7]. It quickly appeared, though, that the basic ingredients of the contamination process could sometimes be masked by deterministic, model-dependent features, and that the reference laminar state could be absorbing only in a weak sense, retaining the global signature of the process but forbidding a detailed comparison with directed percolation.

In order to give a more definite answer to Pomeau's conjecture, and encouraged by the early work of Kaneko[8], a "minimal" coupled map lattice model was built to study the transition in detail. It was shown that the transition to spatiotemporal intermittency is best described as a phase transition, but, contrary to what happens in directed percolation models, it can be discontinuous (first-order) as well as continuous (second-order)[9]. Moreover, in this last case, the scaling properties are not universal and depend on the details of the local dynamics of the system considered. Hence, the equivalence proposed by Pomeau does not hold *stricto sensu*, but it sets the correct stage for studying spatiotemporal intermittency. Clearly, concepts borrowed from the field of critical phenomena are the best suited to describe the statistical properties of the corresponding systems. In the "thermodynamic" limit where the number of elementary units composing these systems is infinite, the complexity of a dynamical-systems-like approach is advantageously replaced by a statistical-mechanics-like description.

Dedicated laboratory experiments have proved that the general picture first unveiled in the numerical study of the models applies to "real" systems such as Rayleigh-Bénard convection[10]. A remarkable example can be found in the disordered regimes of a meniscus submitted to the so called "printer's instability"[11] (see the contribution of Couder in this volume). As a matter of fact, many experimental illustrations of spatiotemporal intermittency can be found, and there remains a number of studies to be made along the lines developed for the models. Nematic liquid crystals are particularly good candidates: a large number of elementary units (rolls, squares) is easily obtained, the time-scales are short, the local dynamics is reasonably understood, and the general phase diagram presents many dynamical regimes akin to spatiotemporal intermittency[12].

2. An Ubiquitous Model: the Complex Ginzburg-Landau Equation

The aim of the present work is two-fold: first, to illustrate the summary presented in the previous section, and second, to stress the ubiquity of spatiotemporal intermittency in extended systems by showing that this phenomenon is present in the most-studied equation of the field. Rather than reviewing experimental situations and models, here the focus is on the complex Ginzburg-Landau, which describes the evolution, following a Hopf bifurcation, of the slowly varying modulations in an extended system. This partial differential equation arises from the analysis of numerous physical situations[13] (hydrodynamics, active media, nonlinear optics, etc.), to the point that it is sometimes asked whether we live in a "Ginzburg-Landau world"...! For reasons of simplicity, the results presented are restricted to one space dimension.

To stay as general as possible, only the "kernel" version of the equation is considered. Depending only on two real parameters b_1 and b_3, it reads:

$$\partial_t A = A + (1 + ib_1)\partial_{xx} A + (-b_3 + i)|A|^2 A$$

where $x \in [0, L]$.

Although many works dealt with various aspects of the model, it is only recently that a systematic, comprehensive study of the different "phases" exhibited in the two-parameter plane has been undertaken[14]. This numerically costly study was achieved in the "thermodynamic" $L \to \infty$ limit, where the dynamics might be understood in terms of concepts derived from statistical mechanics and critical phenomena.

In order to set the problem of spatiotemporal intermittency regimes correctly, a few major features of the equation have to be reviewed. The equation admits travelling wave solutions. The line $b_1 = b_3$ is the so-called Benjamin-Feir or modulational instability line, beyond which all travelling wave solutions are unstable. For $b_3 > b_1$, there exists a band of (Eckhaus) stable wavenumbers. For $b_1 > b_3$, a phase turbulence regime is expected[15]; it is actually observed up to the lines L_1 and L_3 in Fig. 1[14]. To the left of L_1 and L_3, the phase dynamics breaks down to give place to an amplitude turbulence regime where A may go to zero (defects). The amplitude turbulent regime has been found to be limited by the lines L_1 and L_2 in Fig. 1. If no theoretical argument is available at this point to account for lines L_1 and L_2, line L_3 seems to correspond fairly well with the limit of existence of convectively stable travelling waves recently found by Kramer *et al.*[16].

This brief overview of the phase diagram already indicates the regions where spatiotemporal intermittency is most likely to be observed. The most obvious is the Benjamin-Feir stable domain below the $b_1 = b_3$ line and to the left of the L_2 line (region I of Fig. 1), where long-wavelength travelling wave solutions are stable and yet amplitude turbulence exists. This coexistence of two states, one regular and linearly stable, the other turbulent, is strongly reminiscent of the prerequisites for observing spatiotemporal intermittency. The laminar, absorbing state would naturally be formed from stable travelling waves. It remains to see whether the amplitude turbulent regimes in this region do present the characteristics of spatiotemporal intermittency.

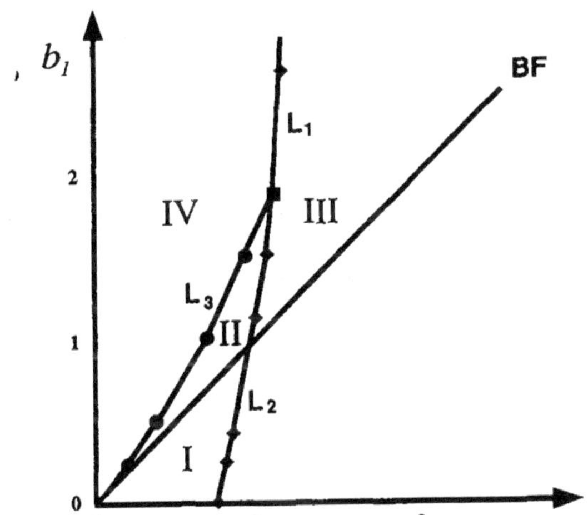

Figure 1: Phase diagram of the one-dimensional complex Ginzburg-Landau equation. The line $b_1 = b_3$, coded "BF", is the Benjamin-Feir stability limit. Amplitude turbulence is observed to the left of lines L_1 and L_2. Phase turbulence occurs in the region above the BF line and below lines L_1 and L_3. For details on the numerical procedure to obtain these results, see[14,17,20].

Another candidate for spatiotemporal intermittency is region II of Fig. 1, where two different turbulent states coexist. In the following section, all regions of the (b_1, b_3) plane where disordered solutions have been observed are considered with respect to the possibility of showing spatiotemporal intermittency.

3. Spatiotemporal Intermittency in the Complex Ginzburg-Landau equation

3.1 In the Benjamin-Feir Stable Region

Fig. 2 presents space-time plots of the amplitude-turbulent regimes of the complex Ginzburg-Landau equation for $b_1 = 0.25$ and b_3 small (region I of Fig. 1). These statistically stationary regimes are reached after a short transient following adequately disordered initial conditions. Propagating localized hole-like structures leave triangular space-time domains in which the amplitude remains constant and equal to its level in the stable standing wave solution at the same parameter values ($|A|_{max} = 1/\sqrt{b_3}$). The phase plots clearly show the standing wave structure of this "laminar" state and the phase jumps accompanying the defect-like objects. A detailed analysis of these objects will be reported elsewhere,[17] but it is already apparent in this representation that actual defects (zeros of A) only occur at discrete space-time points, even though the objects can be seen as long-lived.

The distributions of sizes and lifetimes of the triangular embeddings are approximately exponential with characteristic scales increasing with b_3. Approaching the L_2 line, one is eventually left with a few, almost non-interacting, space and time periodic objects with an effective low-dimensional dynamics. These structures, which provide the elementary scales on which the comparison with directed percolation should be built, will be shown to be "close" to the family of exact solutions found by Nozaki and Bekki[18].

From the spatiotemporal intermittency point of view, although the characteristic scales of the triangular regions increase when approaching L_2, they do not diverge, so that one cannot easily determine the nature of the transition. Nevertheless, the regimes presented are strictly of the spatiotemporal intermittency type. The laminar state is indeed a linearly stable solution and its absorbing character is insured. The visual analogy with the famous diagrams of class 3 cellular automata[19] is not fortuitous: there, the triangular embeddings are formed from the absorbing state.

Most of the amplitude turbulence regimes observed in the Benjamin-Feir stable region are of the type shown in Fig. 2. But other features appear when $b_1 \sim 0$. In addition to the propagative objects already described, long-lived, very slowly moving, hole-like structures superpose onto the dynamics and eventually dominate when approaching the transition line L_2 (Fig. 3). The regular laminar regions then consist most often of patches of travelling waves separated by hole-like sinks and shock-like sources. Although details about these objects and their relationship with the Nozaki-Bekki solutions will be reported elsewhere[17], it is worth mentioning here that they are *not* strict defects ($A = 0$) subsisting for a finite time. The transition region is thus even more complicated near the $b_1 = 0$ axis than elsewhere around the section of the L_2 line crossing the Benjamin-Feir region. This, of course, hinders the comparison with simple systems such as directed percolation.

2a

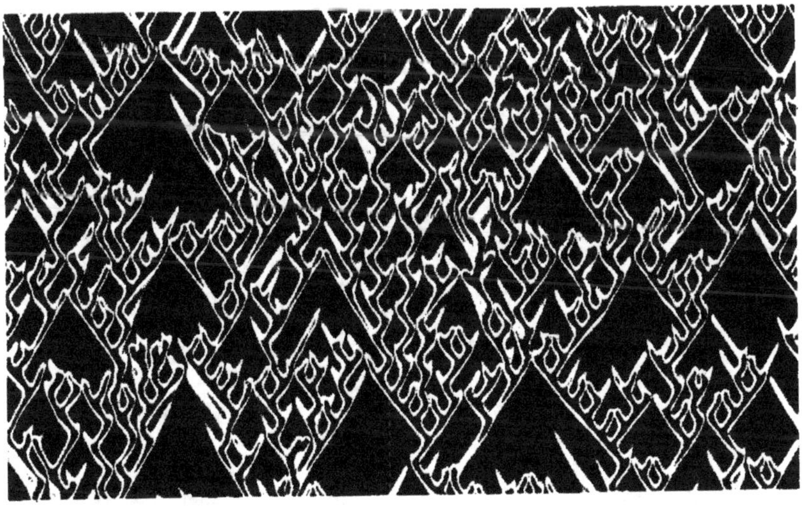

3a

Figures 2a and 3a: Spatiotemporal plots of the amplitude of the solution in region I. Time is running upward during $\Delta T = 200$; $L = 500$; periodic boundary conditions; transients were discarded. Color scale from red ($|A|_{\max} = 1/\sqrt{b_3}$) to purple ($|A| = 0$) via yellow, green and blue. (2a): $b_1 = 0.25$, $b_3 = 0.6$. (3a): $b_1 = 0$, $b_3 = 0.6$.

2b

3b

Figures 2b and 3b: Spatiotemporal plots of the phase ($\phi = arg(A)$) for the same runs as in Figs. 2a and 3a. Grey scale modulo 2π from black ($\phi = 0$) to white ($\phi = \pi$). Time is running upward.

3.2 In the Bi-Turbulent Region

The triangular region of the (b_1, b_3) plane situated between the Benjamin-Feir, L_2 and L_3 lines (region II of Fig. 1) is characterized by the bistability of the dynamics: both phase and amplitude turbulent regimes are observed, depending on the initial conditions. No intermediate regime is stable, so that these two spatiotemporally disordered phases can be thought of as being at a finite distance from each other, defining thus the two local states necessary to any description in terms of spatiotemporal intermittency.

Regimes of the two coexisting turbulent states are shown in Fig. 4. It appears that in the amplitude turbulent regime, the domains between the defect-like objects are similar to those in the phase turbulence regime. The small variations of amplitude above its basic level ($\sim 1/\sqrt{b_3}$), which characterize the phase turbulence regime, are here nothing but the shock-like objects of the amplitude turbulence regime. Although this point will be made more quantitative in a detailed analysis of the phase turbulence regimes to be reported elsewhere[20], it is already clear that the amplitude turbulence observed is a genuine spatiotemporal intermittency regime whose corresponding laminar state is the phase turbulent one coexisting at the same parameter values. The absorbing character of this laminar state is insured by the stability of phase turbulence in the region of interest.

3.3 In the Phase Turbulence Region

It has been well known since the early papers by Kuramoto[15] that the phase turbulence regimes of the complex Ginzburg-Landau equation can be described by a phase equation. At fourth order, this equation is the famous Kuramoto-Sivashinsky model with a few other terms, some of which are actually stabilizing[21]. It is thus reminiscent of the modified Kuramoto-Sivashinsky models first shown to exhibit spatiotemporal intermittency[7]. There, additional stabilizing terms, such as a damping term allowed the study of the transition from Kuramoto-Sivashinsky turbulence to a more regular dynamics.

Here, the coefficients of all the terms depend upon b_1 and b_3 in a complex manner. Moreover, a third (discrete) parameter has to be considered: the winding number (i.e. the integral of the phase gradient), a constant of motion which classifies the solutions. A comprehensive study of the phase turbulence regimes or of the dynamics of the phase equation is thus a difficult task. Preliminary results[20] indicate that spatiotemporal intermittency-like regimes are indeed observed in the phase turbulence region of the (b_1, b_3) plane (regions II and III of Fig. 1). Fig. 5 shows a typical case, where large patches of almost-regular trains of solitons emerge out of more disordered domains where the dynamics is strongly reminiscent of the "classic" Kuramoto-Sivashinsky model.

From the spatiotemporal intermittency point of view, the laminar state appears to be the propagative state. Similar regimes are observed in the transition from

4a

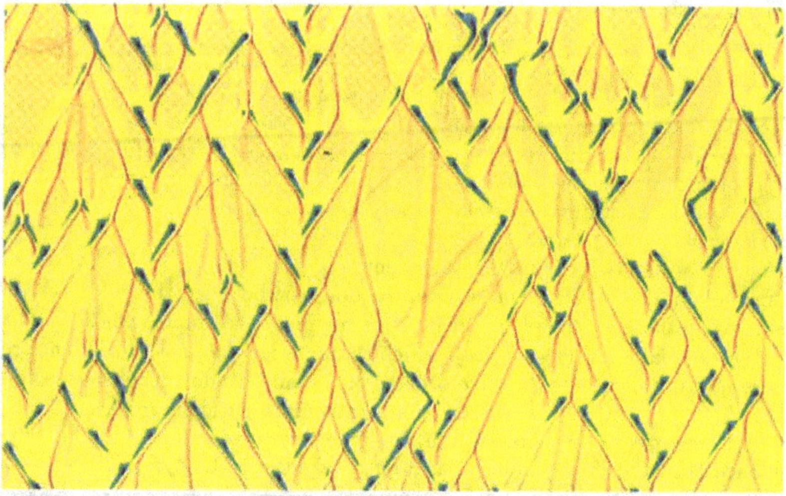

4b

Figure 4: Spatiotemporal plots of the amplitude of the solution for two turbulent regimes coexisting at the same parameter values (region II); $b_1 = 1$, $b_3 = 0.8$, $L = 500$. (a): amplitude turbulence; color scale from 1.15 (red) to 0 (purple); $\Delta T = 200$. (b): phase turbulence; grey scale from 1.15 (white) to 1.09 (black); $\Delta T = 2000$.

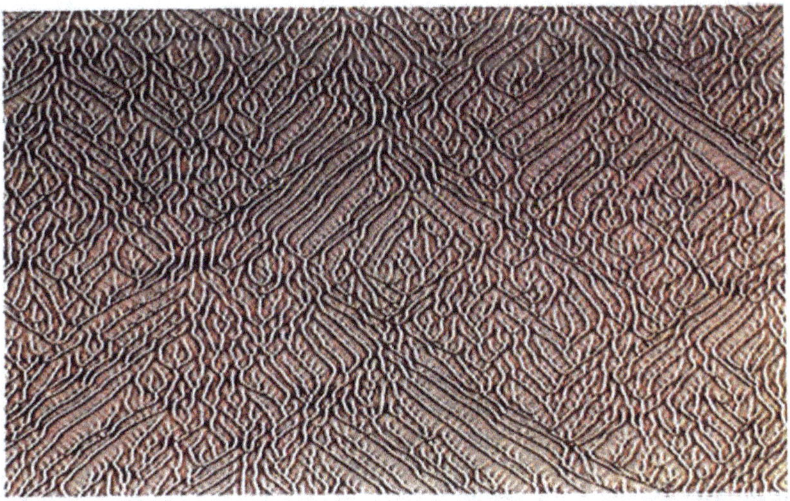

Figure 5: Spatiotemporal plot of the amplitude of the solution in the phase turbulence region (II and III); $b_1 = 2$, $b_3 = 1.03$, $L = 1000$, $\Delta T = 2000$. Grey scale from 1.12 (white) to 0.78 (black). The winding number is null.

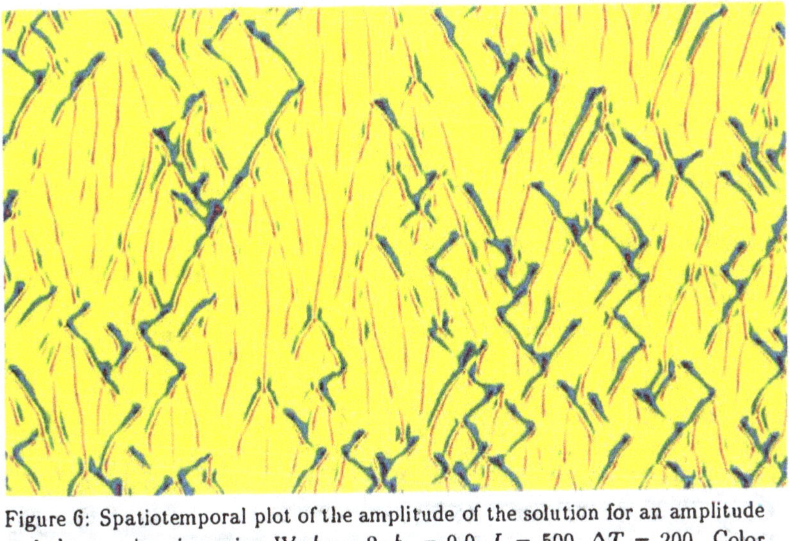

Figure 6: Spatiotemporal plot of the amplitude of the solution for an amplitude turbulent regime in region IV; $b_1 = 2$, $b_3 = 0.9$, $L = 500$, $\Delta T = 200$. Color scale from 1.26 (red) to 0 (purple).

Kuramoto-Sivashinsky turbulence to Korteweg-de-Vries integrable soliton dynamics[22]. Of course, it is not clear, at this point, whether the observed regimes strictly qualify as spatiotemporal intermittency ones. In particular, the precise definition of two local states remains to be done, followed by an assessment of the absorbing character of the laminar state.

3.4 Elsewhere

In the remaining part of the (b_1, b_3) plane where disordered regimes are observed (region IV in Fig. 1), there are, strictly speaking, no spatiotemporal intermittency regimes. Nevertheless, close enough to the transition lines L_1 and L_3, one can observe regimes which fulfill the most general definition of spatiotemporal intermittency (intermittent character of the solutions in both space and time). This is due to the fact that the "laminar" state is absorbing only in a weak sense due to the Benjamin-Feir instability in this region. The regimes studied by Gil[23] are of this type and cannot be compared to directed percolation-like processes, not for the reasons invoked, but simply because they lack an absorbing state. Fig. 6 presents a typical case where the spontaneous nucleation of a defect ("turbulent" state) in a "laminar" domain can be observed.

4. Conclusion

Ample evidence has been given that spatiotemporal intermittency does occur in the one-dimensional complex Ginzburg-Landau equation. Of course, many points still require a deeper understanding, in particular the transition lines L_1 and L_2, and the role of the Nozaki-Bekki solutions in the dynamics of the spatiotemporally disordered regimes. Also, the case of higher space dimensions should be carefully considered since defects then possess a topological nature. These points will be addressed in future publications.

At any rate, the results strengthen the belief that spatiotemporal intermittency is a generic mode of spatiotemporal disorder in extended systems. At a more general level, they underline the need to go to a formalism inspired by statistical mechanics and critical phenomena if one is to correctly describe "extensive chaos"[24] situations.

Acknowledgements

This work stems out of collaboration with P. Manneville, P.C. Hohenberg, B. Nicolaenko, A. Pumir, W.Van Saarloos, and B.I. Shraiman, to whom I am greatly indebted.

References

1. For a recent review see, e.g.: H. Chaté, in *Spontaneous Formation of Space-Time Structures and Criticality*, eds. T. Riste and D. Sherrington, (Kluwer, Dordrecht, 1991).

2. P. Manneville, *Dissipative Structures and Weak Turbulence*, (Academic Press, Boston, 1990).

3. Y. Pomeau, *Physica* **D23** (1986)3.

4. See, e.g., M. Van Dyke, *An Album of Fluid Motion*, (Parabolic Press, Stanford, 1982).

5. W. Kinzel, in *Percolation Structures and Processes, Ann. Israel Phys. Soc.* **5** (1983) 425.

6. W. Kinzel, *Z. Phys.* **B58** (1985) 229.

7. H. Chaté and P. Manneville, *Phys. Rev. Lett.* **58** (1987) 112; H. Chaté and B. Nicolaenko, in *New trends in Nonlinear Dynamics and Pattern Forming Phenomena: the Geometry of Nonequilibrium*, eds. P. Huerre and P. Coullet, (Plenum, New York, 1991).

8. K. Kaneko, *Prog. Theor. Phys.* **74** (1985) 1033.

9. H. Chaté and P. Manneville, *Physica* **D32** (1988) 409; *Europhys. Lett.* **6** (1988) 591; *Physica* **D37** (1989) 33.

10. S. Ciliberto and P. Bigazzi, *Phys. Rev. Lett.* **60** (1988) 286; F. Daviaud, M. Dubois and P. Bergé, *Europhys. Lett.* **9** (1989) 441; F. Daviaud, M. Bonetti and M. Dubois, *Phys. Rev.* **A42** (1990) 3388.

11. M. Rabaud, S. Michalland and Y. Couder, *Phys. Rev. Lett.* **64** (1990) 184.

12. S. Nasuno, B. Janiaud, and Y. Sawada, private communications.

13. M.C. Cross and P.C. Hohenberg, *Rev. Mod. Phys.*, to be published.

14. B.I. Shraiman, A. Pumir, W. van Saarloos, P.C. Hohenberg, H. Chaté and M. Holen, "Spatiotemporal chaos in the one-dimensional complex Ginzburg-Landau equation", submitted to *Physica D.*

15. Y. Kuramoto, *Prog. Theor. Phys. Suppl.* **64** (1978) 346; *Chemical Oscillations, Waves and Turbulence*, (Springer, Berlin, 1984).

16. A. Weber, L. Kramer, I.S. Aranson and L.B. Aranson, "Stability limits of travelling waves and the transition to spatiotemporal chaos in the complex Ginzburg-Landau equation", preprint.

17. H. Chaté, P.C. Hohenberg, A. Pumir, W. van Saarloos and B.I. Shraiman, "Spatiotemporal intermittency regimes of the one-dimensional complex Ginzburg-Landau equation", in preparation.

18. N. Bekki and K. Nozaki, *Phys. Lett.* **A110** (1985) 133.

19. See, e.g., S. Wolfram, *Rev. Mod. Phys.* **55** (1983) 601.

20. H. Chaté, P.C. Hohenberg, A. Pumir, W. van Saarloos and B.I. Shraiman, "Phase turbulence in the one-dimensional complex Ginzburg-Landau equation", in preparation.

21. H. Sakaguchi, *Prog. Theor. Phys.* **83** (1990) 169; **84** (1990) 793.

22. H. Chaté, unpublished.

23. L. Gil, *Nonlinearity*, **4** (1991) 1213.

24. P.C. Hohenberg and B.I. Shraiman, *Physica* **D37** (1989) 109.

Intermittency effects of energy cascade and passive scalar fields

M.H. Jensen[1], G. Paladin[2] and A. Vulpiani[3]

[1] *NORDITA, Blegdamsvej 17, DK 2100 Copenhagen Ø, Denmark*
[2] *Dipartimento di Fisica, Università dell'Aquila, Via Vetoio, I-67010 Coppito L'Aquila, Italy*
[3] *Dipartimento di Fisica, Università "La Sapienza", P.le Aldo Moro 2, I-00185 Roma, Italy*

Abstract

We review some recent work on scaling laws in turbulence. The Navier-Stokes equations and the equation for the transport of passive scalar fields are formulated in terms of shell models in Fourier space, which incorporate the appropriate conservation laws. We find that the Kolmogorov scaling for the energy cascade and the Obukhov-Corrsin scaling for the passive scalar cascade are dynamically unstable. This fact gives rise to corrections to classical values of the scaling exponents. We relate these corrections to the dynamical behavior of the intermittency bursts present in the shell models. Our numerical results are in good quantitative agreement with the experimental results in fluids.

1. Introduction.

It is of great current interest in turbulence research to understand whether intermittency effects influence properties like the energy spectrum for the turbulent flows, the probability distribution functions for velocities and temperatures (and their gradients), the scaling behavior of passively and actively carried scalar fields, etc. Many experimental data indicate that energy spectra follow quite closely the classical Kolmogorov[1] law and that the scaling of passive scalar fields are well described by the Obukhov-Corrsin law.[2] On the other hand, small deviations from these laws have been reported quite often over the last ten years and it has been proposed that these small deviations are due to the very non-homogeneous structures which are observed particularly on small scales, the so-called intermittency bursts.[3−11] Furthermore, it has become evident, that for turbulent flows, the probability distribution functions of various quantities do not follow standard Gaussian forms but are instead exponentials, or even stretched exponentials, indicating that large excursions are more likely to happen with respect to the Gaussian prediction.[7] Again, these effects have been interpreted by some authors as a clear sign of the presence of strong intermittent behavior.

One fundamental problem in experimental and numerical investigations is that the inertial range for the energy spectrum is short and it is really hard to extract reliable exponents for the scaling behavior and even more so for eventual corrections to standard scaling. For this reason it is useful to formulate approximate models of the Navier-Stokes equations which allows one to extract more accurate information about intermittency effects. This is the motivation for the work presented in this contribution. We use the technique of formulating a partial differential equation by ordinary differential equations (O.D.E.) which describe the evolution of shells in the Fourier k-space, limiting the number of degrees of freedom to 2 for each shell. Each shell interacts with various other shells and we can formulate a closed set of O.D.E. which respects the conservation laws of the original equations. Since we have a limited degrees of freedom ($O(10^2)$), we can study the dynamical behavior of energy bursts and intermittency structures in all details and thereby get a deeper insight into what might happen in real turbulence.

2. Equations and shell model.

Let us first recall the equation for the time evolution of a scalar field Θ passively carried by the fluid:

$$\partial_t \Theta + (\mathbf{u} \cdot \nabla)\Theta = D\Delta\Theta + f_\Theta \tag{1a}$$

where $\mathbf{u}$ is the velocity field, given by the Navier-Stokes equations:

$$\partial_t \mathbf{u} + (\mathbf{u} \cdot \nabla)\mathbf{u} = -\frac{\nabla P}{\rho} + \nu\Delta\mathbf{u} + \mathbf{f}_u \tag{1b}$$

Here $\mathbf{f}_u$ and f_Θ are external forcings, D is the diffusion coefficient, ρ is the fluid density, P is the pressure and ν is the viscosity. A first characterization of the statistics of passive scalars is given by the correlation function

$$C(\mathbf{r}) = < \Theta(\mathbf{x})\ \Theta(\mathbf{x}+\mathbf{r}) >, \tag{2}$$

where $< \cdots >$ means spatial average. The theory of Obukhov-Corrsin[2] predicts that its power spectrum

$$\Gamma(k) = \int_{|\mathbf{k}|=k} d\mathbf{k} \int C(\mathbf{r}) e^{-i\,\mathbf{k}\cdot\mathbf{r}}\, d\mathbf{r}, \tag{3}$$

exhibits the power law decay $\Gamma(k) \sim k^{-5/3}$ in an inertial convective subrange, where there is a nonlinear transfer of the energy and molecular diffusion is negligible. For Prandtl number $Pr = \nu/D \geq 1$, this range coincides with the usual inertial range where inertia forces dominate and there exists an energy cascade towards small scales with a power law scaling for the energy spectrum $E(k)$. Hereafter we shall assume $Pr = 1$, for simplicity. The inertial range extends down to scales where the energy is dissipated by molecular viscosity. In the classical Kolmogorov theory[1] K41, which neglects fluctuations in the energy transfer, one has $E(k) \sim k^{-5/3}$ and the viscous cut-off happens at the so-called Kolmogorov length $\ell_K \sim (\nu^3/\epsilon)^{1/4}$ (ϵ is the mean energy dissipation).

Shell models have been introduced in order to study the intermittent behavior of the energy dissipation.[12–20] In spite of the few degrees of freedom considered, these models are able to reproduce the main features of the small scale statistics of the velocity field at very high Reynolds number Re. The success of the approach stems from the fact that detailed structures of the Navier-Stokes equations are not necessary to describe the appearance of power laws and intermittent behavior.

Our shell model for the passive scalar advection is a generalization of the model introduced by Ohkitani and Yamada[16–17] for the evolution of the velocity field. First, one divides the Fourier space in N shells. Each shell k_n ($n = 1, 2, .., N$) consists of the wavenumbers with modulus k such that $k_0 2^n < k < k_0 2^{n+1}$. The velocity and Θ increments on scale k_n^{-1} are given by the complex variables u_n and θ_n. The evolution equations are obtained according the following criteria:

(a) the linear term for u_n and θ_n are given by $-\nu k_n^2 u_n$ and $-Dk_n^2\theta_n$

(b) the non-linear terms for u_n are combination of the form $k_n u_{n'} u_{n"}$; for θ_n they are combination of the form $k_n \theta_{n'} u_{n"}$

(c) n' and $n"$ are nearest and next nearest neighbors of n

(d) in absence of forcing and damping one has conservation of volume in phase space (Liouville theorem) and the conservation of energy $\frac{1}{2}\sum_n |u_n|^2$ and of $\sum_n |\theta_n|^2$

(e) the ('non-intermittent') scaling laws of Kolmogorov $E(k) \sim k^{-5/3}$, i.e. $u_n \sim k_n^{-1/3}$, and of Obukhov-Corrsin $\Gamma(k) \sim k^{-5/3}$, i.e. $\theta_n \sim k^{-1/3}$, are unstable fixed points of the inviscid unforced evolution equations.

The properties (a), (b) and (d) are also valid for the Eq.s (1a) and (1b) in Fourier space, while (c) is an assumption on the locality of interactions among modes.

(e) is a central ingredient of the model: it seems necessary to obtain scaling laws close to those derived by dimensional arguments. The instability of the fixed point is also very important. In fact, intermittency and corrections to the classical exponents of the scaling arise as a consequence of the motion of the system on a chaotic attractor in the $4N$ phase space.

It is important to remark that u_n is a complex scalar so that the vectorial structure of Navier-Stokes equations is lost. That is the weak point of the model but one might hope that the coherent structures widely observed in turbulent fluids do not affect the very nature of scaling, at least on a qualitative level. More sophisticated vectorial shell models have been introduced[20] but they have a much larger number of degrees of freedom than the model of Ohkitani and Yamada.[16,17]

The equations of our cascade model with N shells can be written as

$$\left(\frac{d}{dt} + \nu k_n^2\right) u_n = i(a_n u_{n+1}^* u_{n+2}^* + b_n u_{n-1}^* u_{n+1}^* + c_n u_{n-1}^* u_{n-2}^*) + f\delta_{n,4} \quad (4a)$$

$$\left(\frac{d}{dt} + Dk_n^2\right) \theta_n = i\,[e_n (u_{n-1}^*\theta_{n+1}^* - u_{n+1}^*\theta_{n-1}^*) +$$
$$g_n (u_{n-2}^*\theta_{n-1}^* + u_{n-1}^*\theta_{n-2}^*) + h_n (u_{n+1}^*\theta_{n+2}^* + u_{n+2}^*\theta_{n+1}^*)\,] + \overline{f}\delta_{n,4} \quad (4b)$$

where u_n and θ_n are complex variables. ν is the viscosity, D is the molecular diffusion and f, $\overline{f}$ are external forcings (here on the fourth mode). The conservation of phase space (for $\nu = D = f = \overline{f} = 0$) is automatically satisfied by the

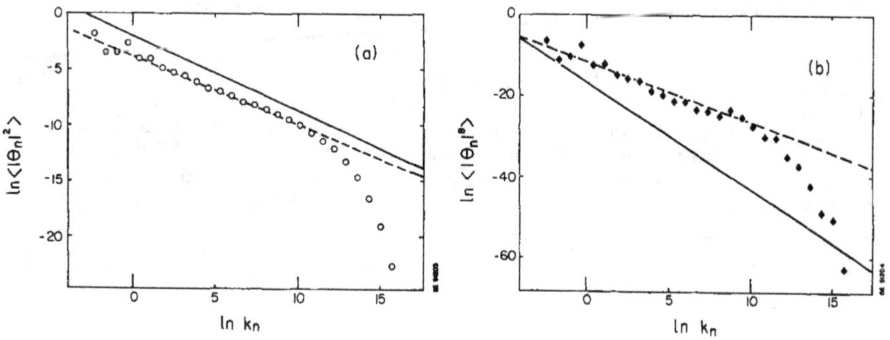

Fig. 1. Plot of the structure functions, $\ln\langle\, |u_n|^p \,\rangle$ versus $\ln k_n$ for a): p=1 and b): p=12 found from simulations with N=27 over 10^4 time units. The full lines have slopes respectively -1/3 and -4 from Kolmogorov scaling. The dashes lines have slopes -0.39 and -3.18, respectively.

absence of diagonal terms proportional to u_n, θ_n in the right hand side of Eq.s (4). The coefficients of the non-linear terms follow from demanding the conservation of $\sum_n |\theta_n|^2$ and of $\sum_n |u_n|^2$ (energy) in the absence of forcing and with $\nu = D = 0$. A possible choice is:

$$a_n = k_n, \qquad b_n = -\frac{k_{n-1}}{2} \qquad c_n = \frac{-k_{n-2}}{2} \tag{5a}$$

$$e_n = \frac{k_n,}{2} \qquad g_n = -\frac{k_{n-1}}{2} \qquad h_n = \frac{k_{n+1}}{2} \tag{5b}$$

with

$$b_1 = b_N = c_1 = c_2 = a_{N-1} = a_N = 0$$

$$e_1 = e_N = g_1 = g_2 = h_{N-1} = h_N = 0$$

The unstable fixed point of Eq.s (4) when $\nu = D = f = \overline{f} = 0$ is given by the Obukhov-Corrsin scaling $u_n \sim \theta_n \sim k_n^{-1/3}$. One observes that the time evolution of the dissipative system (4) is chaotic and confined on a strange attractor in the $4N$ dimensional phase space.

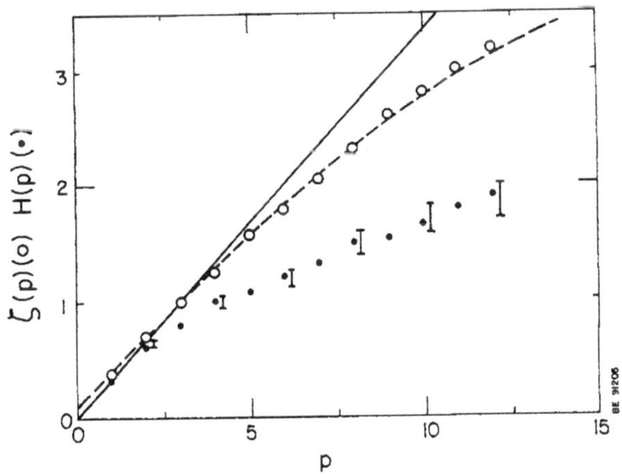

Fig. 2. The structure functions exponents $\zeta(p)$ (open circles) and $H(p)$ (dots) plotted versus p. The data are obtained by an integration of Eq.s (4) with $N = 27$. For larger p, the experimental data of $H(p)$ from Ref. [4] are indicated by vertical bars, with a slight shift in p. The error bars of $H(p)$ computed in the shell model are comparable with the experimental errors (vertical bars). The solid line is the Obukhov-Corrsin prediction $H(p) = \zeta(p) = p/3$; the dashed line is the random β model fit for the $\zeta(p)$.

3. Results for structure functions.

Let us consider the scaling of the velocity differences over small length scales (but larger than the Kolmogorov inner length):

$$\delta u(\ell) \equiv |\mathbf{u}(\mathbf{x}+\ell) - \mathbf{u}(\mathbf{x})| \sim \ell^h.$$

From the Kolmogorov result $h = \frac{1}{3}$ we find for the structure functions

$$< \delta u(\ell)^p > \propto \ell^{\zeta_p} , \qquad \text{with} \qquad \zeta(p) = p/3 \tag{6}$$

where $< \cdot >$ is a spatial average. In the shell model the velocity differences are instead averaged over time and we have obtained structure functions up to the order $p = 12$, by a numerical integration of Eq.(4) with N=27 shells. Fig.1 shows a plot of $\ln < |u_n|^p >$ versus $\ln k_n$ for the two cases, p=1 and p=12. We find a good scaling over about five decades between the shell of the energy input and the shell of the Kolmogorov inner scale which defines the inertial range. The slope is $-\zeta(p)$ and is different from the Kolmogorov result $< |u_n|^p > \propto k_n^{-p/3}$ which is indicated by the full drawn line. Fig.2 collects the results for the velocity structure function

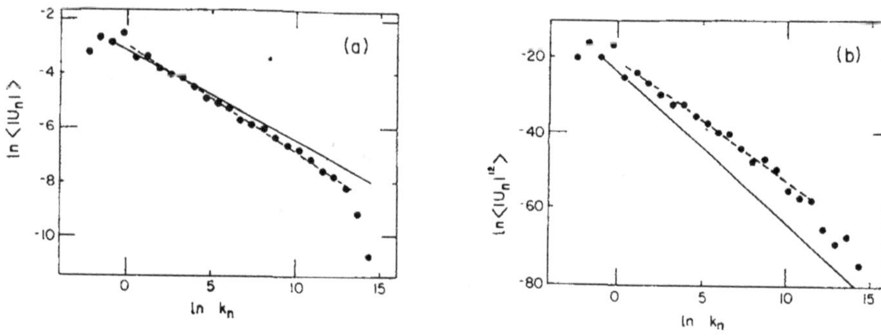

Fig. 3. Scaling of the structure functions for passive scalar. Plotted are $\ln < |\theta_n|^p >$ versus $\ln k_n$ for $p = 2$ (Fig. 3a) and $p = 8$ (Fig. 3b) again with $N = 27$. The solid lines with slopes respectively $-2/3$ and $-8/3$ indicate the Obukhov-Corrsin scaling laws. The dashed lines have slopes -0.61 and -1.5

exponents with N=27 and $\nu = 10^{-9}$. We note that the exponent $\zeta(p)$ is not linear in p. The results are in very good agreement with the experimental results by Anselmet et al obtained from velocity measurements in a channel flow [6]. Our data suggest an underlying multifractal structure of the energy dissipation since $\zeta(p)$ can be obtained by the random β model [21], as shown by the dotted line in Fig.2.

Similarly, we have calculated the higher order structure functions for the passive scalar field, Fig.3. The scaling $< |\theta_n|^p > \sim k_n^{-H(p)}$ is found not to follow the Obukhov-Corrsin prediction $H(p) = p/3$ (see next section), even for small p's, as shown in Fig.2. In fact, $H(p)$ is a more strongly non-linear function of p than $\zeta(p)$. Fig. 2 also shows the experimental data of $H(p)$. The agreement between the experimental values and the ones obtained by our model is impressive. In particular, for $p = 2$ we find $H(2) = 0.61 \pm 0.01$ and $\zeta(2) = 0.70 \pm 0.01$ so that the correction to the Obukhov-Corrsin exponent 5/3 for the power spectrum $\Gamma(k)$ has the opposite sign of the correction to the exponent for the energy spectrum $E(k)$, since $\Gamma(k) \sim k^{-(H(2)+1)}$ and $E(k) \sim k^{-(\zeta(2)+1)}$. This result has been also observed in real fluids.[22]

We have also computed the probability distribution function (PDF) of the passive scalar increments at different scales - i.e. the PDF of θ_n at varying n., as shown in Fig. 4. For k_n at the beginning of the inertial range, that is Θ differences on large scale, the PDF is quite close to a gaussian form. On the contrary at the end of the inertial range, the PDF exhibits the typical exponential tails of velocity and Θ gradients in turbulence. Note that in our model, for n at the end of the inertial range, k_n^{-1} is order of the Kolmogorov length and one can identify $k_n\theta_n$

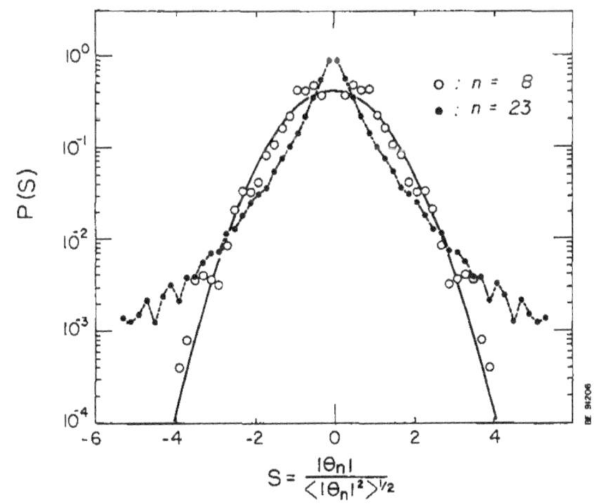

Fig. 4. PDF of $s = |\theta_n| / < |\theta_n|^2 >^{1/2}$ for $n = 8$ (open circles) and $n = 23$ (dots). The solid line is the standard gaussian. The PDF for negative s is drawn as a help to the eye and it is taken to be equal to the PDF of the corresponding positive s values.

with the gradient of Θ. The PDF of $\theta_{n=23}$ in Fig. 4 should thus be compared with the experimental PDF for the Θ gradients.[7]

4. Scaling laws for passive scalars

Let us briefly recall the predictions of the Obukhov-Corrsin theory[2] which makes use of arguments similar to those of the Kolmogorov K41 theory.[1] By dimensional arguments, it is assumed that the time for the spectral transfer at $k \sim r^{-1}$ is $\tau(r) \sim r/\delta u(r)$ where $\delta u(r) = |\mathbf{u}(\mathbf{x}+\mathbf{r}) - \mathbf{u}(\mathbf{x})|$ is the velocity increment on scale r. It follows that the transfer rate of Θ on scale r is

$$N(r) = \frac{\delta\Theta^2(r)}{\tau(r)} \sim \frac{\delta u(r)\,\delta\Theta^2(r)}{r} \tag{7a}$$

as in the K41 theory the energy transfer rate is

$$\epsilon(r) = \frac{\delta u^2(r)}{\tau(r)} \sim \frac{\delta u(r)^3}{r}. \tag{7b}$$

The classical theories ignore the intermittency in $N(r)$ and $\epsilon(r)$, which are assumed to be constant and equal to the correspondent dissipation (and injection) rates, N and ϵ, taken uniform in space and time. Under this hypothesis, the structure functions are easily derived to be

$$< \delta u^p > \sim r^{\zeta(p)} \sim \epsilon^{p/3} r^{p/3} \qquad \zeta(p) = p/3 \tag{8a}$$

$$< \delta\Theta^p > \sim r^{H(p)} \sim N^{p/2}\, \epsilon^{-p/6} r^{p/3} \qquad H(p) = p/3 \tag{8b}$$

However, there is clear evidence of strong fluctuations in the dissipation rates.

According the second revised similarity hypothesis of Kolmogorov,[23] the fluctuations of $\delta\Theta$ and δu are determined by the fluctuations of ϵ and N so that the scaling laws are given by

$$< \delta u^p > \sim < \epsilon(r)^{p/3} > r^{p/3} \tag{9a}$$

$$< \delta\Theta^p > \sim < N(r)^{p/2}\, \epsilon(r)^{-p/6} > r^{p/3} \tag{9b}$$

Intermittency has been described by cascade models by several authors.[21,24,25] Basically the exponents $\zeta(p)$ are expressed in terms of the generalized fractal dimensions D_q, related to the scaling of $\epsilon_x(r)$, which is the coarse grained probability measure given by the integral of the energy dissipation over a sphere of radius r around x. The D_q's are defined via the relation[21,25]

$$< \epsilon_x(r)^q > \sim r^{(q-1)\,(D_q-3)} \tag{10}$$

so that from (9a) and (10) one has

$$D_q = \frac{\zeta(3q) + 2q - 3}{q-1} \tag{11}$$

In this way, the D_q are related to the multifractal spectrum of singularities of the velocity field. In fact, one can consider the sets $S(h)$ of points where the velocity increments scale as $\delta u_x(r) \sim r^h$. The fractal dimensions of these sets are the multifractal spectrum $D(h)$ related via the Legendre transformation

$$\zeta(p) = \min_h [ph - D(h) + 3] \tag{12}$$

to the structure function exponents. The problem is much more complicated for the exponents $H(p)$, even if one introduces the generalized dimensions d_q for the probability measure given by the passive scalar dissipation via the scaling

$$< N_x(r)^q > \sim r^{(q-1)(d_q-3)} \tag{13}$$

It is then simple to construct multifractal cascade models for N . However, the knowledge of the two multifractal spectra is not sufficient to determine the scaling exponents $H(p)$ of the structure functions, without taking into account the correlation between N and ϵ.

In order to continue our argument we should introduce a so-called joint multifractal distribution for N and ϵ. This goes beyond the scope of this presentation but we refer to the discussion in our paper, Ref. 26. We however give a simple expression that relates $H(p)$ to d_q and D_q, at least for not too large q's. Since $\epsilon(r)$ does not appear in combinations of the kind $\delta\Theta(r)^{2q}\delta u(r)^q$, we have

$$< \delta\Theta(r)^{2q} \delta u(r)^q > \sim < N(r)^q > r^q \sim r^{(q-1)\,(d_q-3)+q}. \tag{14}$$

Under the assumption that $\delta\Theta$ and δu are uncorrelated, i.e.

$$<\delta\Theta(r)^{2q}\delta u(r)^q> \sim <\delta\Theta(r)^{2q}><\delta u(r)^q>, \tag{15}$$

the structure function exponents become

$$H(2q) = q - \zeta(q) + (q-1)(d_q - 3) \tag{16}$$

We have then computed the fractal dimensions d_q from the time signal of the Θ^2 dissipation. These exponents are also in good agreement with the experimental data.[5] Our data indicate that the approximate relation (16) between $H(q)$, d_q and $\zeta(q)$ works rather well. We find small discrepancies (0.01 ± 0.01 for $q = 1$ and 0.2 ± 0.1 for $q = 4$) from the value of $H(2q)$ and the estimates obtained by inserting the values of d_q and $\zeta(q)$ into (16).

5. Dynamical Information: Lyapunov Exponents and Eigenvectors.

In this section, we shall study some of the dynamical information and the relation to intermittency effects, which can be obtained from integrating the equations of motion. To answer these questions we have ONLY studied the velocity equation in (4a) and do not consider the passive scalar at all. An appropriate way to characterize the dynamics is by means of the Lyapunov exponents $\lambda_1 \geq \lambda_2 \geq \ldots \geq \lambda_{2N}$. They have been calculated by Yamada and Ohkitani[16] and Fig.5 shows the Lyapunov spectrum, $\frac{1}{H}\sum_{i=1}^{j}\lambda_i$ versus $\frac{j}{D}$, where H is the Kolmogorov entropy and D is the Kaplan-Yorke dimension. The spectrum is plotted for three different values of the viscosity indicating a similarity scaling structure. We note a very interesting characteristic of the $N \to \infty$ limit of the spectrum: There is a small fraction of positive exponents, then a large fraction of marginal (i.e. zero) exponents, followed by the negative ones of order $-\nu k_i^2$ which correspond to the viscous damping and are responsible for the strongest contraction rates. The appearance of many almost zero Lyapunov exponents might be connected to the existence of spatial power law scaling for the structure functions. In fact, it seems difficult to maintain the spatial power laws if there were many exponentially diverging directions in the $2N$ phase space for orbits. A finite density of zero Lyapunov exponents, a sort of "weak chaos",[27] has been obtained by Ruelle for the discrete spectrum of operators that linearizes the Navier-Stokes equations[28] and appears also in Hamiltonian systems near integrability.[29]

Whereas the Lyapunov exponents give information about the relevant time scales the corresponding Lyapunov tangent vectors can resolve the interplay between various length scales during time evolution. Actually, as shown by Ohkitani and Yamada[16,17] the time average of the i'th Lyapunov vector gives a measure of the localization on the various shells related to the instability of the i'th Lyapunov exponent.

In particular, there is a strong correspondence between Lyapunov eigenvectors of the last negative Lyapunov exponents and dissipative modes following the end of the inertial scaling range, since the viscous damping is responsible of the strongest contraction rates, so that $\lambda_{2i} \approx \lambda_{2i-1} \propto -\nu k_i^2$ for $i \approx N$. On the other hand, the eigenvector corresponding to the small Lyapunov exponents are concentrated, although in a less sharp way, in the inertial wavenumbers. This weak correspondence, as well as the existence of a large number of small negative γ_i of

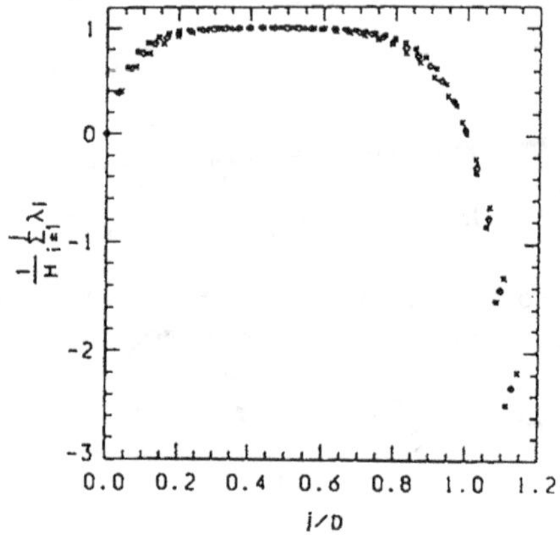

Fig. 5. The Lyapunov spectrum from Ref. [16] for $\nu = 10^{-7}$, $10^{-8}, 10^{-9}$. Plotted is $\frac{1}{H}\sum_{i=1}^{j} \lambda_i$ versus $\frac{j}{D}$, where H is the Kolmogorov entropy and D is the Kaplan-Yorke dimension. Note the similarity of the spectrum for different values of ν. A small fraction of the spectrum are positive exponents but the dominating part consists of marginal exponents.

the jacobian matrix at the Kolmogorov fixed point, suggest that the power scaling law are related to the presence of a large number of (almost) zero Lyapunov exponents. In this picture, the largest Lyapunov exponent - and its eigenvector - should be responsible of the intermittency.

With the following notation:

$$\frac{du_n^R}{dt} = F_{2n-1}(\mathbf{u}) \qquad \frac{du_n^I}{dt} = F_{2n}(\mathbf{u}), \tag{17}$$

where u_n^R (u_n^I) denotes the real (imaginary) part of u_n, we have integrated the linearized equation

$$\frac{dz_i}{dt} = A_{i,j} \cdot z_j \qquad i\ ,\ j\ =\ 1, ..., 2N \tag{18}$$

for the time evolution of an infinitesimal velocity increment $\mathbf{z} = \delta \mathbf{U}$, where $A_{n,j} \equiv \partial F_n / \partial U_j$ is the jacobian matrix of Eq.(4a), and $\mathbf{U} = (u_1^R, u_1^I, ..., u_N^R, u_N^I)$. The solution for the tangent vector $\mathbf{z}$ can thus be formally written as $\mathbf{z}(t_2) = M(t_1, t_2) \cdot \mathbf{z}(t_1)$, with $M = \exp \int_{t_1}^{t_2} A(\tau) d\tau$. We introduce a stability basis $\mathbf{e}_i$ given by the eigenvectors of the matrix M so that a generic tangent vector $\mathbf{z}(t)$ is projected by the evolution along $\mathbf{e}_1$ (i.e. $\mathbf{z}(t) = c\ exp(\lambda_1\ t)\ \mathbf{e}_1$) a part corrections $O(exp - |\lambda_1 - \lambda_2| t)$. The time average of the first eigenvector spreads in the whole inertial range and there seems to be no correspondence with the large scale.[16,19]

It is thus interesting to compute the time evolution of the stability eigenvector $\mathbf{e}_1(t)$. In fact, we observe that its projection on the different vector shells varies quite violently in time. We estimate the fraction of the stability vector in the shell k_n as $p(n) \equiv |\mathbf{e}_1(k_n)|^2/\sum_j |\mathbf{e}_1(k_j)|^2$, where $|\mathbf{e}_1(k_n)|^2 = |\mathbf{e}_1(2n-1)|^2 + |\mathbf{e}_1(2n)|^2$. In particular, we see that during an intermittent burst there appears to be a concentration on the components related to the dissipation (Kolmogorov) scale. To make this idea more precise, we have computed the response after a time τ to an infinitesimal perturbation, defining an instantaneous maximum Lyapunov exponent as

$$\chi_\tau(t) \equiv \frac{1}{\tau} \ln \, \left|\frac{\mathbf{z}(t+\tau)}{\mathbf{z}(t)}\right| \tag{19}$$

Note that $\lambda_1 = \lim_{T\to\infty} (1/T) \int_0^T dt \; \chi(t)$. The value of χ is an indication of the global chaoticity of the system at a given instant. We now focus in particular on the dissipation scale which for the case of N=19 and $\nu = 10^{-6}$ is at the wave number $k_D = k_{15}$. In Fig.6 we show time series of for three quantities: a). The energy dissipation E_D estimated by $|u_{15}|^2$; b). The instantaneous Lyapunov exponent χ; c). The eigenvector fraction on k_D, $p_D \equiv p(k_{15})$. First of all, we note the strongly intermittent nature of energy dissipation. Furthermore, the spikes in E_D are connected to spikes in χ and p_D. In the same way during a laminar period of low E_D values, also χ and p_D are extremely small. The correlation can be quantified by calculating the time correlation function between each two of the quantities. We find that the correlation is very strong at zero time difference τ but is quickly decaying at increasing τ. Moreover, the time correlation function between E_D and χ becomes negative after $\tau \sim 1$.[19] This effect can be directly seen in the time series of Fig.6b. If the value of χ suddenly undergoes a big positive burst, it is immediately followed by a big negative burst. This suggests the following picture: By a big chaoticity burst, the system is suddenly "kicked" away from the normal energy cascade corresponding to a system motion in the phase space close to the Kolmogorov fixed point of Eq.(4a). However, the system is immediately obliged to invert the excursion and contracts back towards the Kolmogorov fixed point, approximately along the same direction (the dissipative shell). This implies that a large value of the instantaneous Lyapunov exponent χ ("a kick") should be followed by a negative χ with large absolute value ("a sudden contraction"). During the bursting event, the Lyapunov eigenvector is found to be localized in the dissipative shell direction in the k-space, indicating that kick and contractions happen in a subspace of very low dimensionality.

The implication of the calculation shown in Fig.6 is that strong energy intermittency is triggered by strong instantaneous chaoticity and a subsequent localization on the dissipation wave numbers and we might be tempted to argue that the intermittency corrections have to do with violent motion (given by the largest Lyapunov exponents) at scales close to the Kolmogorov length. This is however too simplified. The maximal Lyapunov exponent λ_1 is a global quantity and when averaging the corresponding eigenvector we find that it spreads all over the inertial range. This has also been observed in Ref.16,17. The intermittency is a complex phenomenon involving the transfer in the whole inertial range. Actually, what appears to happen is that the distribution in the eigenvector switches from the dissipation wave number during intermittent bursts to the inertial range during more laminar periods.

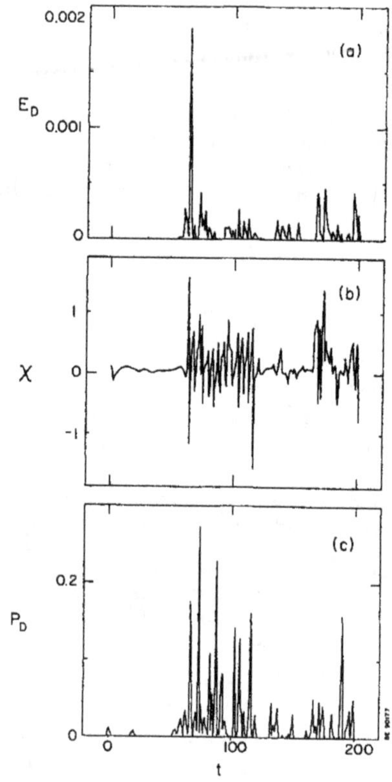

Fig.6. Temporal sequences (200 time units) of E_D (Fig.6a), the instantaneous Lyapunov exponents χ (Fig.6b), and P_D (Fig.6c) for $N = 19$. The sequences show that the energy bursts are in correspondence with large deviations of χ and localization of its eigenvector on the dissipative modes. Note the laminar phase (very small E_D values) during the first 50 time units corresponding to almost vanishing p_D and χ.

References.

[1] A.N. KolmogorovC.R. Acad. Sci. USSR **30** 301 (1941)

[2] A. M Obukhov Izv. Acad. SSSR, Ser. Geogr. Geofiz. **13** 58 (1949)
S. CorrsinJ. Appl. Phys. **22** 469 (1951)

[3] C.W. Van Atta Phys. Fluids **14** 1803 1971

[4] R.A. Antonia and C.W. Van Atta J. Fluid. Mech. **84** 561 (1978)

[5] R.A. Antonia and K. R. Sreenivasan Phys. of Fluids **20** 1800 (1977)

[6] F. Anselmet, Y. Gagne, E.J. Hopfinger and R. Antonia J. Fluid Mech. **140** 63 (1984)

[7] R.A. Antonia, E.J. Hopfinger, Y. Gagne and F. Anselmet Phys. Rev. **A30** 2704 (1984)
[8] H.G.E. Hentschel and I. Procaccia Phys. Rev. **A28** 417 (1983)
[9] C.M. Meneveau and K.R. Sreenivasan, P. Kailasnath and M.S. Fan Phys. Rev. **A41** 894 (1990)
[10] L.C. Andrews and B. K. Shivamoggi Phys. of Fluids **A2** 105 (1990)
[11] V. R. Chechetkin, V.S. Lutinov and A. Yu. Turygin J. Stat. Phys. **61** 589 (1990)
[12] A.M. Obukhov Atmos. Oceanic Phys. **7** 41 (1971);
[13] V.N. Desnianskii and E.A. Novikov J. of Appl. Mathem. and Mech., PMM **38** 468 (1972);
E.B. Gledzer Sov. Phys. Dokl. **18** 216 (1973)
V.N. Desnianskii and E.A. Novikov Atmos. Oceanic Phys. **10** 127 (1974);
[14] E.D. Siggia Phys. Rev. **A15** 1730 (1977);
E.D. Siggia Phys. Rev. **A17** 1166 (1978);
R.M. Kerr and E.D. Siggia J. Stat. Phys. **19** 543 (1978)
[15] R. Grappin, J. Leorat and A. Pouquet J. Physique **47**1127 (1986)
[16] M. Yamada and K. Ohkitani J. Phys. Soc. of Japan **56** 4210 (1987);
M. Yamada and K. Ohkitani Progr. Theo. Phys. **79** 1265 (1988);
M. Yamada and K. Ohkitani Phys. Rev. Lett. **60** 983 (1988)
[17] K. Ohkitani and M. Yamada Progr. Theo. Phys. **81** 329 (1989)
[18] R. Ruiz and D.R. Nelson Phys. Rev. **A 23** 3224 (1981)
[19] M.H. Jensen, G. Paladin and A. Vulpiani Phys. Rev. **A43** 798 (1991)
[20] J. Eggers and S. Grossmann Physics Lett. **A156** 444 (1991);
[21] R. Benzi, G. Paladin, G. Parisi and A. Vulpiani J. Phys **A17** 3521 (1984);
[22] R.M. Williams and C.A. Paulson J. Fluid Mech. **83** 547 (1977)
[23] A. M Obukhov J. Fluid Mech. **13** 77 (1962)
A.N. Kolmogorov J. Fluid Mech. **13** 82 (1962)
[24] B. B. Mandelbrot J. Fluid Mech. **62** 331 (1974);
U. Frisch, P. Sulem and M. Nelkin J. Fluid Mech. **87** 719 (1978)
[25] C.M. Meneveau and K.R. Sreenivasan Phys. Rev. Lett. **59** 1424 (1987)
[26] M.H. Jensen, G. Paladin and A. Vulpiani Phys. Rev. A in press (1991)
[27] K. Chen, P. Bak and M.H. Jensen Phys. Lett. A **149** 207 (1990)
[28] D. Ruelle Comm. Math. Phys. **87** 287 (1982)
[29] R. Livi, M. Pettini, S. Ruffo and A. Vulpiani J. Stat. Phys. **48**530 (1987)

MECHANISM OF VORTEX RECONNECTION

Ryuji TAKAKI
Tokyo University of Agriculture and Technology,
Fuchu, Tokyo 183, Japan

and

Yoshitaka KAKIZAKI
Tokyo University of Agriculture and Technology,
Fuchu, Tokyo 183, Japan

ABSTRACT

Process of the vortex reconnection is investigated theoretically, experimentally and numerically. In the theory a simplified model is constructed based on the vorticity equation and several assumptions, which are considered valid in a local region where the vortex filaments are being reconnected. In the experiment the process was observed in two vortex rings ejected in the air. The numerical simulation is made for these vorte rings with the same initial condition as in the eperiment. These three results agree well and the mechanism of the process proposed in the theory was proved to be appropriate for explanation of this mysterious fluid motion.

1. Introduction

The reconnection of vortex filaments in a viscous fluid is observed since many years, but its mechanism is not yet understood well and this process remains one of the most mysterious phenomena in the fluid mechanics. It is observed when two vortex rings approach each other, or a single elliptical vortex ring is ejected.[1-3] The lines of magnetic force within plasma also make reconnection, where the finite electric conductivity is considered to be responsible.[4] Such a reconnection process is assumed to exist in other physical systems, such as the quantized vortex filament in the superfluid Helium and the disclination in the liquid-crystal. These phenomena are especially interesting because the vorticity lines or the magnetic lines suffer a sudden topological change, and supply interesting problems of dynamical pattern formation.

In spite of a lot of experiments and computations, theoretical studies of the reconnection mechanism are very rare, analyses by Yeh & Axford for the magnetic lines[4] and by the present author for the vortex filaments[5,6] being only examples. The reason of difficulty is that this fluid motion is essentially three-dimensional, time-dependent (impulsive) and that the viscosity plays an essential role.

On the other hand, numerical computations of this phenomenon have been made recently by several researchers.[7-11] Their results show certainly an occurrence of the reconnection process and reveal a complicated behavior of vorticity associated with the reconnection. However, a true understanding of the mechanism is not yet obtained. A theoretical explanation is considered to be the final goal for it.

Now, in order to get a better understanding of such a complicated phenomenon, it would be necessary to compare results of three ways of researches, i.e. theoretical analysis, experiment and computation. The purpose of this paper is to report the results of such researches by the present authors and to provoke interests in this peculiar phenomenon.

2. Theoretical Analysis

The theoretical analysis[5] is confined only to the region where the fluid is moving vigorously and the reconnection process is occuring. The analysis is limited to the case with the filaments with the equal circulations but with the opposite sense of rotation. They are assumed to have a symmetric configulation as shown in Fig. 1(a). The far parts of the filaments are assumed to lie on the x,y-plane and their central parts are lifted up and pressed each other owing to the induced velocity. The reconnection process is ignited in these parts (called an interaction region) to produce the final reconnected state (Fig.1(b)).

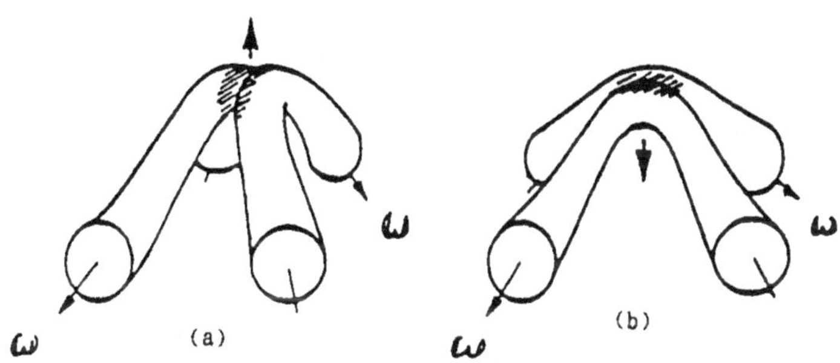

Fig. 1. Sketchs of the initial and the final states.

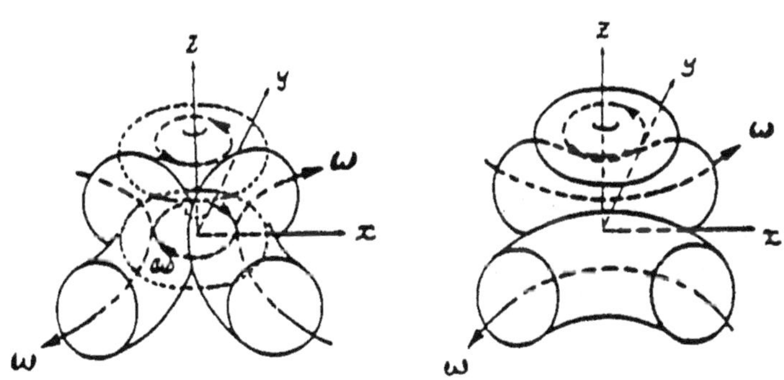

Fig. 2. A model to explain the process of reconnection.

The flow field within this interaction region is estimated based upon the following assumptions:

(i) Each component of the velocity or the vorticity is assumed to have a relatively simple polynomial function of the spatial coordinates because the effect of viscosity is expected to surpress rapid spatial variation of the velocity field.

(ii) The flow state is assumed nearly stationary both before and after reconnection, which are called the initial and the final states, respectively. The final flow state is expressed by a superposition of the initial state and an additional time-dependent with configuration of a vortex ring, as shown in Fig. 2, and grows rapidly.

(iii) In the initial and the final flow fields, the vorticities are directed to y-direction and x-direction, respectively, so that the flow fields does not depend on y and x, respectively.

(iv) The flow field has inversion symmetries with respect to the x,z-plane and the y,z-plane, and the superposing ring has an axisymmetry with respect to z-axix. These symmetries give limitations to possible functional forms of velocity and the vorticity components.

(v) The transition from the initial to the final states is described by a single function of time T(t), which has the following asymptotic conditions

$$T(t) \to 0 \text{ for } t \to -\infty \quad , T(t) \to 1 \quad \text{for} \quad t \to \infty. \tag{1}$$

Therefore, the superposing ring field changes in proportion to this function T(t).

Based upon these assumptions, the velocity and the vorticity components are written as follows:

$$u = -m(1-T)x + k(1-T)xz, \qquad v = mTy - kTyz,$$

$$w = w_0(0-) - \delta T + m(1-2T)z - \frac{\ell}{2}(1-T)x^2 + \frac{\ell}{2}Ty^2 - \frac{k}{2}(1-2T)z^2, \tag{2}$$

$$\omega_x = (k+\ell)Ty - nTyz, \qquad \omega_y = (k+\ell)(1-T)x - n(1-T)xz, \qquad \omega_z = cxy, \tag{3}$$

$$\Delta\omega_x = g_x Ty, \qquad \Delta\omega_y = g_y(1-T)y, \qquad \Delta\omega_z = 0, \tag{4}$$

where m, k, ℓ, δ, n, g_x, g_y, c are constants. In these expansions the terms up to the second order are retained. The terms with Laplacian, given in Eq.(4), are necessary because the vorticity equation has the viscous term.

Now, substituting Eqs.(2)-(4) into the x- and y-components of the vorticity equation and picking up terms of the first order of coordinates, we obtain after some manipulations the following differential equation for T(t):

$$\frac{dT}{dt} - t^{*-1}T + t^{*-1}T^2 = 0, \tag{5}$$

where

$$t^* = \frac{\delta n}{k + \Omega} = \nu(g_x - g_y) - 2m, \tag{6}$$

and ν is the kinematic viscosity. In deriving Eq.(5) the asymptotic condition (1) has been used in order to reduce the number of unknown constants.

The constant t^* is the characteristic time of the reconnection process. According to an order estimation based on the circulation Γ and the core radius σ of the vortex filament, we have $t^*=O(\sigma^2/\Gamma)$.

The solution of Eq.(5) with the condition (1) is written as

$$T = \frac{e^\tau}{1 + e^\tau}, \qquad \text{with} \quad \tau = \frac{t}{t^*}. \tag{7}$$

This solution shows that the superposing flow field, i.e. the vortex ring, grows exponentially and asymptotes to a stationary flow field, and that its time scale is given by t^*. If the solution (7) is substituted into Eqs.(2)-(4), we obtain the desired expression for the flow field in the interaction region, which describes a motion of vortex reconnection as illustrated in Fig. 1.

Thus, the mechanism of reconnection is understood in terms of a relatively smooth 3D unsteady flow.

3. Experiment

The experimental method was to eject two equal vortex rings side-by-side into the air in the same direction, and to observe what happened.[12] The reconnection process was visualized by adding a smoke into the air and illuminating the vortices by a storoboscopic light. Fig. 3 shows an event of experiment. The two vortex rings were seen to merge soon after they were ejected from the two nozzles, where at the contact point of the two rings two vortex filaments should have been reconnected.

Fig.3. Merger of two vortex rings.

Velocity distribution within the interaction region was measured by the use of Laser-Doppler velocimeter. In order to obtain the spatial distribution, a number of events needed to be measured with enough reproducibility. Reproducibility of the velocity signals was assured within a standard deviation of 10 %.

The distributions of the vertical velocity component w(z) at several instants are shown in Fig.4 along with the theoretical ones. The theoretical curves were fitted to the experimntal data by adjusting unknown parameters. However, the qualitative agreement is satisfactory. The flat distribution of experimental data in the later stage is considered to be caused by a turbulent diffusion which is not considered in the theory.

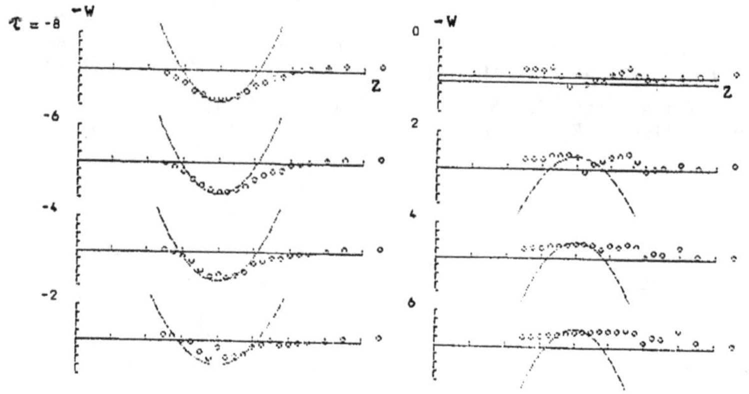

Fig. 4. Distributions of w(z) on the central vertical axis.
o: experiment, line: theory with w(0-)=3.5, δ=5.5, m=0, k=6.

4. Numerical Computation

A numerical computation was made for the same configuration as the present experiment.[12] Finite difference schemes were constructed from the Navier-Stokes equation for an incompressible viscous fluid and the continuity equation, which were solved by the high-speed computer.

Velocity vectors based on results of computation are shown in Fig. 5 along with the results of experiment. The velocity distributions at the three instants show respectively, the initial vortex ring just after ejected, the state in the interacting stage and the final state with a reconnected single ring vortex. The locations of vortex cores are indicated by allows.

The agreement of experiment and computation is remarkable.

5. Discussion

In this paper the mechanism of vortex reconnection is discussed based on the three ways of researches, i.e. theory, experiment and computation. The results are shown to agree each other, and the mechanism suggested in the theory is confirmed. In short, the reconnection process is explained by an impulsive 3D fluid motion with spatial symmetry, where the interaction region is described by polynomials of relatively lower order.

From the process of theoretical analysis it can be also shown (not stated in this paper) that the reconnection process does not occur without viscosity.

Finally, a significance of the present work is suggested on the stand point of the science of pattern formation problem. There are many physical systems where two objects with thin rod-like structures interact each other

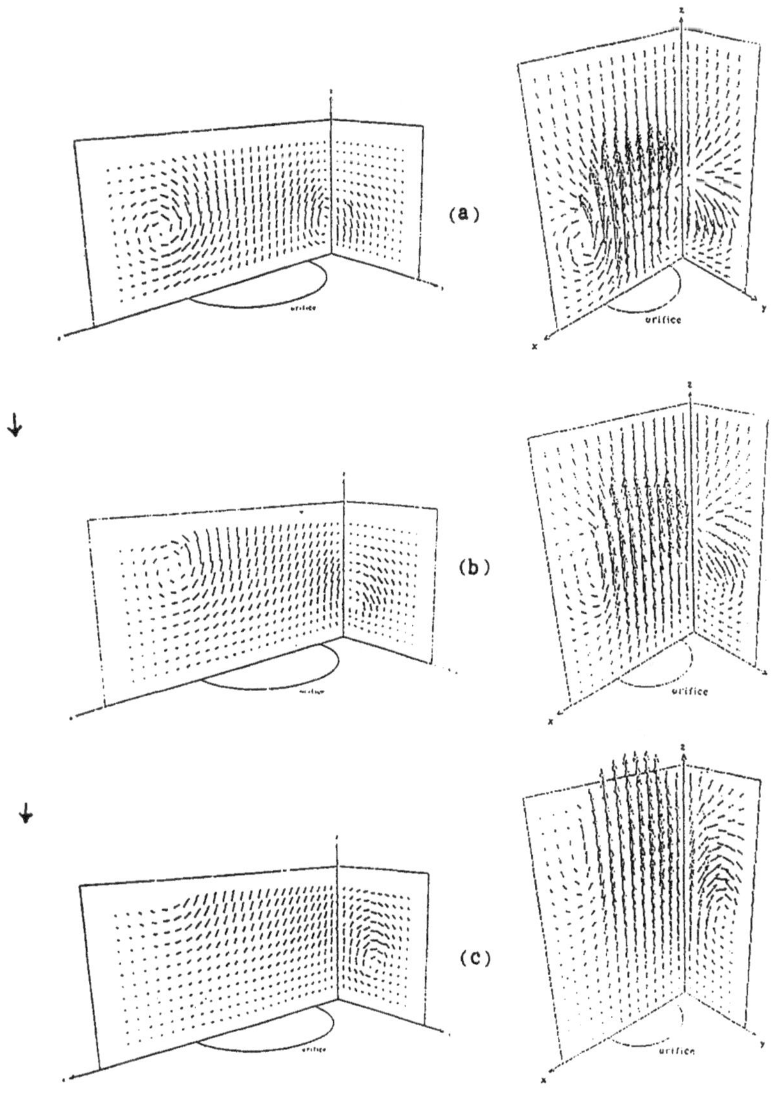

Fig. 5. Results of experiment(left) and computation(right). Velocity vectors in the x,z-plane and the y,z-plane are shown.
(a) before, (b) during, and (c) after reconnection process.

and undergo a reconnection process. In such a case a problem arises whether this process is governed by an dynamical equation, how such a topological change of structure comes out, whether such a change is described by a smooth field quantity, etc.

The present analysis gives a hint to these problems. It states that owing to the presence of viscosity the local interacting region behave regularly, and that the governing equation should have a higher derivative term (the viscous term including the Laplacian) and a nonlinear term to allow a smooth but a rapidly changing behavior. Such an idea would be applied to understandings of similar reconnection processes in other physical systems.

Acknowledgement

The present authors would like to express their cordial thanks to Professor S. Kai for his interest in this work. The part of this work is made in the U.S.-Japan Joint Research Program (1987-88) supported by Japan Society for Promotion of Science, where the chiefs were R. Takaki and Professor A.K.M.F. Hussain in the University of Houston.

References

1. T. Kambe and T. Takao, J. Phys. Soc. Jpn. 31(1971) 591.
2. T. Fohl and J.S. Turner, Phys. Fluids 18(1975) 433.
3. Y. Oshima and J. Asaka, J. Phys. Soc. Jpn. 42(1977) 708.
4. T. Yeh and W.I. Axford, J. Plasma Phys. 4(1970) 207.
5. R. Takaki and A.K.M.F. Hussain, Acad. Proc. Eng. Sci. (ed. R. Narasimha, Indian Academy of Science), 10(1987) 389.
6. R. Takaki and A.K.M.F. Hussain, Fluid Dyn. Res. 3(1988) 251.
7. M.V. Melander and A.K.M.F. Hussain,
8. J.P. Chamberlain and C.H. Lie, AIAA J. 23(1985) 868.
9. M.I. Aksman and E.A. Novikov, Fluid Dyn. Res. 3(1988) 239.
10. M.V. Melander and N.J. Zabusky, Fluid Dyn. Res. 3(1988) 247.
11. S. Kida and M. Takaoka, Fluid Dyn. Res. 3(1988) 257.
12. Y. Kakizaki, Master thesis, Tokyo Univ. Agr. Tech. (1990).

MULTIAFFINITY IN SURFACE GROWTH AND TURBULENCE

TAMÁS VICSEK

Department of Atomic Physics, Eötvös University, Budapest
P.O. Box 328, 1445 Hungary and
Institute for Technical Physics, Budapest, P. O. Box. 76, 1325 Hungary

ABSTRACT

In this paper the relation of the formalisms commonly used to describe self-affine surfaces and turbulent flows is demonstrated. In particular, it is shown that at least in some surface growth models and in a model for the velocity distribution in fully developed turbulence the moments of the corresponding functions are characterised by the same kind of scaling which we call multiaffinity.

1. Introduction

The complex spatial or temporal behavior of a large class of phenomena in physics can be interpreted in terms of self-affine functions[1-3]. These sigle-valued functions $h(\vec{x})$ are continuous, but nowhere differentiable, and satisfy the affine scaling

$$h(x_1,...,x_n) \simeq \lambda^{-H_1}...\lambda^{-H_n} h(\lambda_1 x_1,...,\lambda_n x_n), \tag{1}$$

where H_i are scaling exponents, λ_i are rescaling factors, and the symbol $\simeq$ expresses the fact that the rescaled version of the function is not completely equivalent to the original, the matching should be understood in a statistical sense only. The above expression in the simplest case of $d = 1+1$ dimensions reduces to the more transparent form

$$h(x) \simeq \lambda^{-H} h(\lambda x). \tag{2}$$

According to (2) the self-affine function $h(x)$ is invariant under the following *anisotropic* rescaling: shrinking along the x axis by a factor of $1/\lambda$, followed by rescaling of the values of the function (measured in the perpendicular direction) by a different factor, λ^{-H_1}. Fig. 1 shows a simple example for self-affine surfaces.

Alternatively, if $h(x)$ is self-affine, the average height (h) difference between points separated by a distance x scales as

$$\langle |h(x'+x) - h(x')| \rangle \sim x^H. \tag{3}$$

Thus, expressions (2) and (3) represent equivalent definitions of self-affine surfaces and the various methods based on these definitions can be looked at as

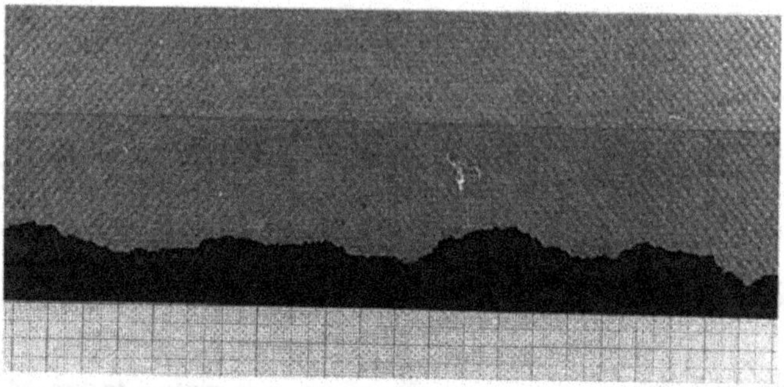

Figure 1. Example for a self-affine surface obtained in a simple experiment. This picture was taken from a paper towel previously immersed into a dish containing coffee. Digital image processing verifies scaling of the height differences according to eq. (3).

analogous approaches. (2) and (3) are given here for a single variable, however, in higher dimensions in most of the cases the same expressions are satisfied (with $\vec{x}$ instead of x, and with a unique exponent H).

In this paper we shall consider two kinds of rather different processes which, in spite of this, produce multiaffine spatial behavior. Multiaffinity will be shown to be analogous to, but even more complex than self-affinity corresponding to (2). The associated formalism will be given in the following section. We we shall discuss a growth process (Section 3) and fully developed turbulence (Section 4) from the point of multiaffine scaling. The results will be discussed in Section 5.

2. Multiaffinity

The concept of multiaffine geometry has been introduced very recently to provide a more complete description of the geometry of complicated functions[4,5]. This has been done in the same spirit as fractal measures can be used to characterize the detailed features of self-similar fractals[6–8]. The two formalisms are closely related, however, on the contrary to affine structures, fractal measures can not be appropriately represented by functions (they are are everywhere discountinuous and do not possess a well defined limiting value for a given x).

The multiscaling properties of the function $h(x)$ of a single variable can be determined by considering the *qth order* height-height correlation function which we define as[4]

$$c_q(x) = \frac{1}{N}\sum_{i=1}^{N}|h(x_i) - h(x_i + x)|^q, \tag{4}$$

where $N \gg 1$ is the number of points over which the average is taken, and only terms with $|h(x_i)-h(x_i+x)|$ strictly larger than zero are considered. This correlation function exhibits a non-trivial multiscaling behavior if

$$c_q(x) \sim x^{qH_q} \tag{5}$$

where H_q changes continuously with q at least for some region of the q values. An assumption analogous to (5) has been used in the studies of fully turbulent flows to describe the scaling nature of the so called velocity structure functions. For self-affine functions defined on a finite interval the above scaling is expected to hold in the $x \ll 1$ limit, while in the discrete case (where there exists a lower cutoff length Δx as, e.g., in the lattice growth models) equation (5) is satisfied in the $x \to \infty$ limit. It can be shown that a continuous spectrum of H_q values is not consistent with the expression $h(x) \simeq \lambda^{-H} h(\lambda x)$ which is valid for standard self-affine functions with a single exponent H. For self-affine functions with multifractal properties this expression is modified since the local scaling of the height differences depends on x in analogy with the local scaling of the measure in a box for fractal measures.

For a continuous surface, which without loss of generality can be considered as defined on $[0,1]$, a possible way to compute $c_q(x)$ is the following. One makes a partition of the $[0,1]$ interval into N equal parts and sums the qth powers of the height differences $\Delta h = |h(x_i) - h(x_i + x)|$ at points separated by $x = 1/N$. However, a more general approach requires that the limits $x \to 0$ and $N \to \infty$ are taken independently[5]. In earlier studies[6] of the velocity structure functions the special case $x = 1/N$ was implicitly assumed.

The significance of the above described concept of multiaffinity depends on its applicability to actual processes. Next we shall consider a computer model of rough surface growth and an iterative construction for the velocity distribution in fully turbulent flows to demonstrate how the above discussed behavior arises in different physical phenomena.

3. Nontrivial H_q spectrum in a surface growth model

Recently it has been proposed[9] that a growth model (Zhang model) with a non-Gaussian distribution of the noise amplitudes should be the appropriate way to describe the non-universal behavior observed in the experiments on the growth of rough surfaces. Let us consider a noise which is independently distributed on each site according to the distribution density

$$P(\eta) = \frac{\mu}{\eta^{1+\mu}} \quad \textit{for} \quad \eta > 1 \qquad \textit{and} \quad P(\eta) = 0 \qquad \textit{otherwise}. \tag{6}$$

According to (6) very large values of η may appear with a finite probability. These are the *rare events* which are supposed to change the behaviour qualitatively.

Perhaps the simplest model in $d = 1 + 1$ dimensions in which the noise (6) can be easily incorporated is represented by the growth rule

$$h(x, t+1) = \max[h(x,t) + \eta(x,t);\ h(x+1,t) + \eta(x+1,t);\ h(x-1,t) + \eta(x-1,t)], \tag{7}$$

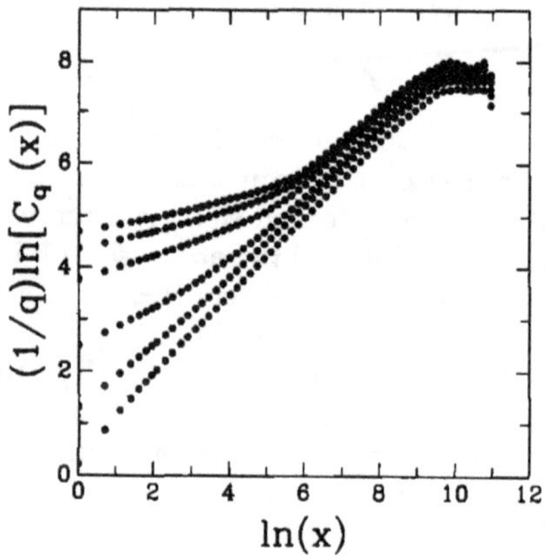

Figure 2. The q-th order correlation functions after the surface width has saturated ($L = 2^{16}$ and at $t = 602890$ sweeps) for q =1, 3, 5, 7 and 9 with q increasing from the bottom to top[10].

where x takes on discrete values along the one dimensional substrate. The above rule means that at every time step a) the local noise $\eta(x,t)$ is added to every site and b) each site then takes a new value equal to the maximum of itself and its two neighbours.

In spite of the fact that the Zhang model is very recent there are already several papers published on the numerical simulation of models related to (7) and the non-universality of its variants has been demonstrated.

Here we shall consider[10] the multiaffine aspects of the Zhang model by investigating the scaling behavior of the qth order height correlation function $c_q(x)$ numerically. For this purpose very large surfaces were grown using the rule (7) and the data were evaluated according to (5).

Fig. 2 shows the results for $L = 2^{16}$ at time $t = 602890$, i.e., deep in the saturation regime; an average over five runs was taken. The important conclusions one can draw from these log-log plots are the following: i) The initial part of the data sets for each q exhibits scaling behaviour with a unique slope depending on q, i.e., multifractal scaling is present, ii) this kind of scaling crosses over into the uniform scaling behaviour for x exceeding some characteristic crossover length $x_\times$.

4. Multiaffine model for the velocity distribution in fully turbulent flows

In this section instead of the actual geometry of a surface the spatial distribution of a rapidly changing quantity will be investigated. We shall consider turbulent flows where, for simplicity, we shall discuss the behavior of the x coordinate of the local velocity $u(x)$ of the fluid along a straight line.

Recent studies of the highly complex nature of fully developed turbulence have revealed that the scaling properties of fluid flows at large Reynolds numbers can successfully be described in terms of fractal geometry and multifractal distributions[6,11,12]. The multifractal description of the singular behavior of the velocity structure functions and the dissipation field represents the most general approach to turbulence and the existing models can be looked at as various realizations cases of the associated formalism. In this picture, places with a given singularity are distributed on fractal subsets of dimension depending on the strength of the singularity.

Here we propose a simple approach[13] in which the the full velocity distribution is directly modelled by a continuous multiaffine function and, in this way, a complete picture capturing the essential features of the geometry underlying turbulent flows is provided. It will be shown that in this approach the relevant quantities can be readily calculated and comparisons can be made with the available experimental data on fully developed turbulence.

As the Navier-Stokes equations in the small viscosity limit ($\nu \to 0$) are invariant under the scaling transformations $\vec{r} \to \lambda \vec{r}$, $\vec{v} \to \lambda^H \vec{v}$ ($t \to \lambda^{1-H} t$, $\lambda > 0$) we are led to assume that the velocity distribution satisfies the usual criterion of self-affinity $u(x) \sim \lambda^{-H} u(\lambda x)$, where in the last expression we only considered the behavior of the x component of the velocity vector. This expression is equivalent to the scaling of the velocity differences of the form $\langle |u(x + \Delta x) - u(x)| \rangle \sim \Delta x^H$. According to the multifractal picture of turbulent flows the scaling exponent H depends continuously on the position in space. This is the property which we called multiaffinity, since it corresponds to an invariance under affine transformations with an infinite hierarchy of rescaling factors depending on the local value of H.

Let us consider the iterative process illustrated by Fig. 3 which is perhaps the simplest realization of a self-affine function satisfying the above outlined criteria. In each step of the recursion the intervals obtained in the previous step are replaced with the properly rescaled version of the generator which has the form of an asymmetrical z made of three intervals (Fig. 3a). During the procedure every interval which is to be replaced is regarded as a diagonal of a rectangle becoming more and more elongated as the number of iterations k increases.

When going to the next level, one can rescale either a) the generator shown in Fig. 3a or Fig. 3b its image inverted around the center of the rectangle. Then the stochastic nature of the turbulent signals can be taken into account by randomly choosing between the possibilities a) and b). The function $u(x)$ generated in this way becomes multiaffine in the $k \to \infty$ limit. Our main statement is that the behavior (geometry) of the one-dimensional projection of the velocity field in the inertial range can be well described by such a multiaffine function. The visual appearance of our model after $k = 5$ iterations is compared in Fig. 4 to a measured time series of the velocity in a turbulent pipe flow behind a grid.

Using the above described velocity distribution we can explicitly calculate

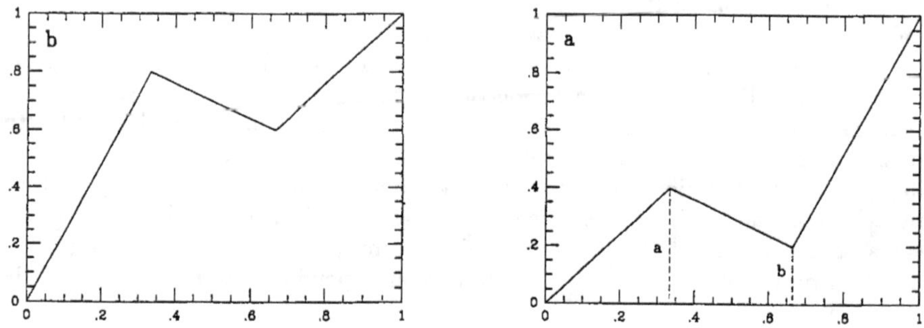

Figure 3. The generators of the multiaffine model. In each step of the recursion one of the the asymmetrical z shaped generators made of three intervals shown in a) and b) is picked randomly and used to replace the single intervals obtained in the previous step (for more details see the text).

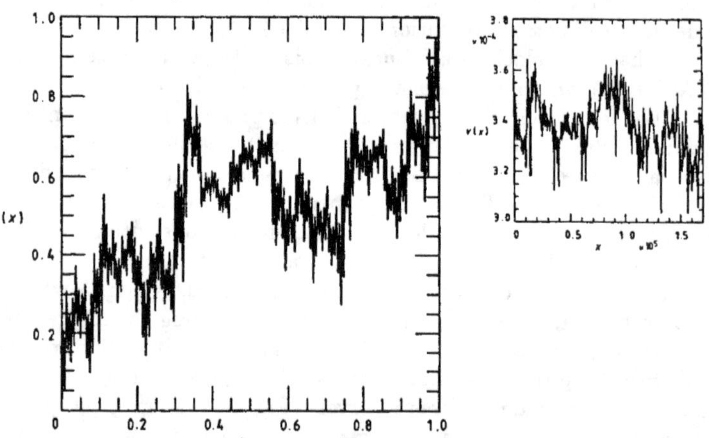

Figure 4. The geometry of the multiaffine model ($k = 5$) for the one-dimensional projection of the velocity field. This figure is for the case $a = 0.67$. The inset shows a typical velocity signal observed in a turbulent pipe flow behind a grid (courtesy of W. Goldburg).

for our model the standard quantities used to characterize fully developed turbulent flows. In such flows the qth order longitudinal velocity structure function $c_q(x)$ is expected to scale as

$$c_q(x) = \langle |u(x' + x) - u(x')|^q \rangle \sim x^{qH_q}, \tag{8}$$

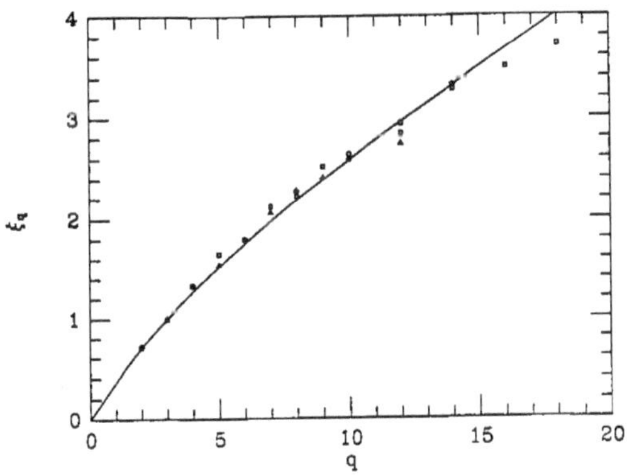

Figure 5. Comparison of the experimentally determined values of the exponent $\xi_q = qH_q$ with the prediction of our model for $a = 0.67$ (continuous line). The experimental data are from Ref. 14.

where the exponent $qH_q = \xi_q$ is one of the major quantities studied in the theory of turbulence. There is firm theoretical result $c_3(r) \sim r$ from which $H_3 = 1/3$ follows. We calculate $c_q(x)$ for our model assuming that the scaling properties are entirely determined by the jumps Δu of the function over intervals of length $\Delta x_k = 3^{-k}$. Thus,

$$c_q(\Delta x_k) = \sum_{m_1=0}^{k} \sum_{m_2=0}^{m_1} \frac{1}{3^k} \binom{k}{m_1} \binom{m_1}{m_2} a^{m_1 q}(a-b)^{(m_1-m_2)q}(1-b)^{(k-m_1)q} \tag{9}$$

Since eq. (9) can be written as $c_q(x_k) = [a^q + (a-b)^q + (1-b)^q]^k/3^k$ we have

$$\xi_q = qH_q = \frac{\ln[(a^q + [a-b(a)]^q + [1-b(a)]^q)/3]}{\ln(1/3)}. \tag{10}$$

In (10) we used $b(a)$ to express the fact that the condition $H_q = 1/3$ has to be satisfied and this, through the corresponding equality $a^3 + [a-b(a)]^3 + [1-b(a)]^3 = 1$, reduces the number of independent parameters to one.

The experimental values[14] for ξ_q can be used to find the best value for the parameter a. Fig. 5 demonstrates that the prediction of eq. (10) for $a = 0.67$ provides a good fit to the experimental results. At this point our multiaffine model has become completely determined and the next important step is to check whether the model provides such a good agreement when other quantities determined in independent experiments are considered.

Thus, we turn to calculating the generalized dimensions D_q associated with the qth moment of the normalized distribution of various quantities related to the nth power of the velocity differences between close points. The motivation for this is that, for example, the energy transfer or dissipation rate on scale l is widely accepted to be proportional to $(\Delta u)^3/l$, while the local viscous dissipation rate can be shown to be proportional to $(\partial u/\partial x)^2 \simeq (\Delta u)^2/(\Delta x)^2$. The fractal measure is constructed for our model by defining the normalized distribution $p_i = |\Delta u_i|^n/\sum|\Delta u_i|^n$, where p_i is the "probability" or weight associated with the ith interval of length Δx. Then the qth moment of the measure is expected to scale as

$$\chi_q(\Delta x) = \sum_N p_i^q \sim \Delta x^{(q-1)D_q} \tag{11}$$

where the summation is taken over $N = 1/\Delta x$ intervals of length Δx. (11) can be evaluated by noting that the sums of the form $\sum_N |\Delta u_i|^{nq}/N$ behave as the nqth order structure functions. After simple algebra we get for the generalized dimensions

$$D_q = 1 + \frac{nq}{q-1}(H_{nq} - H_n) \tag{12}$$

The predictions of expression (12) are in agreement with the experimental results of Meneveau and Sreenivasan[12] for the energy dissipation rate.

From the D_q spectrum we can obtain an estimate for the fractal dimension D of the set of points at which the essential part of the dissipation takes place in a three-dimensional turbulent flow. According to the standard formalism of multifractals we assume that the dominant contribution to the total dissipation $\chi_1(x)$ comes from a fractal subset of dimension f_1 of regions of size δx in which the dissipation is proportional to $(\delta x)^{\alpha_q}$. Then from the well known relation $(q-1)D_q = q\alpha_q - f_q$ it follows that $f_1 = D_1$, and using (10), (12) and $a = 0.67$ we obtain the estimate

$$D = 2 + D_1 \simeq 2.88, \tag{13}$$

where we have taken into account that the fractal dimension of a one-dimensional cut of a D-dimensional object embedded into three dimensions is equal to $D-2$. This is consistent with the related earlier theoretical considerations and the experimental results of Ref. 15 which gave estimates close to $D = 2.9$.

5. Conclusions

In this paper the relation of the formalisms commonly used to describe self-affine surfaces and turbulent flows has been demonstrated. In particular, we have shown that at least in some surface growth models and in a model of turbulence the associated functions are characterised by the same kind of scaling, which we called multiaffinity.

As concerning our studies of the moments of the height correlation function in kinetic roughening[10], on the basis of Fig. 2 we can conclude that in the Zhang model, in addition to the usual characteristic length t_c always present in surface growth before the system saturates, a new characteristic length $x_\times$ appears. For $x < x_\times$

the q-th order correlations show multifractal scaling behavior. For $x_\times < x < t_c$ conventional scaling sets in while no correlations are present for $x > t_c$.

In our approach to the description of the multiscaling aspects of turbulence we have used the simplest multiaffine function[13] to simulate the space (or time) dependence of the velocity and found good agreement with the experimental results concerning various aspects of fully developed turbulent flows. This analytically treatable model allows the calculation of many more quantities, including the scaling functions entering the spectra of such quantities as the energy, the thermal or the velocity fluctuations.

Acknowledgements: The author thanks A. L. Barabási, R. Bourbonnais, J. Kertész and P. Szépfalusy for collaboration and W. Goldburg and K. R. Sreenivasan for helpful correspondence. The present research was supported by the Hungarian Scientific Research Foundation Grant No. 693.

References

1. B. B. Mandelbrot, *The Fractal Geometry of Nature* (Freeman, San Francisco, 1982)
2. T. Vicsek, *Fractal Growth Phenomena* (World Scientific, Singapore, 1989)
3. F. Family and T. Vicsek, *Dynamics of Fractal Surfaces*, (World Scientific, Singapore, 1991)
4. A.-L. Barabási and T. Vicsek, *Phys. Rev.* **A44** (1991)
5. A.-L. Barabási, P. Szépfalusy and T. Vicsek, to appear in *Physica A*
6. U. Frisch and G. Parisi, in *Turbulence and Predictability in Geophysical Fluid Dynamics and Climate Dynamics* edited by Ghil, M., Benzi, R. and Parisi, G. (North-Holand, Amsterdam, 1985)
7. T. C. Halsey, M. H. Jensen, L. P. Kadanoff, I. Procaccia and B. Shraiman, *Phys. Rev.* **A33**, 1141 (1986)
8. B. B. Mandelbrot in H. E. Stanley and N. Ostrowsky, editors, *Random Fluctuations and Pattern Growth* H. E. Stanley and N. Ostrowsky, editors, (Kluwer, Dordrecht, 1988)
9. Y.-C. Zhang, *J. Physique* **51**, (1990) 2129
10. A.-L. Barabási, R. Bourbonnais, M. Jensen, J. Kertész, T. Vicsek, and Y.-C. Zhang, preprint
11. R. Benzi, G. Paladini, G. Parisi and A. Vulpiani, *J. Phys.*, **A17**, 3521 (1984)
12. C. Meneveau and K. R. Sreenivasan, *Phys. Rev. Lett.* **59**, 1424 (1987)
13. T. Vicsek and A.-L. Barabási, *J. Phys.* **A24**, (1991) L845
14. F. Anselmet, Y. Gagne, E. J. Hopfinger and R. A. Antonia, *J. Fluid Mech.* **140**, 63 (1984)
15. K. R. Sreenivasan and C. Meneveau *Phys. Rev.* **A38**, 6287 (1988)

Stochastic Patterns of Wave Turbulence

Grigory E. Falkovich
Department of Nuclear Physics, Weizmann Institute of Science, Rehovot 76100, Israel

Alexander V. Shafarenko
Department of Electronic & Electrical Eng., University of Surrey, Guildford GU2 5XH, U.K.

Summary.

Nonstationary regimes of the wave turbulence evolution are considered in the framework of isotropic kinetic equation. It is predicted analytically and confirmed by numerical experiment that there is a class of wave systems in which any initial distribution of the turbulence energy in k-space comes into a universal, Kolmogorov type spectrum in a finite time. Before and after the formation of the Kolmogorov spectrum, two different self-similar patterns of evolution occur: the first one is responsible for explosively forming the universal spectrum and the second one determines energy dissipation of decay turbulence.

1 Introduction

In this paper we present the theory of nonstationary wave turbulence. We consider weak turbulence approximation when linear effects expressly predominate over nonlinear ones: dispersion of phase velocity has a stronger effect than interaction between waves [1]. We will analyse both the free evolution regimes that appear when no external source is in effect and the processes of forming the Kolmogorov spectra after switching the source on.

We shall discuss two physical concepts playing an important role in the analysis of nonstationary turbulence. One of these is the concept of cascade spectra of the Kolmogorov type understood as stationary spectra of the developed turbulence. In nonstationary regimes, which are the subject of this paper, the Kolmogorov spectrum is realized either as an asymptotically stable state (i.e. one that an arbitrary distribution tends to in the presence of a source and a sink) or as an intermediate asymptotics (for freely decaying turbulence). The other concept is self-similar regimes of evolution, which can be of two kinds:

1. those that form for the long time (formally at $t \to \infty$), and
2. quickly forming and living for a finite time.

The latter case corresponds to explosion regimes of evolution, which proceed in self-accelerating manner. As we shall show, bi-self-similar regimes, where one of the asymptotics of the distribution forms in a self-accelerating (explosion) way for a finite time, and then the evolution of the distribution as a whole goes on by a decelerating self-similar law, are also possible.

2 Kinetic Equation and Dimensional Analysis

The various occurrences of turbulence in conservative or slightly dissipative weakly nonlinear media can be described from a single viewpoint in terms of the double correlator of the wave field [2]:

$$n(\vec{k},t)\delta(\vec{k}-\vec{k}') = \langle c(\vec{k},t)c^*(\vec{k}',t)\rangle \ .$$

Here $c(\vec{k},t)$ is the amplitude of the wave with the wave vector $\vec{k}$, the averaging is over the ensemble of random phases. In the simplest case where the three-wave interaction is allowed, the pair correlator $n(\vec{k},t)$ evolves according to the following kinetic equation:

$$\begin{aligned}\frac{\partial n(\vec{k},t)}{\partial t} &= \int |V_{\vec{k}\vec{1}\vec{2}}|^2\,\delta\left(\omega_{\vec{k}}-\omega_{\vec{1}}-\omega_{\vec{2}}\right)\delta(\vec{k}-\vec{k}_1-\vec{k}_2)\,\times \\ &\times \left[n_{\vec{1}}n_{\vec{2}}-n_{\vec{k}}\left(n_{\vec{1}}+n_{\vec{2}}\right)\right]\,d\vec{k}_1 d\vec{k}_2 - \\ &- 2\int |V_{\vec{1}\vec{k}\vec{2}}|^2\,\delta\left(\omega_{\vec{1}}-\omega_{\vec{k}}-\omega_{\vec{2}}\right)\delta(\vec{k}_1-\vec{k}-\vec{k}_2)\,\times \\ &\times \left[n_{\vec{k}}n_{\vec{2}}-n_{\vec{1}}\left(n_{\vec{k}}+n_{\vec{2}}\right)\right]\,d\vec{k}_1 d\vec{k}_2 \ . \end{aligned} \tag{1}$$

Here $\omega_{\vec{k}}$ is the wave frequency, $V_{\vec{1}\vec{2}\vec{3}}$ is the coefficient of three-wave interaction. The kinetic equation (1) is valid in the limit of small nonlinearity, where the characteristic time of interaction $t_a \propto \omega\left(|V|^2 k^d n\right)^{-1}$ is much greater than the time of the wave packet diffusion determined by dispersion $t_D \propto \left(k^2\omega_k''\right)^{-1}$.

Equation (1) makes sense if the conditions imposed by δ-functions are compatible, i.e., the dispersion law $\omega(\vec{k})$ enables the spatio-temporal resonance:

$$\omega\left(\vec{k}_1\right)+\omega\left(\vec{k}_2\right)=\omega\left(\vec{k}_1+\vec{k}_2\right) \ . \tag{2}$$

Such a dispersion law is termed decay, and we restrict ourselves to that case only. For considering nonstationary wave turbulence with a nondecay dispersion law see [4].

The number of stationary solutions (precisely, the number of independent parameters in the general solution) is equal to the number of independent integrals of motion. Generally speaking, the three-wave kinetic equation (1) has only one integral of motion: the energy $E = \int \omega_k n_k \, dk$. There is one and only solution that corresponds to a constant energy flux in the k-space. It is called the *Kolmogorov spectrum* and can be derived analytically for the case of scale invariant wave systems where $\omega(k) = ck^\alpha$ and $V\left(k, k_1, \left|\vec{k} - \vec{k}_1\right|\right) = k^m V_0 f(k_1/k)$ (here f is a dimensionless structured function). In that case $n_k \propto k^{-s_0}$, $s_0 = m+d$ and d is the dimensionality of the k-space [1]. If the Kolmogorov distribution falls off faster than the equilibrium one $n_k \propto \omega_k^{-1}$, i.e. if $m + d > \alpha$, then the energy flux is directed towards large k.

The validity of kinetic equation should also be checked for the Kolmogorov solutions. For example, substituting $n(k) \propto k^{-s_0}$ into the estimate of the characteristic interaction time gives $t_s \propto k^{s_0+\alpha-d-2m}$. Comparing it with the dispersion time $t_{rmD} \propto k^{-\alpha}$, one can see that the validity of weak turbulence approximation improves while energy cascades towards small scales if $s_0 + 2\alpha > d + 2m$.

It should be noted that all the subsequent analysis will be as general as possible, the results to be applied to the following physical systems [1]:

1. Capillary waves on deep ($d = 2$, $\alpha = \frac{3}{2}$, $m = \frac{9}{4}$) and shallow ($d = 2$, $\alpha = 2$, $m = 2$) water.
2. The 3-dimensional sound with a positive dispersion $\omega(k) = ck(1 + a^2k^2)$, $ak \ll 1$ ($d = 3$, $\alpha = 1$, $m = \frac{3}{2}$)
3. The 2-dimensional sound with a positive dispersion, e.g., gravitational-capillary waves on shallow water ($d = 2$, $\alpha = 1$, $m = 1$).

We shall start considering the evolution of wave distributions with simple dimensional estimations, which clear out the dependence of the characteristic evolution time on the mean wavelength. Let us consider the cascade transfer of waves from large scales to smaller ones. The character of evolution dramatically depends on the sign of the exponent $h = \alpha + d - s_0$. To demonstrate that is true, let us first consider the evolution of small perturbations against the background of the stationary spectrum $n_k \propto k^{-s_0}$ carrying an energy flux. Such perturbation evolve according to the kinetic equation linearized with respect to $\delta n(k,t)$:

$$\frac{\partial \delta n(k,t)}{\partial t} = \hat{L}(k, k_1)\delta n(k_1, t) \ . \quad (3)$$

Here the operator $\hat{L}(k, k_1)$ is expressed in terms of the frequency, interaction coefficient and spectrum $n(k)$, thus being scale invariant (see [1] for details). The scaling exponent of this operator could be established by the following simple derivation. The exponent of the right hand side of the complete kinetic equation with $n(k) \propto k^{-s_0}$ is equal to $-\alpha - d$, since after multiplying by ω_k and integrating over $d\vec{k}$ it should ensure that the energy flux does not dependend on k. Since the exponent of the complete equation differs from that of the linearized operator by s_0, then the latter is equal to $s_0 - \alpha - d = -h$. So if the distribution deviates from the stationary one at some k, the typical time of variations will be

$$t_{\rm NL}^{-1} \propto k^{-h} \quad (4)$$

When external influence is absent, the energy must conserve, therefore the characteristic time may be evaluated from the kinetic equation as follows:

$$t_{\rm NL}^{-1} \propto k^{-2h} E \ , \quad (5)$$

where $E = \int \omega_k n_k d\vec{k}$ is the total energy of the distribution. In particular, in both cases the process of wave transfer to large k accelerates or decelerates depending on the sign of h. It is appropriate to remind here of one more role of the h value: it points out the edge of the Kolmogorov distribution at which the major part of the energy of the turbulence will concentrate:

$$E = \int \omega_k n_k^0 d\vec{k} \propto k^h \ . \quad (6)$$

For example, at $h > 0$ the major part of energy of the Kolmogorov spectrum is confined in the region of large k. In this case, as we can see from ((4) and (5)), the evolution slows down, with the distribution drifting towards greater wave numbers.

3 Analysis of Self-Similar Substitutes

We start with the three-wave kinetic equation (1). For the sake of simplicity we assume the distribution to be isotopic. In this case, there is only one integral of motion, *viz.* the energy. We shall assume that $m + d > \alpha$, i.e., after establishing itself, the Kolmogorov distribution transfers energy to the short-wave region. Two physically different statements of the problem will be discussed:

P1: Expansion of the Kolmogorov distribution into the region of large k, and

P2: Decaying turbulence: free evolution of the initially long-wave packet which, owing to a tendency to equilibrium distribution, should also spread all over the k-space.

It would be natural to presume that the evolution will in some time become self-similar a long way from the source or the initial localization site of a wave packet. Let us discuss possible self-similar substitutions in Eq. 1, i.e. such which have the following form:

$$n(k,t) = t^{-q} f(kt^{-p}) = t^{-q} f(\xi) \ . \tag{7}$$

We shall consider here the ξ variable to be dimensionless, assuming that k is measured in k_0 units (where k_0 is the location of the source or the initial location of the packet), and t in the units of

$$t_N = V_0^2 k_0^{2m+d} n^2(k_0)/\omega(k_0) \ ,$$

where t_N is the characteristic time of the nonlinear interaction in the region $k \simeq k_0$ and V_0 is the dimensional constant associated with the interaction coefficient.

Substituting (7) in the three-wave kinetic equation

$$\frac{\partial n(k,t)}{\partial t} = I(k) \ , \tag{8}$$

we obtain

$$-(qf + p\xi f') = I(\xi) t^{p(2m-\alpha+d)-q+1} \ .$$

It is evident from this that the solution of the form (7) can exist only if a condition

$$p(2m - \alpha + d) - q + 1 = 0$$

is satisfied.

To obtain another relationship between the parameters p and q, we should specify which of the above cases *P1* and *P2* we are going to address.

Case P1. The quasistationary Kolmogorov distribution $n_k^0 \propto k^{-m-d} = k^{-s_0}$ must occur between the source and the relaxation front. This means that $f(\xi) \propto \xi^{-m-d}$ as $\xi \to 0$ and the condition of steadyness gives

$$p(m+d) = q \ . \tag{9}$$

From (3) and (9), we find $q = (m+d)/(\alpha - m) = s_0/h$ and

$$p = (\alpha - m)^{-1} = 1/h \ , \tag{10}$$

which are consistent with the estimate (4). The boundary of the Kolmogorov distribution corresponds to $\xi \simeq 1$ and moves in the k-space by the law $k_b \propto t^p = t^{1/h}$. So the solution (7) in this case describes the motion of the Kolmogorov spectrum towards large k only if $h > 0$. The logarithmic front velocity,

$$\frac{d \ln[k_b(t)/k_b(0)]}{dt} \propto (ht)^{-1} \ ,$$

decreases with time, in agreement with the general idea that the process of the distribution expansion is a slowing-down process when expanding the energy-containing region.

As $h \to 0$, the front velocity dramatically increases and becomes infinite at $h = 0$. This implies that at $h < 0$ the rate of formation of the Kolmogorov spectrum increases with time so quickly that the relaxation front reaches infinity for a finite time. Indeed, at $h < 0$, in order to obtain a self-similar relaxation front moving to large k, one should replace t by $\tau = t_0 - t$ in the right-hand side of (7):

$$n(k,t) = \tau^{-q} f(k\tau^{-p}) \ . \tag{11}$$

The equations (3–10) for p and q will retain their forms, but now the right boundary of the Kolmogorov distribution corresponding to $k_b\tau^{-p} \simeq 1$ moves by the explosion law $k_b \propto \tau^{1/h}$ and reaches infinity for the finite time t_0 determined by the initial distribution (we shall explain how a little later). Qualitative understanding of t_0 being finite lies with the fact that wherever $h < 0$ the stationary spectrum energy is concentrated in the long-wave region, which means that an interval $[k, \infty]$, $k \neq 0$, contains finite energy. This finite energy obviously takes finite time to become redistributed over the interval.

The universal set-in regimes of the Kolmogorov spectra in the cases $h > 0$ and $h < 0$ are realized with absolutely different time scales. At $h > 0$ the self-similar front (7) is formed for the time much larger than the typical stabilization time for the occupation numbers in the pump region. In the case $h < 0$, the self-similar front appears for a small period of time ($t_0 - t \ll t_0$), being too small for any essential changes in the occupation numbers of long waves to occur.

Case P2. We shall start analysis of the free evolution with the case $h < 0$. Since the front (11) is self-accelerating

$$\frac{1}{k_b}\frac{dk_b}{dt} \propto \frac{1}{\tau} \ ,$$

the behaviour of the short-wave part of the distribution should be insensitive to the presence or absence of a long-wave source that changes the occupation numbers not faster than exponentially. So until $t = t_0$ the free expansion of the distribution to large k proceeds in the same way as it does with the source: the self-similar front $n(k,t) = \tau^{-q} f(k\tau^{-p}) = \tau^{-q} f(\xi)$ with $q = s_0/h$ and $p = 1/h$ moves explosively, leaving behind the Kolmogorov distribution, $f(\xi) \propto \xi^{s_0}$ at $\xi \ll 1$. Ahead of the front, the occupation numbers should quickly diminish with ξ (and, accordingly, k) (see [4] and Fig.1).

Since for $t < t_0$ the asymptotics of the distribution at $k \to \infty$ falls off steeper, the total energy conserves. What happens at $t \geq t_0$, i.e., after the right boundary of the Kolmogorov distribution has reached infinity? From that moment on, $n_k \propto k^{-s_0}$ at $k \to \infty$, the energy flux at $k = \infty$ being nonzero, therefore the energy has to go down. Any real system, of course, has a sink at a finite k_m and the energy will start to decrease somewhat earlier than at the moment

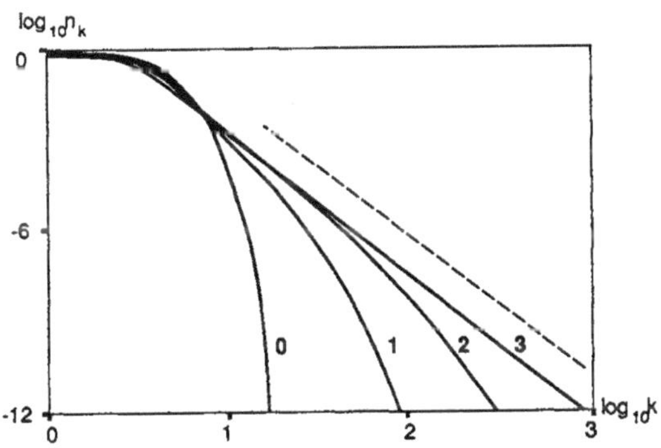

Figure 1: Distribution of occupation numbers n_k in k-space at different moments in time for the case of 3d sound turbulence. Curves 0–3 represent $\ln n(k)$ at the moments in time $0, 10^{-4}, 3 \cdot 10^{-4}$ and $5.3 \cdot 10^{-4}$, respectively. The dashed line represents the Kolmogorov power law $n(k)^0 = Ak^{-9/2}$.

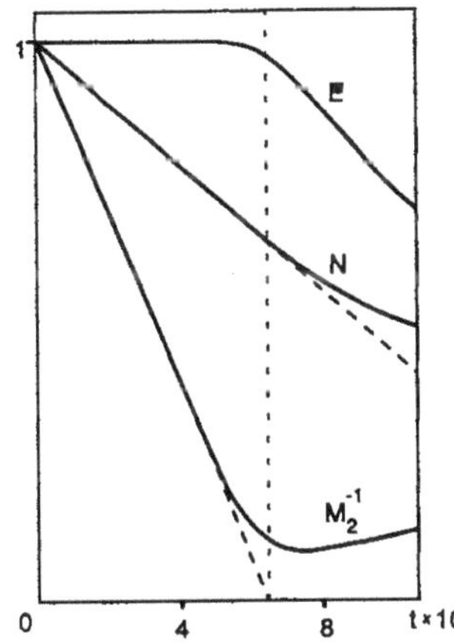

Figure 2: Time dependence of all three moments of the distribution function for Var. 2, L =1000. The vertical dashed line marks the time of the explosion.

t_0, namely, when the relaxation front reaches the sink. It would be natural to presume that after a large enough period of time ($t \gg t_0$), the process of attenuation will come into a second self-similar regime with the energy falling off by a power law. Such a behaviour should be described by a self-similar solution of the form (7), the relation between q and p being given by (3). To determine the exponent p, one has to solve a nonlinear eigenvalue problem stemming from the kinetic equation in the self-similar variable, with an additional requirement $f(\xi) \geq 0$ at $0 < \xi < \infty$ (see [4]). Thus, at $h < 0$ the behaviour is bi-self-similar.

The free evolution of a system with $h > 0$ in the large k region also has at some time to come into a self-similar solution of the type (7). But now the energy has to conserve, therefore the exponent p is determined unambiguously: $p = (2h)^{-1}$ which is consistent with the estimate (5). The positiveness of p implies that the distribution drifts to the short-wave region. Such a solution doesn't have the Kolmogorov asymptotics anywhere, since the energy flux is equal to zero at $k \to 0$ and $k \to \infty$.

Now we would like to emphasize the profound difference between the systems with $h > 0$ and $h < 0$. In the former case, the energy of freely decaying turbulence conserves all the time provided there is no dissipation. On the contrary, for the case $h < 0$ the finite rate of energy dissipation arises after a finite time even without any external dissipation mechanism (formally, the role of a sink is played by the point $k = \infty$). Strong hydrodynamic turbulence of incompressible fluid also belongs to this type, since the energy containing region for the Kolmogorov spectrum is at small k, too. That is why the explosion regime of forming the spectrum is also possible in that type of turbulence [5]. In hydrodynamics, it is the way to resolve the D'Alembert paradox about the formal absence of drag in inviscid fluid: dissipation is still finite as viscosity tends to zero.

The intermediate case $h = 0$ corresponds to the self-similarity of exponential rather than power type [4].

Thus, according to the character of evolution, the wave systems with the decay dispersion law are subdivided into two classes: one with $\alpha > m$ ($h > 0$) and the other with $\alpha < m$ ($h < 0$).

4 Method of Moments

It turns out that the most interesting property of the evolution, the explosion character of the Kolmogorov asymptotics formation, may be given a rigorous mathematical proof for a particular case of weak acoustic turbulence. The proof is based on the analysis of moments of the distribution $n(k,t)$:

$$M_i(t) = \int k^i n(k,t) d\vec{k} \ .$$

If at $k \to \infty$ the power asymptotics $n(k) \propto k^{-s}$ establishes for a finite period of time, then the moments M_i for

$i > s - d$ have at some instant to become infinite. The evolution of three-dimensional isotropic distributions of weak acoustic turbulence is described by the following equation in dimensionless variables

$$\frac{\partial n(k)}{4\pi\partial t} = \int_0^k (k-k_1)^2[n(k_1)n(k-k_1) - n(k)n(k_1)$$
$$- n(k)n(k-k_1)]k_1^2dk_1 - 2\int_k^\infty (k_1-k)^2[n(k)n(k_1-k) -$$
$$- n(k_1)n(k_1-k) - n(k_1)n(k)]k_1^2dk_1 \ . \qquad (12)$$

As we consider the weak dispersion limit $\omega_k \approx k$, the energy is the first moment of the distribution function

$$E = M_1 = \int kn(k,t)d\vec{k} = 4\pi\int_0^\infty k^3n(k,t)\,dk \ ,$$

which conserves if function $n(k,t)$ diminishes faster than by the Kolmogorov law $n(k) \propto k^{-9/2}$ at $k \to \infty$. For the case in question, $\alpha = 1$, $m = \frac{3}{2}$ and $h = -\frac{1}{2}$.

Let us consider other moments of the distribution function. For zeroth moment, which is the total number of waves $N(t) = 4\pi\int_0^\infty n(k,t)k^2\,dk$, we compute from (12)

$$\frac{dN}{dt} = -(4\pi)^2\int_0^\infty k^2dk\int_k^\infty k_1^2(k-k_1)^2\times$$
$$\times[n(k)n(k_1-k) - n(k_1)n(k) - n(k_1)n(k-k_1)]dk_1$$
$$= (4\pi)^2\int_0^\infty k^2n(k)\,dk\int_0^\infty k_1^2n(k_1)\left[(k-k_1)^2 - (k+k_1)^2\right]dk_1$$
$$= -4E^2 \ . \qquad (13)$$

In deriving this equation, we used a rearrangement of integration limits, which is admissible provided the integral $\int_0^\infty k^3n(k)dk$ converges. Thus, (13) is valid if $n(k)$ falls off more rapidly at $k \to \infty$ than k^4 does. Note that we assume no singularities of $n(k)$ at $k = 0$. It follows from (13) alone that a distribution with finite and nonzero N and E that diminishes quickly at $k \to \infty$ initially, can not remain such for an unlimitedly long time.

Equation (13) prompts the following scenario of evolution: during the time $t_0 = N(0)/4E^2$, the initially long-wave packet spreads throughout the k-space, N becomes zero, and the energy-containing scale $k_E = E/N$ becomes infinite. This scenario corresponds to the pumping-over of all the energy to infinity (actually to large k) for a finite period of time.

Actually, the evolution does not follow it.

Let us consider the evolution of the second moment $M_2 = 4\pi\int_0^\infty k^2n(k)k^2dk$:

$$\frac{dM_2}{dt} = 6M_2^2 + 8\pi E\int_0^\infty k^3n(k)k^2dk +$$
$$+ 32\pi^2\int_0^\infty u^2n(u)du\int_0^u v^4(u-v)^2n(v)dv \geq 8M_2^2 \ . \qquad (14)$$

Here we have used a simple inequality

$$\int_0^\infty k^5n(k)dk\int_0^\infty k^3n(k)dk \geq \left(\int_0^\infty k^4n(k)dk\right)^2 \ .$$

The inequality (14) is valid provided that the integral

$$M \propto \int_0^\infty k^4n(k)dk$$

converges, i.e., only if $n(k)$ diminishes faster than k^{-5}.

The equation $dx/dt = 8x^2$ has an explosion solution $x(t) = x(0)/(1 - 8x(0)t)$. Consequently, (14) demands that M_2 become infinite by the time $t \leq t_1 = [8M_2(0)]^{-1}$. It should be noted that the same method of moments could also be applied to prove that the blow-up exists in some model system suggested for describing hydrodynamic turbulence [5].

It is easy to show that $t_1^{-1} = 8M_2(0) \geq 2t_0^{-1} = 8E^2/N(0)$, i.e., that the explosion associated with the second moment M_2 becoming infinite occurs at least twice earlier than N becomes zero. We shall see below (see also Figs. 1 and 2) that M_2 increases with the Kolmogorov asymptotics $k^{-9/2}$ explosively forming in the large k region.

In the distribution $n(k) \propto k^{-9/2}$, the energy-containing region is that of small k: $\varepsilon \propto k^{-3/2}$. Therefore only small part of the initial energy of turbulence appears in the region of large k until the Kolmogorov asymptotics has established itself. The effect of explosive formation of the Kolmogorov distribution in the short-wave region may be compared with a weak collapse (see [3]) when the value of an integral of motion captured into the singularity (in our case, into $k \to \infty$) tends to zero.

Thus, the evolution of a weak acoustic turbulence distribution will be two-stage. During the first, explosion, stage the energy conserves, the number of waves decreases by a linear law and the second moment increases explosively. As a result, a finite energy flux sets on at $k \to \infty$ to initiate the decrease in the full energy. Then at $t \gg t_1$, the system comes into a type (7) self-similar regime with $p = 0$: $n(k,t) = t^{-1}f(k)$, with the energy decreasing by a power law: $E \propto t^{-1}$. The short-wave asymptotics in this case is of the Kolmogorov type: $n(k) \propto k^{-9/2}$, so that (13) remains valid and the full number of waves monotonically decreases (this time not by the linear law though, see Fig. 2).

One can see from (12) that the reverse time of nonlinear interaction $t_{NL}^{-1} \propto kE$ (cf. (5)) grows with k slower than the dispersion correction to frequency $\delta\omega = a^2k^3$. This means that with the distribution drifting towards large k, the applicability criterion of the weak turbulence approximation, $\delta\omega t_{NL} \gg 1$, will be satisfied increasingly better. Consequently, the formation of the power Kolmogorov asymptotics $n(k) \propto k^{-9/2}$ in an infinite k-space will proceed up to the absorption region or to $k \simeq a^{-1}$ where the dispersion will cease being small and (12) will become inapplicable.

Also note that the presence of an external wave source, i.e., the addition of positive terms like F_k (external force) or $\gamma_k n_k$ (increment) into the right-hand side of (12) would not violate (14). Therefore, forming the stationary Kolmogorov spectrum will take a finite period of time as well.

5 Numerical Simulation

Any real experiment, be it natural or numerical, deals with a finite number of modes. This is due to either finiteness of the system observed and discreteness of the medium, or strong attenuation of high harmonics. Accordingly, we carried out two variants of numerical simulation of (12):

Var. 1: for a closed system of L modes,

$$\frac{\partial n(k,t)}{\partial t} = \sum_{l=1}^{k} l^2(k-l)^2 \times$$

$$\{n(l)n(k-l) - n(k)[n(l)+n(k-l)]\} - 2\sum_{l=k}^{L} l^2(l-k)^2 \times$$

$$\{n(k)n(l-k) - n(l)[n(k)+n(l-k)]\} = W_k \; ; \qquad (15)$$

Var. 2: for an open system with the waves going out (we set $n(k) \equiv 0$ at $k > L$),

$$\frac{\partial n(k,t)}{\partial t} = W_k - 2n(k)\sum_{l=L}^{L+k} l^2(l-k)^2 n(l-k) \; . \qquad (16)$$

The last term in (16) corresponds to sweeping waves towards the region $k > L$ due to confluence processes. It plays the role of nonlinear damping and provides for effective dissipation of energy. The evolution in the closed system must lead to forming the equilibrium distribution $n(k) = T/k$ at $k \gg 1$, with (15) conserving the total energy

$$E = \sum_{k=1}^{L} k^3 n(k) \; .$$

For the total number of waves $N = \sum_{k=1}^{L} k^2 n(k)$, we have from (15):

$$\frac{\partial N}{\partial t} = -4E^2 + \sum_{k=1}^{L} k^2 n(k) \sum_{l=L-k}^{L} l^2(k+l)^2 n(l) \; .$$

The last term has the meaning of the "overlapping integral" over the regions $(1, L/2)$, $(L/2, L)$ and, consequently, as the waves are distributed over the whole interval, the law of N decrease will increasingly deviate from the linear one.

In the open system (16), the energy monotonically diminishes

$$\frac{\partial E}{\partial t} = -2\sum_{k=1}^{L} k^3 n(k) \sum_{L-k}^{L} l^2(k-l)^2 n(l) \; ,$$

also owing to a similar "overlapping integral".

Thus if a wave packet was placed initially in the region $k \ll L$, the evolution in both cases (15) and (16) will proceed identically until finiteness of the k-space manifests itself. In the closed system, the waves will be gathering near the right edge. In the open one, on the contrary, the occupation numbers will be fading away (in contrast to the evolution in an infinite system), because of nonlinear damping. In the moments of time and in the regions of the k-space where the solutions (15) and (16) are close, one can deem the numerical experiment a good simulation of wave behaviour in an infinite medium.

The initial distribution was chosen to be localized in the region of small k

$$n(k,0) = \exp[-k^2/k_0^2] = \exp[-k^2/10] \; .$$

Equations (15,16) were solved numerically, the time derivative being approximated by the first difference

$$\frac{\partial n(k,t)}{\partial t} \approx \frac{n(k,t+\Delta t) - n(k,t)}{\Delta t} \; ,$$

where Δt was made sufficiently small $(5 \cdot 10^{-5} \div 2 \cdot 10^{-8})$ to provide for stability of the numerical procedure. The number of modes was chosen to be equal to 200, 400 or 1,000.

Fig. 1 presents the evolution of the $n(k)$ value up to the moment in time from which the solutions (15) and (16) start diverging. As we see, by the time $t^* = 5.3 \cdot 10^{-4}$, the Kolmogorov distribution transferring the constant energy flux to large k spans almost the whole of the interval of 1,000 points.

Now let us look again at Fig. 2. One can see that there are two clear-cut stages of evolution. Approximating the dependence $M_2^{-1}(t)$ by a dashed line, we get $t^* = 5.32 \cdot 10^{-4} \approx [8.7M(0)]^{-1}$. The fact that the $M_2^{-1}(t)$ value decreases without reaching zero and that the energy starts going down earlier than at $t = t^*$ is, of course, due to finiteness of the system, see Fig. 3.

At $t \gg t^*$, the law of energy E decrease approaches the power law $E \propto t^{-\beta}$, which may be seen in Fig. 4, which displays the function

$$\beta(t) = \frac{\partial \ln E}{\partial \ln t} \; .$$

According to our supposition, $\beta(t)$ must tend to unity with increase in time. Figure 4 is consistent with this supposition.

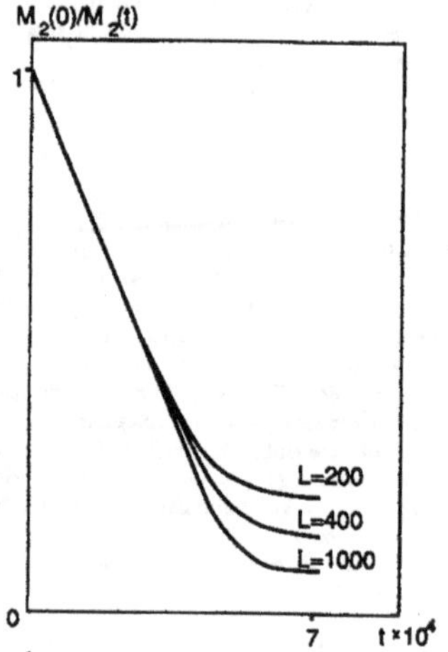

Figure 3: $M_2^{-1}(t)$ for different numbers of modes L.

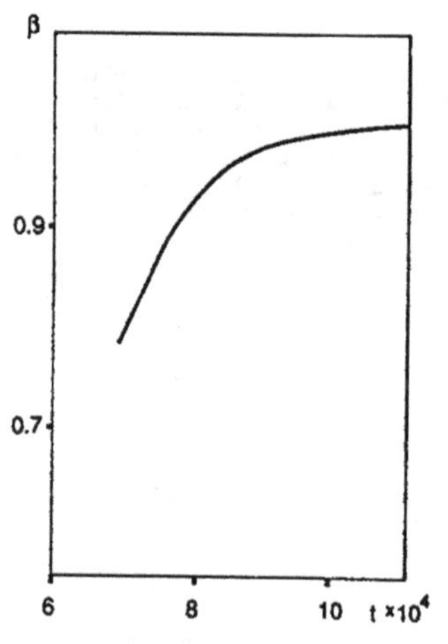

Figure 4: Energy exponent β versus time.

References

[1] Zakharov, V.E., L'vov, V.S., Falkovich G.E.: Kolmogorov Spectra of Turbulence. 1 Wave Turbulence. Springer-Verlag, Heidelberg (1991)

[2] Zakharov, V.E., Musher, S.L., Rubenchik, A.M.: Physics Reports 129 (1985) 285–366.

[3] Zakharov, V.E.: in "Handbook of Plasma Physics" 2 North-Holland, Amsterdam (1984)

[4] Falkovich, G., Shafarenko A.: Journal of Nonlinear Science, 1991 December

[5] Lesieur, M.: Turbulence in Fluids. Kluwer, London (1990)

SELF-SIMILAR STRUCTURE OF AN ACCUMULATED EXPANSION RATE IN THE TACONIS OSCILLATION SYSTEM

K. FUKUSHIMA
Faculty of Education, Kumamoto University,
Kumamoto, 860 Japan

T. YAMADA
Division of Electronic Physics, Kyushu Institute
of Technology, Kitakyushu, 804 Japan

and
T. YAZAKI
Department of Physics, Aichi University of Education,
Kariya 448, Japan

The local expansion rate of nearby orbits on the critical golden torus is calculated from the experimental data obtained in the Taconis oscillation system. The time series of an accumulated expansion rate has a self-similar structure characterized by the Fibonacci sequence. The self-similar structure can also be clearly seen in the power spectrum.

1. Introduction

Direct transition from quasiperiodicity to chaos occurs for the case in which the rotation number is equal to the golden mean, $\rho_G=(\sqrt{5}-1)/2$. The torus with rotation number ρ_G at the critical point of the onset of chaos is called critical golden torus. Similarly to the Rayleigh-Benard convection system,[1] Yazaki and others realized the critical golden torus in the Taconis oscillation by applying the external force.[2] The Taconis oscillation is a spontaneous oscillation of the gas column in a tube with steep temperature gradient.[3,4] The $f(\alpha)$ spectrum was calculated to show that the experimental results can be understood as the system based on the sine circle map.[2]

Theoretically, Mori and his group presented a new manner to dynamically capture the geometrical structure of the critical attractor.[5] These authers showed for several systems that the time series of the n-step average of the local expansion rate has a remarkable structure rather than that of the local expansion rate itself.[6,7] According to them,[6] for a critical golden torus derived from the sine circle map,

$$\begin{aligned}\theta_{i+1} &= f(\theta_i) \\ &= \theta_i + \omega_c - K \sin \theta_i \quad (\mathrm{mod}\ 2\pi)\end{aligned} \tag{1}$$

with $\omega_c=3.811764\cdots$ and $K=1$, the sum of the local expansion rate,

$$S_n = \sum_{i=1}^{n} \lambda_i = \sum_{i=1}^{n} \log |f'(\theta_i)| , \tag{2}$$

fluctuates with n, while the envelope increases as logn. The time series of the following exponent,

$$\beta_n = S_n/\log n , \tag{3}$$

has a self-similar structure characterized by the Fibonacci sequence $\{F_m\}$, where $F_m=F_{m-2}+F_{m-1}$ with $F_0=F_1=1$.

2. Calculation of the Local Expansion Rate

We calculate the local expansion rate from the experimental data by the following manner. See Ref. 2 and 4 for the experimental information. A method of determining the Lyapunov exponent from the experimental time series has been presented by Wolf et al..[8] We apply their method to the present data. Let X_i be a point on the Poincare section. We take k points, $X_i^{(j)}$ (j=1, 2, ···, k), which make smaller angle θ with X_i than the value $|\delta|$, but larger angle than the value $|\varepsilon|$ which corresponds to the noise amplitude. These points map by one step of time evolution as $X_i^{(j)} \rightarrow X_{i+1}^{(j)}$ and $X_i \rightarrow X_{i+1}$. When we denote the angle between X_i and $X_i^{(j)}$ as $\Delta\theta_i$, the individual local expansion rate $\lambda_i^{(j)}$ is written as

$$\lambda_i^{(j)} = \log \left| \frac{\Delta\theta_{i+1}^{(j)}}{\Delta\theta_i^{(j)}} \right| . \tag{4}$$

Thus, λ_i is given by the average of $\lambda_i^{(j)}$'s,

$$\lambda_i = \frac{1}{k} \sum_{j=1}^{k} \lambda_i^{(j)} . \tag{5}$$

We can successively obtain λ_i's by repeating the same procedure for next steps.

Near the critical point the fluctuation of β_n may be written as

$$\beta_n = (S_n - n\,\Lambda^\infty)/\log(n+1) , \tag{6}$$

where Λ^∞ is the Lyapunov exponent and the argument of logarithm is shifted by 1 for the avoidance of divergence for n=0. Though Λ^∞ is equal to zero just at the critical point, it is difficult for experiment to realize $\Lambda^\infty=0$ completely. Therefore, determining Λ^∞ from whole data, we calculate Eq. (6) by subtracting the linear part of S_n with respect to n.

3. Results

We show the time series of β_n, Eq. (6), in Fig. 1. The lower graph in Fig. 1 is a partially magnified figure of upper one. The step intervals between peaks are coincident with the Fibonacci numbers as they are indicated by F_m's. It is found from Fig. 1 that the time series of

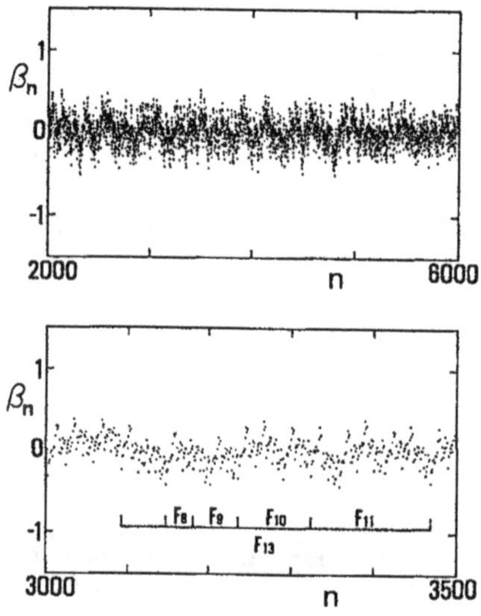

Fig. 1 The time series of β_n. A lower graph is a magnified figure of an arbitrary part in upper one. The Fibonacci numbers are indicated by F_m's.

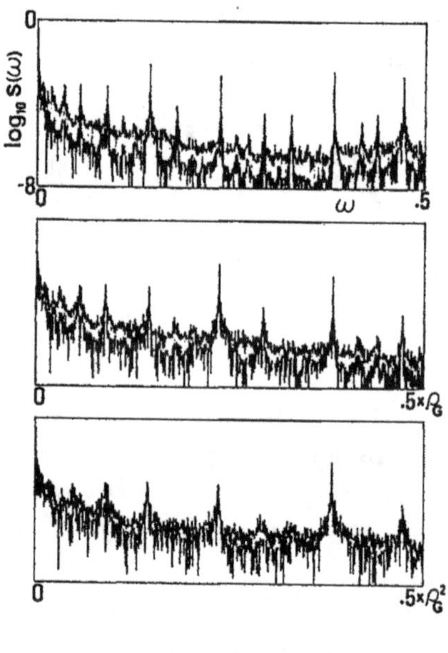

Fig. 2 The power spectra of β_n. We take $N_T=2^{13}$ in Eq. (7). A maximum value of peak is set to be 1.

β_n has a self-similar structure charcterized by the Fibonacci sequence $\{F_m\}$. This self-similar structure is recognized more clearly in the power spectrum of β_n,

$$S(\omega)=\frac{1}{N_T}\sum_{n=1}^{N_T}|\beta_n \exp(-2\pi i\omega n)|^2 , \tag{7}$$

which is shown in Fig. 2 together with the magnified figures. The transverse axes of middle and lower graphs in Fig. 2 are magnified by ρ_G^{-1} and ρ_G^{-2} times, respectively. The peaks appear at the same position even by such a magnification. This means that self-similar structure exist in the power spectrum.

4. Conclusion

We have shown from the experimental data that the time series of n-step average of the local expansion rate has a self-similar structure characterized by the Fibonacci sequence. It is considered that the decomposition process of the critical golden torus, which is regarded as a fusion of the infinite number of attractors with the rotation number $\rho_m=F_m/F_{m+1}$ $(m=1, 2, \cdots, \infty)$, is reflected in such a time series. The self-similar structure in β_n is universal sal for the systems described by the sine circle map. Therefore, we expect that the similar results may also be observed in the other systems.

Acknowledgements

We would like to thank Professor H. Mori for valuable discussions. This work was partially performed in the Instrument and Cryogenic Centers, the Institute for Molecular Science.

References

1 A. P. Fein, M. S. Heutmaker, and J. P. Gollub, Phys. Scr. T9, 79 (1985); J. Stavans, F. Heslot, and A. Libchaber, Phys. Rev. Lett. 55, 596 (1985); M. H. Jensen, L. P. Kadanoff, A. Libchaber, I. Procaccia, and J. Stavans, Phys. Rev. Lett. 55, 2798 (1985).

2 T. Yazaki, S. Sugioka, F. Mizutani, and H. Mamada, Phys. Rev. Lett. 64, 2515 (1990).

3 N. Rott, Z. Angew. Math. Phys. 20, 230 (1969); 24, 54 (1973).

4 T. Yazaki, A. Tominaga, and Y. Narahara, J. Low Temp. Phys. 41, 45 (1980), and Phys. Lett. 79A, 407 (1980); T. Yazaki, S. Takashima, and F. Mizutani, Phys. Rev. Lett. 58, 1108 (1987).

5 H. Mori, H. Hata, T. Horita, and T. Kobayashi, Prog. Theor. Phys. Suppl. No. 99, 1 (1989).

6 T. Horita, H. Hata, H. Mori, and K. Tomita, Prog. Theor. Phys. 81, 1073 (1989).

7 H. Hata, T. Horita, and H. Mori, Prog. Theor. Phys. 82, 897 (1989).

8 A. Wolf, J. B. Swift, H. L. Swinney, and J. A. Vastano, Physica 16D, 285 (1985).

Using Thermal Noise to Measure the Correlation Lengths of Electroconvection

Frank Hörner and Ingo Rehberg

Physikalisches Institut, Universität Bayreuth
W-8580 Bayreuth, Germany

Abstract

The correlation lengths of electroconvection in a nematic liquid crystal both parallel and perpendicular to the director are measured by analyzing the correlation functions of thermally induced director fluctuations in the subcritical regime.

1 Introduction

Pattern forming instabilities in anisotropic systems can be described by an amplitude equation in the neighborhood of their critical point [1]:

$$\tau_0 \partial_t A = \epsilon A + \xi_{\|}^2 \partial_{xx}^2 A + \xi_{\perp}^2 \partial_{yy}^2 A$$

which contains two length scales $\xi_{\|}$ and $\xi_{\perp}$. Within this linear approximation, the correlation lengths parallel and perpendicular to the convection rolls diverge near the transition ($\epsilon = 0$) according to $\xi_P^2 = \xi_{\|}^2/(-\epsilon)$ and $\xi_N^2 = \xi_{\perp}^2/(-\epsilon)$, respectively. An example for an anisotropic system is electroconvection in a thin layer of nematic liquid crystal. It offers the possibility to study the influence of thermal noise on a nonequilibrium phase transition directly [2, 3]. This fact should make it possible to address a number of fundamental questions experimentally [2], but in addition it also offers a convenient tool to study the linear response of the hydrodynamic system to small perturbations. The applicability of this noise method to measure the decay rate τ_0 has already been demonstrated in Ref. [4]. In the present paper the thermal noise is used to measure the correlation lengths $\xi_{\|}$ and $\xi_{\perp}$.

2 Experimental Setup and Procedure

The experimental setup is similar to the one described in Ref. [2]. We used a thin layer of the nematic liquid crystal N-(p-methoxybenzylidene)-p-butylaniline (MBBA). A lower bound for the thickness is given by the use of $d = 8\,\mu$m mylar spacers between the transparent electrodes. The actual thickness was measured by optical inspection with a microscope, a procedure which is believed to be reliable within an error of $\pm 2\,\mu$m at best, which gave a cell thickness of $d = (14 \pm 2)\,\mu$m. The director alignment is enforced by rubbing of the polymer-coated glass slides which confine the sample. The temperature is controlled to (21 ± 0.01) °C. The mean-square value of the AC-voltage applied to the transparent electrodes serves to define the control parameter $\epsilon = V^2/V_c^2 - 1$, where V_c is the frequency dependent critical voltage for the onset of convection.

A light bulb illuminates the sample through a polarizer. The microscope objective forms an image of the layer located 0.5d below the upper boundary of the liquid crystal, where the linear approximation of the optical response is expected to have the largest range of applicability [5]. This image is detected by a charge-coupled device array (CCD)-Camera of 768×512 quadratic pixels thermostatted at (5 ± 0.05) °C, with a spacing of 0.011 mm. The exposure time for the diodes is 0.04 seconds. The light intensity is digitized with 8 bits, which corresponds to 256 gray levels.

The camera is aligned such that the camera lines are located parallel to the director of the nematic liquid crystal, i.e. perpendicular to the convection rolls, and the camera columns parallel to the rolls. The camera works in an interlaced modus, thus only the even lines are used to ensure simultaneous illumination. When measuring the correlation length parallel to the director, the lines over one entire image are added in order to restrict the measurement to long wavelength modes perpendicular to the director, as explained in Ref. [3]. When measuring the correlation length perpendicular to the director, one central column of the camera is used.

We determine the structure function of 2048 lines for measuring $\xi_{\|}$, and 2048 columns for measuring $\xi_{\perp}$, as explained in detail in Ref. [5]. The correlations lengths ξ_P and ξ_N are obtained by fitting exponential decays to the structure functions.

3 Experimental Results

In order to characterize the liquid crystal sample used, Fig. 1 presents the phase diagram of our convection cell. The cutoff-frequency can be seen to be at about 200 Hz. Above this frequency the motionless ground state goes unstable with respect to the dielectric rolls.

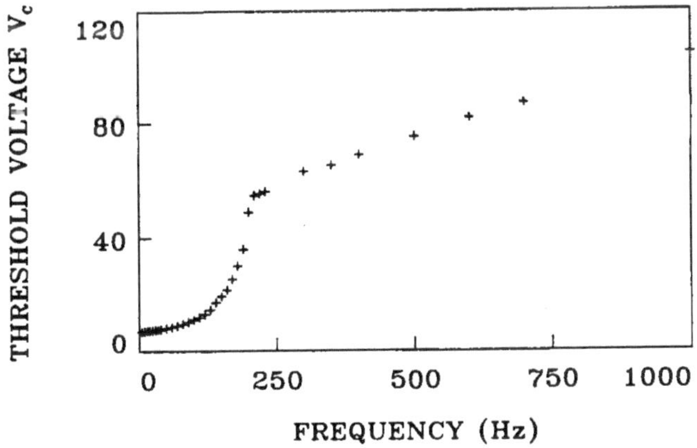

Figure 1: *The phase diagram. The crosses indicate the critical voltage for the onset of convection.*

Further characterization of the sample is provided by measuring the wave number of the rolls in the conduction regime. This has been done by measuring fully developed convection rolls slightly above threshold. The results are presented in Fig. 2. The data are larger than the numbers presented in Ref. [1], at least when comparing with the limit of thick cells or high electrical conductivity. The fact that the numbers do not match might be attributed to the fact that our sample is not within this limit. It is not known whether this assumption can also explain the qualitative disagreement observed in the curvature of the line, when comparing our Fig. 2 to Fig. 3 of Ref. [1]. In order to make better contact with the linear stability analysis it might be advantageous to measure the critical wave number by analyzing subcritical fluctuations. That problem is currently under investigation.

For determining the correlation lengths, we measure spatial structure functions at different subcritical values of ϵ and at a frequency of the driving voltage of 25 Hz. For determining ξ_P we fit the structure function obtained along the director orientation to $S(\Delta x) \sim \exp(-\Delta x/\xi_P)\cos(k\Delta x)$ similar to Ref. [2], and the results are shown in Fig. 3. The structure function obtained along the roll orientation is fitted by $S(\Delta y) \sim \exp(-\Delta y/\xi_N)$, and the resulting values for ξ_N are shown in Fig. 4. From the slope of the fitted straight lines in Figs. 3, 4 we get $\xi_{\|} = 0.26$d, and $\xi_{\perp} = 0.09$d. Both numbers are smaller than the ones which can be read of Fig. 15 of Ref. [1] to be

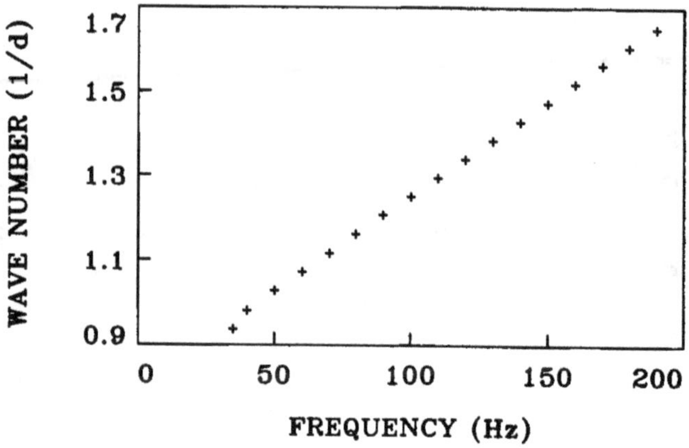

Figure 2: *The crosses indicate the wave number as measured slightly above the onset of convection.*

$\xi_{\parallel} = 0.28d$ and $\xi_{\perp} = 0.11d$. The ratio $\xi_{\parallel}/\xi_{\perp}$ of the measured correlation lengths is larger by about 10% than the theoretically expected ratio. A possible explanation for these discrepancies might be the fact that we are dealing with a relatively pure and thin sample, which tends to change the theoretically expected numbers as indicated by Fig. 12 of Ref. [1]. However, a detailed calculation of the correlations lengths as a function of the cell thickness does not seem to be available.

4 Summary and Discussion

In summary, we have shown that correlation lengths can be measured by means of analyzing subcritical hydrodynamic fluctuations caused by intrinsic thermal noise. An elegant deterministic alternative for measuring these numbers is to analyze the neighborhood of a defect [6]. One might argue, however, that this procedure is not a direct method, but it rather relies on the applicability of the amplitude equation in the neighborhood of the defect, and on a linear response of the optical signal, which can be guaranteed only in a very small ϵ – interval [5]. The noise method offers a fairly large working interval free of these restrictions, because the fluctuations are so small that the optical response to director fluctuations remains linear, and also the applicability of a linear stability analysis cannot be questioned.

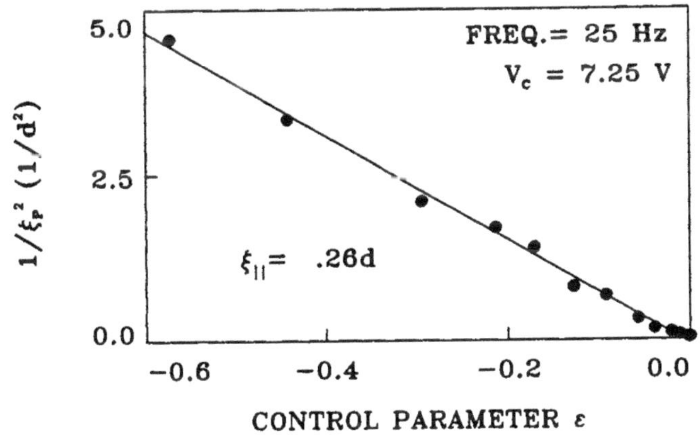

Figure 3: *The correlation length as measured parallel to the director orientation at a driving frequency of 25 Hz. The slope of the fitted straight line yields the value of* $\xi_{||} = 0.26d$.

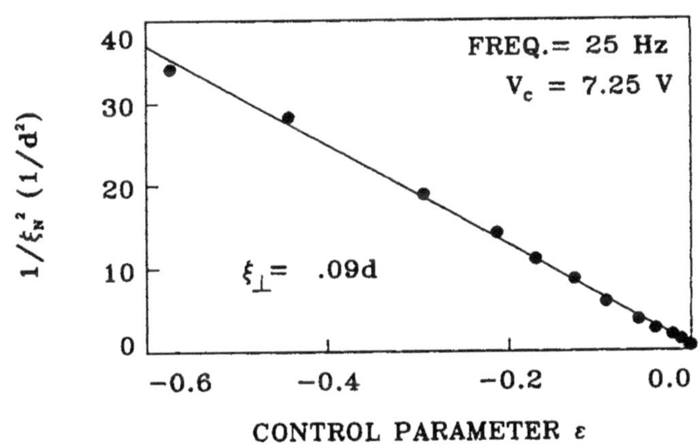

Figure 4: *The correlation length as measured perpendicular to the director orientation at a driving frequency of 25 Hz. The slope of the fitted straight line yields the value of* $\xi_\perp = 0.09d$.

5 Acknowledgements

It is a pleasure to thank G. Ahlers and D.S. Cannell for a helpful discussion. The experiments were supported by Deutsche Forschungsgemeinschaft, and by the NATO Collaborative Research Grant 910112.

References

[1] E. Bodenschatz, W. Zimmermann, and L. Kramer, J. Phys. France **49** (1988) 1875.

[2] I. Rehberg, S. Rasenat, M. de la Torre Juárez, W. Schöpf, F. Hörner, G. Ahlers, and H.R. Brand, Phys. Rev. Lett. **67** (1991) 596.

[3] I. Rehberg, F. Hörner, L. Chiran, H. Richter, B. L. Winkler, Phys. Rev. **A44** (1991).

[4] B.L. Winkler, W. Decker, H.Richter, and I. Rehberg, submitted to Physica D.

[5] I. Rehberg, F. Hörner, G. Hartung, J. Stat. Phys. **64** (1991) 1017.

[6] S. Rasenat, V. Steinberg, and I. Rehberg, Phys. Rev. **A42** (1990) 5998.

Transient Multimodality, and Frequency and Noise Dependences in T1-T2 transition

Shoichi KAI Masanori ANDOH Shigenori YAMAGUCHI

Department of Electrical Engineering, Kyushu Institute of Technology
Tobata, Kitakyushu 804 Japan

Abstract

In electrohydorodynamic convection, turbulence 1-turbulence 2 (DSM1-DSM2) transition is observed when high voltage (typically 4-5 times of onset voltage V_c for convection) is applied. This transition is local , occurs via production of nulei of new phase and shows dynamic hysteresis which means hysteresis due to the characteristic relaxation time scale of the system. As a frequency f is increased the width of hysteresis is shrinked and disappears at ceratin frequency f_c. The shape of turbulent nuclei at the transition changes into deformed one with increase on f. This probably indicates a change of type of transition. The transient properties during turbulence 1-turbulence 2 transition are also studied and the temporal development of distribution probability is obtained. while far from transition threshold the distribution shows always single peak, close to the threshold it shows double peaks in the late stage of transient process, *i.e.* a transient multimodality. This indicates that the instability of T1-T2 transition is due to the symmetry breaking by the intrinsic randomness in a system, *i.e.* DSM1-turbulence. Under superimposing a noise field, the transient properties are affected by noise, *e.g.* that is, the boundary of nuclei is less pronounced.

1 Introduction

Turbulence-turbulence transition is not well known phenomenon and a few examples are found out such as convective states in Helium (He) gas [1], in superfluid He [2] and in the electrohydrodynamic convection (EHC) in liquid crystals [3]. Turbulence 1-turbulence 2 (T1-T2) transition in superfluid He 3 is induced by a topological instability, such as due to the formation and the characteristic density change of voltex filaments by increase in the counterflow between the superfluid and the normal fluid of liquid He. Unfortunately direct visual observation is not possible in liquid

He, unlike in EHC where the system shows a similar phenomenon. Therefore there is more advantage for the detailed investigation of spatial structures in EHC because of easy direct observation.

In the T1-T2 transition (traditionally called as DSM1-DSM2 transition [3]) in EHC we found that the topological disclinations by a director oritentation play an impotatant role. We have also found the following facts in this transition.

The transition is local via nucleation process of new phase and however there is no hysteresis [3]. The nucleation rate follows a conventional nucleation theory [3]. Additionally several uncommon facts are observed: The relaxation time tends to diverge at the T1-T2 threshold V_2. The growth velocity of nulei of T2 does not show classical dependence proportional to $\epsilon_2^{1/2}$ and indicates nonclassical dependence, $\epsilon_2^{0.7}$ [3]. Here $\epsilon_2 = (V - V_2)/V_2$,and V and V_2 is the applied voltage and the threshold for the onset of turbulence 2. An index is recently obtained from the amplitude equation containing higher terms of amplitude in different systems [4]. In this article we report further details observed in T1-T2 transition in EHC such as external noise called multiplicative noise effects, frequency dependences and their transient properties. In special, we discover the transient bimodality phenomenon which has been described in combussion system by Nicolis and Baras [6]. In the system with spatial degrees of freedom, however, there is no report and this is the first time to describe here.

2 Preparation

The sample used in the present study was the room temperature nematic liquid crystal MBBA[3]. The thickness of the enclosed cells was ranging from 25 to 100 μm of which surface was coated by transparent electrodes In_2O_3. The four samples are prepared whose threshold voltage V_2 for the onset of T2 turbulence and whose critical frequency f_c for the dielectric regime are as follows;(A)V_2 = 37.8 V (at f = 50 Hz) and f_c = 100 Hz, (B)V_2 = 27.5 V and f_c = 250 Hz , (C)V_2 = 77.7 V and f_c = 70 Hz, and (D)V_2=27 V and f_c = 500 Hz. All samples show qualitatively similar results. The experimantal set-up was conventional one. The external noise was also prepared by special noise generator (NF co.) which produces a gaussian white noise (correlation time is changed from 2 ms to 5 ms) [7]. The observation was acheived in visual observation and in light transmittance under a cross-polarized microscope, and recorded on a magnetic tape and a disket through a video camera and a personal computer. The details on these experimental procedure have been already reported in Refs. 3-5 and 7.

3 Results

Figure 1 shows the hysteresis in the light-transmittance I/I_{max} near DSM1-DSM2 (T1-T2) transition point V_2 which is related to the onset of the disclination due to turbulence [3]. Here I and I_{max} are the transmitted light intensity and the intensity without convection through a sample respectively. The changing rate (the ramp rate:r) of applied voltage is 25 mV/s. T1-T2 transition can be observed near 38 V for

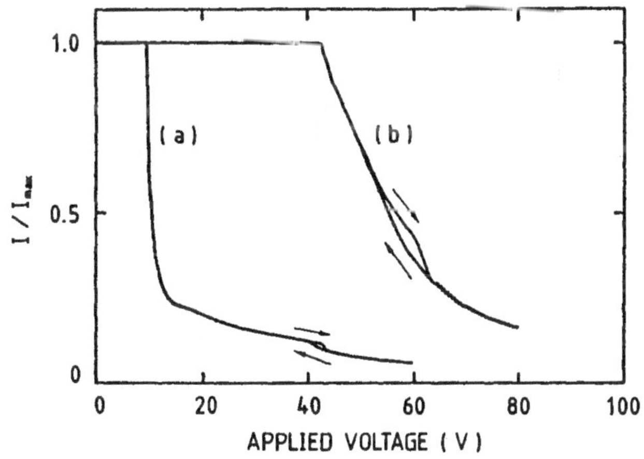

Fig.1. Hysteresis in light-transmittance near T1-T2 transition point V_2. Ramp rate:25 mV/s (a)f=50Hz, (b)150Hz. $\Gamma = 800$, $d = 25\mu$m.

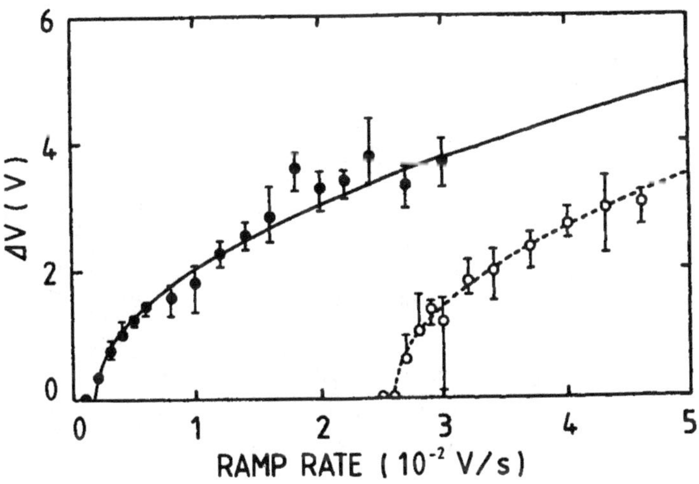

Fig.2. Ramp rate dependence of hysteresis. •:50 Hz, o:80 Hz.

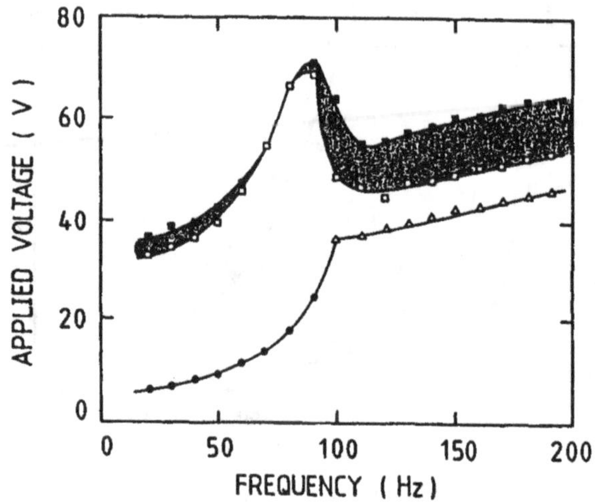

Fig.3. Frequency dependence of hysteresis. The shaded region shows the hysteresis width at the ramp rate $r = 25$ mV/sec.

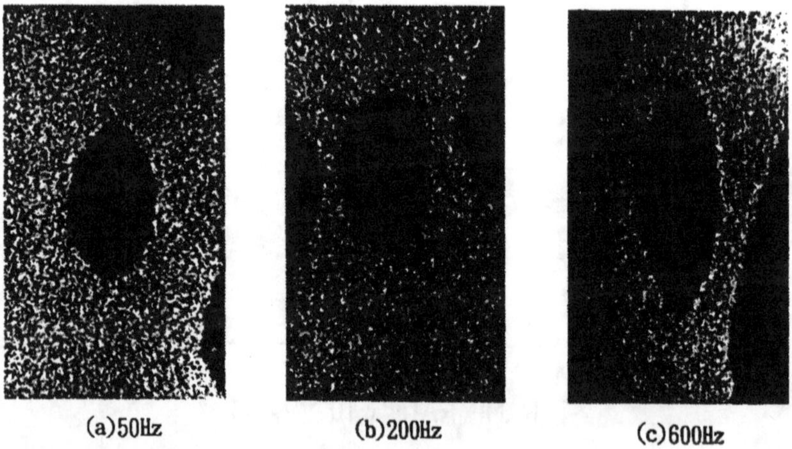

Fig.4. Frequency dependence of T2-nuclei shape. A critical frequency f_c is 500 Hz.

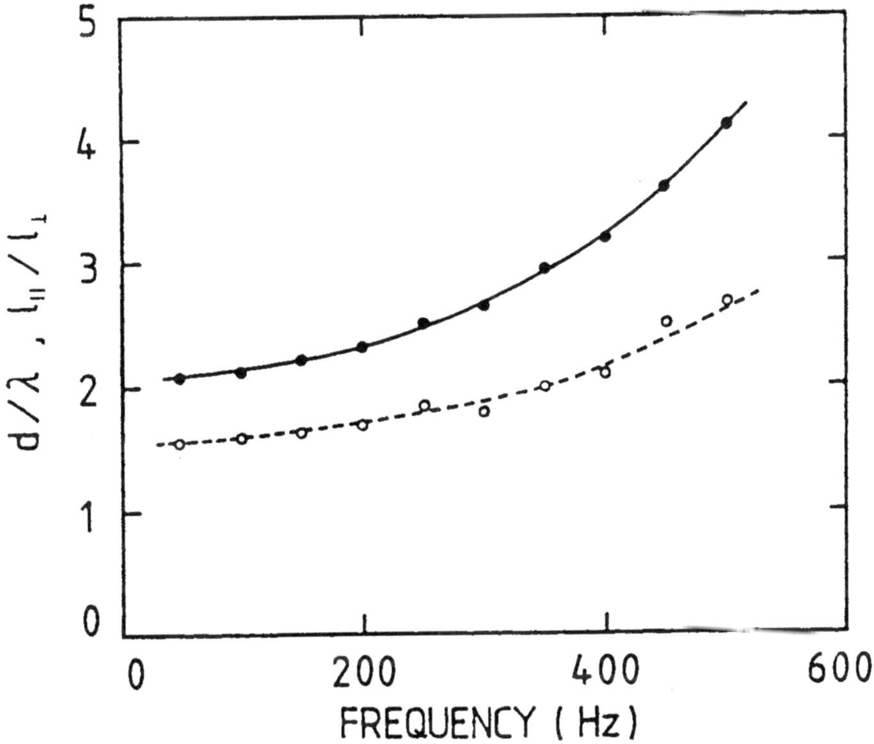

Fig.5. Frequency dependence of geometric ratio of nuclei shape.

50 Hz and 60 V for 150 Hz. The width of hysteresis for 150 Hz is wider than that for 50 Hz obviously. As already reported however the width depends on the ramp rate[3], because it is due to dynamic response of the system. To check it we have measured the ramp rate dependence of the width of hysteresis. As seen in Fig.2 decreasing r the width of hysteresis decreases and finally it disappears. This indicates that the transition is not subcritical. When the frequency is increased to 80 Hz no hysteresis is already observed for $r = 2.6 \times 10^{-2}$ V/s which indicates also that the dynamic response of the system for higher frequency becomes faster. This shows that the transition here has a possiblity similar to the transcritical bifurcation. Figure 3 shows the frequency dependence of the width of hysteresis for the fixed r (=25 mV/s) which is indicated by the shaded mask. Below f_c($\approx$ 100 Hz) as f increases the width decreases and finally disappears. On the other hand above f_c the width is independent of f except frequency region close to f_c. According to the detailed investigation of the ramp rate dependence of hysteresis above f_c, there is no real hysteresis in the high f region. Therefore in whole reagion the type of bifurcation would be either supercritical or transcritical. In order to understand the actual difference of the transition for different frequency, we measured the shape change of T2 nuclei in the transition. Figure 4 shows the shape difference for three different frequencies, 50, 200 and 600 Hz. Increasing the frequency, the shape becomes flater and needle-like. To characterize an anisotropy of the shape, Figure 5 shows the ratio (○) of the lengths between the long axis $l_{\parallel}$ and short axis $l_{\perp}$ for f together with the ratio (•) between a cell thickness d and half-wave length λ of the Williams domain rolls (WDR). Both dependences are very similar and the ratio of both ratios is constant ($\approx$ 0.71). Therefore the shape of nuclei is strongly related to the width of the first pattern, *i.e.* WDR.

In Fig.6, the temporal development of the probability ditribution of DSM1-DSM2 area is shown when the step-like voltage is applied from the below V_2 to the above in the small observation area (200 μm in diameter). For small deviation from ϵ_2 it shows transiently two peaks in the distribution in (b), that is bimodality, while only single peak is observed in (a) of Fig.6 for large deviation. In Fig.6(a), starting with a distribution centered on a high transmission state , its peak is gradually dispersed and move into a low transmission state (*i.e.* the high density state of disclination). As shown in Fig.6(b), on the other hand, starting with a distribution centered on a high transmission state, the system gradually develops a long tail and subsequently a transient bimodality before collapsing again to a unimodal distribution centered on the stable attractor that is a low transmission state. This probably indicates exsistence of two different time scales of dynamics. The profiles and the variances of this transient distributions depend on the observation area as seen in the following figures.

Figure 7 shows the temporal developments of the statistical variance σ_0 and the total statistical mean value P_0 ,and those of two peaks P_F and P_S [5];

$$P_0(t) = \frac{1}{N} \sum_{i=1}^{N} P_i(t), \tag{1}$$

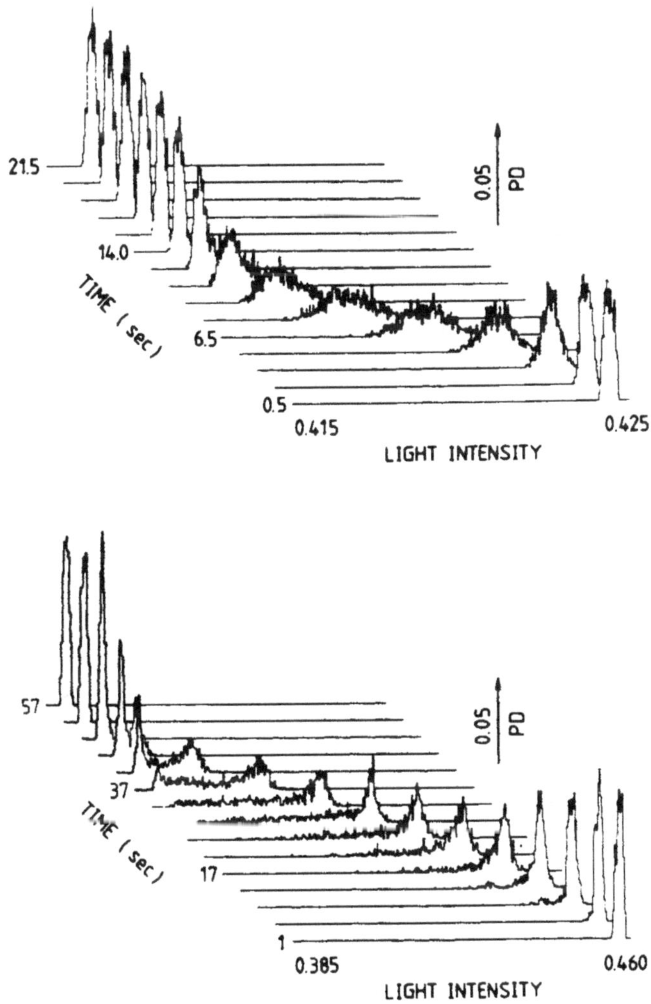

Fig.6. Stochastic distribution of transient processes obtained by the previous method. (a)$\epsilon_2=4$ (b)$\epsilon_2=3$. Close to threshold ϵ_2, the transient multimodulity can be observed.

$$\sigma_0^2(t) = \frac{1}{N-1}\sum_{i=1}^{N}(P_i(t) - P_0(t))^2. \tag{2}$$

Here $P_i(t) = I_i(t)/I_{max}$ and I_i and I_{max} is respectively the transmitted intensity of a light with turbulence at i–th repetition and without convection through an EHD cell. In Fig.7(a), conventional aspects for P_0 and σ_0 can be observed similar to thoes which have been already discussed in Ref.[5].

The original peak (P_F:the initial peak) in Fig.7(b) is observed in the whole region with temporal development and without disappearance. But the temporal dependence of its development is sharper after a second peak (P_S) reaches at fixed location. Before reaching steady state location of P_S in the late stage of a transient phenomenon the temporal change of P_F is rather slow. Alter the vanishment, however, it starts to move very quickly into the low transmission state. The temporal development of σ_0 is a common one [5], that is , it increases once then decreases in the late stage. P_0 shows a monotonical change as expected from a simple evolution equation [5]. Figure 7(b) shows P_0, P_F, P_S and σ_0 for $\epsilon_2 = 3$ where transient bimodality is observed.

Figure 8 shows P_0, P_F, P_S and σ_0 for a large observation area of 1.5 mm in diameter. (We also achieved experiments for the observation area 6 mm in diameter. The results are almost the same as ones in Fig.8.) The transient bimodality is more pronounced and the centered peaks P_F and P_S do not move from their initial locations and instead both heights are changed with time. Then P_F is finally replaced by P_S completely. This bimodal distribution starts to appear around half the time to rearch a steady state, for instance $t \approx 20$ s when the system reaches a steady state at $t \approx 40$ s. In other words, the bimodality appears slightly before $t_{1/2}$ at which the order parameter develops until half the value of a steady state value. Nevertheless P_0 and σ_0 shows a common aspect. Therefore this kinds of behavior cannot be obtained only when the mean aspects are measured. These characteristic features suggest the exsisitence of two different dynamics in T1-T2 transition and probably explained similarly by Nicolis and Baras's theory [6]. Also here the symmetry breaking is due to the intrinsin randomness already existing DSM1 which is nonperiodic (turbulent) flow. This also very new phenomena [6]. Details will be described elsewhere in near future [4].

Figure 9 shows an expected director orientation of disclination in T2 state. This illustration can be obtained from consideration according to the photograph after the applied field is switched off. However it is not yet quantitative and just within qualitative explanation because it is not the observation in the flow state of T2.

When the system is sustained by superimposing a noise field, the threshold V_2 is increased as shown in Fig.10. Here V_2^* means the threshold for the onset of T1-T2 transition at superimposing external noise. This aspect is also understandable in a viewpoint of multiplicative effect as already reported in Ref.7.

When V is increased further from V_2, new transition can be observed. We call this T2-T3 transition. Figure 11 shows the T2-T3 transition. Turbulence 3 appears under very high intensity of V. The flow is very complex and fast. Therefore there might be isotropic liquid by heating effects (ohmic heating). As a detail is unknown

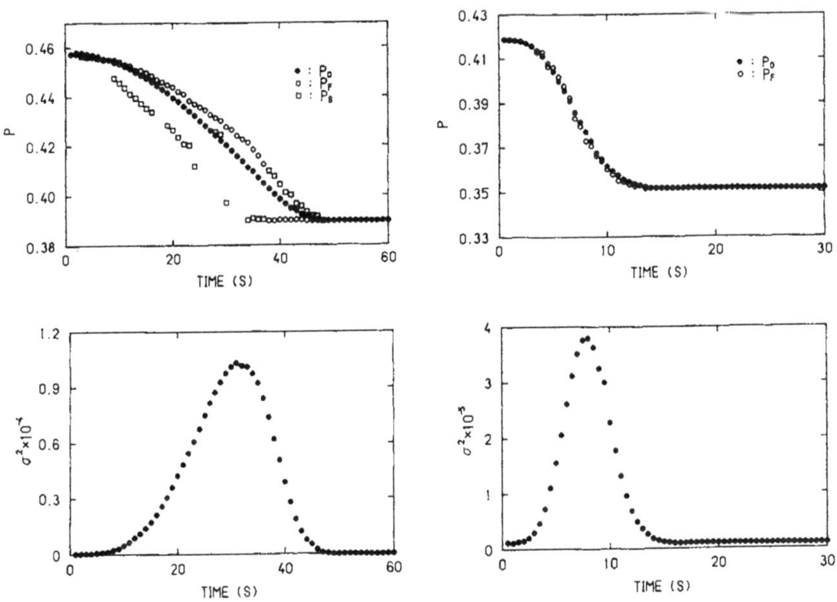

(a) (b)

• **Fig.7.** Temporal developments of two peaks (∘, □) and statistical mean values (•), and ones of variance in small observation area (200 μm in diameter) for $\epsilon_2 = 3$ (a) and 4 (b). Upper figure: peaks Lower figure: variance

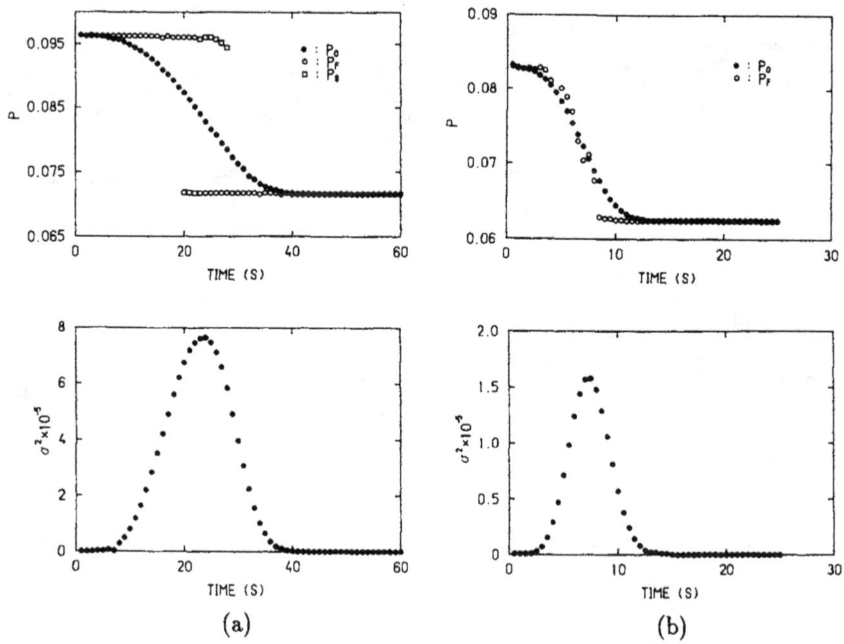

• **Fig.8.** Temporal developments of two peaks (o, □) and statistical mean values (•), and ones of variance in large observation area (1.5 mm in diameter) for $\epsilon_2 = 3$ (a) and 4 (b). Upper figure: peaks Lower figure: variance

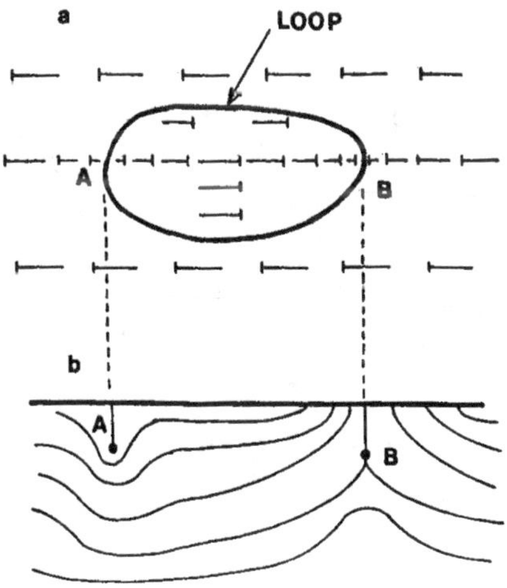

Fig.9. Schematic drawing of director orientation of disclination line. The head of nail of director shows inclination parpendicular to the plane of this page. It indicates that the strong shear due to the up- and down-flows take place at the disclination.

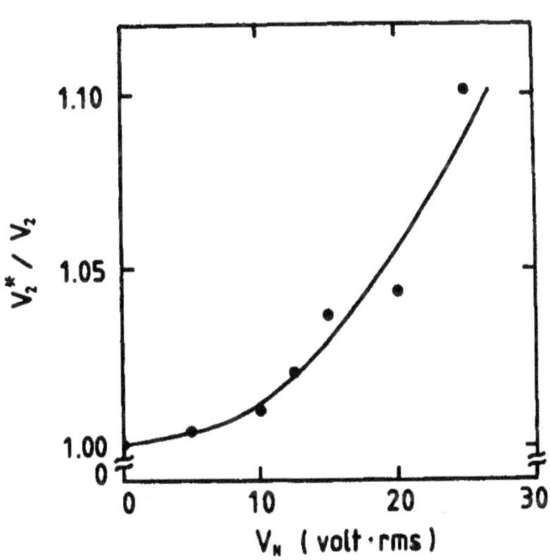

Fig.10. External noise effect of T1-T2 transition point. ($r = 25$ mV/s).

Fig.11. T2-T3 transition at further voltage. (30 Hz, 100V, 30°C, 50 μm)

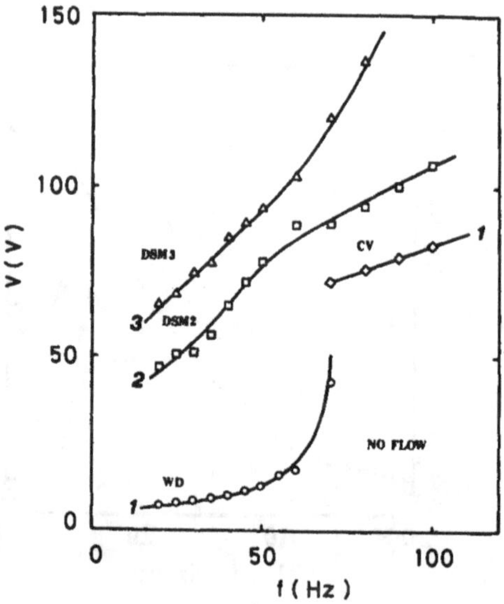

Fig.12. Phase diagram for T1-T2-T3 transition. ramp rate r = 10 mV/s, $\Gamma = 600$.

yet here, we stop to describe only facts of the exsistence of T3, with phase diagram as shown in Fig.12.

4 Conclusion

The obtained results are summarized as follows. Bifurcation seems to change at a certain critical frequency because histeresis vanishes there as changing the shape of T2 nuclei. The higher the freqency, the larger the aspect ratio of the nuclei shape. This might lead to the change of type of bifurcations. Therefore we may say that pattern dependent instability which may be multi-dimensional. The observation of a transient bimodality is the first time for the convective systems with spatial degrees of freedom. The transient bimodality indicates two different dynamics, fast and slow ones, This kind of phenomena is observed in the system where the symmetry breaking is induced by the intrinsic randomness of a system [6]. The deatiled analysis is in progress and will be reported in near future [4]. External noise can also induce the higher threshold for the onset of T1-T2 transition. Turbulence 3 is observed above T2 phase. Turbulence 3 might be turbulence of an isotropic fluid by ohmic heating because the applied field is sufficiently high and also the flow scale is sufficiently small which realizes the complete disturbance of director orientation.

Acknowledgements

This work is in part supported by the Grant-in-Aid for Scientific Research (No.02452047) and The Kajima Foundation. We would like to thank Professor F.Sagues for his valuable suggestion and discussion on a transient bimodality.

References

[1] F. Heslot, B. Castaing and A. Libchaber, Phys.Rev. A36, 5870(1987).

[2] J.T. Tough, *Superfluid Turbulence in Low Temperature Physics, edited by D.F.Brewer(North-Holland, Amsterdam, 1982), Vol.8, p134.*

[3] S. Kai, W. Zimmermann, M. Andoh and N. Chizumi, Phys.Rev.Lett. 64, 1111(1990). 64, 1111 (1990); S. Kai and W. Zimmermann, Prog.Theor.Phys. Supp.99, 458 (1989).

[4] S. Kai et al. , in preparation for publication

[5] S. Kai, S. Wakabayashi and M. Imasaki, Phys.Rev.A. 33, 2612(1986).

[6] G. Nicolis and F. Baras, J.Stat.Phys. 48, No.5/6, 1071 (1987).

[7] S. Kai, H. Fukunaga and H.R. Brand, J.Stat.Phys. 54, 1133(1989).

DYNAMICS OF ANNIHILATION PROCESS OF +1 AND -1 DISCLINATIONS

TOMOYUKI NAGAYA, HIDEKI HOTTA, HIROSHI ORIHARA
and YOSHIHIRO ISHIBASHI
Synthetic Crystal Research Laboratory,
School of Engineering, Nagoya University
Nagoya 464-01, JAPAN

ABSTRACT

Dynamics of interacting disclinations in a nematic liquid crystal cell, which can be regarded as the two-dimensional XY system, has been investigated from the view point of statistical physics. We made numerous wedge disclinations of the strength ±1 and observed the annihilation process of disclinations. By analyzing microscope image, it was found that the number of disclinations is inversely proportional to time. Furthermore, the spatial correlation function of the pattern was calculated, and then the dynamical scaling law was found to hold in the system.

1. Intruduction

Dynamics of topological defects has been investigated in various fields of physics. After the symmetry breaking phase transition, in which a system changes to a lower symmetry phase quickly, many topological defects are created and the system goes to the new equilibrium state. The coarsening dynamics of topological defects, which depends on the dimensions of the order parameter N and of the system D, becomes an extremely important subject in various fields of physics.

The system considered here is a non-conserved order parameter system. Theoretically, Toyoki[1-4] investigated such a coarsening dynamics of topological defects for any dimensions of N and D from the viewpoint of statistical physics. Experimentally, the disclination is one of the best objects to study the dynamics of topological defects, because we can observe the disclinations very easily with a polarizing microscope due to the optical anisotropy of liquid crystals and the motion of disclinations is slow enough to be observed in real time. In some works the collective motion of disclinations[5-12] was investigated, while in others attention was paid to the annihilation process of disclination pairs[13,14]. Orihara and Ishibashi[5,6] studied the two-dimensional coarsening dynamics in the twisted nematic cell. This system can be regarded as the two-dimensional Ising system, namely $N=1$ and $D=2$. The dynamical scaling law was found to hold in the system, the scaling index being 0.5 for the time evolution of the correlation length, which agrees with the theoretical value 0.5. Hashimoto *et al.*[7-9] investigated the annihilation process of a great number of disclinations in the thin film of polymer liquid crystals. They regarded the system as $N=2$ and $D=2$ and obtained the scaling index 0.35 for the time dependence of the average distance between two neighboring disclinations. But the index deviates from the theoretical value 0.5. For the case of three dimensional system, $N=3$ and $D=3$, Yurke *et al.*[10-12] studied the coarsening dynamics of disclinations in the nematics subject to a rapid pressure-jump quenching. They classified the disclinations topologically and investigated the elementary process of the dynamics in detail. They measured the time evolution of disclination density, in which the system consists of a number of $\pm 1/2$ disclinations and some ± 1 disclinations and obtained the scaling index 1.02, which agrees with the theoretical value 1. Furthermore, they also investigated an analogy between the disclinations and the cosmological defects.

These studies are concerned with the statistical properties of the coarsening dynamics of a number of disclinations. On the other hand, Nagaya *et al.*[13] investigated the individual motion of a few disclinations for the case of $N=2$ and $D=2$ on the basis of the equation of motion. They developed a new experimental method to generate only the +1 and -1 disclinations. This technique is very important, since the interaction between the disclinations depends on the strength of each one.

The purpose of the present study is to investigate the coarsening dynamics of a number of disclinations for the case of $N=2$ and $D=2$ by using the above method[13]. Our system corresponds to the polymer system studied by Hashimoto *et al.*[7-9] In contrast to the polymer system, which has two kinds of disclinations, namely a number of $\pm 1/2$ disclinations and some ± 1 disclinations, our system consists of only ± 1 disclinations.

2. Experiments

The liquid crystal used in the present study is the mixture of MBBA and a little amount of cholesteryl chloride with the weight ratio of 99.5:0.5. The dielectric anisotropy of the mixture is negative. The pitch of the mixture was 30μm at 24°C. The mixture was confined between two ITO coated glasses. The boundary condition is a homeotropic alignment. The cell gap is 16μm. As the result of the competition between the chiral force of the cholesterics and the anchoring force on the glass surfaces, the mixture has the bubble domain structure[15] under no field. When we applied an electric field of 5kHz and 25V to the bubble domain cell, a great deal of disclinations, the type of which is +1 or -1, appear from the bubble domain.[1] In such a system, the molecules align perpendicular to the field due to the negative dielectric anisotropy, namely, parallel to the glass surfaces, and thus the disclination lines are perpendicular to the glass plates. Due to the interaction between the disclinations, the pair annihilation of the +1 and -1 disclinations occurs.

3. Results and discussion

Figures 1 shows the time evolution of disclination pattern. The director is parallel to the paper. The time evolution of disclination texture was recorded in a V.T.R. The image was digitized to 256 × 256 pixels of 6 bit number. We analyzed this coarsening process by means of image processing technique.

This coarsening process can be characterized by following three quantities. The first one is the time evolution of the total number of disclinations $N(t)$ in the field of microscope ($1.0 \times 0.8\text{mm}^2$). The next important quantity is the spatial correlation function of the bright and dark pattern. The intensity of the image at $\boldsymbol{r}$ is given as

$$I(r,t) = I_0 \sin^2\left(\frac{\pi\, d\, \Delta n}{\lambda}\right) \sin^2(2\varphi(r,t)) \quad , \qquad (1)$$

where d is the cell gap, Δn the anisotropy of optical indices, λ the wave number, I_0 the intensity of incident beam and φ the angle between the polarizer and the director. Since the values I_0, d, Δn and λ are constant, the intensity depends only on the angle φ. We defined a two-dimensional spatial correlation function as

$$C(r,t) = \frac{\langle (I(r,t) - \langle I \rangle)\,(I(0,t) - \langle I \rangle) \rangle}{\langle I(r,t) - \langle I \rangle \rangle^2} \quad , \qquad (2)$$

which is obtained by the inverse Fourier transformation from the two-dimensional power spectrum. Since the disclination pattern is isotropic, we use a circularly averaged correlation function $C(r,t)$. The last important quantity is the correlation length. We defined the correlation length

[1] The applied voltage is much larger than the critical value of the Fredericks transition and the electric coherence length is only 0.4μm, which is much smaller than the cell thickness 16μm. Therefore, this system can be approximately regarded as the two-dimensional system except for the vicinity of glass plates.

$\xi(t)$ as

$$C(\xi(t),t) = \frac{1}{e}\,, \tag{3}$$

that is, the length in which the correlation decreases to the inverse of e.

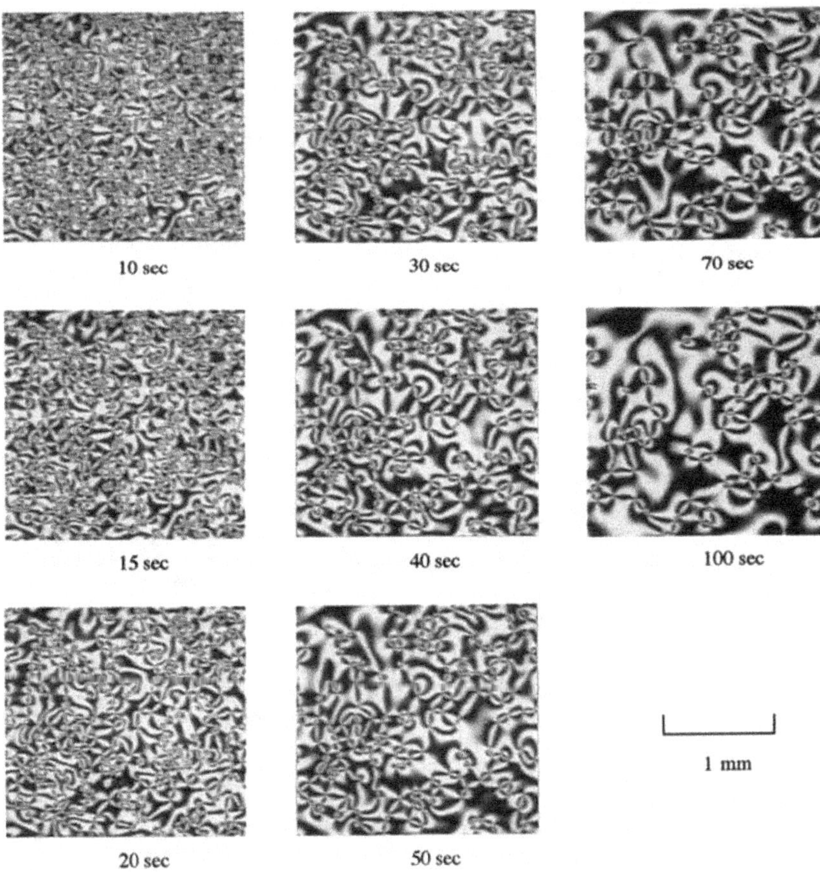

Fig.1 Annihilation process of interacting disclinations.

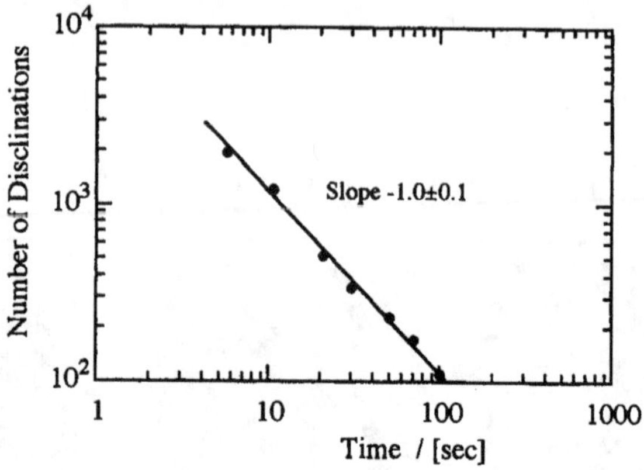

Fig.2 Time dependence of the number of disclinations.

Figure 2 shows the time dependence of the number of disclinations. These data indicate that the power law holds. The scaling exponent is -1.0±0.1.

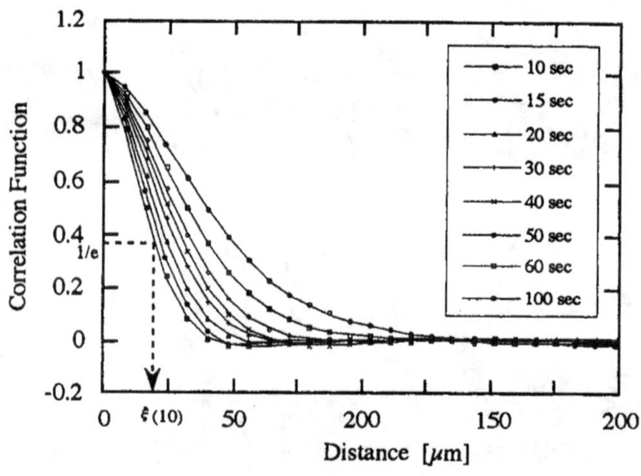

Fig.3 Time evolution of the radial correlation function.

Figure 3 shows the time evolution of spatial correlation function. The correlation length, for example, at t=10[*sec*] corresponds to the dotted line in Fig.3. The correlation length increases with the time.

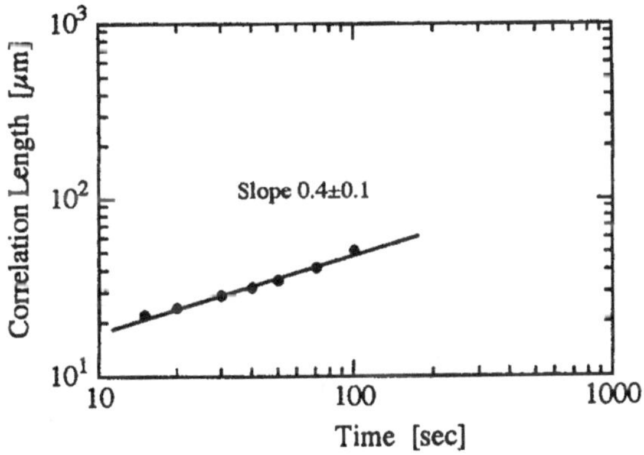

Fig.4 Time dependence of the correlation length.

Figure 4 shows the time dependence of the correlation function. This figure also shows that the power law holds, where the scaling index is 0.4±0.1. Then, we tried to scale the distance by the correlation length and replotted the correlation function in Fig.5. All the data at each time fall on a universal curve. This figure shows that the dynamical scaling law holds in this coarsening process.

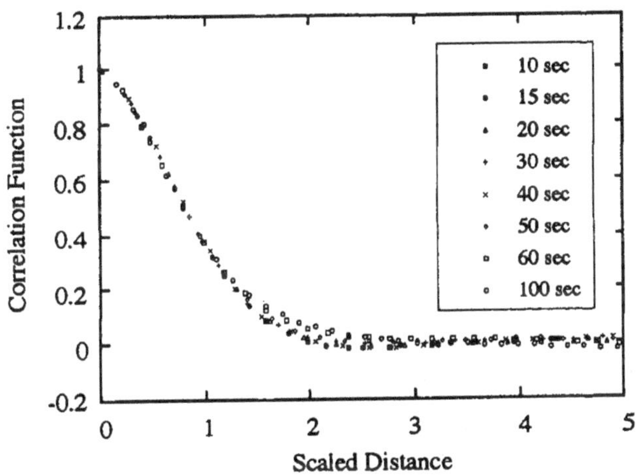

Fig.5 Scaled distance dependence of the correlation function.

Let us discuss the scaling exponents briefly by applying the dimensional analysis. In the previous study[13], we found that the motion of disclination at the position r_i is well described by the following equation of motion,

$$\gamma \frac{dr_i}{dt} = 2\pi S_i K \sum_{i \neq j} S_j \frac{r_i - r_j}{|r_i - r_j|^2} , \tag{4}$$

where S_i is the strength of the i th disclination, γ the viscosity coefficient, and K the averaged Frank elastic constant. If the annihilation process has only one characteristic length l, a dimensional analysis leads to

$$l\,(t) \sim t^{\,0.5} \tag{5}$$

and

$$N\,(t) \sim t^{\,-1} . \tag{6}$$

In conclusion, we have investigated the coarsening dynamics of a number of +1 and -1 disclinations and the dynamical scaling law was found to hold in this system. The experimentally obtained scaling exponents agree with the theoretical ones with in the accuracy of experimental errors.

References

1. H. Toyoki, *Dynamics of ordering process in condensed Matter*, ed. S. Komura and S. Furukawa (Plenum, 1988).
2. H. Toyoki, Phys. Rev. A**42**(1990)911.
3. H. Toyoki and K. Honda, Prog. Theor. Phys. **78**(1978)3086.
4. H. Toyoki: to be published.
5. H. Orihara and Y. Ishibashi, J. Phys. Soc. Jpn. **55**(1986)2151.
6. T. Nagaya, H. Orihara and Y. Ishibashi, J. Phys. Soc. Jpn. **56**(1987)1898, *ibid.* **56**(1987)3036, *ibid.* 59(1990)377.
7. T. Shiwaku, A. Nakai, H. Hasegawa and T. Hashimoto, Polymer Commun. **28**(1987)174.
8. T. Hashimoto, A. Nakai, H. Hasegawa, S. Rojstaczer and R. S. Stein, Macromolecules **22**(1989)422.
9. T. Shiwaku, A. Nakai, H. Hasegawa and T. Hashimoto, Macromolecules

23(1990)1590.
10. B Yurke, A. N. Pargellis, I. Change and N. Turok, Physica B to be published.
11. I. Change, R. Durrer, N. Turok and B Yurke, Science **251**(1991)1336.
12. I. Change, N. Turok and B Yurke, Phys. Rev. Lett. **66**(1991)2472.
13. T. Nagaya, H. Hotta, H. Orihara and Y. Ishibashi: J. Phys. Soc. Jpn. **60**(1991)1572.
14. O. D. Lavrentovich and S. S. Rozhkov: Pis'ma Zh. Eksp. Teor. Fiz. **47**(1988)210 [LETP Lett. **47**(1988)254].
15) S Hirata, T. Akahane and T. Tako: Mol. Cryst. & Liq. Cryst. **75**(1981) 47.

EXPERIMENTS ON PATTERN-FORMING SEMICONDUCTOR SYSTEMS

JÜRGEN PARISI

Physics Institute, University of Zurich,
Schönberggasse 9, CH-8001 Zurich, Switzerland

ABSTRACT

Nonequilibrium dissipative systems from semiconductor physics have prevailed as a paradigmatic testing field for complex nonlinear dynamics during the last decade. Especially, low-temperature impact ionization breakdown in extrinsic semiconductor crystals (germanium, gallium arsenide, and indium phosphide) displays a variety of interesting nonlinear phenomena, such as spontaneous oscillations and filamentary patterns of the current flow. We report on recent experimental results concerning the self-organized formation and subsequent growth behavior of current filaments in n-GaAs and p-Ge bulk semiconductors. These patterns can be visualized in high-resolution imaging experiments, taking advantage of conventional scanning electron microscopy combined with a liquid-helium cryostage. We develop an appropriate power balance model capable to quantitatively evaluate the characteristic limiting sizes that govern the nucleation and annihilation process of the spatial structures resolved experimentally (namely, a minimum and a maximum existence value of the filament current).

1. Introduction

At low temperatures, the electronic transport in a slightly doped bulk semiconductor becomes highly nonlinear, and the relation between electric current and voltage strongly deviates from Ohm's law. One of the most important mechanisms underlying such nonlinear behavior is the process of charge carrier multiplication via impurity impact ionization, giving rise to the avalanche breakdown of the semiconductor /1/. The autocatalytic nature of this breakdown effect as transition from a low to a high conducting state can lead to a bistability region in the current density versus electric field characteristic, favoring the generation of spatio-temporal current flow in the semiconductor /2/. Here, a common phenomenon is the formation of current filaments which represent high conducting channels embedded within a much lower conducting semiconductor medium. These high conducting channels nucleate above a distinct threshold value of the applied electric field and, in many ways, are similar to the lightning discharge in a gaseous atmosphere.

Recently, we have investigated the phenomenon of current filamentation in p-Ge /3/ and n-GaAs /4/. Besides the pronounced nonlinear behavior of the current-voltage characteristic (CVC) of the samples, we looked at the filamentary spatial current flow, using the well-established method of low-temperature scanning electron microscopy (LTSEM).

In the present paper, the following points will be addressed: (i) There exists a minimum filament diameter. (ii) With increasing electric field applied, the growth of existing filaments takes place only up to a distinct maximum size beyond which an additional filament is created. In order to provide a theoretical description explaining the above experimental facts both qualitatively and quantitatively, we propose a phenomenological model based on the power balance in the semiconductor system.

2. The Experiment

Our studies were either performed on homogeneously doped p-type Ge single crystals (shallow acceptor level of about 12 meV) or n-type GaAs epitaxial layers (shallow donor level of about 6 meV) having different impurity concentration and the typical dimensions of about 0.3 mm x 3 mm x 6 mm. The samples investigated carry ohmic contacts on one of the two largest surfaces. The electric circuitry consists of a series combination of the specimen and a load resistor biased by a constant voltage source. The resulting current I and the sample voltage V_s were measured at the load resistor and the surface contacts, respectively. For two-dimensional imaging of spatial current patterns, the semiconductor sample was mounted in the cryostage of the scanning electron microscope with good thermal contact to the liquid-helium bath (kept at T = 4.2 K). The specimen surface was scanned by the electron beam, and the beam-induced changes in the conductivity of the sample were recorded as a function of the coordinates (x,y) of the beam focus. Due to the low beam power applied, usually a high response signal is obtained only from the boundaries of the current patterns where the local conductivity of the sample reacts most sensitively to the beam-injected hot charge carriers. Therefore, brightness-modulated LTSEM images cause current filaments to appear as dark channels within bright regions indicating their boundaries. The experimental procedures are discussed elsewhere in detail /3,4/.

In the following, two essential features plainly observed in GaAs samples are pointed out. First, we present experimental evidence for the existence of a minimum filament cross-section. Referring to the CVC of Fig. 1, breakdown occurs at a voltage of about 0.34 V, when the bias increases monotonically from zero. For decreasing voltage supply, the current flow vanishes at 0.266 V. This way, the detailed CVC magnified from the section marked in the inset of Fig. 1 uncovers a variety of vertical steps that could be attributed to the gradual disappearance of existing current filaments with decreasing voltage bias /4/. On the other hand, the continuous decrease of current with the voltage between these jumps was found to be caused by the simultaneous shrinking of a constant number of filaments each having about the same cross-sectional size /4/. It turned out that the filament proportions cannot decrease continuously to zero, but that at a distinct minimum diameter each filament abruptly decays to zero. The image of three current filaments with minimum size obtained at point A on the CVC is shown bottom left in Fig. 1, clearly demonstrating the existence of a lower limit in the filament diameter that amounts about 15 μm. From the height of the vertical steps in the CVC, we see that each minimum-size filament carries a current of the value I_0 = 0.310 mA. This result has proved to be independent of the whole sample length, and it was also confirmed under current controlled measurements.

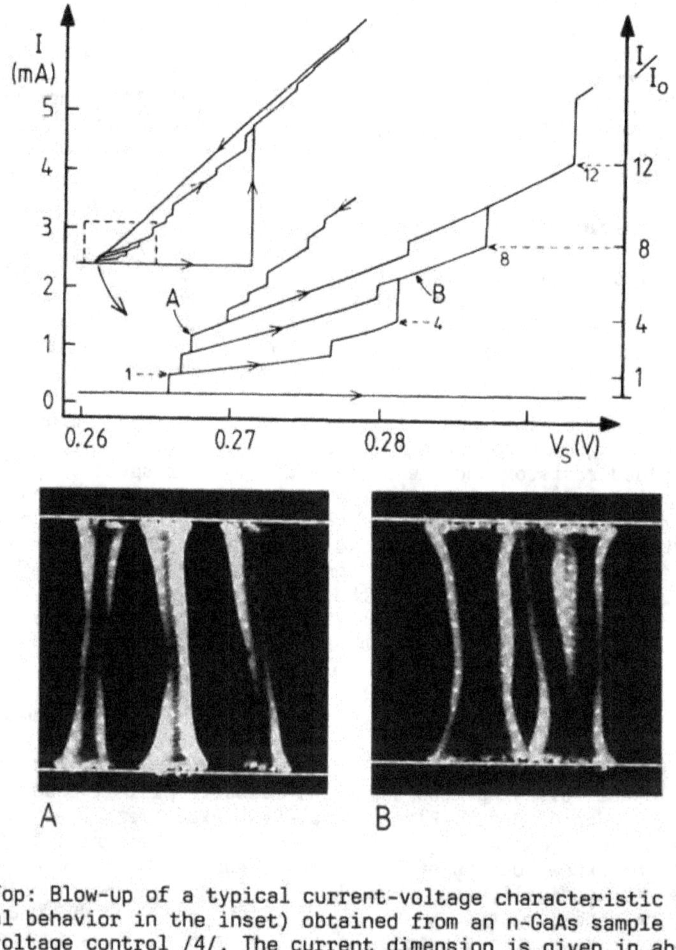

Fig. 1. Top: Blow-up of a typical current-voltage characteristic (global behavior in the inset) obtained from an n-GaAs sample under voltage control /4/. The current dimension is given in absolute values (left axis) as well as in units of the finite size of the last minimum filament (right axis). Bottom: Two-dimensional images of different filamentary states corresponding to point A and point B in the CVC. The contact edges are indicated by the horizontal white lines that bound the active sample area (with the current filaments extending in vertical direction) to about 350 µm distance.

Second, we consider the experimental fact that, for samples with a parallel contact geometry, there exists also an upper limit of the filamentary cross-section, making it impossible, for example, for one current filament to grow continuously with increasing voltage, until it occupies the whole specimen. Instead, additional filaments are nucleated at distinct voltage levels. To gain information about the growth of filaments, we started with a particular configuration (one, two, three, or more filaments) and increased the voltage again. We found a continuously growing filament size correlated with the continuous increase of current as a function of the voltage applied. However, we also observed jumps in the current due to the creation of additional filaments. Such a behavior can be recognized by the detailed shape of the inner hysteresis branches in the CVC of Fig. 1. At bottom right, a two-filament structure grown by a large amount from its minimum size is illustrated. Note that one filament has split partially into two branches, showing up as little step in the CVC. A more systematic analysis of the variation in the filament diameter with the voltage or the current can be found elsewhere /4/.

The nucleation process of new filaments apparently follows a particular scheme. Starting from a one-filament state, we find that the bifurcation point at which an additional filament appears is given by the total current I_1, corresponding to four times the current step I_o caused by annihilation of the last filament on the descending branch of the CVC (see Fig. 1). The resulting two-filament state that develops after this bifurcation coincides with the one obtained from two minimum filaments by increasing the voltage. Moreover, further bifurcations to three- and four-filament states take place at $2I_1$ and $3I_1$, respectively, on the ascending branch of the CVC. It appears that, besides the existence of a minimum value I_o in the filament current, there is also a maximum value I_1 exceeding of which gives rise to the creation of one additional filament portion. In our case, we have $I_1 = 4I_o$.

3. The Model Approach

To develop a theoretical model capable to describe the experimental phenomena observed, we have to start with a brief survey recapitulating the physics of avalanche breakdown in semiconductors. At low temperatures, most of the charge carriers are frozen out at the impurities, and the sample becomes an electric insulator. However, a few mobile carriers generated via thermal excitation are heated by the external electric field to energies sufficient to ionize the neutral impurities. The system of mobile carriers can be characterized by an effective temperature T^* which is higher than the temperature of the embedding lattice. At the onset of breakdown, T^* ranges in the order of the ionization energy E_i (in case of GaAs, about 6 meV), i.e., $(3/2)kT^* = E_i$ and, thus, $T^* = 45$ K. This temperature is reached by the hot charge carriers just at the critical electric field F_c (of about 8 V/cm) where the formation of current filaments sets in and establishes a phase separation between a low and a high conducting state.

The filament boundary derives from a region of space charges that provide the dynamical equilibrium between the diffusion of hot carriers from the inside to the outside of the filament and a drift current flowing in the inverse direction. The typical length scale of the boundary width l_b (about 0.1 µm) is governed by the effective Debye length associated with the dielectric relaxation time. To end up with the characteristic length of the minimum filament diameter, we look at the electric power balance of our semiconductor system taking into account three channels of dissipation: (a) The power loss within the filament (to the lattice through the weak electron-phonon coupling and inelastic impurity scattering). (b) The power loss across the filament boundary (decreasing the density of hot charge carriers). (c) The power loss in the ohmic contacts (reducing the electric field available to the bulk impact ionization process). The only process that contributes to the avalanche carrier multiplication and, hence, to the maintenance of the current filament is (a). Filamentation dies out if too much power goes into channels (b) and (c). A lower limit for the filament diameter can be expected at some critical surface to volume ratio (i.e., if on the whole (a) and (b) being proportional to the filament volume and surface, respectively, are of the same order). Since particle numbers must be conserved at the filament boundaries, process (b) is only possible when the carriers outside the filament are colder than inside. Denoting this effective temperature difference by dT*, the power balance above yields the following expression for the minimum filament diameter: $d_o = (6/l_b F_c^2)(kT^*/e)(kdT^*/e)$. This equation fits our experimental results, if dT* is taken as 2% - 3% of T*, and the parameter values discussed are used. For the case of GaAs and Ge samples, we find about $d_o = 10$ μm and $d_o = 100$ μm, respectively.

Our experimental observation of a maximum filament diameter can be understood again via the power balance argumentation. Process (b) favors a configuration with a minimum surface to volume ratio, i.e., where the total current is carried by a single filament with a large size (instead of many filaments having much smaller size). On the other hand, process (c) prefers a homogeneous current flow, i.e., a subdivision of the total current into a large number of filaments with the smallest possible diameter. By minimizing the sum of both power losses, one obtains for the maximum current per filament $I_1 = (R_o/2R_c)^{2/3} I_o$. Here, I_o and R_o denote the current and the ohmic resistance, respectively, of a single minimum-size filament, R_c the resistance of the ohmic contacts. We note that the total power dissipation inside the filaments is independent of the specific subdivision of the total current and, hence, does not affect the minimization leading to our result. The maximum current I_1 estimated encourages the experimental bifurcation pattern shown in Fig. 1 where the first three ramifications have been found to occur at equal current intervals.

4. Conclusion

We have demonstrated experimental and model evidence that characteristic lower and upper limiting sizes dominate the filamentary spatial current structures during low-temperature semiconductor breakdown.

5. References

/1/ J. Parisi, U. Rau, J. Peinke, and K.M. Mayer, Z. Phys. B - Condensed Matter **72**, 225 (1988).

/2/ J. Parisi, in: Nonlinear Evolution of Spatio-Temporal Structures in Dissipative Continuous Systems, eds. F.H. Busse and L. Kramer (Plenum, New York, 1990), p. 343.

/3/ K.M. Mayer, J. Peinke, B. Röhricht, J. Parisi, and R.P. Huebener, Phys. Scr. T **19**, 505 (1987).

/4/ K.M. Mayer, J. Parisi, and R.P. Huebener, Z. Phys. B - Condensed Matter **71**, 171 (1988).

BLOCH-TYPE MODELING AND PERIODIC-ORBIT ANALYSIS OF LASER DYNAMICS

JÜRGEN PARISI

Physics Institute, University of Zurich,
Schönberggasse 9, CH-8001 Zurich, Switzerland

ABSTRACT

The conventional Bloch-Kirchhoff description of nuclear magnetic resonance (NMR) laser activity is shown to necessitate the inclusion of a nonlinear relaxation term for the transverse magnetization, in order to account for experimental observations. We confirm validity of the modified equations via comparing Poincare cross-sections and periodic-orbit structures in an extended parameter space for both model and experiment. All unstable orbits up to order nine have been extracted and employed to obtain an approximation to a generating partition. The logic tree representation of their symbolic encoding clearly unveils an overall accordance of the extended Bloch-type description with the experimental data. For both cases, a precise quantitative characterization of the temporal scaling behavior near a heteroclinic tangency crisis has been carried out, accentuating the predictability power of our laser model.

1. Introduction

In dissipative, nonlinear dynamical systems, it is often found that small variations of a control parameter p lead to discontinuous structural changes in chaotic attractors. These transitions, called crises /1/, are determined by the collision between the unstable manifold of some periodic orbit A and the stable manifold either of the same orbit A (homoclinic crisis) or of another orbit B, with the same period as A (heteroclinic crisis). Both phenomena display a characteristic widening of the strange attractor which coincides with (the closure of) the unstable manifold of A over a region of phase space that was empty for p smaller than the critical crisis value p_c. Just above crisis, the dynamics consists of an intermittent "bursting" between the two subsets of phase space. The average time elapsing between consecutive bursts of this so-called crisis-induced intermittency /1/ obeys the power law

$$\tau(p) = \text{const.}\,|p - p_c|^{-\beta} \qquad (1)$$

in the vicinity of p_c, where the critical exponent β has been predicted to encompass values in the interval between 1/2 and 3/2, corresponding to the strongly dissipative and conservative limits, respectively.

However, a precise definition of the two domains in phase space, i.e., the "inner" core region and the "outer" transient region, is not easy to achieve when the equations of motion (and, therefore, the invariant manifolds) are not known. This is particularly evident with experimental data, so far analyzed just by taking an arbitrary partition in embedding space. We resolve that ambiguity by constructing an approximation to a generating partition Z and identifying the "outer" region by the element of the refinement of Z with the same label as the "new" symbolic sequence appearing in the dynamics. In this way, we have a clear cut definition of the times spent in the two regions. Our approach, thus, affords a criterion for the onset of attractor widening due to the collision of the stable and the unstable manifold belonging to different unstable periodic orbits. We apply the above technique to experimental data produced by a parametrically modulated NMR laser. Moreover, a generalized Bloch-type laser model is well adapted to the peculiarities of the experiment and investigated accordingly. The conjectured scaling behavior of crisis-induced intermittency via heteroclinic tangency will be confirmed quantitatively for both model and experiment. The accuracy of the present symbolic-dynamical method stems from its ability to resolve neighborhoods of high-order periodic orbits.

2. The Experiment

The NMR-laser activity derives from nuclear ^{27}Al spins in a ruby crystal subject to a static magnetic field at liquid-helium temperature. The spin population inversion is obtained by means of dynamical nuclear polarization, the resonance by enclosing the crystal in a cavity: in our case, an LC circuit tuned to single-mode selection which provides the feedback radiation field $\underline{B}$ (proportional to the current in the circuit) necessary for coherent spin-flip behavior. Furthermore, the cavity is forced to operate with a sinusoidally varying quality factor Q of modulation amplitude p, the actual control parameter, and constant frequency f_m. The laser output relates to the voltage across the LC circuit, proportional to the amplitude of the transverse nuclear magnetization, $M_t = M_t(M_x,M_y)$, where $\underline{M} = (M_x,M_y,M_z)$ means the total magnetization /2/.

3. The Model

An adequate description of the laser dynamics in the rotating frame is given by the extended Bloch-type model,

$$\begin{aligned} dx/dt' &= s(y - x/f(t')) \\ dy/dt' &= -y(1 + ay) + rx - xz \\ dz/dt' &= -bz + xy \ , \end{aligned} \qquad (2)$$

where the variables x, y, and z are proportional to B_t, M_t, and (M_z-M_e), respectively (B_t denotes the rotating field amplitude, M_e the pump magnetization). Their derivative is taken with respect to a rescaled adimensional time t'. The proportionality factors, as well as the parameters $s = 4.875$, $r = 1.807$, $a = 0.2621$, and $b = 2\text{x}10^{-4}$, depend on various physical constants. The function f(t') describes the periodic forcing with amplitude p and frequency f_m. Equations (2) reduce to the Lorenz system for $a = 0$ and $f(t') = 1$ (i.e., $p = 0$). Under these conditions, the asymptotic attractor represents nothing else than the fixed point $x = y =: x^* = (b(r - 1))^{1/2}$, $z =: z^* = r - 1$.

The difference to the conventional model /2/ consists of the term $-ay^2$ which stands for a nonlinear damping of the transverse magnetization y. This term has been introduced to account for the observed decay of the laser output to a stable fixed point in absence of the modulation (p = 0), after switching of the quality factor Q (relaxation oscillation /2/). One can easily show that the correction term is actually very small: in fact, by setting $(X, Y, Z) = (x/x^*, y/x^*, (z - z^*)/x^*)$ and rescaling time t' to $T' = t'x^*$, we end up with the expression $-Y(1 + a'Y)/x^*$ for the damping of Y, where $a' = 3.33 \times 10^{-3}$ and Y becomes an element of the interval (0,3) for all values of p considered. This transformation also demonstrates that X can be eliminated adiabatically. Since the new parameter s' reaches a large value of about 384, X can be approximated by Yf(T'), and Y never changes sign.

4. The Symbolic-Dynamical Approach

In order to characterize our results in a more general framework, we take advantage of methods delivered by the symbolic dynamics theory. This approach includes all relevant metric and topological features of the system under investigation /2,3/. Once the attractor has been reconstructed by embedding the time series in a suitable auxiliary space, all unstable periodic orbits up to some maximum order are localized. A Poincare cross-section is then chosen in such a way that one obtains the same number n of intersections for all period-N orbits (with $N = n/f_m$). Finally, a generating partition is approximated by attributing different symbolic sequences to each unstable periodic orbit (and to their neighborhoods).

5. Experimental and Model Results

We have collected in total 14 different experimental time series via sampling the laser output with a frequency $f_s = 4f_m$. They consist of approximately 10^6 12-bit integers. The modulation frequency, fixed at f_m = 120 Hz, corresponds to about twice the intrinsic relaxation frequency. The modulation amplitude was varied in the range between p = 0.01800 and p = 0.01870 (where the former value is slightly below the crisis point p_c = 0.01802 and the latter well above it). The experimental data have been reconstructed in a 3- to 16-dimensional embedding space (in fact, the dimension of the attractor is around 2.5). The unstable periodic orbits of the system were located by looking at portions of reconstructed trajectories recurring in certain spherical regions of radius R within distinct time intervals (chosen to be integer multiples of the period of the external forcing term). The value of R gives the precision by which the unstable periodic orbits are shadowed. In order to minimize the relative error in the search, R is taken to be proportional to the local density of points in phase space and to the square root of the embedding dimension. For comparison, the unstable periodic orbits of the model were extracted by a straightforward application of the Newton method. Up to order nine, we found a complete correspondence with the experimental results: not only all periodic points have the same symbolic representations, but also the shapes of the orbits are in close resemblance.

Since the dynamics is non-autonomous, Poincare cross-sections can be readily constructed for both experimental and numerical data via synchronizing the trajectorial flow to the external forcing (see Fig. 1). The intersection points with all extracted unstable limit cycles are assigned a different symbolic encoding through a binary partition the elements of which were labelled by the two names 0 and 1 /3/. Recall that a generating partition associates a unique point in phase space to each infinitely long symbolic sequence (among which are the periodic ones). Hence, the present partition constitutes a reasonable approximation to a generating one, because it distinguishes all periodic points up to order nine (of course, this represents a necessary, although not, in general, sufficient, condition for the partition to be generating). Below crisis (e.g., at p = 0.01800), all points belonging to the first subset of the partition are mapped onto the second subset in one iteration. Therefore, the string 00 is forbidden and the most condensed description of the dynamics is given by a binary tree over the two primitive /3/ words $w_1 = 1$ and $w_2 = 01$. After the onset of crisis (e.g., at p = 0.01850), string 00 no longer remains forbidden and, consequently, the additional primitive word $w_3 = 001$ needs to be introduced. Attractor widening shows up through emergence of the "new" unstable period-3 orbit B with label 001 (cf. Fig. 1 (a',b')), the "old" orbit A having label 011 (cf. Fig. 1 (a,b)). The symbolic signal now consists of combinations of w_1, w_2, and w_3. Besides two-dimensional projection of the Poincare cross-sections together with the curves defining the partition employed, Fig. 1 plainly distinguishes the intersection points of the two prevailing orbits, numbered in order of appearance. The full coincidence found between measured and simulated data clearly demonstrates that our model accurately reproduces the experimental observation up to the available resolution.

Knowledge of the symbolic dynamics yields a precise criterion for the transition determined by the collision between the unstable manifold of the old period-3 orbit A (011) and the stable manifold of the new period-3 orbit B (001). If sequence 00 occurs, the second 0 is part of the new branch, i.e., the crisis has taken place. Hence, all points of phase space contained in element 00 are mapped onto such a region. They can be reached only from a small vicinity of the point marked by a 1 in Fig. 1 (a',b'). Accordingly, we have estimated the average time spent in the old region as a function of the distance the actual control parameter is away from the crisis point and calculated the characteristic critical exponent via Eq. (1) from the slopes of the corresponding log-log plots, displayed in Fig. 2. The values of β obtained from experiment and model are 1.02 ± 0.05 and 1.10 ± 0.05, respectively, proving strong accordance within the uncertainty limits. Moreover, we have confirmed the theoretically predicted scaling behavior of a heteroclinic tangency crisis for the dynamics of an NMR-laser system.

6. References

/1/ C. Grebogi, E. Ott, F. Romeiras, and J.A. Yorke, Phys. Rev. A **36**, 5365 (1987).

/2/ L. Flepp, R. Holzner, E. Brun, M. Finardi, and R. Badii, Phys. Rev. Lett. **67**, 2244 (1991).

/3/ R. Badii, Europhys. Lett. **13**, 599 (1990).

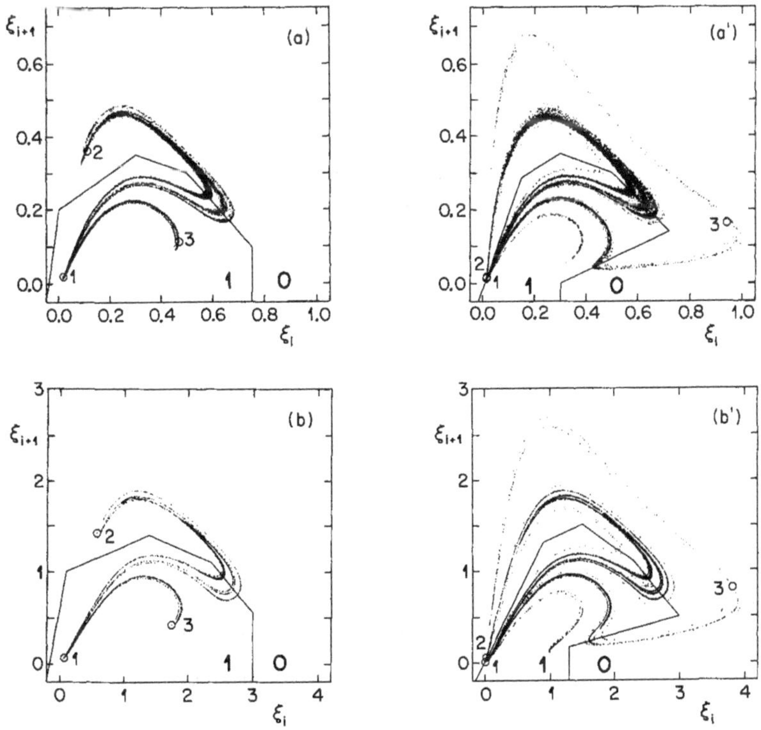

Fig. 1. Poincare cross-sections obtained from experiment (a,a') and model (b,b') at the control parameter values p = 0.01800 (unprimed) and p = 0.01850 (primed). The solid curves indicate an approximation to a generating partition with elements 0 and 1.

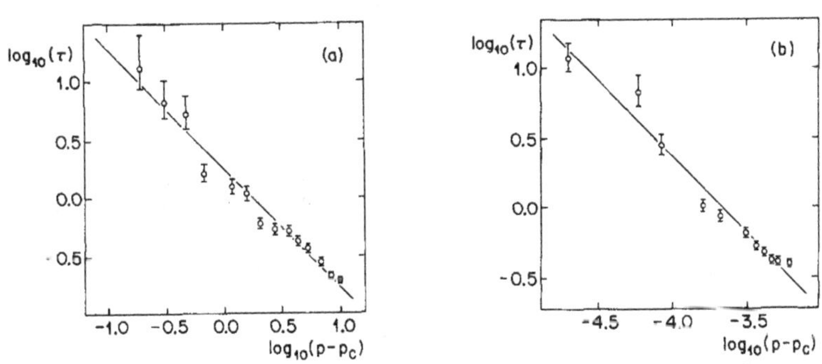

Fig. 2. Average time between successive intermittent bursts versus distance of the control parameter from its critical value (doubly-logarithmic scale) obtained from experiment (a) and model (b).

Coupled Map Lattice Model for Boiling

Tatsuo YANAGITA *

The Graduate University for Advanced Studies
The Institute of Statistical Mathematics
4-6-7 Minami-Azabu, Minato-ku, Tokyo 106, JAPAN

ABSTRACT

A simple model for boiling is proposed. With increasing the temperature of a bottom plate, our minimal model shows three successive phases; conduction, nucleate and film boiling. In the nucleate regime the heat flux increases with the temperature of the bottom plate, while it starts to decrease in the film boiling regime. At a boiling state, the maximum Lyapunov exponent is positive, implying that the boiling phenomena are spatiotemporal chaos.

1 Boiling Transition

The dynamics of boiling is known to show the transition from nucleate to film boiling [1]. More than two centuries ago, Liendenfrost noted the fact that a droplet evaporates slowly on a very hot metal plate [2]. The rate of vaporization is much lower than those occurring at smaller temperature differences between the hot plate and the droplet. This behavior is the result from the thermally insulating action on a thin vapor layer between the plate and the liquid. S.Nukiyama characterized this boiling transition experimentally, so-called boiling characteristic curve [3]. He measured the temperature of heater as a function of the heat flux fed into the system. With this curve, he found that heat flux increases with the heating temperature and reaches a maximum value (critical heat flux) at burnout point. Above the burnout point, the heat flux decreases with the increasing temperature of heater.

2 Model

Boiling phenomena occur during the combined transport of energy, momentum, and matter in the two phase flow of boiling media. In principle, physical transport phenomena are described by Navier-Stokes equation. However, it is difficult to simulate the boiling phenomena based on Navier-Stokes equation because the boundary of two phases is complex and varies in time. To describe the nature, there are two categories of models:(1) models which describe the phenomena quantitatively :(2) models which are used as aids to understand conceptual problems, need not quantitative agreement. Navier-Stokes equation should be used to capture the phenomena quantitatively. To

*E-mail address: yanagita@ism.ac.jp

represent the qualitative feature of dynamical phenomena in spatially extended system, Coupled Map Lattice (CML) is one of the most powerful strategies. A CML is a dynamical system with a discrete time and space, and continuous field variables. It has been investigated by K.Kaneko extensively in various contexts [4] [5]. Use of CML as simulators for various physical phenomena are proposed [6]. CML is successfully applied to phase-separation dynamics by Y.Oono and S.Puri [7]. Analyzing boiling phenomena, we introduce a minimal CML model which exhibits the transition from nucleate to film boiling. Here we study two dimensional boiling phenomena, taking a two dimensional lattice (x, y) and temperature field $T^t_{x,y}$ on it at time t. Our dynamics of boiling is decomposed into independent units by diffusion, floating motion of bubbles and latent heat. We replace each unit by the following simplest parallel dynamics on the lattice

1. Thermal Diffusion

$$T'_{x,y} = T^t_{x,y} + \frac{\varepsilon}{4}\{T^t_{x+1,y} + T^t_{x,y+1} + T^t_{x-1,y} + T^t_{x,y-1} - 4T^t_{x,y}\} \tag{1}$$

2. Floating motion of bubbles

$$T''_{x,y} = T'_{x,y} - \frac{\sigma}{2}T'_{x,y}\{\rho(T'_{x,y+1}) - \rho(T'_{x,y-1})\} \tag{2}$$

$$\rho(T) = \tanh(\alpha(T - T_c)) \tag{3}$$

3. Latent Heat Effect

   if $T''_{x,y} > T_c$ and $T^t_{x,y} < T_c$ then $T^{t+1}_{n(x,y)} = T''_{n(x,y)} - \eta$

   if $T''_{x,y} < T_c$ and $T^t_{x,y} > T_c$ then $T^{t+1}_{n(x,y)} = T''_{n(x,y)} + \eta$

   where $n(x, y)$ shows nearest neighbor of the position (x, y).

These three unit dynamics ("procedures") are separated and successively carried out. The first procedure is thermal diffusion process, given by the discretized Laplace equation. The second one represents generation and floating motion of bubbles. The bubbles floating upward by buoyancy which depend on the inhomogeneity of density. We represent density $\rho(T) = \tanh(T - T_c)$, assuming local equilibrium. The hyperbolic tangent represents the sudden change of density by the phase transition. The cell whose temperature beyond the parameter T_c is regarded as a bubble and is transferred upward due to inhomogeneity of density $\rho(T'_{x,y+1}) - \rho(T'_{x,y-1})$. The third one represents the latent heat effect. When the phase transition from liquid to gas occurs at a cell, the temperature of its nearest neighbors decreases. We use the parameters $\varepsilon, \alpha, \sigma, \eta$ and T_c describe this model. We set these parameters as $\varepsilon = 0.5, \alpha = 10, \sigma = 0.3, \eta = 0.5$ and $T_c = 10$. The boundary condition of horizontal direction is periodic. The top (cooling surface) and the bottom (heating surface) boundary is fixed at T_{Top}, T_{Bottom} respectively. The model is initialized $T_{x,y} = (T_{Bottom} + T_{Top})/2 + \delta$ where δ represent small uniform random number.

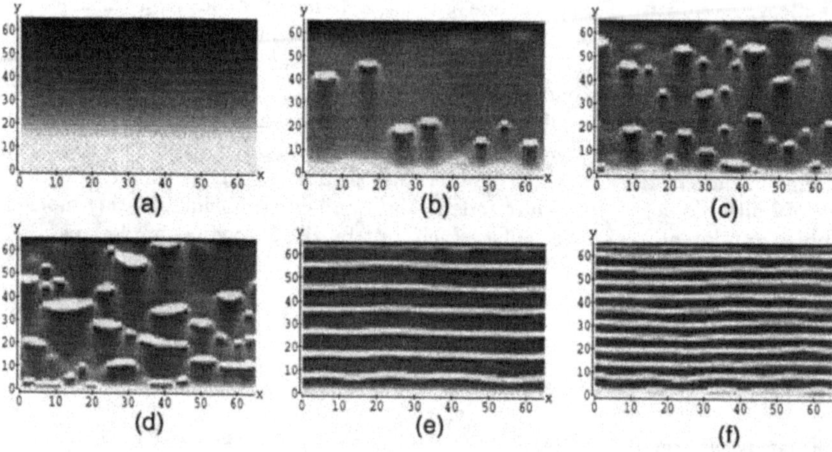

Figure 1: Snapshot pattern

Lattice size is 64 × 64. White color means high temperature and Black means low one. All parameter is same, except for the bottom temperature T_{Bottom}. Above burnout point $T_{B.O.} \sim 9.91$ film vapor appears and comes off simultaneously. (a) $T_{Bottom} = 9.75$ (b) $T_{Bottom} = 9.8$ (c) $T_{Bottom} = 9.85$ (d) $T_{Bottom} = 9.8875$ (e) $T_{Bottom} = 9.9125$ (f) $T_{Bottom} = 9.95$.

3 Simulations

We investigate the pattern of this model increasing the temperature T_{Bottom}.

(A) Below some critical value $T_{B.P.}(\sim 9.77) < T_{Bottom}$, heat is mainly transported by the diffusion (procedure 1) from the bottom. If the temperature gradient $\Delta T = T_{Bottom} - T_{Top}$ is sufficiently small, mean temperature field obeys the Fourier law for the heat conduction.

(B) Above boiling point $T_{B.P.}$, heat is no longer transferred efficiently by the diffusion process only and a large temperature gradient appears near the bottom. Thus the convective procedure 2 is relevant to form and bubbles (i.e. patches where $T > T_c$) appear. Slightly above boiling point, the bubbles is not strong to grow. Then the bubbles eventually disappear before floating. As the increase of T_{Bottom}, bubbles are released from the bottom constantly. By the thermal diffusion, most of them diminish in size and eventually disappear (Fig. 1-b,c,d).

(C) At a burnout point $T_{B.O.} \sim 9.91$, bubbles are jointed to form a film vapor. In our model, this film vapor detaches from bottom plate simultaneously. Thus we observe a stripe pattern in Figure 1-e,f. This pattern is stable because our model has no hydrodynamic and surface effect. To include such an effect, we must add a variable for density on the lattice and also consider the dynamics of the phase transition. H However,in our experient, We found the bubbles are created periodically in film boiling.

We calculate the distribution of bubble size. For simplicity, we define the size of a bubble by the length of x-direction. We observed that the bubble size is distributed exponentially in the nucleate regime since bubbles are generated independently. **Mixing suppresses boiling**

Boiling is suppressed by mixing in experimental observation. In our simulation, we define the mixing as following. We choose a cell randomly and change place with its upside cell with probability P_{mix}. Increasing the probability P_{mix}, the strength of boiling B decrease

$$B = \frac{1}{L_x L_y} \langle \sum_{x=1,y=1}^{L_x,L_y} \Theta(T_{x,y}^t - T_c) \rangle_t \tag{4}$$

where Θ is step function definition.

S.Nukiyama measured the temperature of heater as a function of the heat flux fed into the system, so-called boiling characteristic curve[3]. In our simulation, we define the heat flux by

$$q = \frac{1}{L_x} \langle \sum_{x=1}^{L_x} \frac{\sigma}{2} (T_{x,1}^t - T_{Bottom}) \rangle_t \tag{5}$$

Through numerical calculate of the heat flux q, we have obtained the boiling characteristic curve (Fig.2-a). Our curve in Figure 2-a is in qualitative agreement with the experimental observation.

Chaotic dynamical systems are usually characterized by Lyapunov exponents [8], which measure the divergence rates of nearby trajectories. Here we calculate the maximum Lyapunov exponent using the method of Benettin et al. [9]. Figure 2-b shows the result which has positive value in boiling regime. Thus boiling phenomena are temporally chaotic. At the burnout point, the maximum Lyapunov exponent shows a large decrease. This decrease is related with the formation of an ordered stripe pattern.

4 Conclusion

We have proposed a minimal CML model for boiling. This model shows the transition from nucleate to film boiling. The model reproduces some qualitative features of this transition. In particular, the boiling characteristic curve agrees with experimental observation qualitatively.

5 Acknowledgements

I am grateful to Prof. Y.Itoh for his continual encouragement and interests on boiling model; to Prof. K.Kaneko and his groups for discussion; to Prof. Y.Iba for his constructive criticisms and collaboration.

I am also grateful to Prof. T.Ogawa and T.Hirata for providing me the opportunity to observe the boiling phenomena experiments, which is very helpful to my study.

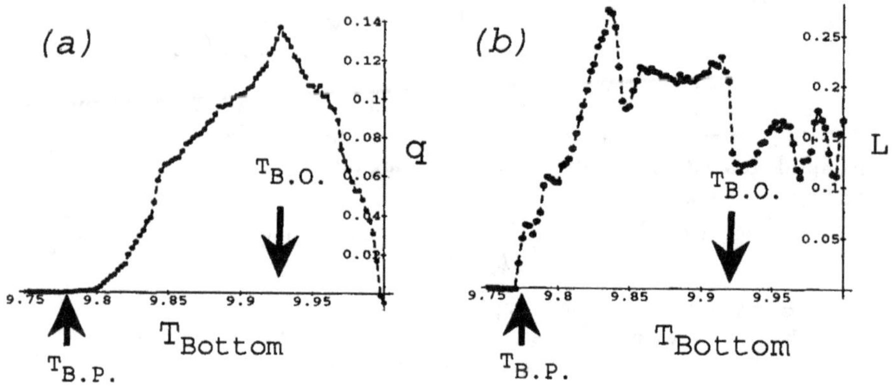

Figure 2: Boiling curve(a) Maximum Lyapunov exponent(b)

References

[1] S. van Stralen and R. Cole. *Boiling Phenomena*, volume 1. Hemishere Washington, 1979.

[2] J.G.Leidenfrost. De aquae communis nonnullis qualitatibus tractatus (1756). *Int. J. Heat Mass Transfer*, 9, 1966. trans. by C.Warea.

[3] S.Nukiyama. The maximum and minimum values of the heat q transmitted from metal to boiling water under atmospheric pressure. *J. Soc. Mech. Eng. Japan*, 37:367–374, 1934. in Japanese.

[4] K.Kaneko. *Collapse of Tori and Genesis of Chaos in Dissipative System.* World Scientific, 1983.

[5] J.P.Crutchfield and K.Kaneko. *Phenomenology of Spatiotemporal Chaos.* World Scientific, 1987. in Directions in Chaos.

[6] K.Kaneko. *Simulating Physics with Coupled Map Lattice*, volume 1 of *Formation,Dynamics and Statistics of Patterns.* World Scientific, 1990.

[7] Y.Oono and S.Puri. Study of phase-sepaeration dynamics by use of cell dynamical systems. 1. modeling. *Phys.Rev.*, 38A, 1988.

[8] I.Shimada and T.Nagashima. A numerical approach to ergodic problems of dissipative dynamical systems. *Prog.Theor.Phys*, 61:1605, 1979.

[9] G.Benettin L.Galgani and J.M.Strelcyn. Kolmogorov entropy and numerical experiments. *Phys.Rev.*, A14:2338–45, 1976.

SOLITON OSCILLATION AND SPATIOTEMPORAL CHAOS IN PARAMETRIC EXCITATION OF SURFACE WAVES WITH LARGE ASPECT RATIO

Makoto UMEKI

Department of Physics, Faculty of Science,

University of Tokyo, Hongo, Bunkyo-ku, Tokyo, 113 Japan

ABSTRACT

Solitary waves of water surface in a narrow rectangular container excited by vertical periodic forcing are considered. Parametric dissipative nonlinear Schrödinger equation derived by weakly nonlinear theory is investigated analytically and numerically. It is found that observed soliton oscillation state is the Hopf bifurcation of a cnoidal or dn-type stationary state. As the forcing parameter increases, the periodic motion changes into temporal and spatiotemporal chaos.

1. Introduction

One of the recent progress in hydrodynamics is understanding of transition from regular to irregular motions. Especially, low-dimensional chaos in parametrically forced surface waves have been studied intensively over this past decade (Ciliberto & Gollub (1985)[1], Simonelli & Gollub (1989)[2], Umeki (1991a)[3]).

In order to understand both temporal and spatial irregularity that appears spontaneously due to homogeneous forcing, spatially extended systems should be studied. For the forced surface-wave system, Wu, Keolian & Rudnick (1984)[4] investigated experimentally the subharmonic excitation of standing solitary waves in a rectangular container with large aspect ratio. They observed that the (0,1) surface-wave mode (one node in the short-side direction) became unstable and changed into a single breather-wave state in a certain region of two parameters, i.e. the amplitude and the frequency of vertical forcing. Moreover, the breather-wave state also changed into a soliton-oscillation state. In order to describe the weakly-nonlinear regime of such surface waves, Miles (1984)[5] derived a nonlinear Schrödinger equation with a complex-conjugate forcing and linear damping. Barashenkov, Bogdan & Korobov (1991)[6] studied the same equation in the context of the easy-plane ferromagnet and discussed the stability of two breather solutions with different amplitudes. Another parametrically forced nonlinear Schrödinger equation investigated by Okamura and Konno (1989)[7] presents a similar behavior to our results.

2. Soliton Oscillation and Spatiotemporal Chaos

Parametric dissipative nonlinaer Schrödinger equation (Umeki 1991b)[8] is given by

$$i(r_\tau + \alpha r) + Br_{XX} + (\beta + A|r|^2)r + A_0 r^* = 0, \tag{1}$$

where r denotes the complex expression of the in- and out-of-phase amplitudes of the modes (0,1), X and τ express the slow modulation of the long-side direction and the time, real parameters A_0, β, α, A, B indicate the normalized amplitude, frequency offset of the external vertical oscillation, phenomenological damping, nonlinear and dispersive coefficients.

Two classes of exact stationary solutions of elliptic-function type are found. One is a cnoidal-wave solution and the other is a dn-type solution. These solutions approach to a solitary-wave solution of sech-type (Miles 1984[5]) as the modulus approaches 1. Dependence of amplitudes of standing waves on the external parameter coincides with that of Wu, Keolian & Rudnick (1984)[4]. The stability of the stationary solutions is examined by solving eigenvalues of the linearized equation. It is shown that Hopf bifurcations occur on both solutions. Numerical simulations by the pseudospectral method illustrate that the Hopf bifurcation leads to the periodic solution similar to the observed soliton-oscillation. Figure 1 shows a numerical soliton-oscillation state with $A = B = 1, \alpha = 0.2, \beta = -0.5, A_0 = 0.3$ and the spatial period $L = 15.5$. Calculated temporal power spectra and Poincaré sections confirm this phenomenon.

Moreover, as the forcing parameter A_0 increases, temporal and spatiotemporal chaotic states emerge. The spatial correlation length decreases when A_0 becomes large. Figure 2 shows a typical spatiotemporal chaotic state with $\beta = -0.3, \alpha = 1$ and other parameters are the same as Figure 1.

3. Conclusion and Discussion

Numerical study of nonlinear Schrödinger equation with complex-conjugate forcing and linear damping shows that there exists a periodic solution similar to the observed soliton-oscillation state and spatiotemporal chaos emerges when the forcing parameter exceeds a critical value. One of the analytical ways to treat chaos in ordinary differential equations is Melnikov's method where a homoclinic orbit plays an important role. Recently Ablowitz and Herbst (1990)[9] constructed homoclinic solutions of unforced nonlinear Schrödinger equation with periodic boundary conditions. Analyses from the viewpoint of homoclinic chaos may be useful and are expected in partial differential equations.

4. Acknowledgements

This work is supported by a Grant-in-Aid (02952085) for Scientific Research from the Ministry of Education, Science and Culture.

5. References

1. S. Ciliberto and J. P. Gollub, J. Fluid Mech. **158** (1985) 99.
2. F. Simonelli and J. P. Gollub, J. Fluid Mech. **199** (1989) 471.
3. M. Umeki, J. Fluid Mech. **227** (1991a) 161.
4. J. Wu, R. Keolian and I. Rudnick, Phys. Rev. Lett. **52** (1984) 1421.
5. J. W. Miles, J. Fluid Mech. **148** (1984) 451.
6. I. V. Barashenkov, M. M. Bogdan and V. I. Korobov, Europhys. Lett. **15** (1991) 113.
7. A. Okamura and H. Konno, J. Phys. Soc. Jpn. **58** (1989) 1930.
8. M. Umeki, J. Phys. Soc. Jpn. **60** (1991b) 146.
9. M. J. Ablowitz and B. M. Herbst, SIAM J. Appl. Math. **50** (1990) 339.

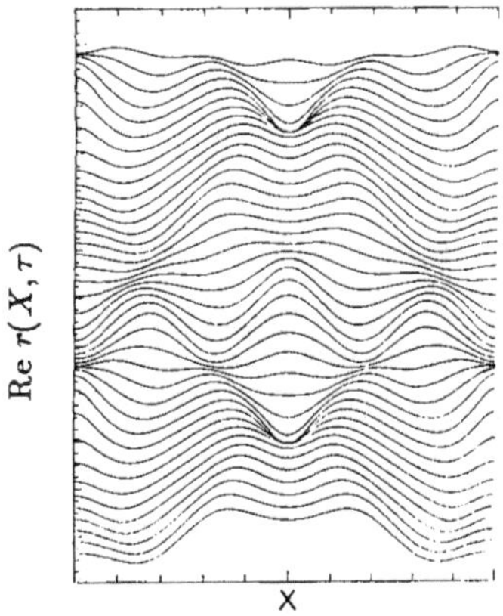

Figure 1. Soliton-oscillation state

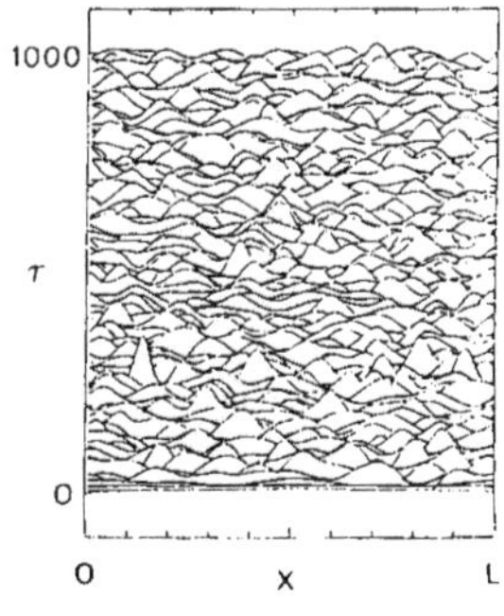

Figure 2. Spatiotemporal chaos state

E: Others

Neural Network Model Carrying Phase Information

YOSHIKI KURAMOTO, TOSHIO AOYAGI, IKUKO NISHIKAWA*
TSUYOSHI CHAWANYA, AND KOJI OKUDA

Department of Physics, Kyoto University, Kyoto 606, Japan
* *Division of Intelligence Science, The Graduate School of Science and Technology, Kobe University, Kobe 657, Japan*

Abstract

A network of periodically bursting model neurons is proposed. Its unique feature is a complex representation of the cell variables and also of the synaptic matrix. In the strong-coupling limit, the model recovers the traditional neural network model of simple on-off units, while in the weak-coupling limit it reduces to the network of smooth phase oscillators. In the special case of all-to-all excitatory coupling, some numerical and analytical evidences are provided for the occurrence of global phase locking. More complicated collective behavior such as clustering is also discovered numerically. Stimulus-evoked collective oscillations as observed in the cat primary visual cortex are explained within the present framework.

Recent electrophysiological studies revealed that neuronal responses with oscillatory modulation of the spiking frequency occur in various parts of mammalian brain[1,2]. It has hence been suggested that multiple sensory inputs with some mutual similarity could be linked through the phase locking of this sort of neural oscillations, and that such linking could be crucial to the early stage neural information processing[3]. Unfortunately, however, traditional neural network models composed of the simple McCulloch-Pitts elements or of their continuous analogues do not seem to provide a suitable framework for testing the above hypothesis nor for evaluating the computational power of oscillatory neural networks in general. This is because such models ignore from the outset the basic fact that a single neural cell itself can behave as an oscillator and hence carry phase information. It is not only meant here that cell membranes are capable of generating a periodic train of action potentials. More importantly, they are known to exhibit quite commonly the periodic bursting i.e. alternation between the period of rapid spiking and that of quiescence[4,5]. This second form of neural oscillation is of our main concern in the present paper.

Recently, some people studied oscillator networks in relation to sensory information processing in neural fields[6,7]. They used relatively simple oscillators such as the phase oscillators which are quite unlike the processing units postulated in the usual network models. It may be wondered if it is not too busy for a neuron to behave in one case like an Ising spin as in the Hopfield model[8] while in the other case like a smooth phase oscillator or even like an XY spin as in Sompolinsky et al's recent model[6]. Isn't there any way of integrating these seemingly conflicting aspects of a neuron into one simple dynamical unit? Once this has been done, one may be able to construct a new network model possessing high dynamical flexibility, yet being simple enough to permit extensive mathematical and numerical analyses. This is exactly what we attempt in this paper. An earlier attempt along the same line was due to Abbott[9]. The main difference between his and the present approaches lies in the way of modeling the internal oscillatory dynamics of a cell. His model is based on a reinterpretation of the FitzHugh-Nagumo dynamics in such a way that the respective stable branches of the sigmoidal manifold are associated with the active and silent phases of a bursting neuron. The bursting mechanism assumed by Abbott is similar to that in the Chay and Keizer's ordinary-differential-equation model[4] for insulin secreting β-cells of the mammalian pancreas. In that case, the slow dynamics of the variable controlling the burst generation is basically relaxational but strongly influenced by the action potential itself. In contrast, what we do in the present paper amounts to an extreme simplification of Plant's model[5] for pacemaker neurons of the Aplisia abdominal ganglion for which the slow subsystem forms an autonomous independent oscillator. As was discussed by Rinzel and Lee[10], the above give the two representative mechanisms for bursting. From a mathematical viewpoint, the model based on the second mechanism seems to be more naturally related to the phase oscillator model, and is easier to analyze especially for collective behavior.

We present our network model by contrasting it with the conventional one. For comparison's sake, the latter is expressed in terms of an auxiliary variable x_i in the form

$$S_i(t) = F(x_i(t) - h) \ , \tag{1a}$$

$$x_i(t+1) = \sum_j J_{ij} S_j(t) \ , \tag{1b}$$

where $S_i(t)$ represents the firing rate of the i-th cell averaged over some short interval about time t. The output function $F(x)$ is an increasing function of x changing abruptly between 0 and 1 near $x = 0$, so that h represents the threshold value of x_i for depolarization. The summation in Eq. (1b) extends over all cells connecting to the i-th cell.

Besides x_i, we postulate in each cell an extra degree of freedom y_i, and this pair of cell variables defines a *complex potential* $z_i(t)$ with amplitude $r_i(t)$ and phase $\phi_i(t)$. Our model then takes the form

$$S_i(t) = F(\mathrm{Re}\ z_i(t) - h) \ , \tag{2a}$$

$$z_i(t+1) = z_i^*(t) e^{i\Omega} + \sum_j C_{ij} S_j(t) \ , \tag{2b}$$

where $z_i^*(t) \equiv r^* \exp(i\phi_i(t))$ is the reduced potential with fixed amplitude r^*. In aforementioned Plant's model[5], x_i may be interpreted as intracellular free calcium concentration, and y_i as the slow conductance for calcium. Intrinsic oscillatory dynamics of the cell is implemented by the first term on the right-hand side of Eq. (2b). The last term in the same equation has the usual form of synaptic coupling, except that the synaptic efficacies C_{ij} are supposed here to be complex i.e. $C_{ij} = | C_{ij} | \exp(i\alpha_{ij})$.

The dynamical rule in Eq. (2b) tells that the transformation $z_i(t) \to z_i(t+1)$ can be decomposed into three elementary steps as $z_i(t) \to z_i^*(t) \to z_i^*(t)\exp(i\Omega) \to z_i^*(t)\exp(i\Omega) + p_i(t) \equiv z_i(t+1)$, where $p_i(t) = \sum_j C_{ij}S_j(t)$ (see Fig. 1). The first step corresponds to the rapid relaxation of z_i to its native attracting cycle which is a circle of radius r^* in the complex plane. Here the phase of z_i is preserved, which is consistent with the general fact that autonomous oscillators lack restoring force against phase disturbances. The second step corresponds to the rotation of z_i along the circle by an angle Ω, reflecting the oscillatory nature of z_i. Finally, the input p_i kicks z_i out of the circle to give $z_i(t+1)$. Note that p_i generally changes the phase of z_i as well as its amplitude.

An alternative way of modeling the internal oscillatory dynamics would be to assume the form

$$z_i(t+1) - z_i(t) = \varepsilon\Big\{(1+i\Omega)z_i(t) - (1+ic) \mid z_i(t) \mid^2 z_i(t) + \sum_j C_{ij}S_j(t)\Big\} \,, \tag{3}$$

or its time-continuous version ($\varepsilon \to 0$), namely, coupled Ginzburg-Landau oscillators. The two models are expected to lead to similar behaviors. In the present study we will work with the first form Eq. (2b).

Our model can be generalized in a number of ways. For instance, h, r^* and Ω may become cell-dependent parameters, and some random perturbations or more general external stimuli may be included in Eq. (2b).

There are two trivial limits in our model. Suppose first that the oscillation amplitude of z_i is vanishingly small or, equivalently, the total input p_i is strong enough to mask the oscillatory nature of z_i. Then, by putting $r^* = 0$, the conventional model is recovered where $x_i(t)$ are identified with $\mathrm{Re}z_i(t)$ and J_{ij} with

$\mathrm{Re}C_{ij}(=|C_{ij}|\cos\alpha_{ij})$. If seen from our ***complex synaptic matrix*** viewpoint, a given cell j is excitatory or inhibitory according to $|\alpha_{ij}|<\pi/2$ or $|\alpha_{ij}|>\pi/2$, respectively. As another limit, consider single-cell dynamics by switching off the coupling. Our model then reduces to

$$S_i(t) = F(r^* \cos\phi_i(t) - h) \quad , \tag{4a}$$

$$\phi_i(t+1) = \phi_i(t) + \Omega \quad , \tag{4b}$$

or $\phi_i(t) = \Omega t$ for Eq. (4b). For simplicity, let $F(x)$ be represented by the unit step function $\theta(x)$, where $\theta(x) = 1$ for $x \geq 0$ and 0 otherwise. Then, under the condition $| h |< r^*$, active phase ($S_i = 1$) and silent phase ($S_i = 0$) alternate. A periodically bursting neuron has thus been modeled in a simplest way. Whether this condition for bursting is satisfied or not, $\mathrm{Re}z_i$ oscillates smoothly like $\cos\Omega t$. Such smooth oscillations of membrane potential are commonly known under the name of ***slow waves***. They can persist even when tetrodotoxin is used to suppress the spike, and hence some people postulated an autonomous oscillator functioning independently of the action potential[5)]. We follow exactly the same idea.

The synaptic coupling is now switched on. Since too strong coupling leads to the usual network model as noted above, our main concern below will be the case of relatively weak coupling such that the oscillatory nature of the cells could hardly be masked. What we expect then is that the cells exchange phase information and, through mutual adjustment of the timing of bursts, a coherent temporal pattern of activity be created in the network. In order to demonstrate this possibility, some numerical simulations were carried out for Eqs. (2a, b). Without loss of generality, we put $r^* = 1$ hereafter. Our numerical study is restricted to the special case of all-to-all excitatory coupling, so that we put $C_{ij} = N^{-1}C(1 - \delta_{ij})$, where $C =| C | \exp(i\alpha), | \alpha |< \pi/2$ and N denotes the population size. The same form of coupling in the conventional model will be of little dynamical interest, except possibly for the occurrence of collective bistability. In the actual simulation, Ω was replaced by cell-dependent quantities Ω_i, the latter being assumed to obey a Lorentz distribution $L(\Omega - \Omega_0; \gamma) \equiv \gamma[2\pi((\Omega - \Omega_0)^2 + \gamma^2)]^{-1}$. We also assumed $F(x) = \theta(x)$. Recent extensive studies on large populations of coupled phase oscillators made it clear that collective oscillations easily arise in the populations of globally coupled oscillators with frequency distribution[11,12]. Our particular concern here is whether the same is true for our network model.

Figure 2 shows a typical case in which a domain of mutually synchronized cells develops out of the state of complete incoherence. Initially, the phases of bursting are totally random, while in a period of a few oscillations a major part of the entire population comes to behave in a coherent fashion.

The degree of global synchronization in sufficiently large populations may be quantified by the average of the reduced potential i.e. $Z(t) \equiv N^{-1}\sum_i z_j^*(t)$. A slightly more convenient measure will be the order parameter R defined by

$$R = \overline{X\dot{Y} - Y\dot{X}} \ , \tag{5}$$

where X and Y are the real and imaginary parts of Z, the bar denoting a long-time average, and the dots should be understood as $\dot{A}(t) = A(t) - A(t-1)$. Since collective oscillations imply a closed orbit in the complex Z-plane, their absence/presence may be indicated by vanishing/non-vanishing value of R ; such a simple property is not shared by Z except for highly symmetric cases. Figure 3 shows how R changes with the synaptic strength $\mid C \mid$. As is expected, a critical value of $\mid C \mid$ for the onset of collective oscillations seems to exist.

There are some striking similarities of our numerical results to what we know about the populations of the phase oscillators studied earlier. We suspect that the present model itself might be reduced to the network of phase oscillators under suitable conditions. The reduction is in fact possible when the synaptic coupling is sufficiently weak, as we see below.

Since the deviation of $\mid z_i \mid$ from $r^*(=1)$ may still be neglected when the coupling introduced is weak, we retain Eq. (4a) as valid. The only remaining problem is therefore to find an approximate form of the coupling term to be added to the right-hand side of Eq. (4b). Comparing the phases of the respective sides of Eq. (2b) with each other, and putting $r_i = 1$, we obtain to the lowest order in the coupling strength

$$\phi_i(t+1) - \phi_i(t) = \Omega + \sum_j \mid C_{ij} \mid \sin(\alpha_{ij} - \Omega - \phi_i(t))S_j(t) \ . \tag{6}$$

Substituting Eq. (4a) for S_j, and using new phase variables ψ_i defined by $\psi_i = \phi_i - \Omega t$, we rewrite Eq. (6) in the form

$$\psi_i(t+1) - \psi_i(t) = \sum_j \mid C_{ij} \mid \sin(\tilde{\alpha}_{ij} - \Omega t - \psi_i(t)) F(\cos(\Omega t + \psi_j(t)) - h) \quad , \quad (7)$$

where $\tilde{\alpha}_{ij} = \alpha_{ij} - \Omega$. Since the time-variation of ψ_i should be sufficiently slow, the left-hand side of Eq. (7) may be approximated by $d\psi_i/dt$ and, moreover, the right-hand side may be time-averaged over the oscillation period $2\pi/\Omega$ under fixed ψ_j's over the same period. In this way, we get

$$\begin{aligned}\dot{\psi}_i &= \sum_j \mid C_{ij} \mid \frac{1}{2\pi} \int_0^{2\pi} d\lambda \sin(\psi_j - \psi_i + \tilde{\alpha}_{ij} - \lambda) F(\cos\lambda - h) \\ &= \sum_j K_{ij} \sin(\psi_j - \psi_i + \tilde{\alpha}_{ij}) \qquad (8)\end{aligned}$$

or

$$\dot{\phi}_i = \Omega + \sum_j K_{ij} \sin(\phi_j - \phi_i + \tilde{\alpha}_{ij}) \quad , \quad (9)$$

where

$$K_{ij} = \frac{\mid C_{ij} \mid}{2\pi} \int_0^{2\pi} d\lambda \cos\lambda F(\cos\lambda - h) \quad . \quad (10)$$

Equation (9) gives the standard form of the phase model. When the natural period of bursting changes from cell to cell as we assumed in our numerical simulations, one may simply replace Ω by Ω_i in Eq. (9). A little careful examination shows that such replacement is consistent with our perturbative reduction only when the width of the distribution of Ω_i is much smaller than the average frequency Ω_0. This condition is assumed to be satisfied below.

Remember that our numerical simulations assumed: (a) all-to-all coupling, (b) $\alpha = \Omega_0$ so that $\tilde{\alpha}_{ij} \simeq \alpha - \Omega_0 = 0$, (c) $F(x) = \theta(x)$, and (d) Lorentz distribution $L(\Omega - \Omega_0; \gamma)$ for Ω_i. Under these conditions, Eqs. (9) and (10) are reduced to

$$\dot{\phi}_i = \Omega_i + \frac{K}{N} \sum_j \sin(\phi_j - \phi_i) \quad (11)$$

and

$$K = \begin{cases} |C|\sqrt{1-h^2}/\pi & (|h|<1) \\ 0 & (|h|\geq 1) \quad , \end{cases} \tag{12}$$

respectively. Previously, the phase model of the form of Eq. (11) with Ω_i obeying the Lorentz distribution $L(\Omega - \Omega_0; \gamma)$ was solved exactly for Z[12]. The result is $Z = \sqrt{1-(2\gamma/K)}\exp(i\Omega_0 t)$ for $K \geq 2\gamma$ and 0 otherwise. Thus R becomes

$$R = \begin{cases} \Omega_0\left(1 - \dfrac{2\gamma}{K}\right) & (K \geq 2\gamma) \\ 0 & (K < 2\gamma) \quad , \end{cases} \tag{13}$$

where K is given by Eq. (12). This approximate result for R is now compared with our numerical curve for the same quantity, which is shown in Fig.3. Their agreement is satisfactory at least not far from the critical point.

Although still preliminary, further numerical study revealed the occurrence of more complicated collective dynamics when the coupling becomes stronger. Figure 4 shows an example of such behavior for a population *without* frequency distribution, where the system is decomposed into two subpopulations with different timing of bursts. The states of three, four and more clusters were also found under different conditions. This kind of clustering behavior was first discovered by Kaneko[13] for globally coupled chaotic units and then by Golomb et al.[14] for globally coupled non-smooth phase oscillators.

The effect of an external stimulus I is now considered, and we add this term to the right-hand side of Eq. (2b). For simplicity, I is assumed to be real positive, as weak as $|C|$, constant in time and cell-independent. We will show below that the collective dynamic is very sensitive to I when the threshold parameter h is near its critical value 1. In such a situation, we are no longer allowed to neglect the small amplitude-deviation of z_i from 1, but instead we have to come back to the original expression Eq. (2a) for $S_i(t)$ with

$$\mathrm{Re}z_i(t) = \cos(\phi_i(t)+\Omega) + \sum_j |C_{ij}|\cos\alpha_{ij}S_j(t) + I \quad . \tag{14}$$

In the 0-th order approximation in which the last two terms were neglected, $S_j(t)$ is vanishing for the most period on account of the assumption $h \lesssim 1$. This implies that in Eq. (14) the coupling term is negligible compared to the input I. Thus, in the first order approximation, we may use in Eq. (6) the expression

$$S_j(t) = F(\cos(\phi_j(t)) + I - h) \quad . \tag{15}$$

On the other hand, the new term $-\sin(\phi_i(t)+\Omega)I$ which appears in the right-hand side of Eq. (6) vanishes on averaging. Consequently, the previous result Eq. (9) without external stimulus remains valid if we change the definition of K_{ij} as

$$K_{ij} = \frac{| C_{ij} |}{2\pi} \int_0^{2\pi} d\lambda \cos\lambda F(\cos\lambda + I - h) \quad . \tag{16}$$

It is clear that the effective coupling K_{ij} is strengthened by I, so that if the system is originally near and below the threshold for collective oscillation, the external stimulus may easily bring it into the oscillatory regime. Stimulus-evoked collective oscillations of neural populations have been observed in the cat primary visual cortex[1]. This phenomenon is thus naturally understood if we suppose our idealized population with all-to-all excitatory coupling to represent one orientational column in the visual cortex. It is also suggested from the experiments that each neuron gives rise to at most a few spikes per oscillation period. This is consistent with our assumption $h \lesssim 1$, namely, the assumption of brief active period.

We conclude this paper with a few additional remarks. First, the reduced dynamics in Eq. (8) bears some resemblance to the relaxational XY spin dynamics. In fact, if $\tilde{\alpha}_{ij} = 0$ and if at the same time the synaptic matrix is symmetric, then Eq. (8) derives from the hamiltonian $H = (1/2)\sum_i \sum_j K_{ij} \cos(\psi_i - \psi_j)$ via $\dot{\psi}_i = -\partial H/\partial\psi_i$. Generalization to finite temperature XY spins would also be easy to achieve by including random noise in Eq. (2b). Secondly, the form of the phase coupling in Eq. (9) implies that a given pair of cells favors in-phase/out-of-phase configurations if the corresponding $|\tilde{\alpha}_{ij}|$ is smaller/larger than $\pi/2$. On account of the difference between α_{ij} and $\tilde{\alpha}_{ij}$, the above condition for $|\tilde{\alpha}_{ij}|$ does not correspond precisely to the excitatory/inhibitory condition. This difference Ω_0 is unimportant, however, as far as the typical period $2\pi/\Omega_0$ of bursting remains, as usual, much longer than the minimum interval between successive spikes which is $O(1)$. Thirdly, our proposed mechanism leading to phase locking seems to be quite efficient energetically because it can work when the synaptic coupling is too weak to change the level of activity to a meaningful extent. In the framework of the conventional neural network model with global coupling, the only way to produce oscillations will be to assume a delayed feedback from inhibitory neurons to excitatory ones[15]. This of course presupposes strong synaptic coupling. Finally, our extension of synaptic efficacies to complex numbers suggests the possibility of a new learning scheme based on the *orientational* plasticity of C_{ij} in the complex plane. In this connection, very suggestive is the fact that $\tilde{\alpha}_{ij}$ (i.e. essentially the phases of C_{ij}) are important parameters in determining phase wave properties associated with a space-continuous version of the phase equation (9)[16]. Possible roles of the phases of C_{ij} in neural information processing are yet to be clarified.

The present work is supported by the Japanese Grant-in-Aid for Science Research Fund from Ministry of Education, Science and Culture (Nos. 03234109 and 03247107).

References

1) C.M. Gray, P. König, A.K. Engel, and W. Singer, Nature **338**, (1989) 334.
A.K. Engel, P. König, C.M. Gray, and W. Singer, Eur. J. Neurosci. **2**, (1990) 588.
R. Eckhorn, R. Bauer, W. Jordan, M. Brosch, W. Kruse, M. Munk, and H.J. Reitboeck, Biol. Cybern. **60**, (1988) 121.

2) W.J. Freeman, Electroencephalogr. Clin. Neurophysiol. **44**, (1978) 586.

3) C. von der Malsburg and W. Schneider, Biol. Cybern. **54**, (1986) 29.

4) T.R. Chay and Keizer, Biophys. J. **42**, (1983) 181.

5) R.E. Plant, J. Math. Biol. **11**, (1981) 15.

6) H. Sompolinsky, D. Golomb, and D. Kleinfeld, Proc. Natl. Acad. Sci. USA **87**, (1989) 7200 ; Phys. Rev. **A43**, (1991) 6990.

7) P. Baldi and M. Ronny, Neural Computation **2**, (1990) 458.
E. Lumer and A. Huberman, Neural Computation (to be published).
D.M. Kammen, P.J. Holms and C. Koch, in *Models of Brain Function*, R.M. J. Cotterill, ed., (Cambridge Univ. Press, 1989), 420.

8) J.J. Hopfield, Proc. Natl. Acad. Sci. USA **79**, (1982) 2554.
D. Amit, *Modeling Brain Function*, (Cambridge Univ. Press, Cambridge, 1989).

9) L.F. Abbott, J. Phys. **A23**, (1990) 3835.

10) J. Rinzel and Y.S. Lee, In *Nonlinear Oscillations in Biology and Chemistry*, H.G. Othmer ed., (Springer, New York, 1980), Lecture Notes in Biomathematics **66**, 19.

11) A.T. Winfree, J. Theor. Biol. **16**, (1967) 15.
Y. Yamaguchi and H. Shimizu, Physica **11D**, (1984) 212.
Y. Kuramoto and I. Nishikawa, J. Stat. Phys. **49**, (1987) 569.
S.H. Strogatz and R.E. Mirollo, J. Stat. Phys. **63**, (1991) 613.

12) Y. Kuramoto, in *Int. Symp. on Math. Prob. in Theor. Phys.*, H. Araki, ed., (Springer, New York, 1975), Lecture Notes in Physics **39**, 420

Y. Kuramoto, *Chemical Oscillations, Waves, and Turbulence*, (Springer, Berlin, 1984).

13) K. Kaneko, Physica D**41**, (1990) 137.

14) D. Golomb, D. Hansel, B. Shraiman and H. Sompolinsky, preprint.

15) S. Amari, IEEE Trans., SMC-**2**, (1972) 643.
H.R. Wilson and J.D. Cowan, Biophys. J. **12**, (1972) 1.

16) Y. Kuramoto, Prog. Theor. Phys. Suppl. **79**, (1984) 223.

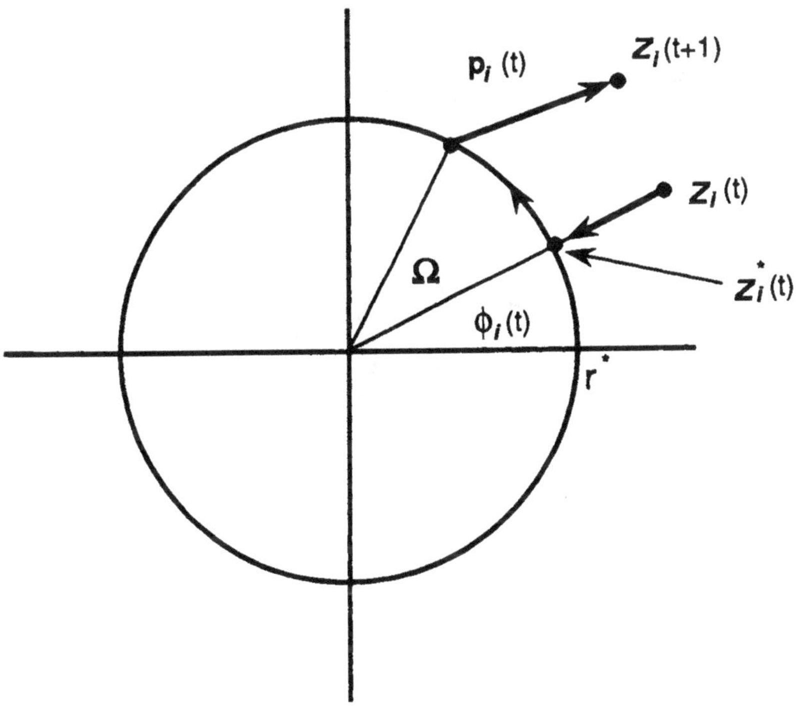

FIG.1 Dynamical rule producing $z_i(t+1)$ from $z_i(t)$ in the complex plane. Three elementary steps indicated by the arrows are involved.

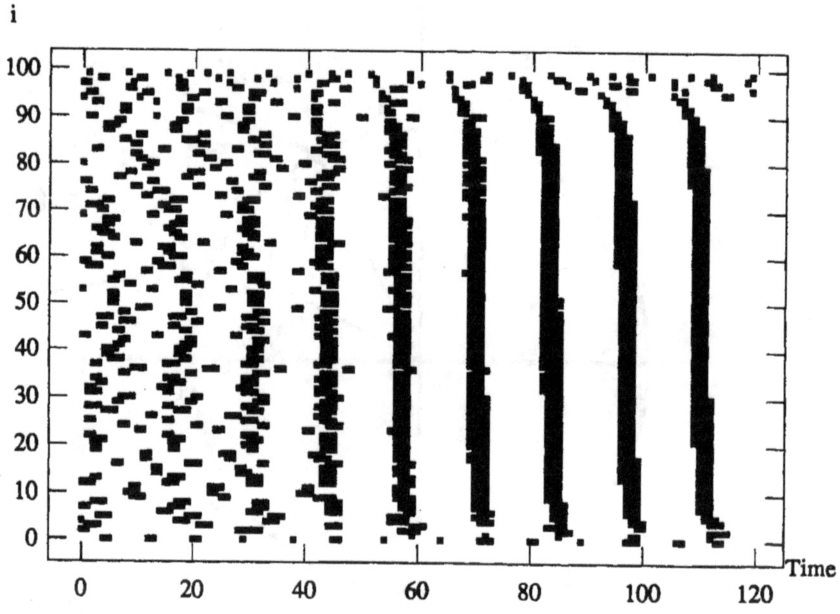

FIG.2 Formation of a phase-locked domain exhibited by 100 globally coupled cells obeying Eqs. (2a, b) with distributed Ω. The cells are numbered vertically in order of increasing natural frequencies. Active phases are indicated by dark horizontal bars. Parameter values : $h = 0.90, \Omega_0 = 0.50, \gamma = 0.02, | C |= 0.70, \alpha = 0.50$.

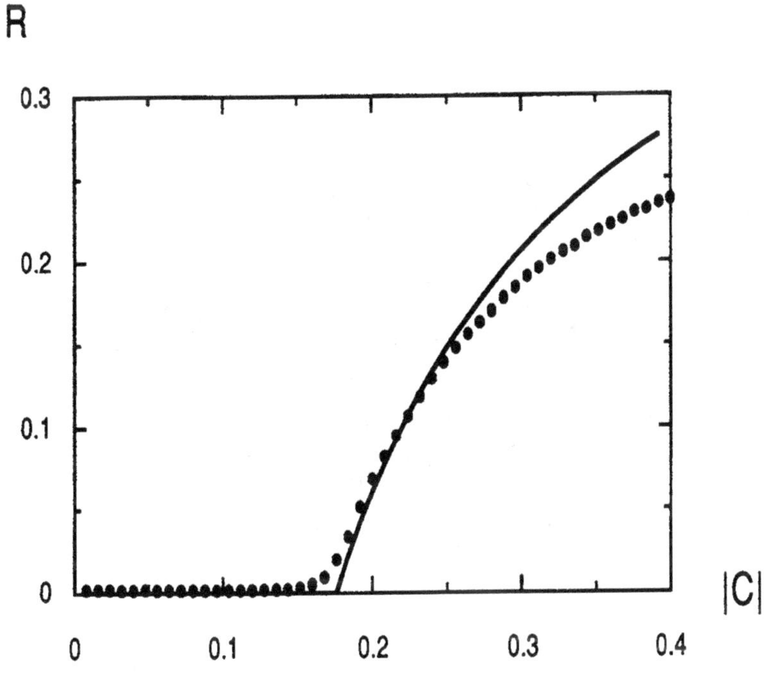

FIG.3 Order parameter R versus coupling strength $|C|$. Circles indicate the data from computer simulation on Eq. (2a, b) with 100 globally coupled cells. Analytical result in Eq. (12) valid in the weak-coupling limit is indicated by a solid curve. Parameter values : $h = 0.7, \Omega_0 = 0.5, \gamma = 0.02, \alpha = 0.5$.

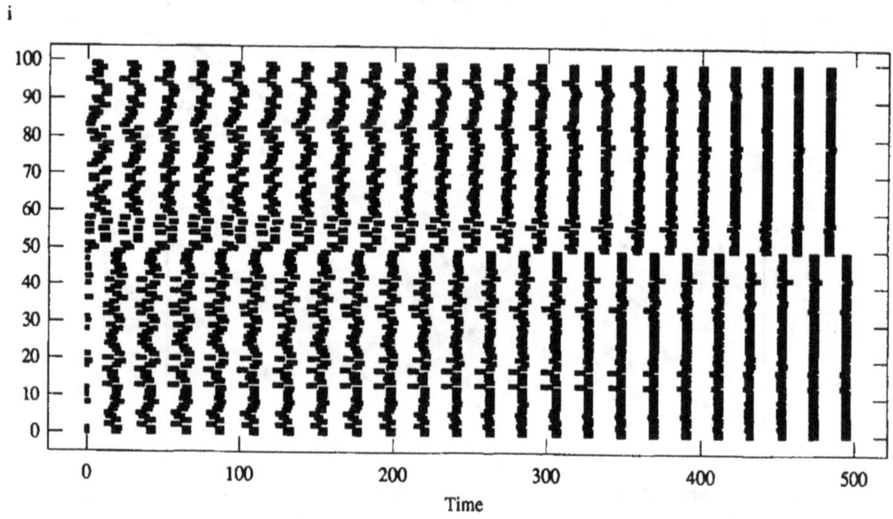

FIG.4 Temporal pattern of bursts exhibited by a population of 100 globally coupled cells without frequency distribution (displayed similarly to Fig.2 but the cells are renumbered suitably). Parameter values : $h = 0.50$, $\Omega = 0.31$, $\mid C \mid = 0.70$, $\alpha = 2.20$.

Nonequilibrium Systems with Self-Organized Boundaries

Yasuji Sawada

Research Institute of Electrical Communication,
Tohoku University, Sendai 980 Japan

ABSTRACT

We report recent progress in our study on regeneration of the structure and information of hydra from dissociated cell aggregate. The process includes cell-sorting period, cavity-formation period and structure -forming period. Simultaneously, nerve network formation takes place, apparently induced by head structure formation. The aggregate changes from initially disordered cell assembly to an ordered animal within a few days. We try to compare the features of the living state thus formed with a self organized pacemaker recently found in the electro-convection pattern of nematic liquid crystal.

1. INTRODUCTION

The bifurcation series of a nonlinear system when it is driven far from equilibrium has been intensively studied theoretically and experimentally in the last decade[1]. The most welknown scheme may be; quiescent homogeneous, steady structure, structure with temporal oscillation, chaos, hyperchaos and finally turbulence. It seems to be very difficult to define a living system as one of the nonlinear system far from equilibrium. There seems to be no place for living state in the bifurcation sequence as it is.

Living state has not been well-defined in physical terms. Therefore it may be helpful for understanding biological system to compare a living state with any dynamical state known so far. To do so we must first point out some features characteristic to living state which are different from those which one can find in the nonlinear physical system driven far from equilibrium.

Let us limit ourselves to the living state of multicellular biological system to avoid discussing the biological structure within cells which are complicated by itself. The multicellular biological structure is composed of cell units with a self-organized free boundary, and internally generated information, which, in turn, controls the structure and produces function to interact with external world.

As an example of such system, we present regeneration phenomena of hydra, and show reconstruction of the body structure and an information-generating system when a dissociated cell aggregate regenerates and how the generated information control the structure.

Some of these features can be found separately in physical systems. Self-organized boundary is seen in crystal growth, and chaos generates information internally. Here, we ask if we can find a nonbiological system where information controls the structure and produce function to maintain the structure itself.

2. REGENERATION OF AN ANIMAL

We describe transition from an artificially created state of a biological system to a normal living state. Since a transition from a totally inanimated state to a living state is technically unrealizable at present, we use a multicellular bilogical system and study the regeneration process from a dissociated cell aggregate to a complete animal. The dissociated cell aggregate is a random disorganized cell ensemble where the original positional information is completely destroyed. Each cell is living in a sense, when it is dissociate and aggregated. But the cluster of cells just aggregated is not living as an animal. Through the course of regeneration the system recovers its life. By studying the process of regeneration one might be able to observe the dynamics of achieving a living state of an animal.

For this purpose we use hydra which is known to have a strong ability for regeneration[2]. Hydra regenerates even from an aggregate of cells dissociated from hydra animals. The body of hydra is simple enough to study the regeneration; essentially a tube-shaped intestines with an opening at the top around which five or six tentacles are regularly arranged. Hydra has relatively few kind of cells; endodermal and ectodermal epithelial cells, which constitute single internal and external cell walls of the body, nerve cells, undifferentiated interstitial cells, nematocytes which are found only in tentacles, and so on.

In this part we describe recent progress of our research on regeneration of hydra from a dissociated cell aggregate carriod out in our laboratory and the readers are referred for the experimental detailes to the literature[3,4,5,6].

2A) Regeneration of Body Structure

The regeneration process from a dissociated cell aggregate to a complete hydra may be divided into three stages[7]. Fig.1 is a schematic drawing of the regeneration process of an aggregate (Fig.1(2)) dissociated from intact hydras (Fig.1(1)).(see reference 4 for real pictures). In the first stage, cell sorting takes place among the ectodermal and endodermal epitherial cells. Through this process the ectodermal cells move to the surface of the aggregate and form a monolayer, which is backed by the mono-cell layer of the endodermal cells, and the surface of the aggregate becomes smooth(Fig.1(3)). The time needed for an aggregte to accomplish this stage is about 6 hrs.

The second stage is the cavity formation period(Fig.1(3),(4)). When the double cell layer is formed, the internal space is isolated from the external world, and some chemical potential difference, such as osmotic pressure, may be established between them. It is often observed that a clump of the cells which have been left unused for forming the double layer are expelled from inside the aggregate(Fig.1(4)). By the end of this stage, 24hrs after regeneration started, the aggregate is an approximately isotropic and homogeneous baloon of double cell layer.

The third stage is a structure formation period. Tentacles are formed after two or three days almost randomly over the surface of the aggregate(Fig.1(5)). Then the aggregate is elongated along an axis which will be the future body axis of a normal hydra(Fig.1(1)). A mouth opening appears later at the apex of the axis. The tentacles which are located far from the opening are absorbed in the body and disappear. The tentacles which happened to be in the vicinity of the opening are gradually rearranged around it.

2B) Regeneration of Control Sytem

It appeared from the measurement[8] that most of nerve cells visualized by an antibody are thrown away from the aggregate in the cell-sorting and cavity-formation stages. Then the observed density starts increasing from about 50hrs. due to differentiation of the interstitial cells into nervre cells. The increment is most steep around 90hrs., then the tendency gradually saturates. Simultaneously, it was observed that the average number of connection of a nerve cells to the others increases. The increment of the

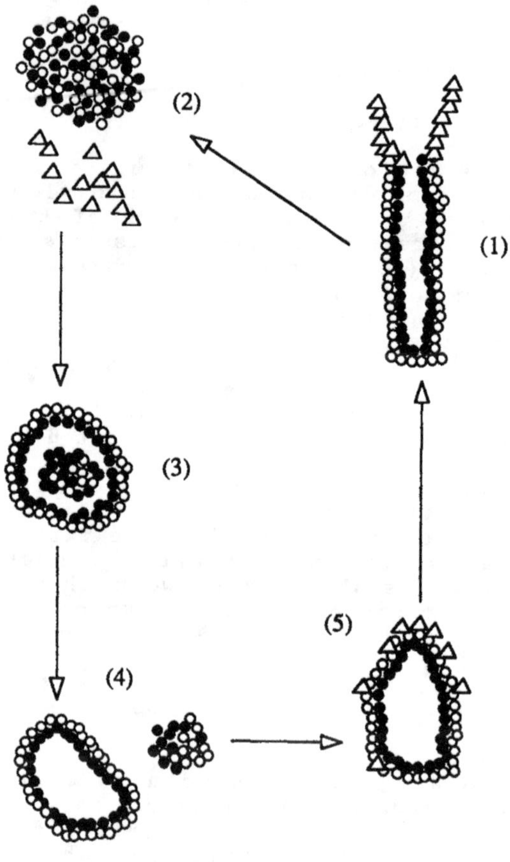

Figure 1 A schematic illustration of regenetaion cycle of hydra; open circles are ectodermal epithelial cells, closed circles are endodermal epithelial cells, and the tiangles are the cells found in the tentacles; 1) intact hydra, 2) dissociated cell aggregate, 3) double cell layer, 4) cavity, 5) pattern formation.

connection is somewhat earlier than that of the density. It might mean that the nerve cells has a tendency to have connection as soon as they are differentiated. The maximum number of connection is about five, suggesting the nearest neighbor interaction as average, due to the two dimensional arrangement of the network.

The electrical activity of the nerve systems and the epithelio-muscular cells were measured by micro glass electrodes.[9] The aggregate does not show any activity clearly distinguishable from noise for about 10 hrs from the starting time of the regeneration. Then the activity increases rapidly. At about 20hrs one finds decrease of the activity, which may be attributed to the the osmotic pressure decrease of the aggregate due to the expulsion of the internal cell clumps. The activity increases again thereafter. Recent two points measurements have revealed that the correlation of the electrical activity at two differents points also has a sharp rise at about 15 hrs and then gradually saturates.

3. LIVING STATE

In hydra, living state as a multi-cellular system is autonomously regenerated from a disordered cell mass. The living state is characterized by self-organized boundary and structures, such as a mouth and a foot in the present case, and by a self-organized information generator used for control of the structure. Study of the first stage of the cell sorting is under way by F.Graner and J.Glazier[10] as a physical phenomena in terms adhesion of dynamics, and bio-chemical studies to identify the adhesion molecules of hydra are now under way by E. Hobmayer, Fujisawa and Sugiyama[11]. No studies have been done for the mechanism of expulsion. The third stage has been extensively studied by M.Sato et. al.[4,6]. Most phenomena in this stage are understood by activator-inhibitor model, although these morphgens have not been identified at present stage.

Together with the construction of structure, there start appearing local excitabilities, which are spatially uncorrelated at early stage and becomes gradually correlated as regeneration proceeds. A percolation model for network formation was investigated by Itayama[12].

There are some similarities and differences between the bifurcation scenario of physical systems and the regeneration scenario of biological systems. In a quasi-one dimensional convection system.[13], for example, steady convectional structure is formed first. Then with increasing parameter, local oscillational activities, weak spatially uncorrelated chaos and strong correlated chaos are successively observed. In this respect, the biological system has similarity with the nonlinear physical systems far from equilibrium. Yet, many features characteristic to living state are manifested by the boundary controled by the internally generated information.

Information in biology is manyfolds over different time scales; genetic information, positional information and nervous information etc.. Genetic information has a long time scale; many many years. Positional information has a time scale of regeneration; days. Nervous information has a time scale of behavior; seconds. All these information are essential for biology. In the present discussion the first kind of information is excluded since it is fixed for the time scale of our observation. It is utilized only to produce necessary chemicals when the cell division takes place or when the second kind of information is autonomously built up in case of regeneration.

The third kind of information which controls the structure may be important for the definition of the living state at this time scale. Any information must, by definition, have a speed of propagation faster than the speed of environmental change. If a piece of metal does not allow any signals to propagate, temperature increment of external world cannot be sensed until the whole body melts. If the heat is conducted by lattice oscillation, or even better if excitable signals or other kind of waves can propagate, it would be possible to fabricate a safety system to avoid a disaster. That is the very reason why information or software is necessary for the survival of an animal to avoid complete destruction of structure by sensing an approaching danger and to send signals to the muscle to move away. It is also much easier to catch food if it is equipped with signals transmitted from the senser to muscles.

Although localized structures are found in binary mixtures and liquid crystals[14], the interaction of the boundary with internally generated information has not well studied in the nonlinear systems in the laboratory.

4. AUTONOMOUS LOCALIZED PACEMAKER IN CONVECTION SYSTEM

We discuss this topics here as an example of a localized autonomous structure whose boundary is controlled by a self-generating signal, like a biological system discussed in the previous section. This topic is discussed in detail experimentally by Sano and theoretically by Sakaguchi and Mizoguchi in this proceeding. I only cites some crucial points which are relevant to the present paper.

The system of our interest is an electro-convection pattern in nematic liquid crystal called grid pattern, in which convection pattern forms regular rectangular oscillating lattice. Some relevant facts observed experimentally[15] are,

1)A pacemaker which sends out phasewaves to its surroundings is created autonomously.(Fig.2)
2)The phasewaves propagate either in a target pattern or a spiral pattern.
3)Expansion of convection lattice is associated in a pacemaker region.

It is evident that two fields are involved in this phenomena. One is an oscillating field of the convection and the other is deformation field of the underlying convection lattice structure. The two fields interact with each other positively by some mechanism which is unknown at present. The purpose of this section is not to discuss the realistic mechanism of the localized pattern in the convection , but to present a possible physical understanding of a localized structure controlled by a self-generating signal. I believe it is useful not only for understanding the biological phenomena which has a direct resemblance, such as the control mechanism of physarum polysephalum[16], but also for physical understanding of fundamental features of living state in relation to the previous section.

The nervous signal sent to muscles to contract in a animal like hydra may seem to be a very specialized phenomena as a physical system. But because the control of the boundary is essential for an animal to survive, one needs to understand how this mechanism arose first in a physical world. The phenomena we wish to describe here is an example to show that control of the boundaries by a self-generating signal is by no means phenomena special only in biological system, but a universal state of a nonlinear system far from equilibrium.

Let us consider a slow variable(a structural variable), the periodic lattice field here, originally homogeneous. The field of this slow variable supports an oscillating field. A Hopf type of instability is assumed. For an example a concentration field may support chemical oscillation, and convectional structure may have an oscillatory flow. The oscillatory field therefore, is sensitive to the fluctuation of the slow field. From the experimental observation the following two mechanism may be essential.

1)The homogeneous slow field is unstable against local deformation in the presence of the oscillation.

2)A local defrmation of the slow field increases the local frequency of the oscillation

Then a following scenario is possible in some parameter space[17]; A local deformation of the slow field will saturate in amplitude and expands slowly in space. The frequency of

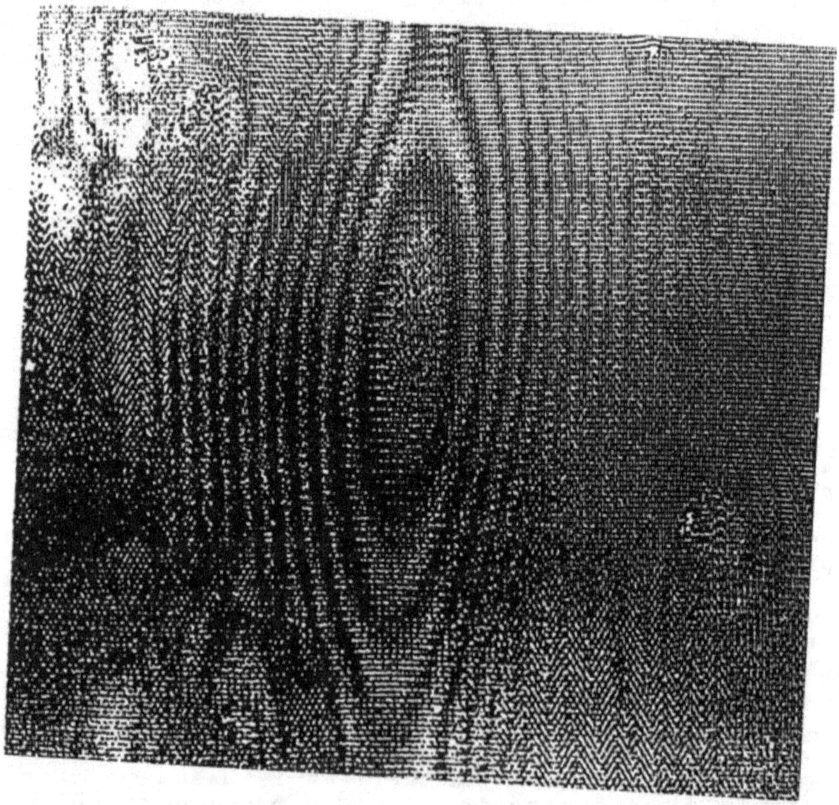

Figure 2 An example of pacemaker sending out phasewaves in the grid convection pattern of nematic liquid crystals (after K.Sato, master thesis, Tohoku Univ.,1991 unpublished).

the oscillation in the deformed domain is higher than the outside region. But if the difference of the frequency is small, the oscillation in outside region would be entrained by the oscillation within the domain. In this case a phasewave would propagate outwards into the outside region.

If a phasedynamics is predominant, that is the phase change is slow in space and time, 1) means that the growth rate of the structure field is sensitive to the curvature of local amplitude of the oscillatory field, which is represented in phasedynamics by the phase gradient squared or the phase curvature. Therefore, in some parameter condition, the homogeneous state of the structure field may be stabilized against local deformation in the presence of phase gradient generated by the phasewave propagation. This stabilization mechanism has two effects. One is to prevent other local instability outside a pacemaker region. The other is to stabilize the pacemaker boundary from expanding further.

As a result, the equation for the slow field of the structure field may be written as:

$$\frac{\partial S}{\partial t} = f(S, (\nabla\phi)^2, \nabla^2\phi, \mu) + D\nabla^2 S \qquad (1)$$

where S is the slow field and ϕ is the phase of the oscillating field. Readers are referred to the complete theoretical treatment[17] for further detail.

Two points may be relevant in relation to biology:

1) The present physical system is a prototype for self-organization for which the activator is a slow field and the inhibitor is the phasewave. The conventional model used for biological self-organization is based on the Turing model[18] for which the inhibitor is a fastly diffusing substance, not an oscillating field. In this case the pattern is repeated with a wavelength comparable to the diffusion length of the inhibitor. Phasewaves can generally propagate over a longer distance than diffusion mechanism, therefore the inhibition length is effectively infinite, and an effectively isolated structure can be constructed by this inhibitor.

Note that the phasewave experimentally observed in liquid crystal is not exactly described in the model written above, but may be more realistically explained by the model by Sakaguchi in case of convection system. However, in chemical case such as found in biology the present explanation may be easier, and it may be related to the pacemaker of heart cells and that of physarum polysephalum[16].

2) The present system suggests that the phasewaves can control the boundary. The equation (1)for the slow field will give the boundary dynamics which is subject to the variation of parameters for the oscillating field. By controlling the parameters it would be possible to move the boundary. In a sense this kind of control may be related to the amoeba motion of dictyostelium discoidium[19] as well as motion of physarum polysephalum[20].

REFERENCE

1. See for example, P.Berge, Y.Pomeau and C.Vidal, Order within Chaos(John Wiley & Sons, 1984)

2. See for example, H.M.Lenhoff(ed.), Hydra:Research Method (Plenum,1983)

3. H.Shimizu and Y.Sawada, Transplantation Phenomena in Hydra:Cooperation of Position Dependent and Structure Dependent Factors Determines the Transplantation Result, Dev. Biol. 122, 113(1987)

4 M.Sato and Y.Sawada, Regulation in Numbers of Tentacles of Aggregated Hydra Cells, Dev. Biol. 133, 119(1989)

5 H.Ando, Y.Sawada, H.Shimizu and T,Sugiyama, Pattern Formation in Hydra Tissue without Developmental Gradients, Dev. Biol. 133, 405(1989)

6. M.Sato, H.Bode and Y.Sawada, Patterning Processes in Aggregates of hydra Cells Visualized with the Monoclonal Antibody, TS19, Dev. Biol. 141, 412(1990)

7. Y.Sawada, T.Itayama and S.Mika, Physics of Living State (Ed. Y.Sawada and T.Musha, Ohmusha to appear)

8. T.Itayama and Y.sawada, Neural Network Formation and Activity of an Aggregate of Diisociated Hydra Cell. In Cooperative Dynamics in Complex Physical Systems (ed. H.Takayama, Springer-Verlag, 1988)p.262

9. T.Itayama and Y.Sawada, Development of Contractile Motion and the Electrical Activity in Regenerating Hydra Aggregates, submitted.

10. F.Graner and J.Glazier, in preparation

11. E.Hobmayer, T.Fujisawa and T.Sugiyama, private

communication

12. T.Itayama, (Doctor thesis, Tohoku Univ. 1991 unpublished)

13. S.Sato, M.Sano and Y.sawada, Phys. Rev. A37,1679(1988)

14. Recent literatures may be found in Advances in Fluid Turbulence, Proc. 8th Int. Conf. Center for Nonlinear Studies, Physica D, 37(1989)

15. S.Nasuno, M.Sano and Y.Sawada, J. Phys. Soc. Jpn., 58, 1875(1989); S.Nasuno and Y.Sawada, Prog. Theor. Phys. Suppl. 99, 450(1989); M.Sano, K.Sato, S.Nasuno and H.Kokubo, submitted

16. K.Matsumoto, T.Ueda and Y.Kobatake, J.Thoer. Biol. 122, 339(1986)

17. A.Mikhailov, Physica D ,to appear in 1992

18. A.M.Turing, Phil. Trans. Roy. Soc. London B237, 37(1952)

19. G.Gerisch, Curr. Top. Devel. Biol., 3, 157(1968)

20. K.Matsumoto, private communication

FILAMENTS IN LIQUID CRYSTALS: STRUCTURE AND DYNAMICS

P. Palffy-Muhoray[1], B. Bergersen[2], H. Lin[1], R.B. Meyer[3] and Z. Rácz[4,5]

1. Liquid Crystal Institute and Department of Physics, Kent State University, Kent, OH 44242 U.S.A.
2. Department of Physics, University of B.C., Vancouver, B.C. V6T 1Z1, Canada
3. Department of Physics, Brandeis University, Waltham, MA 02254 U.S.A.
4. Institute for Theoretical Physics, Eötvös University, Budapest 1088, Puskin u. 5-7, Hungary
5. Physics Department, Clarkson University, Potsdam, New York 13676 U.S.A.

ABSTRACT

On cooling the isotropic phase of certain liquid crystalline materials, the low temperataure smectic A phase first appears in the form of filaments. We discuss the results of experiments, and propose a simple model for the structure and dynamics of these filaments. We compare our observations with the predictions of this model and with computer simulations.

1. Introduction

When condensed matter is cooled across phase boundaries, the emerging low temperature phase can be accompanied by the formation of patterns with a wide variety of morphologies. Isotropic fluids condensing from a supersaturated vapor tend to form simple spherical domains. Solid crystals growing in an undercooled melt or vapor can exhibit strikingly beautiful and complex structures, as exemplified by snowflakes and ice crystals. The anisotropy of the low temperature phase is believed to play a key role in determining the pattern morphology. Liquid crystals are materials with phases which are intermediate between those of isotropic liquids and anisotropic solids. It may be expected, therefore, that the emerging low temperature phases of liquid crystals incorporate features of both types or materials.

Nematic liquid crystals consist of anisotropic molecules with long range orientational order. The smectic A phase also exhibits a layer structure in which the average molecular orientation is parallel to the layer normal. A wide variety of other liquid crystalline phases exist. In many materials, the S_A phase appears directly on cooling the high temperature isotropic liquid phase. The S_A phase usually appears via the formation of small compact domains, which are elongated and irregular in shape. These 'bâtonettes' grow in size, and eventually coalesce to form the focal conic structure[1] of the S_A phase.

In certain materials, the S_A phase first appears in the form of filaments instead of bâtonettes as the isotropic liquid is cooled below the isotropic (I) - S_A transition temperature. Single filaments, which grow rapidly in length but not in diameter, have been observed. During growth, the filaments continuously buckle to take on a serpentine form, and subsequently become extremely convoluted and nearly space-filling. The envelope of the region occupied by the filament exhibits instabilities similar to those seen in other pattern forming systems with tip-splitting and dendritic growth. The filaments themselves are metastable, and eventually transform to compact domains whose texture is similar to the characteristic focal conic texture of S_A phases. The filament lifetime is strongly dependent on undercooling. Moreover, the filaments are associated with other small structures such as spheres, and these complex structures exhibit a variety of unusual dynamic behavior.

We present a simple model for the structure and growth of these filaments, which incorporates aspects of both perfect faceted crystals and of isotropic liquid droplets, but with additional features which originate from the unique properties of smectic liquid crystals.

2. Experiments and Observations

The first observation of filamentary growth in liquid crystals was made by R.B. Meyer and F. Jones[2,3] in mixtures of EEBAC and silicon oil. The chemical structures of these materials are shown below.

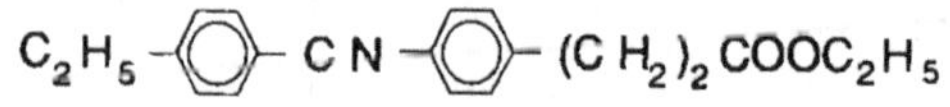

ethyl (p-ethoxybenzylidene amino cinnamate) (EEBAC)

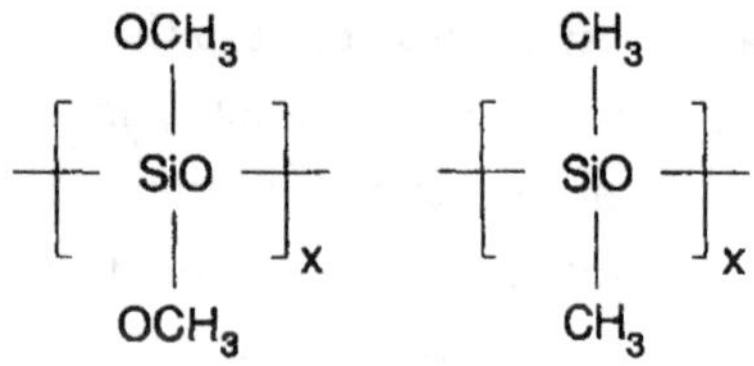

Silicon oil

For a wide range of concentrations, the EEBAC-silicon oil mixture undergoes an I - S_A transition on cooling, and the S_A phase first appears in the form of filaments[3].

Filaments have been observed in ternary mixtures of nematic, cholesteric and smectic liquid crystals[4,5], in mixtures of smectogens and alcohols[6,7], as well as in mixtures containing phospholipids[8]. Filaments have also been observed in pure liquid crystalline materials[9-13]. The chemical structure of the recently synthesized 4'-decoxyphenyl 4-(4-cyano-4-oxybiphenyl) butanoate[11] (DCOB) and its phase sequence are shown below.

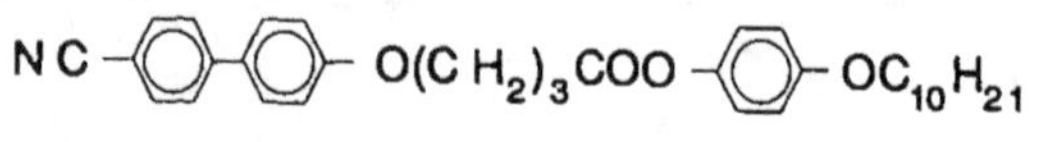

4'-decoxyphenyl 4-(4'-cyano-4-oxybiphenyl) butanoate (DCOB)

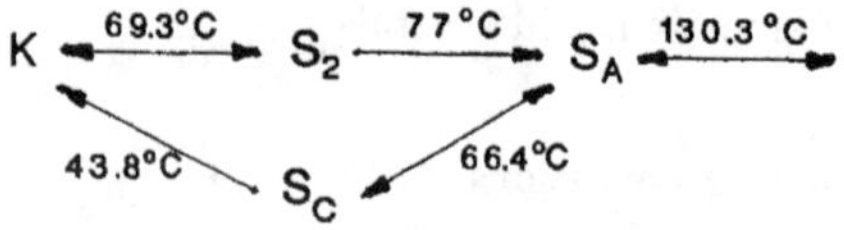

The phase sequence of DCOB.

Filaments were observed when DCOB was cooled from the isotropic phase into the S_A phase. The DCOB samples were contained in cells consisting of parallel glass plates separated by mylar spacers. The cells were placed in a Instec microscope stage, with ± 0.1°C temperature control. Using a Nikon polarizing microscope, the samples were observed in transmission using white light illumination. As the sample temperature was slowly lowered to values below the I - S_A transition, with typical sweep rates $\sim$1mK/s, filaments formed which were videotaped. A sequence of pictures showing the growth of a typical DCOB filament is shown in Fig. 1.

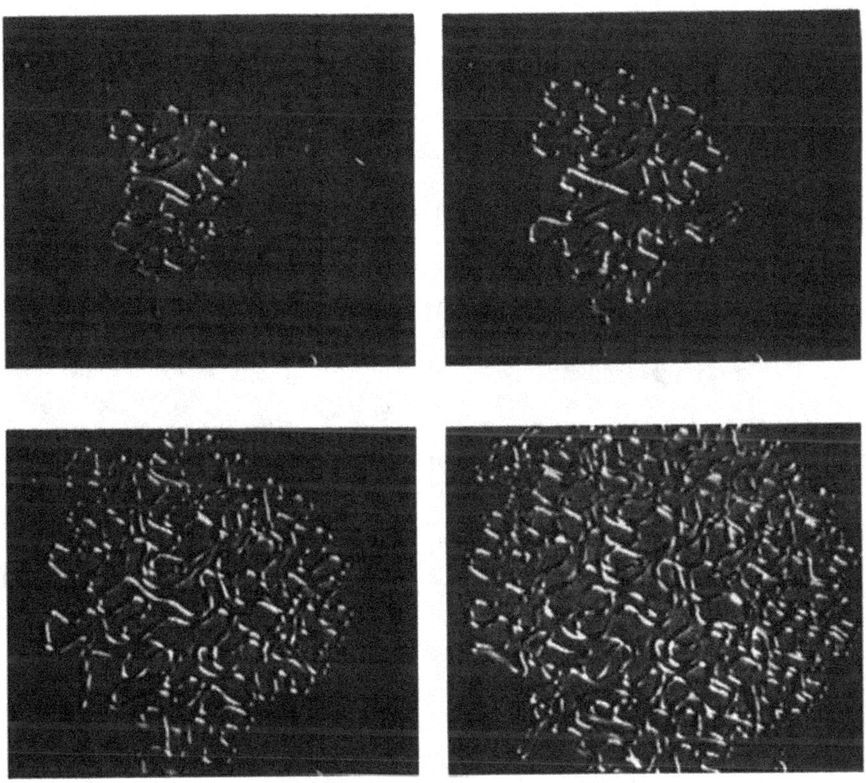

Fig. 1 Growth sequence of a S_A filament from the istropic phase of DCOB. The diameter is $\sim 2\mu$m; the cooling rate is 2mK/s.

The EEBAC-silicon oil and DCOB filaments exhibit similar behavior. The filaments grow rapidly, usually with no apparent change in thickness. Typical filament diameters are in the range of a few microns, but thicknesses both smaller and greater than this can also be found. During growth, the filaments buckle, and take on a serpentine form. The loops thus formed lengthen and further buckle, indicating that the growth takes place by local elongation, rather than by growth at the ends. When two filament segments approach each other during growth, the velocity of approach usually decreases as they get closer. As the filament grows and buckles, it eventually occupies an approximately disk shaped region in the cell. The filaments grow more rapidly near the edge of this region than close to the center. The outer envelope of the region containing the filament can be unstable during the growth of the filament, and its shape may resemble branching or dendritic patterns. These observations are consistent with diffusion limited growth. The filaments are apparently metastable, and eventually coalesce into compact domains, as shown in Fig. 2.

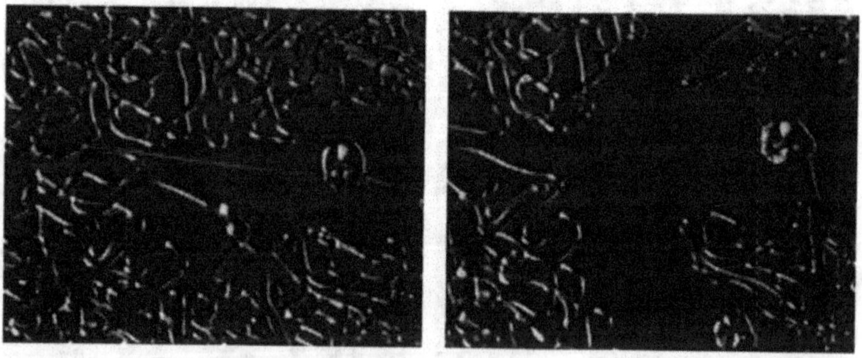

Fig. 2 The appearance of compact domains in DCOB which absorb the filament.

3. Microstructure and growth of filaments

The filaments show uniform birefringence colors. Our observations indicate that the optic axes are parallel and perpendicular to the filament axis, and that the ordinary refractive index is associated with polarization along the filament axis. The S_A phase is characterized by parallel layers, where the director $\hat{n}$, the unit vector along the direction of average molecular orientation, is parallel to the layer normal. The observed cylindrical symmetry leads us to assume that the filaments consist of concentric cylindrical smectic layers[2]. Since the elastic strain associated with splay in the center of the filament diverges, it is likely that the filaments have an isotropic core[10].

Our proposed model[3,10] of growth of a filament from the undercooled isotropic phase relies on the similarity of the S_A phase to both liquids and solids. We propose that, as in the case of a droplet growing in a condensing liquid, the filament absorbs molecules through its outer surface, at a rate which is proportional to its surface area. As in a perfect faceted crystal, there exists a nucleation barrier against the formation of new layers, and hence we assume that, after the initial formation, no new layers are added during the growth of the filament. Since all layers elongate at the same rate there must be a net transport of matter from the outer to the inner layers. This behavior is unlike that of either ordinary liquids or crystals.

If no new layers are formed, the outer radius r_o of the filament stays constant and the growth of the filament can be related to the particle 'solidification' current J_s through the filament surface. The rate of change of volume for a filament of length L is

$$\frac{\partial V}{\partial t} = 2\pi r_o L v_o J_s = \pi r_o{}^2 \frac{\partial L}{\partial t} \tag{1}$$

where v_o is the volume of a liquid crystal molecule. This gives at once $L = L_o e^{t/\tau}$, where the time constant for the growth is $\tau = \frac{r_o}{2v_o J_s}$. We further assume that the current of 'freezing' particles is controlled by the diffusion of latent heat in a perfectly pure material, or of impurities or solvent in a mixture. In this case, as in other models of solidification[14], the current is given by

$$J_s = -\frac{D}{v_o} \nabla u \tag{2}$$

where u is the dimensionless diffusion field, and D is the diffusion constant. The dimensionless field u satisfies the diffusion equation, with the boundary condition that at $r=r_0$,

$$u = \Delta = u_{SI} - d_o\kappa - \tfrac{1}{2}d_o d_1\kappa^2 \tag{3}$$

where u_{SI} is the dimensionless undercooling or supersaturation, κ is the curvature of the interface and d_o is the capillary length. (In the thermal model, $u_{SI} = \frac{T_{SI}-T_\infty}{L_H/c}$ and $d_o = \frac{\gamma_{\|} c T_{SI}}{L_H^2}$ where $\gamma_{\|}$ is the interfacial energy of the interface in the plane of the smectic layer, c is the specific heat, T_{SI} is the S_A- I transition temperature and L_H is the latent heat.) The terms on the r.h.s. of Eq. 3 involving κ are correction terms due to curvature; in addition to d_0 in the Gibbs-Thompson term, a new capillary length[16] $d_1 = \frac{K}{\gamma_{\|}}$, associated with the splay elastic constant, also enters the problem.

If the diffusion field is known, the current can be calculated from Eq. 2. Assuming steady state, and ignoring the Gibbs-Thompson corrections, we obtain $|\nabla u| \simeq -\frac{\Delta}{r_o}$, hence $J_s = \frac{D\Delta}{v_o r_o}$ and $\tau = \frac{r_o^2}{2D\Delta}$. We thus expect that thin filaments will grow fast, and that large undercooling will also result in rapid growth. Fig. 3 shows the logarithm of the length of a typical DCOB filament as a function of time[16], showing exponential growth over two decades.

Inside the filament, there exists a radial permeation current J_p of particles, transporting material to the inner layers. Since all layers must elongate at the same rate, the current is proportional to distance r from the filament axis. According to Darcy's law,

$$J_p = -\frac{k\rho}{\mu}\nabla P \tag{4}$$

where k is the permeability, ρ is the number density and μ the viscosity and P is the pressure. The permeation current at the surface of the filament is

$$J_p = \frac{k\rho P_0}{\mu r_0} \tag{5}$$

where $P_0 \sim \gamma/l_m$ is the permeation pressure and l_m is the molecular length.

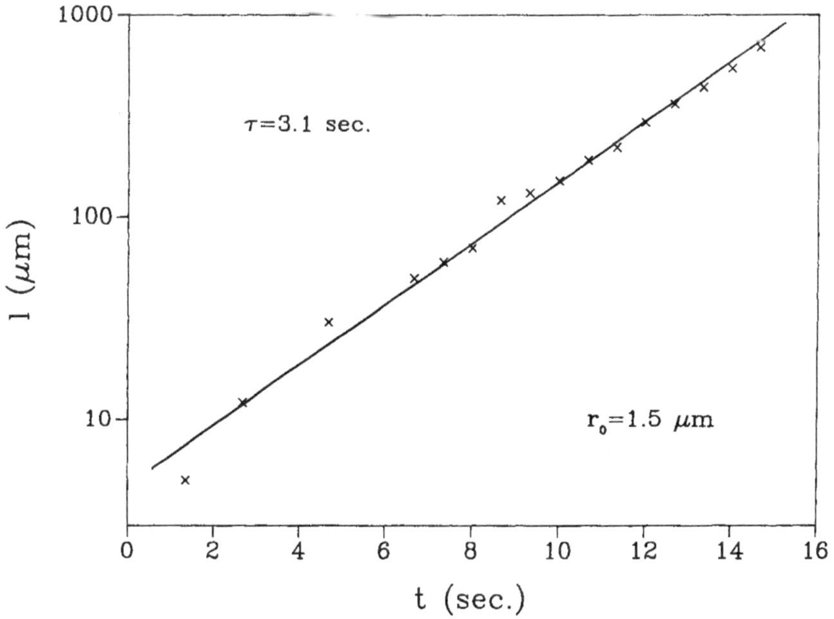

Fig. 3 Lenght of a filament in DCOB as function of time.

The solidification current is

$$J_s = \frac{D\rho\Delta}{r_0} \tag{6}$$

Equating the two currents at the filament surface gives $\Delta = \frac{kP_0}{\mu D}$ or, from Eq. 3, in the thermal model,

$$\frac{T_{SI}-T\infty}{L_H/c} - \frac{d_0}{r_0} - \frac{d_0 d_1}{r_0{}^2} = \frac{kP_0}{\mu D} \tag{7}$$

Eq. 7 suggests a possible selection mechanism for the filament diameter. During the early stage of filament formation, the filament radius r_0 adjusts itself until the solidification and permeation currents are equal. Since to first order both of these are proportional to $1/r_0$, the adjustment is via the correction terms containing the capillary lengths. From Eq. 7, if the r.h.s. is constant, if the temperature T_∞ is decreased, the filament radius of must also decrease. This is what is typically observed experimentally.

4. Shape and time evolution of a filament

To characterize the shape of a filament, we regard it as a space curve, of length L, defined by a position vector $\mathbf{R}(s)$ where s is the arc length measured along the curve. The unit tangent to the curve at point s is given by $\hat{\mathbf{T}}(s) = \frac{\partial \mathbf{R}}{\partial s}$, and we define the correlation function

$$g(r) = \langle \hat{\mathbf{T}}(s) \cdot \hat{\mathbf{T}}(s+r) \rangle \tag{8}$$

where the brackets <> denote the average along the curve. The characteristic wavelength of the filament undulations may be estimated from the first zero of g(r).

We next propose a simple model for the time evolution of a filament. The free energy contains contributions from elastic strain due to bending of the filament, and from the free energy change associated with the (non-equibrium) I - S_A transition.

If the filament consists of concentric smectic layers, $\nabla \times \hat{\mathbf{n}} = 0$, and the splay energy per length associated with the bending of the filament is $AK\kappa^2$ where $A=\pi r_0{}^2$ is the cross-sectional area of the filament, K is the splay elastic constant and κ is the curvature of the space curve representing the filament. If the free energy density difference between the isotropic and the S_A phase is Γ, then the free energy is given by

$$\mathcal{F} = A \int_0^L \{K(\frac{\partial^2 \mathbf{R}}{\partial s^2})^2 - \Gamma(\frac{\partial \mathbf{R}}{\partial s})^2\} ds \tag{9}$$

The thermodynamic force acting on an element of the filament is $-\delta\mathcal{F}ds$. We assume that inertial effects are negligible, and that this force is balanced by viscous drag on the filament. The viscous drag on a cylinder of length L moving through a fluid with velocity v and Reynolds number R_e is[17] $F_{drag}= \frac{4\pi\mu vL}{\ln(7.4/R_e)} \simeq \mu_e vL$ where μ_e is an effective viscosity. Equating the thermodynamic and viscous forces gives the equation of motion[18]

$$-\frac{\mu_e}{2AK}\frac{\partial \mathbf{R}}{\partial t} = \frac{\partial^4 \mathbf{R}}{\partial s^4} + \left(\frac{\Gamma}{2K} + \frac{3}{2}<\kappa^2>\right)\frac{\partial^2 \mathbf{R}}{\partial s^2} \tag{10}$$

where $<\kappa^2>$ is the mean squared curvature averaged over the filament length. Measuring length in units of $\sqrt{A}$ and time in units of $\mu_e A/2K$ gives the dimensionless form of Eq. 10

$$-\frac{\partial \mathbf{R}}{\partial t} = \frac{\partial^4 \mathbf{R}}{\partial s^4} + \beta\frac{\partial^2 \mathbf{R}}{\partial s^2} \tag{11}$$

where $\beta = \left(\frac{\Gamma A}{2K} + \frac{3A<\kappa^2>}{2}\right)$. In our case, $K\kappa^2/\Gamma<<1$, and the second term can therefore be neglected. Eq. 11 describes the time evolution of a filament. It is worth noting that it is similar to the dynamical equations describing the evolution of cosmic strings and superstrings[19]. It differs from these in that it contains a viscous rather than an inertial term, and curvature is neglected for superstrings, but not always for cosmic strings.

We next consider the fastest growing mode of an initially straight filament. Letting x=s we obtain the equation for the y-component of **R**

$$-\frac{\partial y}{\partial t} = \frac{\partial^4 y}{\partial x^4} + \beta\frac{\partial^2 y}{\partial x^2} \tag{12}$$

If $y=y_0 e^{\omega t - iqx}$, this gives

$$-\omega = q^4 - \beta q^2 \tag{13}$$

The wave number of the fastest growing mode is $q_{max}=\sqrt{\frac{\beta}{2}}$. Since Γ and hence β in general depend on concentration and temperature gradients and are not directly determined experimentally, it is useful to express β in terms of ω which gives $\beta^2 = 4\omega$. If the elongation is exponential in time with time constant $\tau = \frac{1}{2\omega}$, we obtain in reduced units $q_{max}= \left(\frac{1}{2\tau}\right)^{\frac{1}{4}}$ and in standard units

$$q_{max} = \left[\frac{\mu_e}{4\pi r_0^2 K\tau}\right]^{\frac{1}{4}} = \frac{1}{r_0}\left[\frac{\mu_e D\,\Delta}{2\pi K}\right]^{\frac{1}{4}} \tag{14}$$

which can be compared with experiment. We estimate $\mu_e = 10^{-2}$Pa·s, K = 10^{-12} N, τ = 1s, $r_0 = 10^{-6}$ m to give $q_{max} \simeq 10^5$ m^{-1} and the wavelength of the fastest growing mode is $\lambda \simeq 37\mu$m in reasonable agreement with observations.

5. Computer Simulations

We have carried out 2-d computer simulations using an algorithm based on Eq. 11. The filament is represented by a set of points. A randomly selected point is displaced by

$$\Delta \mathbf{R} = -\Delta t\left[\frac{\partial^4 \mathbf{R}}{\partial s^4} + \beta \frac{\partial^2 \mathbf{R}}{\partial s^2}\right]. \tag{15}$$

When the distance between adjacent points is greater than some threshold value, a new point is added to keep the density constant. The results of simulations are shown on Fig. 4.

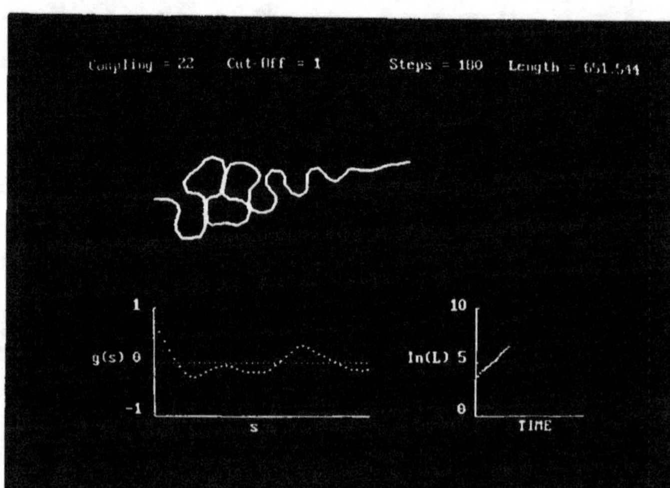

Fig. 4. Simulation of filament evolution. g(s) is the correlation function; ln(L) is the logarithm of the filament length.

Detailed results of simulations will be presented elsewhere. Here we have assumed that β is constant; that is, we have not taken into account the effect of spatial variation the external diffusion field. Work to include this in the simulations is currently under way.

6. Discussion and Conclusions

Experimental observations of filaments are consistent with the 'myelinic'[2] structure consisting of concentric smectic layers; in addition, we suppose the filaments to have an isotropic core. Our model for the growth is a process of 'solidification' frustrated by permeation, where newly forming smectic material 'freezing' from the isotropic phase onto the surface of the existing filament is absorbed into the filament interior. The surface tension associated with the interface separating the isotropic and S_A phases is anisotropic; the anisotropy is $\Delta\gamma = \gamma_{\parallel} - \gamma_{\perp}$ where $\gamma_{\parallel}$ and $\gamma_{\perp}$ are the surface tensions for interfaces parallel and perpendicular to the smectic layers. For a number of materials[20] $\Delta\gamma > 0$; but for the filament to be stable and for the permeation process to take place, we must have $\Delta\gamma < 0$. This has been confirmed for DCOB[21], and we conjecture that negative surface tension anisotropy is a necessary condition for the formation of filaments. The growth rate is limited by the diffusion of heat or impurities, exponential growth has been observed. Qualitatively, the time constant decreases as the undercooling is increased, in agreement with our prediction that $\tau = \frac{r_o^2}{2D\Delta}$. In addition, for one pair of coexisting filaments we have verified that $\frac{\tau_1}{\tau_2} = \left(\frac{r_{o1}}{r_{o2}}\right)^2$. The selection mechanism for the filament outer radius r_o is likely dynamical and r_o may depend on the capillary lenghts in the problem as suggested by Eq. 7. The details of this selection mechanism are not well understood. It is straightforward to estimate the radius of the inner isotropic core by minimizing the free energy; this gives $\sim 0.1\mu$m in the limit of both large and small undercooling which is not inconsistent with the typically observed filaments with outer diameters of a few microns. The shape of the growing filament is the result of competition between elongation, which takes place everyhwere along the length of the filament, and viscous drag as the filament moves trough the surrounding isotropic fluid. The dynamics of this process appear to be well described by the viscous string dynamics of Eq. 11.

Since β depends on temperature and concentration, we expect it to vary in space. This spatial variation is likely responsible for the observed instabilitics of the boundary of the region surrounding a long filament, as well as for the different growth rates observed near the center and at the edge of such a region. In our computer simulation we have neglected any spatial variations of β.

In addition to simple growth of filaments, a variety of complicated dyamical phenomena have been observed[3-13]. These include the rapid ejection of a thin filament from the end of thicker one, the ejection of filaments by compact domains, the flattening and splitting of filaments into ring-like structures covered with a smectic film and the appearance of 'beads' on filaments. All of these instabilities may be related to a single underlying effect arising out of the permeation process. The pressure inside the filament increases with distance from the axis, as suggested by Eq. 4. If this pressure is sufficiently large; that is, the surface area of the outer layers is 'too large' for the volume they enclose, an instability may arise which enables these layers to expand, relieving this pressure. The ejection of a filament by another[3] or by a compact domain[10] is such an instability; as is the flattening and splitting of a filament to form a ring. Conversely, if the temperature is suddenly increased, the permeation pressure outside the filament is reduced, and the unbalanced pressure inside leads to the formation of beads in a manner somewhat similar to aneurisms. The notion that the sudden enlargement of a closed surface enclosing a fixed volume leads to instabilities is well established in the case of biological systems. When spherical lipid bilayer vesicles are caused to expand in water, they exhibit 'budding' phenomena[22,23], where the parent vesicle ejects small spherical vesicles, tethered to the parent vesicle by filaments[24]. It is likely that the birth of S_A filaments is related to such an instability. Similar to the ejection of a filment by a compact domain, an initially spherical S_A germ consisting of concentric spherical layers could become unstable during growth as the pressure in the outer layers increases, giving rise to the instability which causes the formation of the filament. The filaments, because of their large surface area are metastable, and they eventually die by collapsing into compact domains. This collapse presumably takes place subject to layer-preserving constraints, but the detailed mechanisms are not yet understood.

Acknowledgements

We acknowledge useful discussions with A. Adamczyk, H.G. Dobereiner, N. Goldenfeld, M.A. Lee, N.V. Madhusudana, B.R. Ratna, W.G. Unruh and M. Wortis. We are grateful to S.L. Arora and R.A. Vora for supplying us with DCOB, to A.P. Dunmur and H. Zou for their help with the computer simulations, and we acknowledge the following grant support: NSF GMR9107013 (R.B.M.) and NSF DMR89-20147 (P.P.-M.).

References

1. G.W. Gray and J.W. Goodby, *Smectic Liquid Crystals: Textures and Structures*, (L. Hill,Glasgow 1984)
2. see P.G. de Gennes, *The Physics of Liquid Crystals* (Oxford University Press, Oxford, 1974) p.304
3. F. Jones, Ph.D. Thesis, Harvard University, 1972.
4. A. Adamczyk, *Mol. Cryst. Liq. Cryst.* **170**, 53, 1989
5. A. Adamczyk, (preprint - IV International Topical Meeting on Optics of Liquid Crystals, Cocoa Beach, Oct. 1991)
6. N.V. Madhusudana, (priv. comm., 1989).
7. R. Prathiba and N.V. Madhusudhana (preprint - paper presented at the Indo-French Conference on Membranes, Monolayers and Multilayers, Chantilly, France, May 1991)
8. S.A. Rudolph, B.R. Ratna and B. Kahn, *Lett. to Nature* **352**, 52, (1991).
9. A.J. Slaney and J. W. Goodby, *J. Mater. Chem.* 1, 5, (1991)
10. P. Palffy-Muhoray, S. Arora, A. Buka and R. Vora, 12th International Liquid Crystal Conference, Freiburg, Abstract IN06, August 1988
11. S.L. Arora, P.Palffy-Muhoray and R.A. Vora, Liq. Cryst. **5**, 133 (1989)
12. P. Palffy-Muhoray, March 1989 meeting of the APS, St. Louis, MO, *APS Bulletin* **34**, 495 (1989); '*Science News*', March 25, 1989
13. P. Palffy-Muhoray, H. Lin, B. Bergersen, R.B. Meyer and Z. Rácz, March 1991 meeting of the APS, Cincinnati, OH, *APS Bulletin* **36**, 820 (1991).
14. J.S. Langer, *Physica* **140A**, 44 (1986)

15. A.J. Simon and A. Libchaber, *Phys. Rev. A*, **41**, 7090 (1990)
16. R.B. Meyer, F. Jones and P. Palffy-Muhoray, 13th International Liquid Crystal Conference, Vancouver, Abstract PAT05 , July 1990.
17. G.K. Batchelor, *An Introduction to Fluid Dynamics* (Cambridge University Press, Cambridge, 1967) p. 247.
18. P. Palffy-Muhoray, B. Bergersen, H. Lin, R.B. Meyer and Z. Rácz, NATO ARW on Nonlinear Dynamical Structures in Simple and Complex Liquids, *Information Guide,* p28, Los Alamos, June 1990.
19. M.B. Gren, J.H. Schwarz and E. Witten, *Superstring Theory,* Vol. 1, Cambridge Monographs on Mathematical Physics (Cambridge University Press, Cambridge, 1987)
20. J.B. Fournier, I. Dozov and G. Durand, *Phys. Rev. A*, **41**, 2252, (1991)
21. X. Huang and P. Palffy-Muhoray (to be published).
22. E. Evans adn W. Ravitz, *Phys. Rev. Lett.* **64**, 2094, 1990
23. M. Wortis, U. Seifert, K. Berndl, B. Fourcade, L. Miao, M. Rao and R.K.P. Zia, in Proceedings of the workshop on '*Dynamical phenomena at interfaces, surfaces and membranes'*, Les Houches, Feb. 1991 (ed. D. Beysens, N. Boccara and G. Forgacs)
24. M. Wortis and H.G. Dobereiner, (private communication).

SOLITON TRAINS IN ANNULAR LONG JOSEPHSON JUNCTIONS: EXPERIMENTAL STUDY

A. V. USTINOV[1]

Institute of Solid State Physics, Russian Academy of Sciences
Chernogolovka, Moscow district, 142432 Russia
and
Lehrstuhl Experimentalphysik II, University of Tuebingen
D-7400 Tuebingen, Germany

Abstract

The annular Josephson junction allows for the study of undisturbed motion of sine–Gordon solitons (flux quanta) in the absence of boundaries and soliton–soliton collisions. The theoretical model based on the perturbed sine–Gordon equation is briefly reviewed. The major discussion in this contribution concerns recent experiments in which the dynamics of a *chain* of solitons trapped in the annular Josephson junction has been studied. The low temperature electron microscopy technique has been used in these experiments in order to trap in a controlled way one by one several solitons in the junction. The comparison of the experimental results with existing multi–soliton perturbation theory is presented. A symmetry–breaking mechanism due to soliton bunching as a relevant explanation for a crossover observed in the dynamical behavior at high velocities is discussed.

1. Introduction

Sine–Gordon solitons play an important role in physics as well–known examples of nonlinear wave motion. In solid state physics the sine–Gordon solitons appear in many different systems (magnetics domains, dislocations, charge density waves, etc.), but the closest correspondence to the pure sine–Gordon dynamics is found for solitons in long (quasi–one–dimensional) Josephson junctions[1]. In a long Josephson junction the soliton is a magnetic flux quantum (fluxon), which is able to move along the junction between its superconducting electrodes. The system is described by the perturbed sine-Gordon equation[2]

$$\varphi_{xx} - \varphi_{tt} = \sin\varphi + \alpha\varphi_t - \beta\varphi_{xxt} + \gamma\,, \tag{1}$$

[1]Present address: Physics Laboratory I, Technical University of Denmark, DK-2800 Lyngby, Denmark

where $\varphi(x,t)$ is a spatially and time dependent superconducting phase difference between the electrodes of the junction, the spatial coordinate x is normalized to the Josephson penetration depth λ_j, the time t is normalized to the inverse plasma frequency ω_0^{-1}, α and β are the dissipation coefficients, and γ is the bias current (driving force).

In case of a conventional topology of a long Josephson junction, a moving soliton cannot avoid collisions with other solitons and junction boundaries. This strongly complicates the analysis and interpretation of the experimental data. Davidson *et al.*[3] have proposed to use experimentally ***an annular, or ring-shaped, Josephson junction geometry.*** In this case the boundary conditions for Eq.1 are periodic

$$\varphi(\ell,t) = \varphi(0,t) + 2\pi n\,, \quad \varphi_x(\ell,t) = \varphi_x(0,t)\,. \tag{2}$$

Here $\ell = L/\lambda_j$ is the normalized length (circumference) of the junction, and n is the number of trapped solitons. Davidson *et al.* have successfully trapped ***a single soliton*** in such a junction and studied its dynamics[3, 4]. An impressive result obtained in Ref.3 was the experimentally-measured force-vs-momentum curve for a soliton, which can be considered as a relativistic particle moving in a solid-state system. Unfortunately, the technique used in Ref.3 and 4 allowed the authors to trap only occasionally a single soliton in the ring, so that it was not possible to study multi-soliton dynamics in their experiment.

The ***multi-soliton states*** for the perturbed sine-Gordon system with periodic boundary conditions have been analyzed theoretically by Marcus and Imry[5]. They calculated analytically the slope of the linear foot of the I-V characteristics in the following form:

$$\gamma = \frac{4\alpha}{\pi}\frac{E(k)}{k}\frac{v}{\sqrt{1-v^2}}\;; \quad L/n = 2kK(k)\;, \tag{3}$$

where $K(k)$ and $E(k)$ are the complete elliptic integrals of the first and the second kind with the modulus $0 < k < 1$, v is the soliton velocity normalized by $\bar{c}$ - the maximum velocity of the electromagnetic waves in the junction (Swihart velocity).

In this contribution I review our recent experiments in which the multi-soliton dynamics of a ***chain*** of solitons trapped in the annular Josephson junction has been studied[6, 7]. In order to trap solitons in the annular junctions the technique of low temperature scanning electron microscopy[8] (LTSEM) has been used.

2. Experimental results

For the measurements we have used annular Josephson tunnel junctions of "Lyngby" geometry proposed for the first time in Ref.3. A sketch of the junction is presented in Fig.1a. The niobium-lead junctions of two slightly different geometries have been used. First[6], we have used a simple two-layer overlap structure equivalent to that one used in Ref.3,4. Later investigations[9] showed a substantial inhomogeneity of the tunnel barrier in the junctions of this geometry. In order

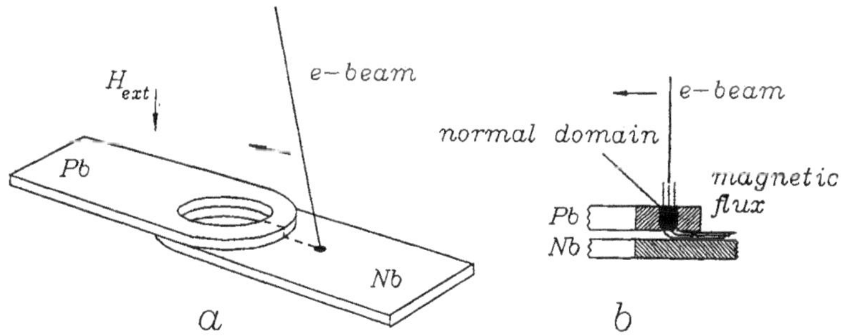

Figure 1: a) The magnetic flux trapping procedure in an annular junction using LTSEM. b) The cross–section of the junction along the dashed line shown in Fig.a at the moment of crossing by the electron beam. The insulating window is not shown.

to improve the homogeneity of our junctions we have added the insulating silicon monooxide layer between two superconductors, so that the edges of the superconducting films were covered by the insulator. The additional insulating layer formed the annular window with the tunnel barrier inside it. This technique has substantially improved the homogeneity of the junctions[7]. The results presented here have been obtained for the junctions of the second (window–type) geometry.

Three annular junctions with the radius of the ring $R = L/2\pi = 100\ \mu m$ and with the width of the tunnel barrier $W = 20\ \mu m$ were made on the substrate. The sample temperature was about 5 K. The Josephson penetration depth λ_j was approximately 70 μm, so that the normalized length of the junction was $L/\lambda_j \approx 9$. The condition $W < \lambda_j$ allowed us to consider the fluxon motion in the tangential direction as a quasi–one–dimensional one in the first approximation.

The sample was attached to the top side of a sapphire disk using high thermal conductivity glue. A compact bifilarily wound heater was also glued on the sapphire disk close to the sample. During the experiments the bottom side of the sapphire disk was in direct contact with liquid helium, whereas the top side carrying the Josephson junctions was exposed to the vacuum of the scanning electron microscope. In order to remove trapped magnetic flux from the sample it was heated above the critical temperature of Nb using the heater at the sapphire. The external magnetic field H_z was applied in the direction perpendicular to the substrate plane. The dc current–voltage (I–V) characteristics were accurately measured by a nanovoltmeter and the data collected by a computer.

The focused electron beam was used as a tool for trapping magnetic flux in the annular junctions. The crossing of the top electrode by the electron beam focus (see Fig.1a) resulted in trapping the magnetic flux in the tunnel barrier. This was due to the local heating of the sample in the focus of the beam and the presence of a small applied magnetic field H_z perpendicular to the plane of the junction. The

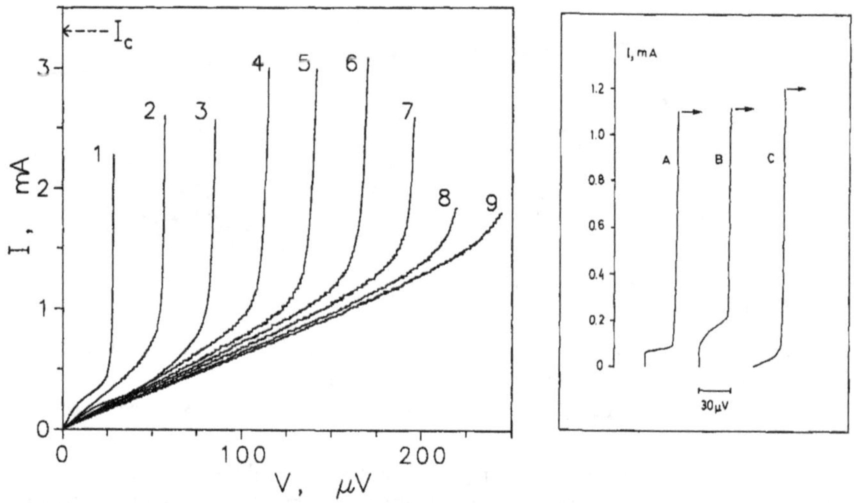

Figure 2: a) The I–V curves corresponding to the sequential trapping of solitons in the annular Josephson junction. The numbers correspond to the numbers of solitons trapped. b) The first soliton step (A) obtained after crossing the junction by the electron beam. The curves B and C is the same step after the irradiation of the sample area around the junction by a defocused beam with lower power.

normal domain carrying the magnetic flux was moved across the lead film together with the electron beam focus as shown in Fig.1b. Once the electron beam focus has reached the interior of the annular junction, the magnetic flux remained trapped inside the superconducting lead ring. Because there appeared no magnetic flux trapped inside the niobium ring, all the flux trapped inside the lead ring was going out through the tunnel barrier of the junction. It means, that *after the electron beam focus had crossed the junction area, a certain number of magnetic flux quanta remained trapped in the annular Josephson junction.* The details of this technique are presented elsewhere[6].

Fig.2a shows the experimental $I-V$ characteristics for the trapping of solitons in the annular Josephson junction obtained by a sequential crossing of the junction with the electron beam. Typically, each crossing resulted in trapping a one flux quantum in the ring. The critical current of the junction without trapped flux was about 3.3 mA. The resonant soliton steps corresponding to different numbers of unipolar solitons n are observed. The $I-V$ characteristics has been found to be very symmetrical with respect to the point $I, V = 0$.

With the properly adjusted electron beam parameters the magnetic flux trapping procedure worked in a quite reproducible way. The electron beam focus was placed on the niobium film outside the junction and moved towards the junction

(Fig.1a). After crossing the junction area the beam was stopped in the opening inside the ring and then blanked. In most of cases such a procedure resulted in trapping one soliton in the junction. When the process was repeated the steps with increasing number n were observed. The reversing of the magnetic field H_z during the trapping sequence resulted in a decrease of the step number. This corresponds to the trapping of solitons of the opposite polarity and their annihilation one by one with those already trapped in the ring. By heating the whole substrate above the critical temperature of the superconducting films it was easy to remove all trapped flux from the junction after each trapping procedure.

The voltages of the steps are quantized as expected, according to the formula[1]:

$$V_n = \frac{n\Phi_0\bar{c}}{L}, \tag{4}$$

where Φ_0 the magnetic flux quantum. At the step voltage V_n the solitons move under the influence of the bias current I as relativistic particles with the velocity close to $\bar{c}$.

If solitons are trapped in an ideally homogeneous annular junction theoretically the non–dissipative critical current I_c should be equal to zero. This is perfectly fulfilled for all the curves in Fig.2a and a good proof for a high homogeneity of our junctions. Sometimes during the trapping procedure a certain critical current (typically $5-10\%$ of I_c) was observed. We suppose, that it was happening due to a redistribution of the magnetic flux occasionally trapped in the superconducting electrodes under the influence of the electron beam heating effect. Fig.2b shows as an example the first step (A) obtained after crossing another junction by the electron beam, which displays a critical current of about $50\,\mu$A. The curves B and C show the same step after the irradiation of the sample area around the junction by a defocused beam with lower power. Obviously, the critical current is strongly sensitive to such an influence. The flux–free critical current I_c of this junction was about 2.8 mA. For the curve C the critical current is at least three orders of magnitude lower than I_c. This corresponds to pinning–free motion of the soliton trapped in the junction.

3. Discussion

We have found the soliton steps in I–V curve to be extremely stable even with respect to the gap voltage state of our tunnel junctions. Due to unavoidably quite high electromagnetic interference and external noise while measuring in the scanning electron microscope, current jumps from the soliton step to the gap voltage and back were occasionally observed only at high currents close to the top of the step. At lower currents we found the soliton steps to be a kind of *globally stable dynamical states* of the system. This inherent property even simplified the measuring procedure considerably, so that highly accurate measurements of the voltage with long–time averaging using a nanovoltmeter were possible. This property in contradiction to conventional long Josephson junctions, where the soliton steps are rather unstable and their measurements require special precautions and shielding.

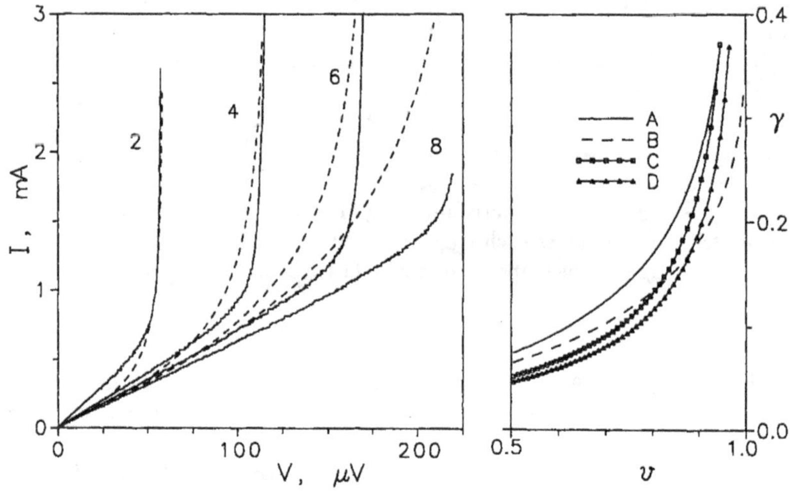

Figure 3: a) The dashed curves represent a fit of Eq.3 to the experimental I–V curves with $n = 2, 4, 6, 8$ from Fig.2a. b) The modification of a simple relativistic I–V curve shape according to Eq.3 with $\alpha = 0.1$ (A) due to taking into account a finite width of the ring $\Delta R/R = 0.2$ (B), β–losses (C) with $\alpha = 0.05$ and $\beta = 0.05$, and both the β-losses and the finite width of the ring (D).

Fig.3a shows the bottom parts of several steps from Fig.2a in slightly magnified current scale. With increasing step number the steps show a tendency to approach asymptotically a slope that most probably corresponds to the McCumber linear slope. This is in agreement[10] with the multi–soliton theoretical formulas by Marcus and Imry[5] already mentioned in the introduction. The dashed curves in Fig.3a show a two–parameter fit of Eq.3 to the experimental I-V curves. The tendency to reach an asymptotic linear slope with the increasing number of solitons n (decreasing spacing $L/n\lambda_j$ between them) is similar both for theory and experiment. However, in the crossover region between the linear slope and the relativistic vertical parts of the steps the deviations of the experimental data from theory are evident.

There may be at least two reasons which to a reasonable degree can explain this discrepancy between theory and experiment. First, due to the finite width of the junction the rotating soliton has a difference in velocity at the outer and the inner radius of the ring. This difference is relatively large (about 20% for our junctions) and is the most important at high velocities. The corresponding correction to the I–V curve can be calculated by averaging the velocity–dependent losses over the width of soliton[7]. Fig.3 shows the shape of the I–V curve (A) modified by taking into account a finite width of the ring $\Delta R/R = 0.2$ (B). Secondly, Eq.3 does not include the surface losses corresponding to the well–known[11] β–term in Eq.1. The curve (C) shows the modified shape of the I–V curve obtained by taking into account the

β-losses. The curve (D) corresponds to the case when both the β-losses and the finite width of the ring are taken into account. One can see, that these corrections of the theoretical curve from (A) towards (D) work in the direction which improves the agreement between theory and experiment.

An unexpected feature which has been predicted by Marcus and Imry[5] for an annular junction but never observed in experiments before was the *increase of the step height I_n with the step number n.* We note a similar tendency for our experimental results shown in Fig.2a. In fact, the real height of these steps was even larger than shown in Fig.2a but the jumps to the gap voltage did not allow us to record the real top of each step. Nevertheless, the step height displays the tendency to increase with n, the highest steps with $n = 4, 5, 6$ are approximately as high as the critical current. To our knowledge this has never been observed experimentally before. In conventional long Josephson junctions the step height always decrease with n. We believe that this difference between linear and annular junctions is due to already mentioned global stability of the soliton steps in an annular system.

Let us now discuss the details of high-current regions of the steps in Fig.2a. Fig.4 shows the voltage per fluxon V_n/n and displays the top parts of the steps in a magnified voltage scale. The value V_n/n is proportional to the average velocity v of fluxons moving in the junction, so from Fig.4 we can compare the dependences of v on the external driving force $I \sim \gamma$ for different densities of fluxons in the system. With increasing the step number a certain crossover in the asymptotic step voltage behavior is clearly seen. At high current I the normalized voltage V_n/n for the steps with $n > 4$ is considerably lower than for the steps with smaller n. In principle, a certain monotonic decrease of v with increasing n is expected for the Marcus and Imry theory[5]. It can be explained as an increase of the effective damping parameter $\alpha E(k)$ as k decrease with n. However, it is evident from Fig.4 that the changes of the curves with increasing n cannot be described by rather smooth one–parameter scaling only. In particular, at certain fluxon velocities (shown by arrows in Fig.4) the steps with $n = 3$ and $n = 4$ display a kind of shift to higher voltages, which is not expected from theory[5].

The following explanation of this crossover at high velocities may be suggested. As known from computer simulations[12], the dissipation due to the β-losses at high velocity generates an effective attraction between unipolar solitons and results in bunching in a localized soliton train moving in the junction. The bunching effect (Fig. 5) breaks the symmetry in the annular system and "helps" the chain of solitons to overcome the dissipative losses. Their average velocity becomes higher than that of a single soliton at the same bias current. The β-damping term in Eq.1 determines the soliton profile at high bias current[11] which can explain the bunching mechanism at high velocities[13, 14]. We suppose, that compared to the one–soliton velocity the higher soliton chain velocity observed in our experiment may be due to the bunching of the trapped solitons at the high current. In this case, the point of soliton bunching in the I-V curve is also the point of a *symmetry breaking in the system with periodic boundary conditions.*

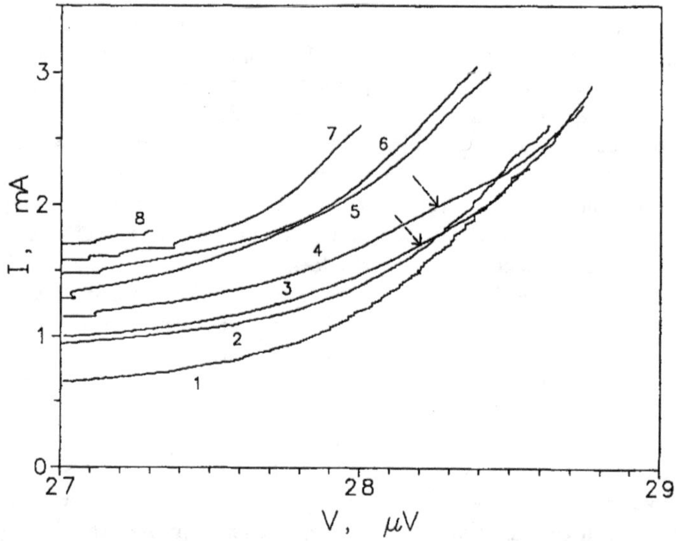

Figure 4: The detailed comparison of the high–current parts of the scaled soliton steps in a magnified voltage scale (the voltage of the n-th step is divided by n). The arrows show approximately regions where the steps $n = 3$ and $n = 4$ start to deviate towards high voltages.

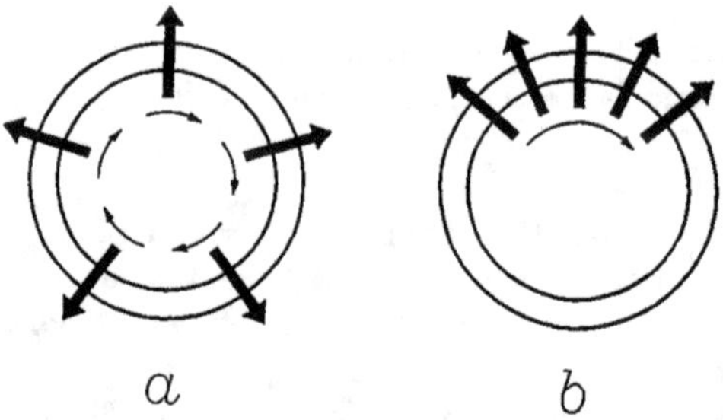

Figure 5: The illustration of the unbunched state (a) and the bunched state (b) of solitons in an annular Josephson junction.

A certain lack of symmetry in the system due to the inhomogeneous biasing and the influence of the superconducting electrode planes should of course influence the bunching. It is not clear, why we do not see a similar crossover for the step with $n = 2$. It might be possible, that bunching does not happen due to the higher degree of symmetry of the state with $n = 2$, namely that in addition to the intrinsic rotational symmetry of the undeformed chain of n solitons with the axis of the n–th order, the electrodes surrounding the junction (see Fig.1a) have also the plane of symmetry which makes the unbunched configuration of two solitons more favorable.

Malomed[15] has recently shown, that in annular system the threshold velocity of soliton bunching is increasing with increasing the soliton number n. Qualitatively, this can be explained in the following way. The dissipation due to the β–losses at high velocity changes in the soliton profile $\varphi(x - vt)$ in a way that an oscillating tail appears behind the moving soliton. The distance Δ from the soliton center of mass to the first local minimum of the tail behind the soliton is decreasing with increasing the soliton velocity. The next soliton in the chain may be trapped in a local magnetic field minimum produced by the tail of the first soliton. The bunching only becomes possible when Δ is smaller then the initial spacing between solitons in the chain $L/n\lambda_j$. It means, that the minimum velocity value where the bunching may appear is increasing with increasing the soliton number n. The experimental curve for $n = 4$ displays the increase of the soliton velocity (which is proportional to the voltage) being at higher current than that one for $n = 3$, which agrees with theoretical prediction. This explanation is relevant for case $L/n\lambda_j \gg 1$ when solitons a well separated. So, this is not a big surprise that we do not see any distinct crossover for $n > 4$, because in those cases $L/n\lambda_j$ is of the order of 1 and the single–soliton approach should not work. It is not clear, whether the bunching is possible at all for very high chain density. Because both coefficients α and β strongly depend on temperature, a more detailed study of trapped solitons in annular junctions at different temperatures seems to be very interesting in further experiments for quantitative comparison with theory.

4. Acknowledgments

I am grateful to N. F. Pedersen for very valuable discussions of the ideas and results presented here already at the stage of preparation of the manuscript. I would like to thank T. Doderer, M. Cirillo, R. P. Huebener, B. A. Malomed, K. Nakajima, S. Pagano, R. D. Parmentier, S. Sakai, and I. V. Vernik for many stimulating discussions. The support of this work by the Alexander von Humboldt Foundation is gratefully acknowledged.

References

[1] N. F. Pedersen, in *Josephson Effect - Achievements and Trends*, World Scientific, Singapore, 1986, edited by A. Barone.

[2] D. W. McLaughlin and A. C. Scott, Phys. Rev. A **18**, 1652 (1978).

[3] A. Davidson, B. Dueholm, B. Kryger, and N. F. Pedersen, Phys. Rev. Lett. **55**, 2059 (1985).

[4] A. Davidson, B. Dueholm, and N. F. Pedersen, J. Appl. Phys. **60**, 1447 (1986).

[5] P. M. Marcus and Y. Imry, Solid St. Commun. **33**, 345 (1980).

[6] A. V. Ustinov, T. Doderer, B. Mayer, R. P. Huebener, and V. A. Oboznov, Europhys. Lett. (submitted).

[7] A. V. Ustinov, T. Doderer, B. Mayer, R. P. Huebener, N. F. Pedersen, and V. A. Oboznov (to be published).

[8] R. P. Huebener, Rep. Prog. Phys. **47**, 175 (1984).

[9] A. V. Ustinov, T. Doderer, B. Mayer, R. P. Huebener, A. A. Golubov, and V. A. Oboznov (to be published).

[10] This fact has been for the first time explicitly pointed out by N. F. Pedersen.

[11] A. Davidson, N. F. Pedersen, and S. Pagano, Appl. Phys. Lett. **48**, 1306 (1986).

[12] W. J. Johnson, Ph. D. Thesis, University of Wisconsin (1968), unpublished.

[13] S. Sakai and N. F. Pedersen, Phys. Rev. B **34**, 3506 (1986).

[14] S. Sakai, Phys. Rev. B **36**, 812 (1987).

[15] B. A. Malomed, unpublished.

NONLINEAR OSCILLATION AT MEMBRANE AND INTERFACE

KENICHI YOSHIKAWA
College of General Education, Nagoya University
Nagoya 464-01, Japan

ABSTRACT

Oscillatory chemical phenomena at interface and membrane were studied on three different experimental systems: (1) Self-pulsing at an oil-water interface, (2) Nonlinear capacitance and resistance of an electrode, (3) Application of nonlinear dynamics to gas sensing.

1. Introduction

Rhythmic phenomenon is one of the most essential aspects in living organisms. There are many studies on rhythms in biological systems, e.g., oscillations in mitochondria, oscillations of cells of nerve and muscle including heart, circadian rhythms. Recently biochemical techniques, especially in molecular biology, have developed rapidly. However, biochemical treatments, such as extraction, purification and identification of various biological molecules, inevitably cause destruction of the system, and therefore destroy the rhythms. Biochemical approach on the bio-rhythms, thus, exhibits a fatal problem. Under these circumstances, it becomes important to study model systems showing various nonlinear characteristics such as excitation and oscillation. In the present paper we would like to describe on three kinds of artificial systems with high "nonlinear characteristics".

1) Self-pulsing phenomenon at an oil-water interface. This is a self-excitatory phenomenon in the presence of various surface active compounds. The driving force of the oscillation is the chemical potential difference of chemical substances, such as surface active compounds and alcohols, between the oil and water phases. In this system, one can observe pulsations of electrical potential. It will be demonstrated that the time-series of the electrical pulses contain abundant information on chemical substances present in the aqueous phase.

2) As for the second subject, a new method to evaluate the nonlinear impedance in an electrochemical system, composed with electrodes and an aqueous solution, will be given together with some experimental results. It was found that the nonlinearity of the impedance, in other words, the voltage dependence of

capacitance and conductance of an electrochemical system, changes greatly depending on the chemical nature of the substance present in the aqueous solution.
3) A novel method of gas sensing is proposed based on the information from the nonlinear dynamic response of a ceramic gas sensor. A sinusoidal voltage is applied to a heater equipped with an SnO_2 gas sensor and the resulting output resistance is analyzed by fast Fourier transformation (FFT). It is found that the intensities of the higher harmonics exhibit characteristic changes in the presence of different kinds of gas.

2. Chemical Sensing by Excitable Artificial Membrane[1]

Self-agitation of the interface is observed in an oil-water system accompanied by mass-transfer across the interface. This phenomenon is well-known as the Marangoni effect.[2] In general, the movement of the interface arising from the Marangoni instability is depressed in the presence of surfactant molecules. In contrast to this, more than a decade ago, Dupeyrat and Nakache[3] found quasi-periodic variations of a relaxation type in the interfacial tension and the electrical potential in an oil/water system in the presence of a cationic surfactant. Recently, we have found sustained rhythmic oscillations are generated in a liquid membrane consisting of water/oil/water phases, where one of the aqueous phases contain a cationic surfactant[4,5] or an anionic surfactant[6,7] or a phospholipid.[8] As the reproducibility of this oscillations in this liquid membrane is excellent, we could quantitatively study the effect of various chemical species on the oscillations.

Experiments were performed in an apparatus in a U-shaped glass tube, 12mm inner diameter .Fig.1 shows the experimental result on the response to various taste-stimuli, i.e., salty, sweet, bitter and sour compounds.[7] In the presence of sodium chloride, small rhythmic oscillations were generated together with giant oscillations of longer period (Fig.1a). When sucrose was added to the aqueous phase, chaotic oscillations were observed, as is shown in Fig.1b. When a bitter compound such as quinine chloride is added, downward pulses are generated (Fig.1c). In the presence of acid, i.e., sour compound, small upward pulses are observed with the large change of the electrical potential for the base line (Fig.1d). It is quite interesting to note that the oscillatory characteristics,

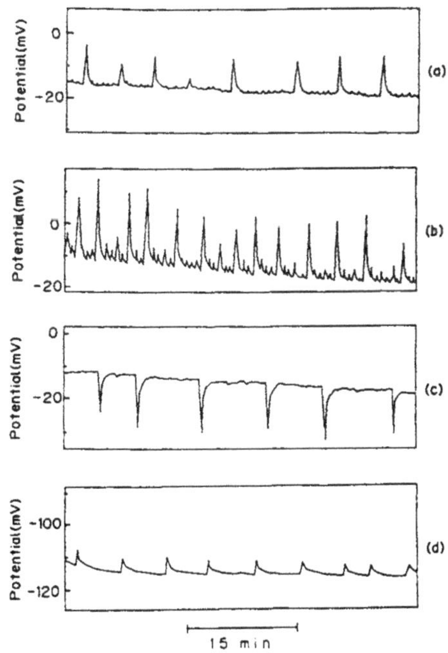

Fig.1 Oscillations of electrical potential by the addition of 0.5 mL of (a) 0.1 M sodium chloride, (b) 1M sucrose, (c) 0.001 M quinine, and (d) 0.03 M hydrochloric acid, to the aqueous phase containing oleate molecules.[7]

such as the amplitude, the frequency and the shape of pulses, change remarkably, depending on the nature of the chemical stimulant. It has been confirmed that other chemical stimulant for taste sensation also induce similar characteristic effects on the behavior of the oscillation, their nature being dependent upon the individual taste-categories, as has been shown in Fig.1.

We have proposed a plausible kinetic model.[7,9] Let X_i, Y_i be the concentration of the surfactant and alcohol near the interface and let Z_i be the concentration of the surfactant-alcohol aggregate or complex at the interface. X_b and Y_b are the concentrations of the surfactant and alcohol, respectively, in the bulk aqueous phase. The system kinetics may be expressed using the following differential equations:[7,9]

$$\frac{dX_i}{dt} = D_X(X_b - X_i) - k_1 Z_i \qquad (1)$$

$$\frac{dY_i}{dt} = D_Y(Y_b - Y_i) - k_2 Z_i \qquad (2)$$

$$\frac{dZ_i}{dt} = k_3 F(X_i\,,\, Y_i) - k_4 G\,(Z_i) \qquad (3)$$

Fig.2 shows the numerical results of the above eqs.(1)(2)(3). The first term on the right-hand side of eqs.(1) and (2) show the rate of diffusion. The second terms correspond to the negative feedback of Z_i. The first term in eq.(3) is the rate of formation of the aggregate of complex between the surfactant and alcohol molecules. As the surface pressure of the monolayer is expected to be proportional to the two-dimensional density of surfactant puls alcohol, the synergetic effect of surfactant and alcohol is assumed to be the simple addition of X_i and Y_i in eq.(3). This assumption is related to the speculation that the second term in eq.(3) also involves the specific pressure-concentration relationship of the monolayer. This term represents the rate of migration of the surfactant-alcohol aggregates from the interface to the bulk organic phase. According

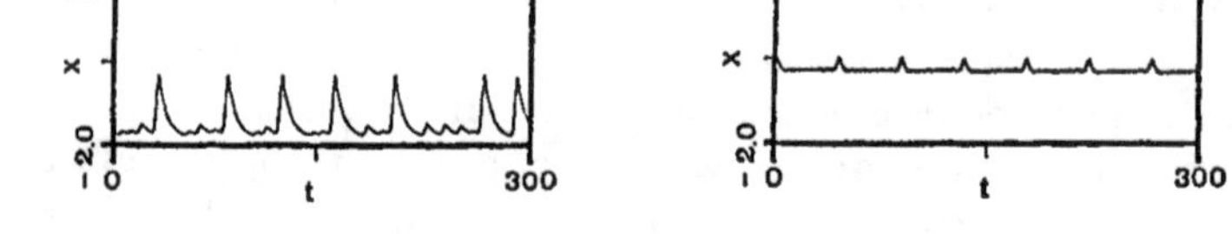

Fig.2 Computer simulation of the electrical oscillation.7,9

to the two-dimensional "van der Waals" equation for the surfactant molecule at the interface, and "N"-shape relationship for the surfactant pressure versus the concentration of surfactant is expected.[9,10,11] It is thus assumed that $G(Z_i)$ exhibits N-type nonlinearity, which is the most essential factor to generate the oscillation at the oil-water interface.

3. Chemical Sensing Based on the Nonlinearity in an Electrochemical System[12,13]

In this section, we shall describe a new sensing system for detecting and quantitating various chemical species, which is based on the information of electrochemical nonlinearity.

Various electrochemical sensors have been developed, such as a pH electrode and selective electrodes for inorganic-metal cations. In these sensors, chemical information, such as structure or activity, is converted into a DC voltage (or DC current) difference. As well as measuring the DC voltage or current, it may be useful to obtain chemical information from the capacitance or resistance of a test solution. In general, the capacitive component in an electrochemical system cen be classified into two components: the dielectroricity of the bulk of the solution and the capacitance on the electrode-surface originating from the electrical double layer of ion and electrosorption at the surface of the electrode. Though there have been many physico-chemical studies on the capacitance in electrochemical systems, there have been only few practical application. This may arise from the unsatisfactory reproducibility of the observed capacitance, caused by the hysterisis and variation of the surface area, including bubble formation on the electrode during its handling. It should be noted that the capacitance usually depends on the electrical potential applied to the electrode.

A purely sinusoidal voltage generated by a function generator, was applied to the electrode through a potentiostat. The AC voltage was applied at a relatively low frequency, 1Hz, because the capacitive component due to the bulk dielectroricity of the solution may be dominated at the higher frequencies. The output AC current was Fourier transformed into the frequency domain by use of a personal computer.

Fig.3 shows the change of the current-voltage, I-V, curve in a 1mM sodium oleate aqueous solution caused by the addition of the following taste compounds, (A) hydrochloric acid as a sour compound, (B) ammonium chloride as a salty compound, (C) sucrose as a sweet compound, and (D) nicotine as a bitter compound.[13] It is interesting to note that the effects of the taste-compounds are apparently different from each other. The caracteristics in the I-V curves can be transformed into the characteristics in voltege-dependent impedance in the electrochemical system. Detailed discussion has been given in a separate manuscript.[13]

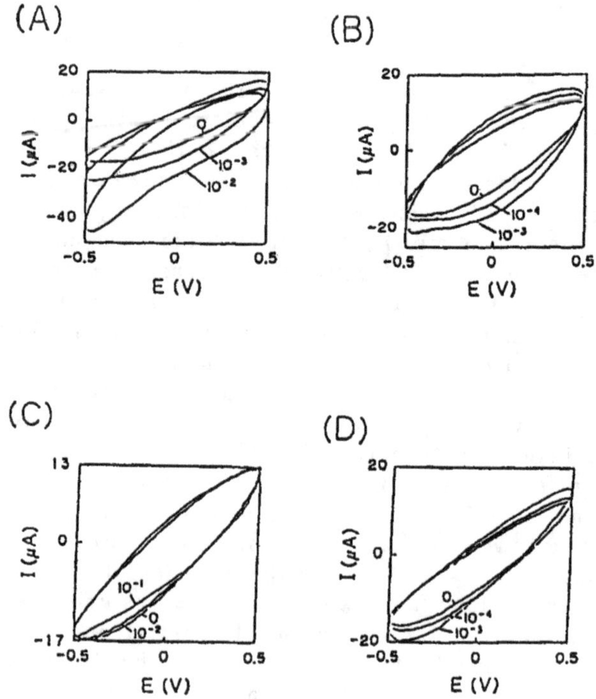

Fig.3 Current-voltage trace for the aqueous solutions, (A) hydrochloric acid, (B) ammonium chloride, (C) sucrose, and (D) nicotine. The numbers given in the figure are the concentrations of the chemical species.[13]

4. Gas Sensing Based on Nonlinear Response[14,15]

We would like to propose a new strategy for the development of gas sensor, which is based on the information of nonlinear, dynamic property of a ceramic gas sensor.In Fig.4, the experimental apparatus is schematically shown. A sinusoidal voltge (amplitude: 2V and frequency: 0.02Hz) is applied to the heater of a tin oxide gas sensor, TGS813, commercially available from Figaro Engineering Inc., Osaka. Fig.5 shows the resistance *vs.* voltage (R *vs.* V) curve, for different odor. The response curve for the different odors are apparently different each other. In order to analyze the characteristics of the response curve in a quantitative manner, time tracing of the resistance was Fourier transformed to the frequency domain. To minimize the experimental error, the intensities of the

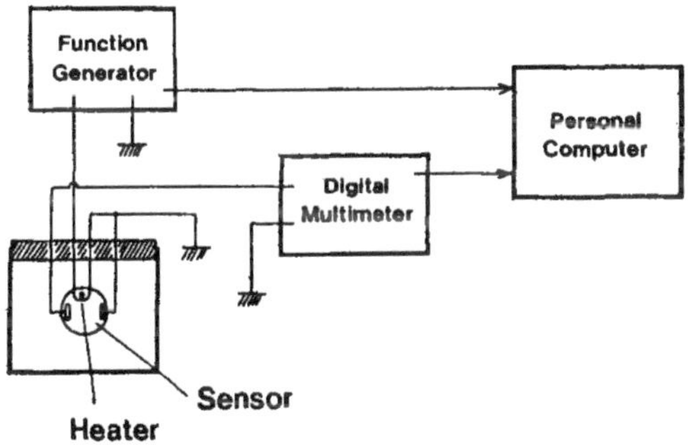

Fig.4 Experimental apparatus

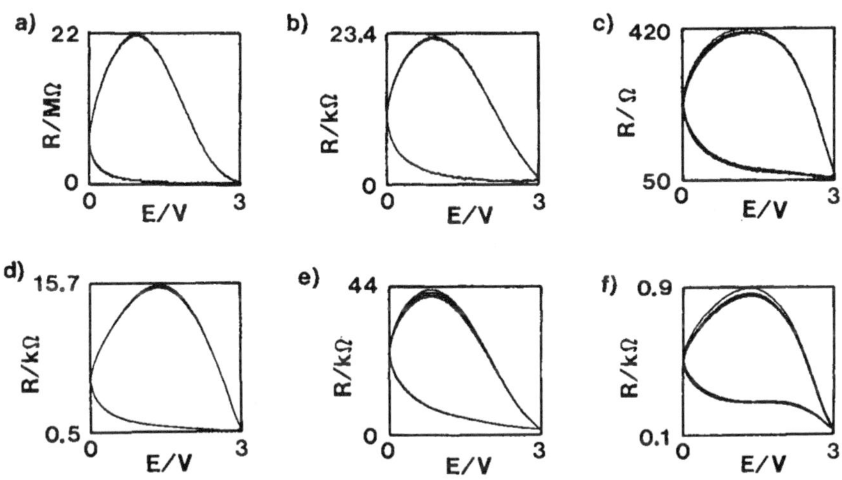

Fig.5 Characteristic response of resistance-voltage curve for a) air, b) whiskey, c) cigarette smoke, d) aromatic perfume, rimonene, e) powder of Colombia coffee bean, f) smoke from cosksec mosquito coil.[15]

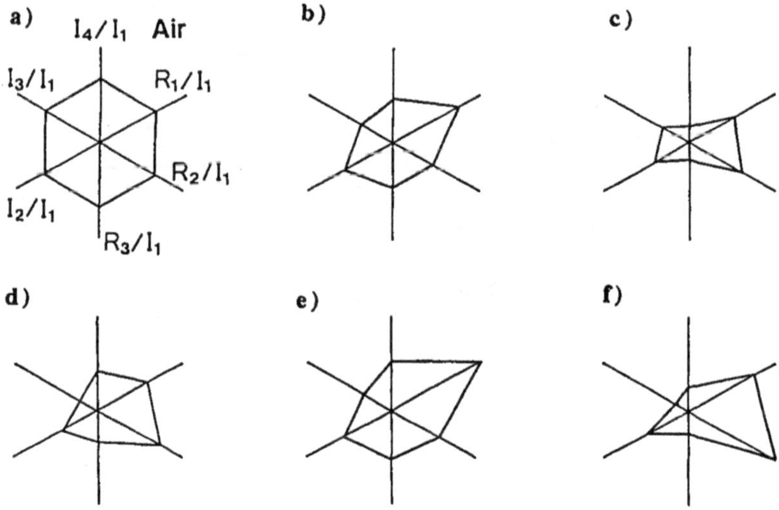

Fig. 6 Characteristic response of the relative intensities of the higher harmonics. R_i and I_i are the real and imaginary components of the ith higher harmonics.[15] a)-f) correspond to those in Fig.5, respectively.

2nd, 3rd, and 4th harmonics were normalized relative to that of the imaginary component of the first harmonics I_1 (Fig.6). It is apparent that discrimination of odor becomes possible by measuring change of the relative intensities.

Currently, gas sensing by ceramic material has been carried out based on the information of the change of the resistance. However, if one withes to distinguish various kinds of odor molecules, solely the change of resistance is insufficient. It is clear that the present method affords us abundant information based on the relative intensities of the higher harmonics. In other words, the nonlinear response gives us higher informational dimension in contrast to the one dimensional information in linear response.

Acknowledgement

The author thanks Prof. H. Kawakami for helpful suggestions. He is grateful to Dr. S. Nakata for his active collaboration. A part of this study was supported by a Grant-in-Aid from the Ministry of Education, Science and Culture of Japan.

References

1. K. Yoshikawa, in *Liquid Membranes: Chemical Applications*, eds. T. Araki & H. Tsukube, (CRC Press, 1990)p.141.
2. For example, C. A. Miller and P. Neogi, in *Interfacial Phenomena*, (Marcel Dekker, 1985)p.184.
3. M. Dupeyrat and E. Nakache, *Bioelectrochem. Bioenerg.*, **5**(1978)p.134.
4. K. Yoshikawa and Y. Matsubara, *J. Amer. Chem. Soc.*, **105** (1983)p.5767.
5. K. Yoshikawa and Y. Matsubara, *J. Amer. Chem. Soc.*, **106** (1984)p.4423.
6. K. Yoshikawa, S. Nakata, T. Omochi, and G. Colaccico, *Langmuir*, **2**(1986)p.715.
7. K. Yoshikawa, M. Shoji, S. Nakata, and S. Maeda, *Langmuir*, **4**(1988)p.759.
8. K. Yoshikawa, and M. Makino, *Chem. Phys. Lett.*, **160**(1989)p.623.
9. K. Yoshikawa, S. Maeda, and H. Kawakami, *Ferroelectrics*, **86** (1989)p.117.
10. K. Yoshikawa, M. Makino, S. Nakata, and T. Ishii, *Thin Solid Films*, **180**(1989)p.117.
11. K. Yoshikawa, M. Shoji, and T. Ishii, *Biochem. Biophys. Res, Commun.*, **160**,(1989)p.699.
12. K. Yoshikawa, in *Far-From-Equilibrium Dynamics of Chemical Systems*, eds. J. Popielawski and J. Gorecki, (World Scientific, 1991)p.329.
13. S. Nakata, K. Yoshikawa, M. Shoji, H. Kawakami, and T. Ishii, T., *Biophys. Chem.*, **34**(1989)p.201.
14. S. Nakata, Y. Kaneda, H. Nakamura, and K. Yoshikawa, *Chem. Lett.* (1991)p.1505.
15. S. Nakata, H. Nakamura, and K. Yoshikawa, *Sensors and Actuators*, in press.

Experimental Evidence of Bekki Nozaki Holes

B. Janiaud[1,2], S. Jucquois[1], J. Lega[3,4] and V. Croquette[1]

(1) *Laboratoire de Physique Statistique, École Normale Supérieure, 24 rue Lhomond, 75031 Paris cedex 05, France*
(2) *Research Institute for Electrical Communication, Tohoku University, 2-1-1 Katahira, Sendai 980, Japan*
(3) *Dept. of Mathematics, University of Arizona, Bldg # 89, Tucson AZ 85721, USA*
(4) *INLN, Université de Nice Sophia-Antipolis, Faculté des Sciences, B.P. 71, 06108 Nice cedex 2, France*

Abstract

We show experimental evidence of traveling hole defects, which are reminiscent of analytical solutions to complex GINZBURG-LANDAU equations. By comparison between experimental results and numerical simulations, we suggest that the appearance of these objects is related to the existence of a local phase instability in the system.

1 Introduction

We present an experimental observation of traveling holes in wave trains. The experiment is Rayleigh Bénard convection, performed with a fluid of low Prandtl number, in an annular cell in which the appearance of two concentric rolls is forced above threshold. Under this condition, convective axisymmetric rolls may then become unstable versus an oscillatory instability [1], that leads to the formation of traveling waves which propagate along the rolls axis. The traveling holes have been observed in an amplitude depression of such a wave train. Though the existence of a region of lower amplitude is likely to be due to a cell inhomogeneity, the presence of such an irregularity is not reflected by the dynamics of the low-amplitude region, which travels along the cell at a quasi-constant speed. Besides, traveling holes appear and disappear randomly inside this amplitude depression. These objects are reminiscent of the traveling holes described by BEKKI and NOZAKI as solutions to the complex GINZBURG-LANDAU equation [2]. These latter solutions connect two different wave trains and are characterized by a localized amplitude drop and a rapid phase variation at their core.

After summarizing the most important features of traveling holes, we shall turn to the experimental observation of such objects. Then, we will explain how, in the amplitude depression, the pattern can be locally driven out of the phase stability band. And we will show that this instability may be responsible for the formation of traveling holes.

2 Complex Ginzburg-Landau equations and traveling hole solutions

2.1 Amplitude equations for a wave pattern

The oscillatory instability of Rayleigh Bénard convective rolls takes the form of waves that propagate along the rolls axis. These waves are easily visualized by the wiggling shape of the cold stream of the roll pattern. As we have already shown [3] [4], they may be described by means of two coupled GINZBURG-LANDAU equations in dimensionless form:

$$\begin{aligned}(\partial_t - c\partial_x)A &= A + (1+ic_1)\partial_x^2 A - (1+ic_2)(|A|^2 + (\beta_r + i\beta_i)|B|^2)A \\ (\partial_t + c\partial_x)B &= B + (1+ic_1)\partial_x^2 B - (1+ic_2)(|B|^2 + (\beta_r + i\beta_i)|A|^2)B.\end{aligned} \tag{1}$$

In this description, the position of the cold stream is given by

$$y = A\exp[i(-k_0x + \omega_0 t)] + B\exp[i(k_0x + \omega_0 t)] + c.c. + \dots \tag{2}$$

where k_0 and ω_0 are the critical wave vector and frequency measured at threshold. The complex quantities A and B are the slowly varying amplitudes of the right and left traveling waves that become unstable at threshold, and c is the dimensionless group velocity of these waves. The temporal frequency above threshold depends on the spatial wave number through c_1, and on the amplitude of oscillations through c_2. Finally, the competition between traveling (($A = 0$ and $B \neq 0$) or ($A \neq 0$ and $B = 0$)) and standing (A and B both finite) waves is ruled by the real part of β.

2.2 Traveling holes

In [2], BEKKI and NOZAKI give analytic solutions to the complex GINZBURG-LANDAU equation (CGL)

$$\partial_t A = A + (1+ic_1)A_{xx} - (1+ic_2)|A|^2 A, \tag{3}$$

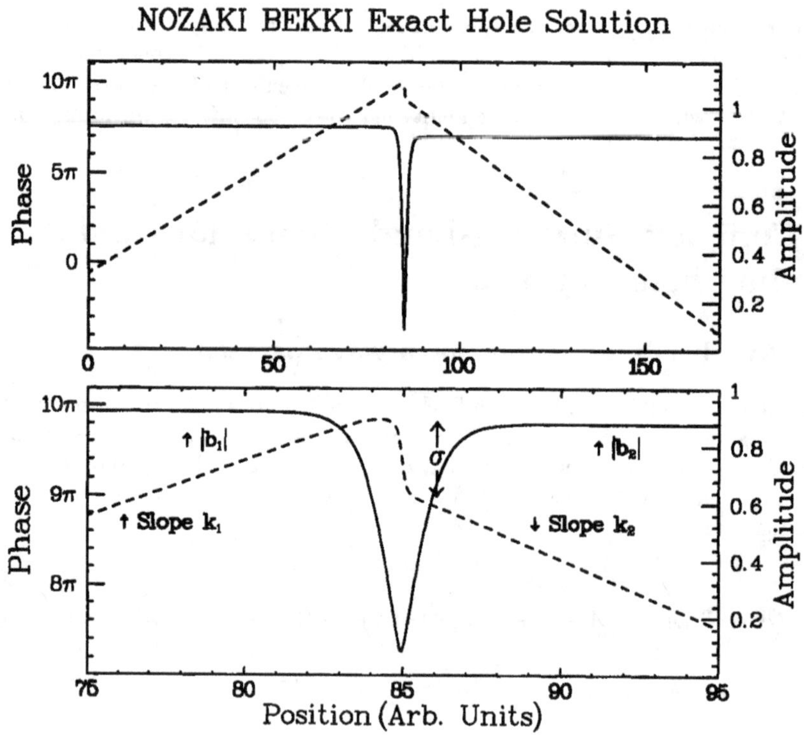

Figure 1: Plot of the amplitude (solid) and the phase (dashed) of a traveling hole solution for $c_1 = 0.5$ and $c_2 = 2.3$. The enlargement about the core (bottom) of the traveling hole demonstrates the smooth behavior of the solution.

that they describe as "traveling holes". Experimentally-relevant characteristics of these objects have been described in detail elsewhere [4]. Here, we only summarize their main features.

Traveling holes asymptotically connect two wave trains of different wavenumbers, say k_1 and k_2. Of course, the asymptotic waves must be solutions to CGL, and therefore, read

$$A_{1,2} = \sqrt{1 - k_{1,2}^2} \exp\left[i\left(k_{1,2}x + \Omega_{1,2}t\right)\right] \tag{4}$$

where

$$\Omega_{1,2} = -c_2 + (c_2 - c_1)k_{1,2}^2.$$

Since the temporal frequency of the phase of the solution does not depend on x, above conditions can only be satisfied if the solution itself travels at a given speed

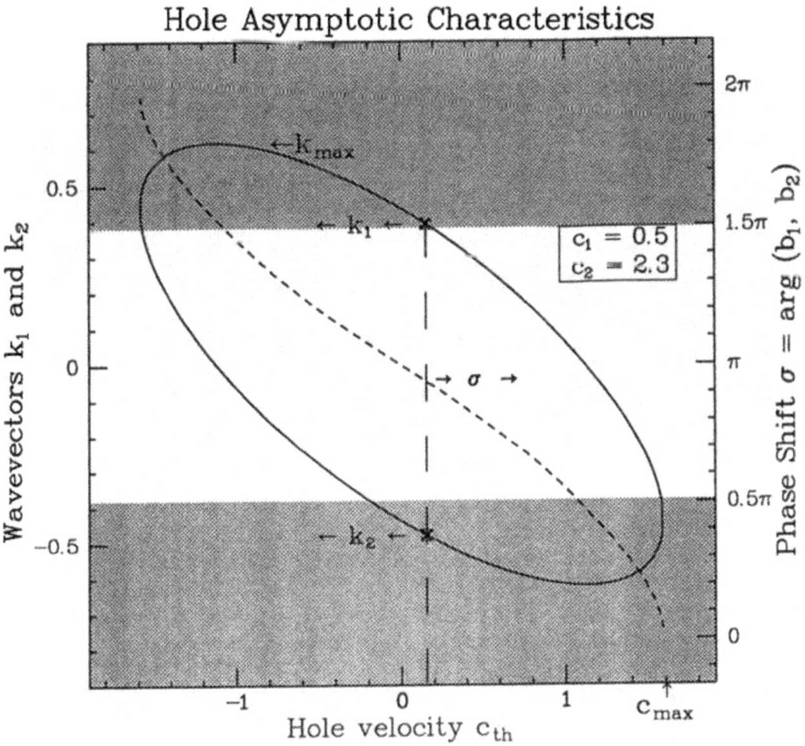

Figure 2: Determination of the caracteristics of the hole solution [2] for $c_1 = 0.5$ and $c_2 = 2.3$. The asymptotic wave-numbers k_1 and k_2 are read at the intersection of the ellipse with the vertical line corresponding to c_{th}. The phase-jump σ is read at the intersection of this line and the S-shaped dashed curve. We have shaded the wavenumbers which are phase unstable.

c_{th}, which is the average of the group velocities of the two asymptotic waves. There is a one-parameter family of solutions that comply with these requirements. It can be parametrized by one, say k_1, of the two asymptotic wavenumbers. Figure 1 is a plot of the amplitude and the phase of one of these solutions, for $c_1 = 0.5$ and $c_2 = 2.3$. At the core of this object, the amplitude of A is minimal and its phase varies rapidly. Besides, the phases of the two asymptotic waves are shifted by a quantity σ, which also depends on k_1. Figure 2 shows how the asymptotic wavenumbers and the phase shift σ are determined when the velocity c_{th} is given.

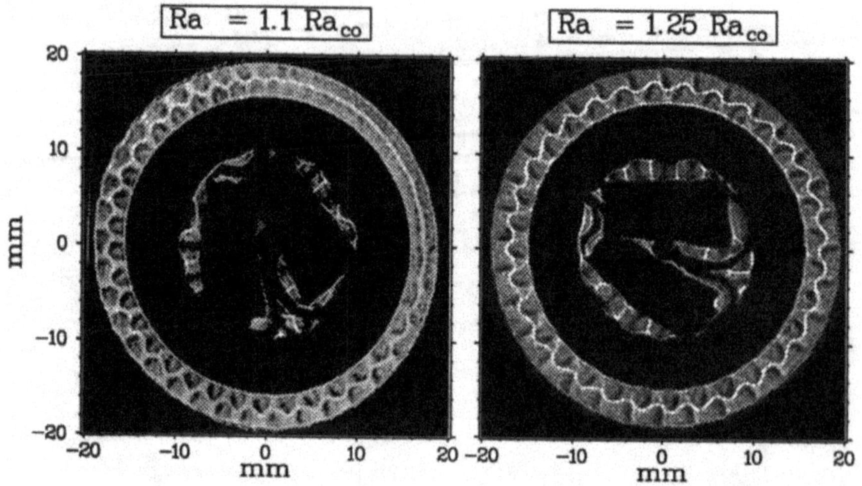

Figure 3: Shadowgraphic pictures of (right) the homogeneous wave pattern arising at high Rayleigh number ($\epsilon = 0.2$), and (left) of the pattern presenting a single amplitude depression ($\epsilon = 0.1$). The white wiggling line in the annular cell represents the cold stream separating the two concentric rolls.

2.3 Traveling holes in wave patterns

We can expect that traveling hole solutions also exist in the case of two coupled GINZBURG-LANDAU equations like 1. Since $|A|$ decreases at the core of these defects, $|B|$ should grow about this point, and therefore exhibit a pulse-shaped envelope. We have numerically checked [4] that such solutions are possible for equations 1. However, depending on the value of β, we can make $|B|$ so small that it can be considered equal to zero everywhere; in this case, its pulse shape disappears. Under these conditions, Eqs. 1 can be reduced to a single equation for A, which is nothing but the complex GINZBURG-LANDAU equation 3, written in a frame moving at speed c. Then, traveling hole solutions to 1 are exactly the traveling hole defects described above. In periodic boundary conditions, a traveling hole is coupled to a shock [2], which connects the two asymptotic waves of wavenumbers k_1 and k_2, and moves at the same speed c_{th} as the hole. Together, these two objects make a persistent localized pattern [2].

3 Description of the Experiment

We have studied the evolution of the wave trains due to the oscillatory instability in an annulus, using a large aspect ratio cell with an outer diameter of 39.7 mm, an inner diameter of 30 mm, and a thickness of 1.18 mm. Except its bigger size, the cell is similar to that of [5]. A resistive wire imbedded inside both circular walls allows to stabilize a two-concentric-roll pattern in the cell, which is visualized by shadowgraphy [6]. The control parameter is the Rayleigh number Ra which is proportional to the temperature difference ΔT across the fluid layer. The radial width of the cell is chosen so that the first instability encountered while increasing Ra is the oscillatory instability. Under its threshold Ra_{co}, the cold stream between the two rolls forms a regular circle, that we use as a reference. Above Ra_{co}, we observe wavy ondulations of the circle as may be seen in Fig. 3, and define the local amplitude of the wave as the deviation from the reference circle. The oscillation frequency being in the $1Hz$ range, we record the time evolution in real-time. The datas are first filtered in time by means of a broad band-pass filter centered at the fundamental frequency of the signal. Then, by keeping only the positive wavevectors of the spatial spectrum, we obtain a complex signal of the wave which make amplitude and phase measurement easy to achieve.

4 Experimental results

4.1 Homogeneous pattern

Above $Ra = 1.2Ra_{co}$, the oscillatory instability always leads to a monotonic evolution with either a clockwise or anti-clockwise wave train. After a transient, during which wave trains propagating in opposite directions compete, an homogeneous wave invades the entire cell (see Fig. 3) and leads to a monoperiodic spectrum. Different wavenumbers (number of wavelengths along the annulus) have been observed, the averaged value being 38. As long as $Ra > 1.2Ra_{co}$, these states, homogeneous in amplitude and in wavenumber, remain stable.

4.2 Dynamics at low Rayleigh number

Lowering the Rayleigh number, we see the appearance of strong modulations of both amplitude and wavenumber. We presume that the building up of these local modulations is related to a cell inhomogeity. However, after creation, these modulations do not reflect any cell inhomogeneity since they are not fixed at some spot but actually slowly turn around the cell. Around $Ra = 1.1Ra_{co}$, these modulations take a simple form: a unique and slowly varying depression of the amplitude of oscillations, and a wavenumber modulation in the region of high amplitude. Inside this depression which typically extends over 10 wavelengths, the amplitude is smoothly reduced at least by an order of magnitude and the local wavenumber varies strongly. It changes nearly by a factor of two between both ends of the region of large amplitude.

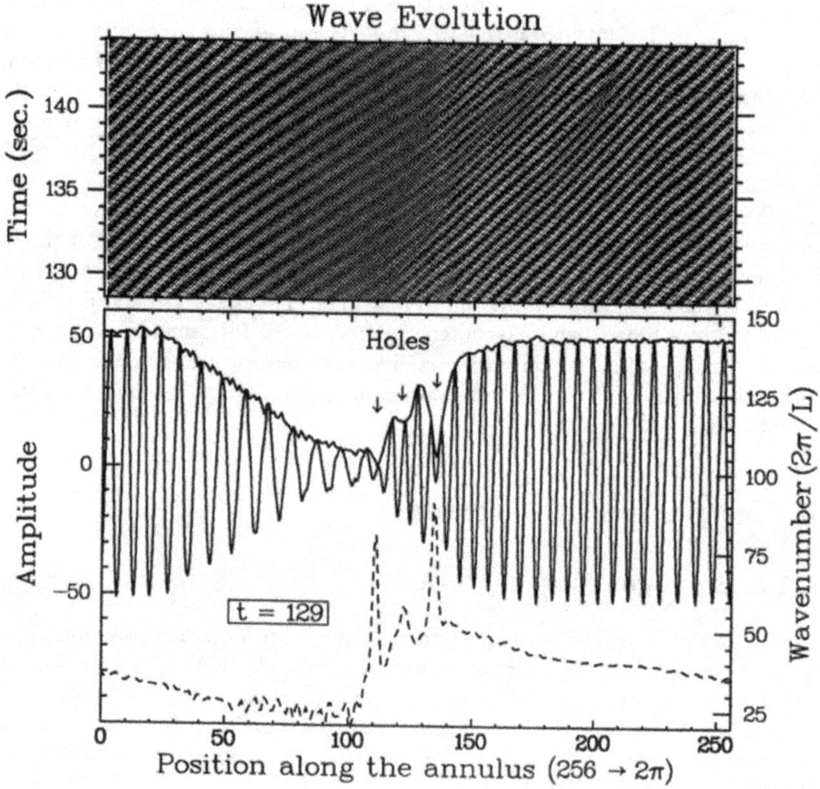

Figure 4: Top: Space time evolution of the wave train shown in grey scale. Black (white) pixels correspond to an inward (outward) displacement of the white line with respect to the reference circle. Bottom : local amplitude and wavenumber (dashed) sampled at a given time.

Lowering even further the Rayleigh number leads to the formation of several extended amplitude depressions, simultaneously present in the cell and separated by wave trains that propagate in opposite directions. Very close to the threshold of the oscillatory instability, the pattern appears as an assembly of small wave trains propagating in both directions and continuously varying in amplitude. If we average the wave dynamics over a long time, no region in the cell is particularly associated with a smaller amplitude of the wave trains. Time spectra measured in this state are chaotic. This chaotic evolution of the wave trains makes difficult the exact determination of the oscillation threshold, Ra_{co}.

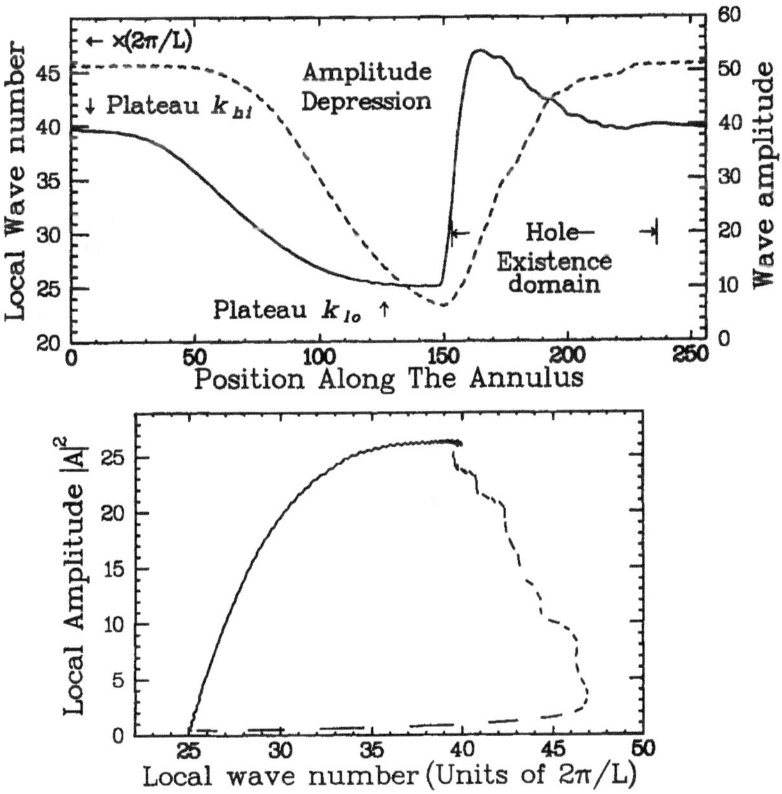

Figure 5: This series of figures represents the wave features obtained by averaging in the frame moving with the amplitude depression. Top : Local wavenumber and wave amplitude (dashed). Bottom : Square of the amplitude versus local wavenumber, obtained from the data corresponding to the leading edge.

4.3 The traveling amplitude depression

We have studied the simplest non stationary pattern having one amplitude depression (see Figure 3) turning around the cell. In the top part of Figure 4, we show the time evolution of the wave train in the entire cell over a short time period. The grey scale corresponds to the local displacement of the cold stream from its reference circle obtained when no oscillation appears along the rolls, i.e. the local wave amplitude. In the bottom part of this figure, we have drawn the amplitude and wavenumber sampled at a given time. We have recorded this signal over a time long enough so that the amplitude depression (appearing as the grey, slightly blurred region) actually makes more than one turn round the cell.

In order to analyze the dynamics of the system inside the amplitude depression, we have averaged the amplitude and the wavenumber in a frame moving at the same velocity as the depression (see fig.5). In the leading edge of the wave train, averaging leads to a noise free pattern with a well defined amplitude and wavenumber. In the trailing edge, which typically extends over one quarter of the cell, we observe a noisier structure, indicating time dependence. In the stationary part of the pattern, the local amplitude and wavenumber present two plateaus (see Fig. 5 top), corresponding to a pattern evolution between two asymptotic wavenumbers: k_{hi} and k_{lo}. This change is smooth and the local wave amplitude adiabatically follows the local wavenumber. Since k_{hi} and k_{lo} are quite different, we could obtain the relation between wavenumber and amplitude. We got the theorically expected parabola (see Fig. 5 bottom).

The trailing edge of the wave train appears to be not stationary. Here, the main feature is the pseudo-periodic nucleation of spatio-temporal dislocations. As it may be seen on figure 6, the formation of new wavelengths occurs roughly at the same place within the amplitude depression; on the other hand, the disappearance of wavelengths follows a more random mechanism. The interesting phenomena that we want to stress here, is that each dislocation nucleation produces a phase jump which propagates at a different velocity from that of the depression and in most cases leads to the occurrence of a dislocation of opposite sign. Although it is not surprising to see a dislocation producing a phase jump, usually such a phase gradient diffuses, and does not propagate. In figure 6, where we have displayed the local wavenumber in grey scale, phase jumps correspond to high values of the local wavenumber (taking as reference the mean wavenumber that is 38) and thus appear as white or dark lines. From figure 5, we know that such a high value of wavenumbers are linearly stable [7], so they cannot come from propagating waves but only from phase jumps. At the location of each phase jump, the wave amplitude drops to a small value over a short distance (typically one or two wavelengths); we shall refer to this structure as a "hole". In figure 4, we present the wave shape along the cell, where both the extended amplitude depression and holes may be seen. The velocity of the holes is typically twice that of the amplitude depression.

Hence, we have seen the spontaneous creation of localized dips, that we called traveling holes by reference to the theoritical solutions. In the experiment, they appear and propagate in a region where the amplitude of the wave is smoothly but strongly decreased. These holes are characterized by their depth and the phase jump at their core which vary as they move.

5 Numerical simulation

In order to make a close comparison with the behaviors observed in the experiment, we need a mechanism to make traveling holes appear and disappear in time. A simple one

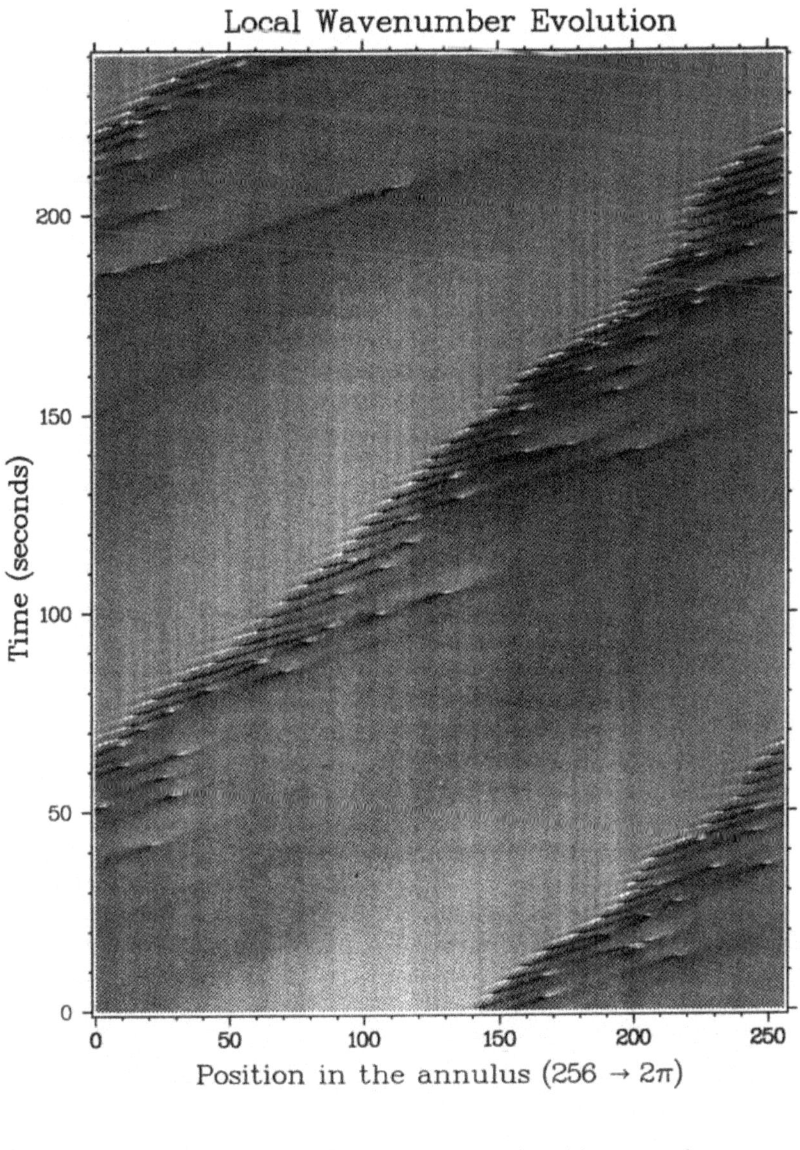

Figure 6: Space-time evolution of the local wavenumber. The grey scale corresponds to 80 for black lines and 0 for white ones. These limits extend far beyong the range of linearly unstable wavenumbers (25 to 50).

consists in the existence of a local phase instability of the wave train. More precisely, we have seen that, in the experiment, the amplitude of the wave train adiabatically follows its phase – at least in the leading edge. If we call $|A|$ the amplitude of the wave, and φ its phase, and if we assume that $|A|^2$ behaves like $1 - |\varphi_x|^2$, then we see that, in a region of low amplitude, $|\varphi_x| = k$ may go above the Eckhaus-Benjamin-Feir stability threshold [8] [9][10]. This rather crude criterion gives an idea of how a wave train can be locally unstable.

We have made a numerical simulation of Eqs.1, with an initial condition that showed a locally unstable amplitude depression – according to above criterion. The first stages of the dynamics that ensues are shown in Fig.7. The point we want to emphasize is the appearance of traveling dips inside the region of low amplitude. They are reminiscent of the traveling holes described above, and characterized by a strong phase variation at their core. Of course, since the amplitude depression is not stable for the amplitude equations we have considered, it quickly disappears. However, holes continue to be created and to move across the box. Eventually, the system relaxes to a flat solution, or to a state with a small number of steady traveling holes, separated by shocks.

6 Conclusion

As suggested by the numerics, the presence of traveling dips may be due to a local phase instability of the wave train. This leads to the existence of an active region in the system, where traveling holes are created and disappear. Outside this low amplitude domain, the wavenumber of the wave train is space-dependent, and varies smoothly between its asymptotic values.

On the other hand, time-averaging of the experimental datas showed the existence of a steady amplitude depression in a wave train otherwise smoothly-varying. Detailed analysis of this region of lower amplitude revealed the presence of many traveling holes, being created and moving across this region. The existence of a local phase instability could then explain the experimental observation of a strongly modulated wave train, both in amplitude and wavenumber. Finally, it was of real importance to observe these traveling holes experimentally, since these defects play a fundamental role in the route towards spatio-temporal disorder.

7 Acknowledgements

We are grateful to D. Bensimon, Y. Couder, P. Coullet, S. Fauve, V. Hakim, Y. Kondo, A. Newell, T. Passot, A. Pumir, M. Rabaud, P. Tabeling, M. Tabor, and H. Willaime for fruitful discusions. B. Janiaud wishes to acknowledge the Monbusho (Japanese Ministry of Education) and J. Lega the Arizona Center for Mathematical

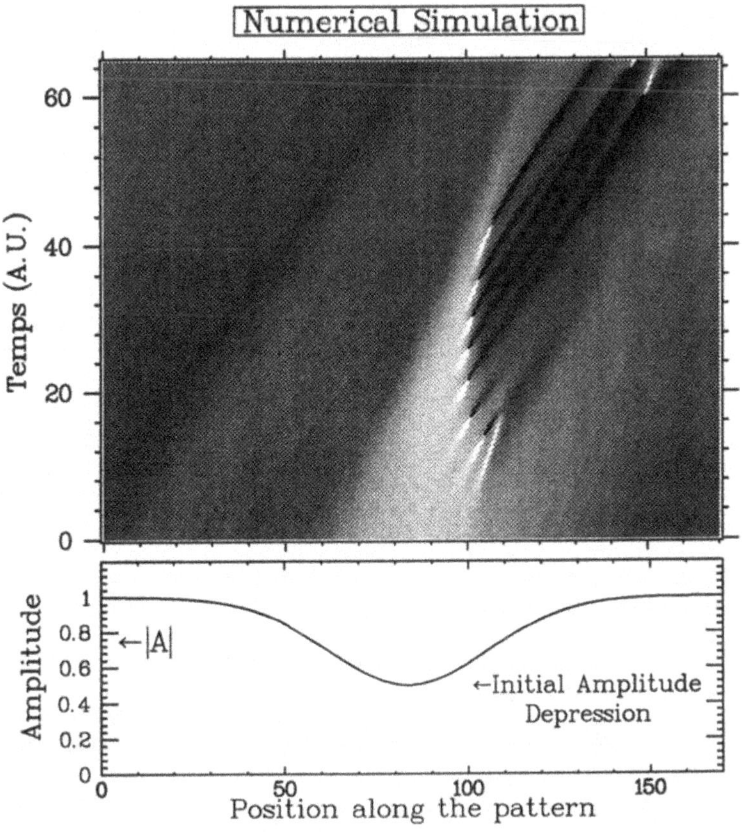

Figure 7: Numerical simulation of Eqs.1, with $c_1 = 0.5$ and $c_2 = 2.3$. The bottom graph correspond to the initial condition. The top image represents the local wavenumber in grey scale.

Sciences (ACMS) for support. ACMS is sponsored by AFOSR contract FQ8671-9000589 (AFOSR-90-0021) with the University Research Initiative Program at the University of Arizona.

References

[1] R.M. CLEVER and F.H. BUSSE, J. Fluid Mech. **65**, 625 (1974).

[2] N. BEKKI and K. NOZAKI, Phys. Lett. **110A**, 133, (1985).

[3] V. CROQUETTE and H. WILLIAMS, Phys. Rev. **A 39**, 2765 (1989); Physica **D 37**, 300 (1989).

[4] J. LEGA, B. JANIAUD, S. JUCQUOIS, V. CROQUETTE, *Localized Phase Jumps in Wave Trains*, to be published in Phys. Rev. A, (1992).

[5] B. JANIAUD, A. PUMIR, D. BENSIMON, V. CROQUETTE, H. RICHTER and L. KRAMER, *The* ECKHAUS *Instability for Travelling Waves*, to appear in Physica D, (Feb. 1992).

[6] W. MESRZKIRCH, *Flow Visualization*, (Academic Press, New York) 1974; S. RASENAT, G. HARTUNG, B.L. WINKLER and I. REHBERG, Phys. Fluids. **15**, 379, (1989).

[7] We call linearly unstable wavenumbers those which naturally grow above the oscillatory instability threshold. In other words, linearly unstable wavenumbers fall inside the parabola of Fig. 5.

[8] A. C. NEWELL, Lectures in Applied Mathematics Vol. **15**, 157, Am. Math. Society, Providence, (1974).

[9] W. ECKHAUS, *Studies in Nonlinear Stability Theory*, Springer-Verlag, New-York, (1965).

[10] J. T. STUART and R. C. DIPRIMA, Proc. R. Soc. Lond. **A 362**, 27 (1978).

DYNAMICS OF PATTERN FORMATION IN BINARY FLUIDS WITH AMPHIPHILIC MOLECULES

TOSHIHIRO KAWAKATSU and KYOZI KAWASAKI
Department of Physics, Faculty of Science,
Kyushu University 33, Fukuoka 812, Japan

ABSTRACT

Pattern formation processes in binary fluid mixtures containing amphiphilic molecules are investigated by simulations on the authors' Hybrid Model and also by Monte Carlo and molecular dynamics simulations. Amphiphilic effects on the phase separation process are discussed.

1.Introduction

Dynamics of pattern formation processes of immiscible binary mixtures caused by phase separation has been extensively investigated both experimentally and theoretically.[1] Final spatial patterns of the domain structures as well as their formation processes are interesting and important problems in physics, chemistry, biology and also in engineering. A lot of experimental investigations on phase separation processes were reported not only on simple alloys and low-molecular liquid mixtures but also on more complex systems like polymer blends.[1,2] One example of such complex systems is microemulsions. Microemulsions are formed in immiscible binary mixtures containing amphiphilic molecules, e.g. oil/water/amphiphile systems.[3,4] Here, the amphiphilic molecule is a molecule composed of two chemically distinct parts, which dissolve into different solvents. It is well-known that such amphiphilic molecules are strongly adsorbed onto interfaces and lower the interfacial tension, which results in complex domain structures on mesoscopic scales.[3,4]

Equilibrium domain structures of microemulsions have been extensively investigated.[5-13] On the contrary, the dynamics of formation processes of such systems have scarcely been studied. This is due to tha fact that the formation process proceeds too fast for low-molecular systems, e.g. oil/water/amphiphile systems, which makes it difficult to measure the process in neutron scattering experiments. One can, however, find suitable examples in polymer systems, where the dynamics of phase separation is slowed down considerably. Immiscible polymer blends containing block copolymers, which consist of different polymer chains connected at the end, show the similar behavior to microemulsions.[14,15] However, a difficulty arises when we try to study such polymer systems analytically because of the complex nature of the system. Recent developments in high-speed supercomputers and the growing use of the computer simulation techniques solved this difficulty. In this paper, we describe our current research on these complex pattern formation processes using microscopic computer simulation techniques.

2.Computer Simulations on the Pattern Formation Processes

2.1 Hybrid Model

We proposed the *Hybrid Model* to study numerically the dynamics of binary mixtures containing amphiphiles. The model is a combination of a continuous scalar field and discrete molecules with spins, the former describing the binary mixture and the latter describing the amphiphilic molecules, respectively.[16–17] We demonstrated that this Hybrid Model can successfully reproduce the main features of amphiphilic systems, e.g. the reduction in the interfacial tension of amphiphile-adsorbed interfaces, equilibrium domain structures and the slowing down of the coarsening in the late stage of phase separation.[16,17]

(a) (b)

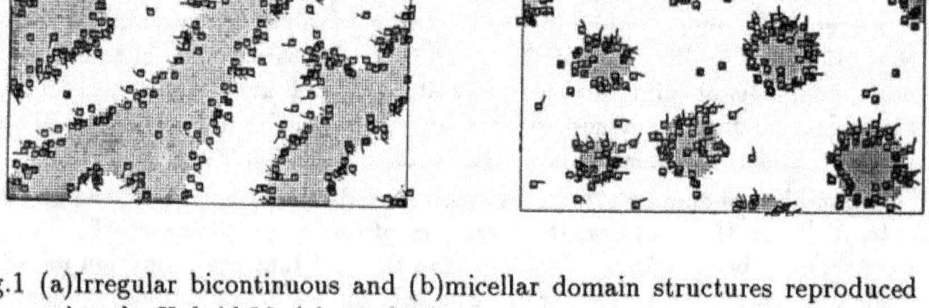

Fig.1 (a)Irregular bicontinuous and (b)micellar domain structures reproduced using the Hybrid Model are shown.

As an example, in Fig.1 we show typical domain structures produced with the use of the Hybrid Model. Fig.1(a) is the so-called irregular bicontinuous structure and Fig.1(b) is the micellar structure. In these simulations, the effects of thermal noise, which were neglected in our previous works,[16,17] is taken into account. These domain structures are formed as a result of self-assembling processes of the binary mixture and the amphiphilic molecules. The effect of thermal noise is more apparent in the micellar case. A regular triangular lattice structure of almost mono-dispersed micelles was observed when the thermal noise is neglected.[16] However here in Fig.1(b), we observe poly-dispersed micelles in irregular configuration, which shows that the regular pattern in the absence of thermal noise is a metastable configuration. Details on the thermal noise effects will be described elsewhere.[18]

Some predictions on the dynamics in the early stage of phase separation processes have also been made on the basis of the Hybrid Model,[19] which have not yet been confirmed experimentally.

2.2 MC and MD Simulations

In order to check validity of our Hybrid Model and its predictions in microscopic level, we designed two types of simulations, one is a Monte Carlo (MC) simulation on spin systems[20,21] and the other is a molecular dynamics (MD) simulation on atomic and molecular systems.[21,22]

In our MC simulation, we adopt an Ising ferromagnetic spin system to describe the binary mixture, and the amphiphilic molecule is modelled as a pair of adjacent spins with opposite spin variables which are connected by a rigid bond. We found that the phase separation is decerelated in the early stage when we add amphiphilic molecules into the phase separating binary mixture, which is caused by the impurity nature of the surfcatant molecule.[20] In this MC simulation, the amphiphilic molecule is small compared to the length scale of the phase separation in the early stage, and therefore the amphiphilic nature is less important than the impurity effects of the amphiphilic molecule.[20]

The amphiphilic effects on the dynamics of phase separation become more apparent when the size of the amphiphilic molecule is comparable to the characteristic length scale of the phase separation. Such a situation can be realized in the case of amphiphilic polymer molecules, i.e. block copolymers. Unfortunately, in the existing experiments, relatively short block polymer chains are used, which turns out to be inadequate to detect the amphiphilic effects in the early stage.[14,15] In order to investigate such amphiphilic effects, we performed a molecular dynamics simulation on the phase separating binary soft-sphere system containing a block copolymer chain.[21,22] Performing MD simulations on the phase separation process from uniformly mixed state, we found that the block copolymer chain induces a large composition fluctuation at later time. Such a composition fluctuation originates from the initial small composition fluctuation associated with the conformation of the block copolymer chain, which acts as a nucleus for the phase separation. An analysis on the Hybrid Model also predicts acceleration of the phase separation due to the amphiphilic effects, and therefore, the MD simulation stated above confirmes this prediction.

3. Conclusions

Our computer simulations on the Hybrid Model and also the MC and the MD simulations described in this paper showed how the amphiphilic nature affects the dynamics of phase separation processes. These results can be tested by real experiments using polymer blends and block copolymers.

Acknowledgements

Computations were performed at the Computer Centers of Institute for Molecular Science, National Laboratory for High Energy Physics, Japan Atomic Energy Research Institute and Kyushu University. The work is also supported by the Scientific Research Fund of the Ministry of Education, Science and Culture of Japan.

References

1. S.Komura and H.Furukawa eds., *Dynamics of Ordering Processes in Condensed Matter* (Plenum, New York, 1988).
2. A.Onuki and K.Kawasaki, eds., *Dynamics and Patterns in Complex Fluids*, (Springer-Verlag, Heidelberg, 1989).
3. S.Safran and N.Clark eds., *Physics of Complex and Supermolecular Fluids* (Wiley, New York, 1987).
4. W.M.Gelbart, D.Roux and A.Ben-Shaul eds., *Modern Ideas and Problems in Amphiphilic Science*, in press.
5. Y.Talmon and S.Prager, *Nature* **267**, (1977) 333.
6. P.G.DeGennes and C.Taupin, *J. Chem. Phys.* **86**, (1982) 2294.
7. B.Widom, *J. Chem. Phys.* **81**, (1984) 1030.
8. S.A.Safran, D.Roux, M.Cates and D.Andelman, *Phys. Rev. Lett.* **57** (1986) 491.
9. K.Chen, C.Ebner, C.Jayaprakash and R.Pandit, *J. Phys. C: Solid State Phys.* **20**, (1987) 1361.
10. G.Gompper and M.Schick, *Phys. Rev. Lett.* **62**, (1989) 1647.
11. R.G.Larson, L.E.Scriven and H.T.Davis, *J. Chem. Phys.* **83**, (1985) 2411.
12. B.Smit, *Phys. Rev.* **A37**, (1988) 3431.
13. J.R.Gunn and K.A.Dawson, *J. Chem. Phys.* **91**, (1989) 6393.
14. R.J.Roe and C.M.Kuo, *Macromolecules* **23**, (1990) 4635.
    D.W.Park and R.J.Roe, *Macromolecules* **24**, (1991) 5324.

15. T.Hashimoto, private communication.
16. T.Kawakatsu and K.Kawasaki, *Physica* **A167**, (1990) 690.
17. T.Kawakatsu and K.Kawasaki, in *Molecular Dynamics Simulations; Proceedings of the Taniguchi International Symposium*, F.Yonezawa ed., (Springer, 1992), in press.
18. T.Kawakatsu and K.Kawasaki, unpublished work.
19. T.Kawakatsu and K.Kawasaki, *J. Colloid Interface Sci.* **145**, (1991) 413, *ibid.* 420.
20. T.Kawakatsu and K.Kawasaki, *J. Colloid Interface Sci.* **148**, (1992) 23.
21. T.Kawakatsu and K.Kawasaki, in *From Phase Transitions to Chaos*, G.Györgyi, I.Kondor, L.Sasvári and T.Tél eds., (World Scientific), in press.
22. T.Kawakatsu and K.Kawasaki, submitted for publication.

HOLE DYNAMICS OF ONE-DIMENSIONAL PLASMA

Kazuo KITAHARA, Nobuo ISHIBASHI and Kohki TANNO
Department of Applied Physics, Tokyo Institute of Technology
Meguro-ku, Tokyo 152, Japan

ABSTRACT

A system of N negatively-charged sheets with uniform positive background is studied numerically by transforming the equations of motion into finite-time mapping of $2N$ degrees of freedom. Evolution of the system from two-stream instability towards large-vortex states is shown.

1. Introduction

Bernstein, Greene and Kruskal[1)] found steady state solutions for the one-dimensional Vlasov equation. However, the stability of the steady state solutions and the time evolution towards the steady states are not well established[2)]. On the other hand, Dupree[3)] and Kadometsev-Pogutse [4)] suggest the existence of highly correlated structures in a turbulent plasma, called "clumps". Therefore it is important to study the time evolution of a system of charged particles with Coulomb interactions.

In this paper, we study numerically Dawson's model[5)] of one-dimensional plasma by casting the dynamics of N electron sheets with positive uniform ion background into $2N$-dimensional map in the phase space and by performing numeically exact computation of the dynamics.

2. The Methods

The equation of motion of the i-th sheet from the left boundary is

$$\frac{d^2x_i}{dt^2} = -\omega_p^2 \left[x_i - \left(i - \frac{1}{2}\right)\Delta\right], \tag{1}$$

where ω_p is the plasma frequency and Δ is the average disntance of neighboring pairs of sheets. Eq.(1) is completely solved for a given initial condition of positions $\mathbf{x}^0 = (x_1^0, ..., x_N^0)$ and velocities $\mathbf{v}^0 = (v_1^0, ..., v_N^0)$. Thus, we have

$$x_i(t) = X_i(x_i^0, v_i^0, t), \qquad v_i(t) = V_i(x_i^0, v_i^0, t), \tag{2}$$

where

$$\begin{cases} X_i(x,v,t) \equiv \Delta\left(i-\frac{1}{2}\right) + \left[x - \Delta\left(i-\frac{1}{2}\right)\right]\cos(\omega_p t) + \frac{v}{\omega_p}\sin(\omega_p t) \\ V_i(x,v,t) \equiv v\cos(\omega_p t) - \omega_p\left[x - \Delta\left(i-\frac{1}{2}\right)\right]\sin(\omega_p t). \end{cases} \tag{3}$$

When the amplitude of the oscillation of sheets is large enough, a pair of neighboring sheets cross each other. Suppose at time $t_c(i)$, the i-th sheet and the $i+1$-th sheet cross each other for the first time from a given initial condition. Then $t_c(i)$ is given as the solution of $X_i(x_i^0, v_i^0, t) = X_{i+1}(x_{i+1}^0, v_{i+1}^0, t)$, which can be expressed analytically. After the crossing the two sheets exchange the velocities and the subsequent evolution is again given by Eq.(2) or Eqs.(3). However, in updating the initial conditions, we should find the first crossing pair i_c and the corresponding crossing time. Thus, we can determine the configuration of N particles in the phase space at a crossing time from the configuration at the previous crossing time. Namely,the configuration $(\mathbf{x}^{(k+1)}, \mathbf{v}^{(k+1)})$ at the $(k+1)$-th crossing is specified completely by the time interval between the k-th and the $(k+1)$-th crossings and by the configuration $(\mathbf{x}^{(k)}, \mathbf{v}^{(k)})$ at the k-th crossing.

We have analyzed the two-stream instability of $N = 1260$ particles in the one-dimensional space of length $L = 4\pi$. The time is scaled by $1/\omega_p$. The initial positions of the particles are randomly distributed and the distribution of velocities is chosen to be narrow ("cold beams"). In fact, we put initially $v(x) = \pm v_0[1 + a\cos(4\pi x/L) + bR]$ where $a \approx 10^{-4}$, $b \approx 10^{-8}$ and R is a random number with $|R| < 1$. We take periodic boundary condition; namely, when we have $x_N(t) - x_1(t) = L$, we consider that the 1-st sheet and the N-th sheet cross each other. A typical evolution of the particles in the phase space is shown in the figures below.

We analyzed also the mean square of relative displacement of two particles, $\langle [x_i(t) - x_j(t)]^2\rangle$ at the later stage ($\omega_p t \approx 2000$). It turns out to be proportional to t^3 for large t, implying that the system is in a turbulent state.

3. Concluding Remarks

It seems that a quasi-steady hole is formed very quickly from the two-stream instability. This type of structure formation was observed in the computer simulation by Ghizzo et. al[6]. It is very surprizing that the system is in a turbulent state as shown in the relative diffusion.Further analysis of the statistical characteristics of the system, for example, spatial and temporal correlations, is necessary.

4. Acknowledgements

The authors thank to Prof. T.Hatori, National institute of Fusion Science, Prof. M. Nambu, Kyushu University, Prof. K. Saeki, Shizuoka University and Prof. H. Irie, Nihon

University for valuable discussions. This work was supported by the Cooperative Research Program of National Institute of Fusion Science.

5. References

1) I. B. Bernstein, J. M. Greene and M. D. Kruskal, *Phys. Rev.* **108**(1957) 546.

2) H. Schamel, *Z. Naturforschung* **42a** (1987) 1167; *Phys. Rep.* **140** (1986) 161; *Phys. Rev. Lett.* **48** (1982) 481.

3) T. H. Dupree, *Phys. Fluids* **15** (1972) 334.

4) B. B. Kadometsev and O. P. Pogutse, *Phys. Fluids* **14**(1971) 2470.

5) J. M. Dawson, *Phys. Fluids* **5** (1962) 445; *Meth. Comp. Phys.* **9** (1970) 1.

6) A. Ghizzo, B. Izrar, P. Bertrand, E. Fijalkow, M. R. Feix and M. Shoucri, *Phys. Fluids* **31** (1988) 72.

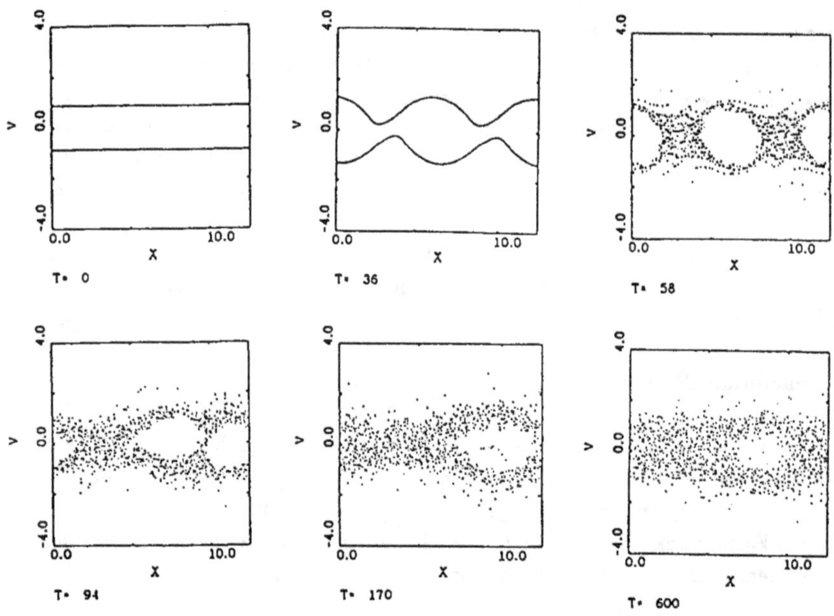

Fig.1. A typical evolution of 1260 sheets in the phase space. The V axis is the velocity axis and the X axis is the position axis. Time T is scaled by the plasma frequency.

Double Hierarchy and Fractal Energy Function of the Discommensuration Structures

Hisashi MITANI and Kenji YAMAMOTO
Department of Physics, Fukuoka University of Education,
Akama, Munakata, Fukuoka 811-41

Abstract

It is shown that discrete sine-Gordon systems form a number of types of hierarchical discommensuration patterns simultaneously, with much different scales and different dimensions. A fractal energy function is obtained in the structural parameter space.

The systems with a coexistence of two systems with different periodicities have solitons or Discommensurations(DCs). Detailed studies have shown the fractal natures in these systems[1,2,3]. In this paper, we study a two-dimensional atomic monolayer system on a periodic substrate potential, where the symmetry of the atomic lattice is generally different from the one of the substrate lattice. Their extensions to various dimensions and symmetries are easy[4].

The Hamiltonian of the systems of a general dimension (n) has the form[3]

$$H = \frac{1}{2}\sum_{i=1}^{N}\sum_{j}^{Z} W(\mathbf{x}_j^{(i)} - \mathbf{x}_0^{(i)}) + V_0 \sum_{i=1}^{N}\sum_{\mathbf{k}} C_{\mathbf{k}} \exp(i\mathbf{g}_{\mathbf{k}} \cdot \mathbf{x}_0^{(i)}) \quad . \tag{1}$$

The first term is the interaction energy between the atoms, where $\mathbf{x}_j^{(i)}$ is on the j-th bond on $\mathbf{x}_0^{(i)}$. The second one is the substrate potential energy. The vectors $\{\mathbf{g}_k\}$ are the reciprocal vectors of the substrate lattice, N is the number of the atoms, and Z is the order of the coordination number of the atomic lattice.

Detailed studies of the one-dimensional model [5] show that the ground state solutions are commensurate (C) or incommensurate (IC) states. We assume that this conclusion is correct in higher dimensions.

The C and IC structures are represented so that the atoms may be modulated to the potential minima from the points of their "averaged lattice". The atomic averaged lattice vectors $\{\mathbf{b}_1,...,\mathbf{b}_n\}$ can be given by linear combinations of the basal vectors of the substrate lattice $\{\mathbf{a}_1,...,\mathbf{a}_n\}$ as

$$\mathbf{b}_j = \sum_{i=1}^{n} \mathbf{a}_i \left[r_{ij}^0 \right] \quad . \qquad (i,j=1\sim n) \tag{2}$$

We have shown[4,6] the two-dimensional competitive systems have generally hierarchical DCs depending on the "inverse matrix expansion" of $\mathbf{r}^{\circ} = [r^{0}_{ij}]$, which is an extension of the continued fractional expansion,

$$\mathbf{r}^{0}_{(1)} = \mathbf{k}^{0}_{(1)} + (\mathbf{k}^{1}_{(1)} + (\mathbf{k}^{2}_{(1)} + \cdots + (\mathbf{k}^{s}_{(1)} + \mathbf{u}^{s}_{(1)})^{-1} \cdots)^{-1})^{-1} \quad . \tag{3}$$

This equation is determined by the following procedure: in each hierarchy $(0 \le l \le (s\text{-}1))$, matrix $\mathbf{r}^{l}$ can be decomposed as $r^{l}_{ij} = k^{l}_{ij} + u^{l}_{ij}$, where $k^{l}_{ij} = \mathrm{Int}\,[r^{l}_{ij}]$, and $\mathbf{r}^{l+1} = (\mathbf{u}^{l})^{-1}$. While the final matrix $\mathbf{u}^{s}_{(1)}$ is singular; $\det(\mathbf{u}^{s}_{(1)})=0$. Each index $\mathbf{k}^{l}$ in Eq.(3) shows that the (l+1)-th hierarchical DC structure $\{\mathbf{a}^{l+1}\}$ can be approximately written by a linear combination of the l-th DC structure $\{\mathbf{a}^{l}\}$, that is,

$$\mathbf{a}^{l+1}_{j} \sim \sum_{i=1}^{n} \mathbf{a}^{l}_{i}\,[k^{l}_{ij}] \quad . \tag{4}$$

In these notations, $\mathbf{a}^{0}$ ($\equiv\mathbf{a}$), $\mathbf{a}^{1}$ ($\equiv\mathbf{b}$), and $\mathbf{a}^{2}$ denote the basal vectors of the substrate lattice, atomic averaged lattice, and the usual (averaged) DC lattice, respectively. Here we define the "DC-dimension", λ, as the rank of the matrix $\mathbf{r}^{l}$. As far as $0 \le l \le (s\text{-}1)$, each DC-dimension is equal to the system dimension, $\lambda=n$; thus, the DC locates as a lattice in the n-dimensional space, which we call "lattice-DC".

The systems in which the symmetry of the atomic lattice is same as the one of the substrate lattice which means that $\mathbf{u}^{s}_{(1)}$ is not existed, has been studied in detail[3,6]. We call such structures "higher symmetrical structures". In these systems, the energy $E(\mathbf{r}^{\circ})$ and the phase transitions of the system $\mathbf{r}^{\circ}(\mu)$, where μ is the external parameter(s), have been shown in "devilish" graphs with hierarchies, self-similarities and breaking of the analysity.

Hereafter, we concentrate on the case except "higher symmetrical structure", i.e., $\det(\mathbf{u}^{s}_{(1)}) = 0$. The rank of $\mathbf{u}^{s}_{(1)}$, λ_2 is lower than the system dimension, i.e., $\lambda_2 < n$. Thus, we find that the matrix $\mathbf{u}^{s}_{(1)}$ shows new DC structures of a lower-dimensional sub-space. This matrix $\mathbf{u}^{s}_{(1)}$ is orthogonally transformed to a non singular $\lambda_2 \times \lambda_2$ matrix, $\mathbf{r}^{0}_{(2)}$, as follows,

$$\mathbf{T}\,\mathbf{u}^{s}_{(1)}\,\mathbf{T}^{-1} = \left(\begin{array}{c|c} \mathbf{r}^{0}_{(2)} & 0 \\ \hline 0 & 0 \end{array}\right) . \tag{5}$$

Note that Eq.(5) shows that there are no components of the DC modulation vectors besides the directions of the λ_2-dimensional sub space. Thus, we see that the matrix $\mathbf{r}^{0}_{(2)}$ in the λ_2-dimensional sub space plays the same role as the matrix $\mathbf{r}^0 \equiv \mathbf{r}^0_{(1)}$ in the n-dimensional space. Therefore, we find that the matrix $\mathbf{r}^0_{(2)}$ is also expanded as Eq.(3),

$$\mathbf{r}^0_{(2)} = \mathbf{k}^0_{(2)} + (\mathbf{k}^1_{(2)} + (\mathbf{k}^2_{(2)} + \cdots + (\mathbf{k}^s_{(2)} + \mathbf{u}^s_{(2)})^{-1} \cdots)^{-1})^{-1} \quad . \tag{3'}$$

The matrix $\mathbf{u}^s_{(2)}$ is also orthogonally transformed into a nonsingular $\lambda_3 \times \lambda_3$ matrix, $\mathbf{r}^0_{(3)}$, as Eq.(5).

Such procedures will be repeated in general. Then, we can obtain hierarchical DC structures on each DC-dimension: such as $\lambda=\lambda_1, \lambda_2, \lambda_3, \cdots$.

We show an example of the case in Fig.1. Here we consider the following matrix $\mathbf{r}^0 \equiv \mathbf{r}^0_{(1)}$ as shown in Eq.(3),

$$\mathbf{r}^0 = \begin{bmatrix} 1 & 1 \\ -1 & 1 \end{bmatrix} + \left[\begin{bmatrix} 4 & 1 \\ -1 & 4 \end{bmatrix} + \left[\begin{bmatrix} 2 & 1 \\ -1 & 2 \end{bmatrix} + \begin{bmatrix} \frac{1}{3} & \frac{1}{3} \\ \frac{1}{3} & \frac{1}{3} \end{bmatrix} \right]^{-1} \right]^{-1} \quad . \tag{6}$$

where the symmetry of the substrate lattice is square for simplicity. The last matrix is singular and orthogonally transformed as follows:

$$\mathbf{T} \begin{bmatrix} \frac{1}{3} & \frac{1}{3} \\ \frac{1}{3} & \frac{1}{3} \end{bmatrix} \mathbf{T}^{-1} = \begin{bmatrix} \frac{2}{3} & 0 \\ 0 & 0 \end{bmatrix} , \tag{7}$$

and $$\frac{2}{3} = 1 - \frac{1}{3} \quad . \tag{8}$$

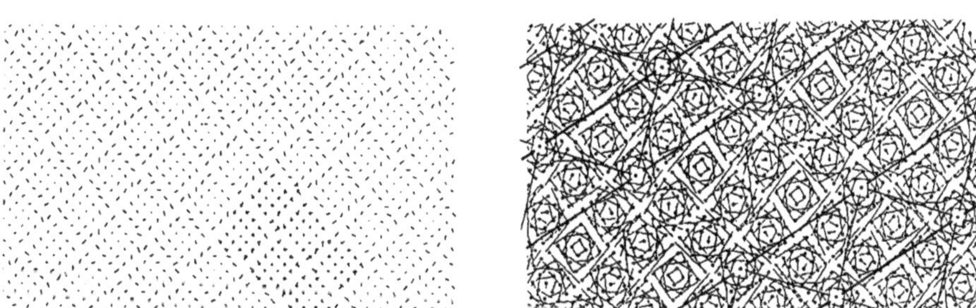

FIG. 1 (a) FIG. 1 (b)

Fig.1 (a) An example of a two-dimensional modulation wave, where each atomic averaged lattice and its neighboring substrate potential minima are connected by a line, for simplicity. (b) The modulation wave is exaggerated, where the atomic averaged lattices are represented by closed circles and the modulation wave is drawn by lines ten times longer than that of (a). The "nest" patterns are lattice-DC, the thin lines represent lattice-DC of lattice-DCs, and the thick lines represent stripe-DC pattern.

Note that eqs.(6), (7), and (8) correspond to Eq.(3), (5), and (3'), respectively. We can see in Fig.1: (i) islands with a high density of the lattice which form a square lattice-DC, (ii) a lattice-DC of lattice-DCs with a large square network, and (iii) a one-dimensional DC (stripe DC), which has a one-dimensional density wave on the lattice-DC networks of (i) and (ii). They correspond to the second matrix of r.h.s. of Eq.(6), its third matrix, and the second term of r.h.s. of Eq.(8), respectively.

Hereafter, we focus upon two-dimensional systems where the substrate lattice has a triangular symmetry and the structure of the atomic lattice has near "2x2 structure" based on the substrate lattice structure [3]. Then the structure matrix $\mathbf{r}^0 \equiv \mathbf{r}^0_{(1)}$ can be represented as,

$$\mathbf{r}^0_{(1)} = \begin{bmatrix} 2 & 0 \\ 0 & 2 \end{bmatrix} + \begin{bmatrix} P_1 & P_2 \\ P_3 & P_4 \end{bmatrix}^{-1} \tag{9}$$

If all of the P_i (i=1~4) are integer, the structure is 1st class and it has 1st hierarchical DC pattern. These structures are represented by points in the four-dimesional parameter space, $\{P_i\}$. The parameter space also has networks in which P_i is moved and the other P_j's $(j \neq i)$ are fixed to be integers. Notice that the parameter space include two characteristic hyper-surfaces; one is "the singular surface" in which the determinant of $\mathbf{P}$ is zero, the other is the "symmetrical surface" which corresponds to the higher symmetrical structures, where the matrix $\mathbf{P}$ is in $\mathbf{R} \times \mathbf{SO}(2)$ group.

We calculcate the energies of the system along one line in the network. Here we use the simple Hamiltonian where the atomic monolayer is rigid and the substrate potential consists of a repeat of parabolic potential [7],

$$H = \frac{V_0}{2} \sum_{i=1}^{N} (\mathbf{x}_i - \mathrm{Int}(\mathbf{x}_i))^2 , \tag{10}$$

where

$$\mathbf{x}_i = \widehat{m}^i_1 \mathbf{b}_1 + \widehat{m}^i_2 \mathbf{b}_2 + \Delta \mathbf{b} ,$$

and $\mathrm{Int}(\mathbf{x}_i)$ = the nearest substrtate lattice point of $\mathbf{x}_i$.

The energy is determined by minimizing H in terms of "phase shift" $\Delta \mathbf{b}$.

We show the energy functions along the P_3 line (with $P_1=0$, $P_2=1$, and $P_4=0$) in Fig.2(a) and Fig.2(b). We see the energy function is a hierarchical one, correspond to the continued fraction of the structural matrix $\mathbf{r}^0$. In the Fig.2(a), the point where $1/P_3=0$, which means that $\{P_i\}$ is the infinite point, corresponds to the singular matrix $\mathbf{u}^s_{(1)}$ where $\det(\mathbf{u}^s_{(1)})=0$. On the other hand, the point $P_3=0$ which is shown in Fig.2(b), is in the singular surface: $\det(\mathbf{P})=0$, which is un-physical. We have also obtained such hierarchical functions along the other lines [7].

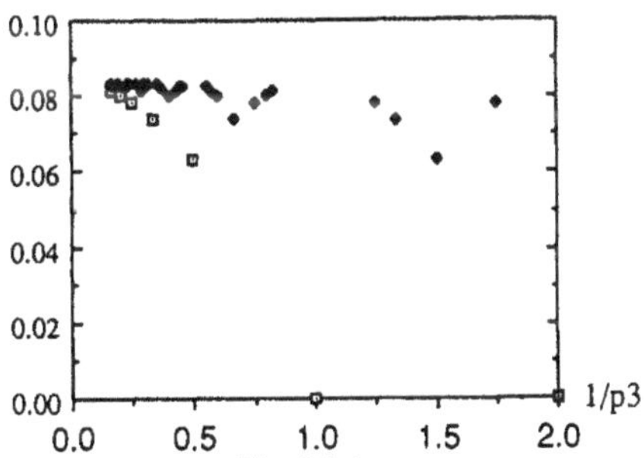

Fig.2(a)

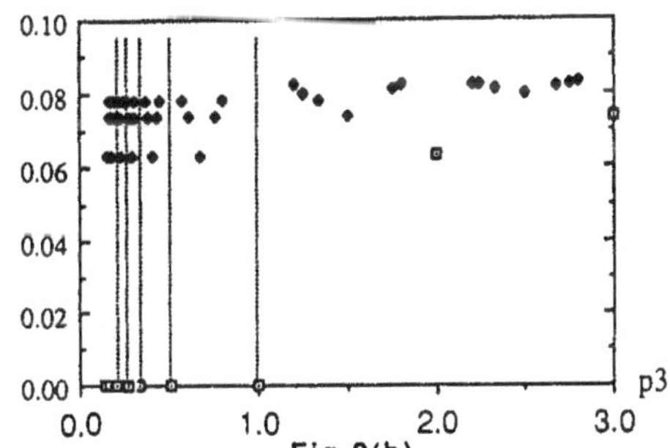

Fig.2(b)

Fig.2 The energy along P_3 line with $P_1=0$, $P_2=1$, and $P_4=0$. (a)The energy versus $1/P_3$. (b)The energy versus P_3. The point $1/P_3=0$ corresponds to $\det(\mathbf{p})=\infty$ which corresponds to $\det(\mathbf{u}^{s}_{(1)})=0$, and the point $P_3=0$ corresponds to $\det(\mathbf{p})=0$.

We expect that the soft atomic lattice systems have the similar energy functions. The ground states and the phase diagrams of the systems of the soft atomic lattice on the periodic substrate potential have been obtained only among the "higher symmetrical structures" [3]. We hope that the true ground states and the true phase diagrams of the soft atomic system will be obtain.

The author would like to thank Professor H.Aoki for valuable discussions.

reference

1) P.Bak: Rep.Prog.Phys 45(82)587.
2) T.C.Halsey: Phys.Rev.B31(85)5758.
3) H.Mitani and K.Niizeki: J.Phys.C21(88)1895.
4) H.Mitani, J.Phys.Soc.Jpn. 60(91)3589.
5) S.Aubry: "Soliton and Condensed Matter Physics" (Springer Series in Sol Stat phys. Vol.8,(78)),264.
6) H.Mitani and K.Niizeki: J.Phys.C20(87)1017.
7) H.Mitani: to be published in J.Phys.Cond.Matt..

SEGREGATION OF MODIFIED MEMBRANE PROTEINS IN PHOSPHOLIPID MEMBRANE INDUCED BY THERMODYNAMICAL CHANGES

Toru YAMADA and Shozo ISHIZAKA
Institute of Biological Sciences, University of Tsukuba, Tsukuba, Ibaraki 305, Japan

ABSTRACT

A membrane protein, bacteriorhodopsin, in reconstituted vesicles could be aggregated by the change of thermodynamical parameters, i.e., decrease in temperature or increase in osmotic pressure by adding poly (ethylene glycol). Also, when bacteriorhodopsin was labeled by fluorescein isothiocyanate or differently given photoemissivity, the change in distance between the two kinds of modified bacteriorhodopsin was observed by the method of excitation energy transfer. When proteins were aggregated in vesicle membranes, the yield of excitation energy transfer between two kinds of molecules was decreased. This implies that two kinds of proteins excluded each other and aggregated in separate regions in the membrane.

1. Introduction

All proteins are synthesized in the same region in living cells, the surface of the endoplasmic reticulum, however, each kind of protein is finally localized to different regions in the cell. These pattern formation of protein concentrations is regarded as a dissipative structure. Proteins are transported in a particular manner called vesicle transport (1); Proteins embedded in the phospholipid membrane of intracellular compartments aggregates in a certain region in the membrane and this region vesiculates and is cut off from the compartment. The transport of these membrane proteins is proceeded by the transport of such vesicles containing the protein and their fusion at the destination compartment.

In these vesicle transport processes, the segregative aggregation among many kinds of membrane proteins would take place preceding the vesiculation. This might be the essential process to form the localized concentration pattern of the membrane proteins. What kind of factors regulate the aggregation of membrane proteins in the phospholipid membrane? And how do they aggregate in the presence of many kinds of membrane proteins?

It has been reported that a membrane protein, bacteriorhodopsin in reconstituted vesicles aggregates with decrease in temperature (2). And we reported that bR aggregation is caused also with increase in the osmotic pressure by adding poly(ethylene glycol) into the vesicle suspension as follows. Thus, bR aggregation can be caused by the change of these thermodynamical parameters. Supposing the bR aggregation to be representing the aggregation of membrane proteins in general, a model experiment by using two types of modified bR, i.e., Fluorescein isothiocyanate labeled bR (FbR) and photoemissive bR (bR*) are performed to approach the question. FbR is intended to simulate a model of the protein modification by oligosaccharides. The excitation energy transfer method was applied to observe the change in the distance between FbR and bR* when they aggregate.

2. Material and Methods

Materials

i)FbR and bR*:

bR was purified from Halobacterium halobium S9 by the methods of Oesterhelt, et al.(3). Fluorescein isothiocyanate isomer I (FITC) was bound with bR by applying a procedure used by Bromer, et al.(4). The Absorbance of FITC at 490nm and bR at 570nm were respectively measured. And the molar ratio of FITC/bR calculated from each absorbance was 0.49.

The treatment of bR in order to give the photoemissivity was carried out by applying the procedure used by Peters, et al.(5). In the sample after the treatment, the peak at 570nm characteristic of native bR disappeared and instead of it, a peak newly appeared at 360nm. The emission spectrum of the suspension excited with 360nm showed a peak at around 470nm.

ii)Reconstituted vesicles:

Reconstituted vesicles containing bR and L-α-dimyristoyl phosphatidylcholine (DMPC) and vesicles containing FbR, bR* and DMPC were prepared by applying the methods of van Dijck, et al.(6). The size distribution of the prepared vesicles in the sample suspension was analyzed by the photon correlation method. The average diameter was around 70nm. The vesicles were thought to be unilamellar.

The contents of protein and phospholipid were analyzed quantitatively by Lowry's and Allen's procedures, respectively. In the case of bR+DMPC vesicles, DMPC/bR molar ratio was 78. In the case of FbR+bR*+DMPC vesicles, DMPC/(FbR+bR*) molar ratio was 73.

Methods

It is well known that a purple membrane or bR aggregation shows a typical band in circular dichroism (CD) spectrum (7),(3). Thus CD spectra measurements were applied to observe the bR aggregation. The measurements were performed with a JASCO spectropolarimeter J-720 using a 2cm path length sample cell.

Excitation energy transfer can occur between the acceptor, the FITC labeled bR (FbR) and the donor, the photoemissive bR (bR*) because the emission wavelength of bR*, 470nm and the absorption wave length of FbR, 490nm is close enough to each other (8). The changes in distance between FbR and bR* in the vesicle membrane was observed from the change in the yield of excitation energy transfer. The emission spectrum was measured with Hitachi fluorescence spectrometer F4010. In both measurements, the temperature of the sample cell was controlled by circulating water into the cell holder.

3. Results

Figure 1 shows the CD spectra of bR+DMPC vesicle suspension at 33, 23 and 2 degrees. As the temperature is decreased, a positive peak and a negative peak appeared at 540nm and 600nm, respectively. This is in accordance with the CD spectrum of purple membrane (7) and of bR trimers formed in the DMPC membrane (3).

As increasing the amount of PEG in the vesicle suspension at a constant temperature of 23 degrees, two peaks, at 540nm and 600nm appeared in the CD spectra as shown in figure 2. These peaks were similar to what was seen at 2 degrees in figure 1,c. This implies that bR formed trimers and aggregated by adding PEG into the vesicle suspension just as it did by decreasing the temperature. In figure 2, the ellipticity was converted to per original concentration before adding PEG. From the above, it was confirmed that bR aggregate in the vesicle membrane by the change in thermodynamical parameters such as temperature or osmotic pressure.

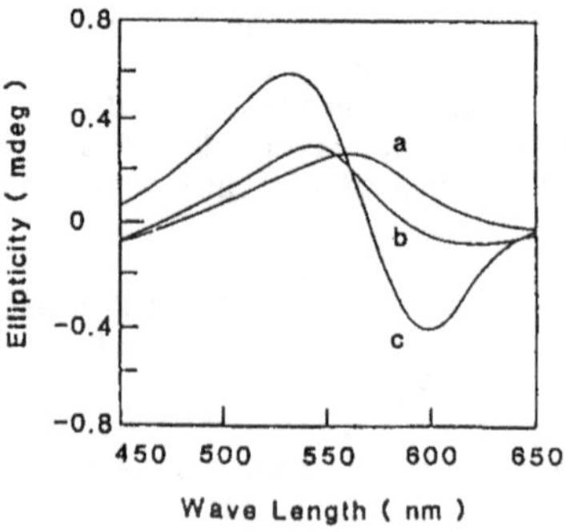

Fig. 1. CD spectrum showing the effect of temperature of bR+DMPC vesicles suspension. Curve a: at 33 degrees; curve b: at 23 degrees; curve c; at 2 degrees.

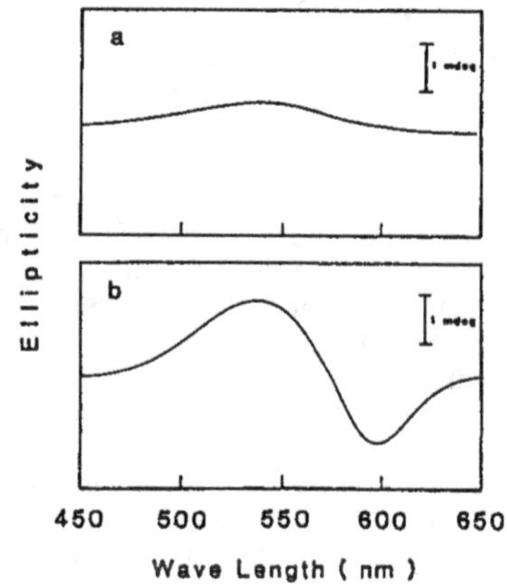

Fig. 2. CD spectra showing the effect of the addition of PEG to bR+DMPC vesicles suspension at 23 degrees. Curve a: without PEG; Curve b: at 80%(w/w) PEG. In curve b, the ellipticity is converted to per original concentration before adding PEG.

Figure 3 shows the emission spectra of FbR+bR*+DMPC vesicle suspension excited at 360nm at 23 degrees with various PEG concentrations in the suspension. Each emission intensity was converted to per original concentration before adding PEG. The spectrum without PEG in Figure 3 shows two emission peaks of bR* and FbR at 470nm and at 520nm, respectively. The spectrum of FbR and bR* directly suspended in the buffer solution without the vesicles, excited at 360nm at 23 degrees only showed the emission peak of bR* at 467nm and no peak at 517nm.

FbR has its excitation wavelength at 490nm, and it can not be excited directly by the excitation light of 360nm. But, if FbR lies as close as several nm to bR* physically, the emission energy of bR*, corresponding to the wavelength of 470nm, can be transferred to FbR and FbR can emit light at 520nm. The diameter of the prepared FbR+bR*+DMPC vesicles was around 70nm and the molar ratio of DMPC/(FbR+bR*) was 73. Supposing the shape of the vesicles to be spherical, in the vesicle membrane, the distance between FbR and bR* was calculated to be around 7nm. So that, two emission peaks indicate the evidence that the excitation energy transfer did occur between bR* acting as an energy donor and FbR acting as an energy acceptor in the DMPC membrane.

By changing the PEG concentration in the vesicle suspension, obvious differences appeared among the spectra as shown in Figure 3. The ratio between the emission from FbR at 520nm and the emission from bR* decreased with increasing PEG concentration and the emission of FbR disappeared entirely at 80%(w/w). This result showing the decrease of the excitation energy transfer implies the increase of the distance between the donor, bR* and the acceptor, FbR.

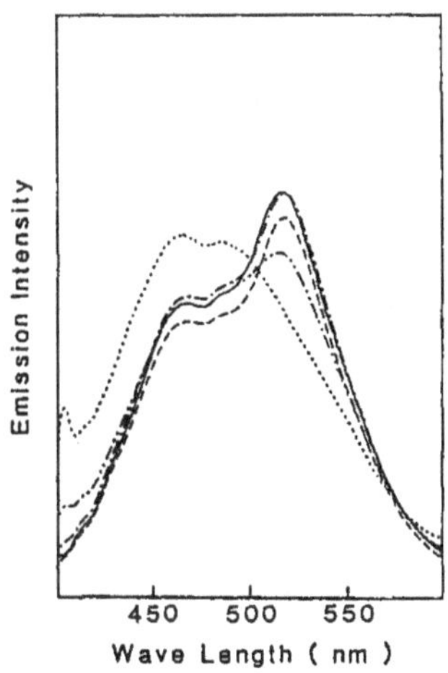

Fig. 3. The emission spectra of the FbR+bR*+DMPC vesicle suspension at various PEG concentrations at 23 degrees. Curve(——): without PEG; curve(– –): at 20%(w/w) PEG; curve(·–·–): at 40%(w/w) PEG; curve(·–··–): at 60%(w/w) PEG; curve(······): at 80%(w/w) PEG. Each emission intensity is converted to the value per original concentration before adding PEG.

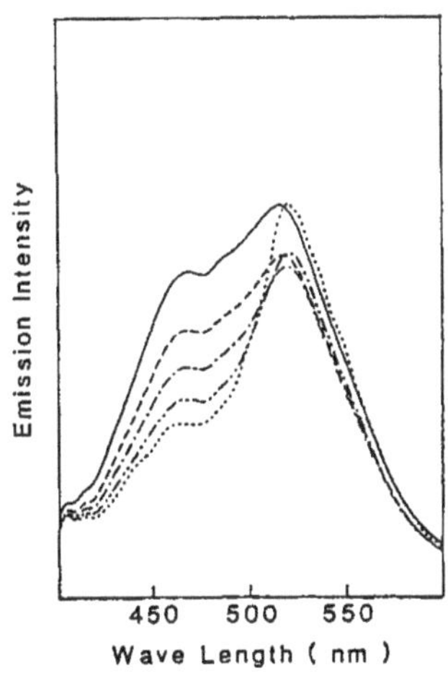

Fig. 4. The effect in the emission spectrum of the FbR+bR*+DMPC vesicle suspension with 80%(w/w) of PEG concentration by increasing temperature. Curve(——): at 23 degrees; curve(– – –): at 33 degrees; curve (·–·–): at 43 degrees; curve(–··–): at 53 degrees; curve(······): at 63 degrees.

Figure 4 shows the change of the emission spectrum of FbR+bR*+DMPC vesicle suspension at 60%(w/w) of PEG concentration by increasing the temperature from 23 degrees to 63 degrees. The peak at 520nm which had disappeared by the addition of PEG recovered with the increase in temperature. The recovery of the peak at 520nm was also observed when PEG was removed by dialysis for 24 hours (data not shown).

4. Discussion

Aggregation of FbR and bR* would occur in FbR+bR*+DMPC vesicles by adding PEG into the suspension just as it would in native bR in DMPC membranes. In FbR+bR*+DMPC vesicles, the distance between FbR and bR* dispersed randomly in the vesicle membrane is calculated to be around 7nm. This is close enough to cause the excitation energy transfer from the donor, bR* to the acceptor, FbR.

The yield of excitation energy transfer is proportional to the negative 6th power of the distance and to the 2nd power of the orientation factor between FbR and bR* (8). If the trimers contain FbR and bR* heterogeneously, the distance between FbR and bR* should be around 3nm. The value of the orientation factor among FbR and the nearest neighboring bR* is hardly changed by the hexagonal coordination of FbR and bR* in heterogeneous aggregation. As a result, the yield of excitation energy transfer should increase. However, figure 5 shows the decrease of the yield of excitation energy transfer by adding PEG into the suspension. It is hard to think that the extrication of FbR and bR* from vesicles is the cause of the decrease, since the energy transfer recovered when PEG was removed by dialysis (data not shown). Again, PEG may not be working to quench the emission of FbR, because in figure 6, the emission of FbR at 517nm recovered in the presence of PEG at 60%(w/w). Hence, the remaining possibility is that each aggregation is composed of homogeneous molecules, containing only FbR or bR*.

Supposing the trimers are composed homogeneously with either one of FbR or bR* and the other dispersed in the vesicle membrane, the average distance between FbR and bR* in the vesicle membrane would increase to several multiples of 7nm, the distance in the case of random dispersion. And in the case where FbR and bR* form homogeneous trimers separately, the average distance between FbR and bR* would increase to about 10 times the distance in random dispersion. The distances in both of the above cases are large enough for the excitation energy transfer between FbR and bR* to vanish. Thus, from the above, it can be said that at least either FbR or bR* aggregates homogeneously excluding the other.

In the case of the sample where FbR and/or bR* has already been aggregated homogeneously by adding PEG, the increase of temperature could cause the aggregations of FbR and/or bR* to disperse into the vesicle membrane and to shorten the distance between FbR and bR*. Thus, the yield of excitation energy transfer would increase. Actually, in Figure 4, the recovery of the emission of FbR at 517nm by the increase in temperature was observed.

5. Conclusion

A membrane protein, bacteriorhodopsin, in reconstituted vesicles could be aggregated by changing thermodynamical parameters such as temperature or pressure. Also, bacteriorhodopsin was labeled by a fluorescent molecule,fluorescein isothiocyanate, or was given photo-emissivity. Even though these modifications are quite small comparing with general chemical modifications, the two kind of molecules excluded each other and aggregated in separate regions in the membrane.

REFERENCES

1. B.W.Wattenberg, W.E.Balch, and J.E.Rothman, J.Biol.Chem., **261** (1986) 2202
2. D.Oesterhelt, and W.Stoeckenius, Meth.Enzymol., **31** (1974) 667
3. R.J.Cherry, U.Mueller, R.Henderson, and M.P.Heyn, J.Mol.Biol., **121** (1978) 282
4. W.W.Bromer, S.K.Seehan, A.W.Berns, and E.R.Arquilla, Biochemistry, **6** (1967) 2378
5. J.Peters, R.Peters, and W.Stoeckenius, FEBS Lett., **61** (1976) 128
6. P.W.M.van Dijck and K.van Dam, Meth.Enzymol., **88** (1982) 17
7. B.Becker and T.G.Ebrey, Biochem.Biophys.Res.Commun., **69** (1976) 1
8. L.Brand and B.Witholt, Meth.Enzymol., **11** (1967) 839

F: Abstracts

TURBULENCE IN A BOX

*A. Libchaber**

The Research Institutes, Department of Physics,
The University of Chicago
5640 S. Ellis Ave., Chicago, IL 60637 USA

Thermal turbulence in He gas at low temperature (5K) allows a broad study of different states of turbulence. By varying the gas density the Rayleigh number can be swept from 10^3 to 10^{15} and the Reynolds number can go beyond 10^7. All the measured quantities, in each turbulent state, scale with Ra or Re. We will describe in detail the characteristics of soft turbulence ($Ra < 10^8$), hard turbulence ($Ra > 10^8$), and the new recent observation at higher Ra ($Ra > 10^{11}$) where the high frequency part of the local temperature spectrum changes. We will present scaling models for various length scales and time scales in the problem.

Visualization in water ($Pr \approx 5$) and in SF_6 in the gas state ($Pr \approx 1$) will be presented and compared to the theoretical models.

References:

D. C. Threlfall J.F.M. 67, 17 (1975).
F. Heslot et al Phys Rev A 36, 5870 (1987).
M. Sano et al Phys Rev A 40, 6421 (1989).
B. Castaing et al J.F.M. 206, 1 (1989).
X. Z. Wu et al Phys Rev Lett 64, 2140 (1990).
G. Zocchi et al Physica A 166, 387 (1990).

* Present address : Physics Department, Princeton University, Princeton, NJ 08540 and NEC Research, 4 Independence Way, Princeton, NJ 08540.

RELATIVE STABILITY IN SINGLE AND MULTIPLE VARIABLE CHEMICAL SYSTEMS

J.Ross

Department of Chemistry, Stanford University

We have formulated a thermodynamic and stochastic theory of chemical systems, far from equilibrium, with multiple stationary states. The theory introduces the concept of the affinity of a system from a stationary state. This excess work is related to changes in thermodynamic functions appropriate for the constrains on the chemical system; it is the thermodynamic potential, the driving force towards the stationary state; it is an extremum at each stationary state; it is a globally valid Liapunov function; its second derivative provides necessary and sufficient criteria of stability of each stationary state; its time derivative is a component of the dissipation in the relaxation to a stationary state; and it determines the stationary distribution of a stochastic birth-death master equation. The excess work is used to define relative stability and coexistence of multiple stationary states under the same external constraints. Experiments are reviewed on relative stability in an optically bistable (one-variable) system, and calculations are presented on relative stability in a two-variable reaction diffusion systems.

TOPOLOGICAL DEFECTS AND MECHANISM OF TRANSITION TO WEAK TURBULENCE

V. Steinberg
Department of Nuclear Physics
Weizmann Institute of Science, Rehovot 76100, Israel

An overview of the experimental results on transition initially ideally ordered state of stationary roll structure (Williams domain) in a electrohydrodynamic convection (EHC) of a nematic liquid crystal to defect-mediated turbulence is presented. A complete quantitative description of the different boundaries of stability of the system is given. There exists three different regimes on the stability diagram: the Eckhaus stable band of normal rolls near the onset of convection, the tilted rolls regime at higher values of the control parameter, and finally, regime of sustain defect nucleation, annihilation and motion which is identified as a weak turbulence. Understanding of the details of the transitions from one to another regime is related tightly with uncovering the underlying symmetry breaking. The transition from Williams Domain to tilted rolls occurs due to spontaneous left-right parity breaking whereas the transition to the defect-mediated turbulence is possible due to fluctuations of soft long wavelength longitudinal modes which are related to translation symmetry. The longitudinal mode is coupled via the mechanism of parity breaking to the transverse modes which are a necessary ingredient leading to the defect nucleation. The Eckhaus instability in two-dimensional patterns occurs through a phase slippage which leads to topological defects although the translation symmetry breaking, which is coupled to long-wavelength longitudinal mode, responsible for it. Since the topological defects play an important role in stability of 2D roll pattern by providing a very effective wavenumber selection mechanism and in destroying spatial coherence, much effort was devoted to studying their dynamics and pair-interactions in both climb and glide motion. It is shown that the mutual interaction of defects in a pair becomes significant only below a characteristic length, which is universal for both climbing and gliding and about an order of magnitude larger than the corresponding coherence length. Thus, the overall picture which emerges from our experimental studies is that the system exhibits a well defined universal route to the weak turbulence which does not depend on the details of the system, i.e. this set-up may serve as a paradigm for anisotropic nonequilibrium pattern forming systems.

THE INTERACTION OF 2-DIMENSIONAL LOCALIZED SOLUTIONS NEAR A WEAKLY INVERTED BIFURCATION

HELMUT R. BRAND ‡,† and ROBERT J. DEISSLER †,*

‡ FB 7, Physik, Universität Essen, D 4300 Essen 1, F.R. Germany

† Center for Nonlinear Studies, MS-B 258, Los Alamos National Laboratory, University of California, Los Alamos, NM 87545, USA

* ICOMP, MS 5-3, NASA Lewis Research Center, 2100 Brookpark Road, Cleveland, OH 44 135

We study the interaction of two-dimensional (2D) solutions as they arise for the envelope equations for a subcritical bifurcation to traveling waves. We show that these 2D localized solutions can collide and reemerge unchanged in size and shape after the collision in contrast to what is found in 2D solitonic systems. Various other types of behavior arise as the impact parameter and the cross-coupling between the waves are varied. We point out that these phenomena should be observable experimentally for hydrodynamic instabilities in anisotropic liquids such as nematic liquid crystals.

We find that behavior occurring in one spatial dimension also occurs in two for the present equations. This is in contrast to solitonic systems for which an increase in the dimensions of the system greatly changes the behavior of the solutions. We find interactions which preserve the form of the solutions (a behavior which occurs with one-dimensional solitons), as well as interactions which cause complete annihilation or a transition from subcritical to supercritical. We also observe behavior which cannot occur in a one-dimensional system such as a change in the perpendicular distance between trajectories as a result of the interaction. It will be most interesting to see whether the phenomena predicted will be observed in electroconvection or thermal convection in nematic liquid crystals or in another hydrodynamic instability in an anisotropic fluid. This would then allow for a detailed comparison of the presented results with the collision behavior of two-dimensional particle-like solutions in physical systems having dissipation and dispersion.

PHASE DEFECTS IN TRAVELLING WAVE

V. CROQUETTE, J. LEGA, B. JANIAUD AND S. JUCQUOIS

LPS-ENS 24 rue Lhomond 75 231 Paris Cedex 05

We have observe traveling holes in wave-trains, reminiscent of those predicted by BEKKI and NOZAKI [1] (see also [2]). Such objects or defects connect wave-trains of different wavenumbers, and are characterized by strong localized amplitude and phase variations at their cores. The experiment is Rayleigh Bénard convection, performed with a fluid of low Prandtl number. Under this condition, convective rolls become unstable versus an oscillatory instability leading to the formation of wave trains propagating along the rolls axis. Using an annular geometry we stufy a pattern with periodic boundary conditions. At low Rayleigh number, a strongly modulated pattern forms, having a region of lower amplitude, where the wave defects have been observed. This phenomenon is generic since once the depressed area is formed, it travels around the annular cell in a quasi-stationary manner, presenting phase jumps whose position is apparently uncorrelated with any special location in the cell.

Although the stability of the traveling holes has not been studied, they are apparently unstable solutions. For this reason, they do not appear spontaneaously in ordinary patterns which display the stable homogeneous travelling wave train. In our experiment, the strong wave amplitude modulation drives the pattern in a phase unstable domain. Apparently the phase instability is associated with the nucleations of the hole defects which propagates over some distance in the pattern with a velocity different from that of the amplitude modulation.

The experimental holes are transients and thus not easily compared with theory, we performed numerical simulations where we impose a strong wave modulation comparable to the one observed in the experiment and then observed similar transient hole defects. This confirms the occurence of hole defects.

References

[1] N. BEKKI and K. NOZAKI, Phys. Lett. **110A**, 133, (1985).

[2] F. CARIELLO and M. TABOR, Physica D **39**, 77 (1989).

FLUCTUATIONS, CORRELATIONS AND NOISE IN A DIFFUSION-REACTION PROCESS

Charles R. Doering

Department of Physics and Institute for Nonlinear Studies

Clarkson University

Potsdam, NY 13699-5820 USA

The reversible single species reaction scheme $X \leftrightarrow 2X$ has long served as an example of a system with nonlinear macroscopic dynamics. In recent years a microscopic version of this interacting particle system has been solved exactly in one spatial dimension.[1] The effects of internal fluctuations, correlations arising in nonequilibrium situations, and external noise can all be studied in this model without approximation. Departures of the macroscopic behavior from the mean field predictions can be dramatic.[2,3,4] Here we present some recent results and discuss the effect of a temporally fluctuating (but spatially homogeneous) externally imposed noise in the growth rate of the process (the rate for $X \rightarrow 2X$ process). We find that the steady state of the system possesses a new microscopic length scale resulting from the nonequilibrium constraints, demonstrating that spatially homogeneous external noise can introduce spatial correlations.

---

[1]D. ben-Avraham, M.A. Burschka, and C.R. Doering, *J. Stat. Phys.* **60**, 695 (1990).

[2]M.A. Burschka, C.R. Doering and D. ben-Avraham, *Phys. Rev. Lett.* **63**, 700 (1989).

[3]C.R. Doering and M.A. Burschka, *Phys. Rev. Lett.* **64**, 245 (1990).

[4]C.R. Doering, M.A. Burschka, and W. Horsthemke, *J. Stat. Phys.* (submitted. 1991)

SPIRAL DYNAMICS IN CHEMICAL AND BIOLOGICAL EXCITABLE SYSTEMS

Stefan C. Müller
Max-Planck-Institut für Ernährungsphysiologie
Rheinlanddamm 201, W-4600 Dortmund 1, FRG

Excitable systems play an important role in the field of spatio-temporal selforganization. In biology, they serve as media for signal transmission, for instance in pulse propagation along nerve fibers or across the chambers of the heart. For the study of traveling waves in two-dimensional extended media, two model systems are especially suited for in depth investigations: a chemical system, namely the excitable Belousov-Zhabotinskii reaction, and a biological system, which is the cellular slime mold Dictyostelium discoideum during early stages of cell aggregation.

We have studied the formation and dynamic evolution of spiral waves in both systems in quantitative detail by applying computerized video imaging techniques. Emphasis was given to elucidate the structure and the properties of the spiral core. Spirals in the Belousov-Zhabotinskii reaction, as observed by light absorption, are approximately Archimedian shaped. Their tip rotates on a circle along the boundary of the core (diameter $\approx$ 0.7 mm) inside which the amplitude of excitation decreases towards the rotation center, which is a singular point (diameter < 10μm) with quasi-stationary chemistry. When decreasing the excitability of the solution, the steady tip rotation undergoes a transition to a compound motion, close to that of an epicycloid, and the spiral shape becomes non-Archimedian.

Archimedian spirals in the slime mold, as observed by dark-field illumination, indicate the pattern of amoebae cells moving towards an aggregation center by chemotaxis. A quantitative velocity analysis by mutual correlation of pixel brightness in microscopic video movies yields information about the distribution of cell velocities. Periodic changes of velocity in a spiral arm are characterized in detail. Close to the spiral center the trajectories of the cells are found to merge on a closed loop (diameter $\approx$ 0.5 mm) both from the outside and from the inside (the spiral core), thus producing a limit cycle in real space.

Chaotic Travelling Waves in Coupled Map Lattices

Kunihiko Kaneko
Department of Pure and Applied Sciences,
College of Arts and Sciences,
University of Tokyo, Komaba, Tokyo 153, Japan

Spatiotemporal chaos (STC) is (temporally) high-dimensional chaos which involves spatial pattern dynamics. Coupled map lattice (CML) has become a powerful tool to analyze the spatiotemporal chaos and pattern dynamics within these 10 years[1,2].

Modelling through CML is carried out as follows: Choose essential procedures for the spatially extended dynamics, and then replace each procedure by a parallel dynamics on a lattice. Coupled map lattice dynamics is obtained by successive applications of each procedure. As an example, assume that you have a phenomenon, created by a local chaotic process and diffusion. This class of examples is seen in convection, chemical turbulence, and so on. In CML approach, we reduce the phenomena into local chaos and diffusion processes. If we choose a logistic map $x'_n(i) = f(x_n(i))$ ($f(y) = 1 - ay^2$) for chaos, and a discrete Laplacian operator for the diffusion, our CML is given by

$$x_{n+1}(i) = (1-\epsilon)f(x_n(i)) + \epsilon/2(f(x_n(i+1)) + f(x_n(i-1))) \qquad (1)$$

The above CML has extensively been investigated as a standard model for spatiotemporal chaos. Studies of CML have clarified some novel qualitative universality classes in spatiotemporal chaos, including spatial bifurcation, frozen random chaos, pattern selection with suppression of chaos, spatiotemporal intermittency, soliton turbulence, and quasistationary supertransients [1-3].

In the present paper we report a novel class of phenomenolgy in CML, which is related with recent experiments in fluids and liquid crystal: (chaotic) travelling waves.

AUTHOR INDEX

www.ingramcontent.com/pod-product-compliance
Lightning Source LLC
LaVergne TN
LVHW010536300926
849162LV00002B/156